The Neurobiology of Schizophrenia

The Neurobiology of Schizophrenia

Edited by

Ted Abel

Thomas Nickl-Jockschat

AMSTERDAM • BOSTON • HEIDELBERG • LONDON
NEW YORK • OXFORD • PARIS • SAN DIEGO
SAN FRANCISCO • SINGAPORE • SYDNEY • TOKYO

Academic Press is an imprint of Elsevier

Academic Press is an imprint of Elsevier
125 London Wall, London EC2Y 5AS, United Kingdom
525 B Street, Suite 1800, San Diego, CA 92101-4495, United States
50 Hampshire Street, 5th Floor, Cambridge, MA 02139, United States
The Boulevard, Langford Lane, Kidlington, Oxford OX5 1GB, United Kingdom

Copyright © 2016 Elsevier Inc. All rights reserved.

No part of this publication may be reproduced or transmitted in any form or by any means, electronic or mechanical, including photocopying, recording, or any information storage and retrieval system, without permission in writing from the publisher. Details on how to seek permission, further information about the Publisher's permissions policies and our arrangements with organizations such as the Copyright Clearance Center and the Copyright Licensing Agency, can be found at our website: www.elsevier.com/permissions.

This book and the individual contributions contained in it are protected under copyright by the Publisher (other than as may be noted herein).

Notices

Knowledge and best practice in this field are constantly changing. As new research and experience broaden our understanding, changes in research methods, professional practices, or medical treatment may become necessary.

Practitioners and researchers must always rely on their own experience and knowledge in evaluating and using any information, methods, compounds, or experiments described herein. In using such information or methods they should be mindful of their own safety and the safety of others, including parties for whom they have a professional responsibility.

To the fullest extent of the law, neither the Publisher nor the authors, contributors, or editors, assume any liability for any injury and/or damage to persons or property as a matter of products liability, negligence or otherwise, or from any use or operation of any methods, products, instructions, or ideas contained in the material herein.

British Library Cataloguing-in-Publication Data
A catalogue record for this book is available from the British Library

Library of Congress Cataloging-in-Publication Data
A catalog record for this book is available from the Library of Congress

ISBN: 978-0-12-801829-3

For Information on all Academic Press publications visit our website at https://www.elsevier.com/

Publisher: Mara Conner
Acquisition Editor: Melanie Tucker
Editorial Project Manager: Kathy Padilla
Production Project Manager: Julia Haynes
Designer: Matt Limbert

Typeset by MPS Limited, Chennai, India

Contents

List of Contributors xv
Acknowledgments xix

PART I INTRODUCTION

CHAPTER 1 Historical and Clinical Overview: Implications for Schizophrenia Research 3
T. Nickl-Jockschat and T. Abel

Introduction 3
A Brief History of the Definition of Schizophrenia 5
Lessons From Therapeutic Approaches 9
The Neurobiology of Schizophrenia 10
References 12

PART II THE GENETIC AND EPIGENETIC BASIS OF SCHIZOPHRENIA

CHAPTER 2 Progress and Prospects for Endophenotypes for Schizophrenia in the Time of Genomics, Epigenetics, Oscillatory Brain Dynamics, and the Research Domain Criteria 17
G.A. Miller and B.S. Rockstroh

Endophenotypes in Schizophrenia: Current Status 22
An Emerging Frontier for Endophenotypes: Epigenetics 26
A New Frontier for Endophenotypes: Brain Connectivity Dynamics 28
Limitations and Future Directions 30
References 31

CHAPTER 3 Insights From Genome-Wide Association Studies (GWAS) 39
S. Cichon and S. Ripke

The Pre-GWAS Era: Linkage and Candidate Gene Association Studies 40
The Developments That Made GWAS Possible 41
The Principle of GWAS 42
GWAS Findings in Schizophrenia 44
Genetic Relationship Between Schizophrenia and Other Psychiatric Disorders 46
Were GWAS of Schizophrenia Successful So Far and Should They Continue? 48
Future Directions 49
References 49

v

CHAPTER 4 Sequencing Approaches to Map Genes Linked to Schizophrenia 51
P.M.A. Sleiman and H. Hakonarson

Introduction and Background 51
Next-Generation Sequencing 52
Exome Versus Whole Genome 53
Sequencing Studies to Date 55
Conclusions 57
Future Directions 57
References 58

CHAPTER 5 Epigenetic Approaches to Define the Molecular and Genetic Risk Architectures of Schizophrenia 61
M. Kundakovic, C. Peter, P. Roussos and S. Akbarian

Introduction 62
An Epigenetic Link Between Environmental Risk Factors and Schizophrenia 64
    Evidence From Human Studies 64
    Evidence From Animal Studies 65
DNA Methylation in Schizophrenia 65
    DNA Methylation and Gene Regulation in the Brain 65
    DNA Methylation Changes in Schizophrenia 66
    Studies in the Peripheral Tissue and Possible Epigenetic Biomarkers in Schizophrenia 66
Epigenetic Approaches to the Molecular and Genetic Risk Architectures of Schizophrenia 67
    Do Epigenetic Mechanisms Contribute to Long-Lasting Alterations in Gene Expression in Schizophrenia Brain? 67
    Functional Neuroepigenomics Could Inform About Disease-Associated Genetic Risk Polymorphisms 69
    Higher-Order Chromatin and the Genetic Risk Architecture of Schizophrenia 69
Epigenetic Mechanisms and the Treatment of Schizophrenia 71
Future Directions 72
    Chromatin-Modifying Drugs: A Future Treatment for Schizophrenia? 72
    Comprehensive, Region-Specific, and Cell Type-Specific Epigenome Mappings in the Brain 73
    Larger and More Ambitious Clinical Studies 73
Acknowledgment 74
References 74

CHAPTER 6 Exploring Neurogenomics of Schizophrenia With *Allen Institute for Brain Science* Resources..................83
M. Hawrylycz, T. Nickl-Jockschat and S. Sunkin

Introduction..83
Introduction to the Portal 84
Mouse Atlas Resources 84
    Common Reference Space in the Mouse Brain..........................86
    Integrated Search and Visualization in the Allen Brain Atlas.............87
    The Mouse Developmental Atlas..91
    Connectivity in the Mouse Brain 93
Human Atlas Resources 97
    The Allen Human Brain Atlas..97
    BrainSpan Atlas of the Developing Human Brain..........................99
    NIH Blueprint Non-Human Primate Atlas 101
Beyond the Atlases...102
Applications of the Allen Institute for Brain Science Resources in Schizophrenia Research...102
Acknowledgments 104
References...104

PART III THE NEUROCHEMICAL BASIS OF SCHIZOPHRENIA

CHAPTER 7 The Dopamine Hypothesis of Schizophrenia: Current Status 109
G. Gründer and P. Cumming

Introduction: History of the Dopamine Hypothesis..................................109
Dopamine Receptor Occupancy Studies..110
Molecular Imaging Studies of Dopaminergic Neurotransmission in Schizophrenia 112
Prefrontal–Subcortical Dopamine Dysregulation......................................116
Ketamine Psychosis 118
Conclusion: Beyond the Dopamine Hypothesis 118
References...119

CHAPTER 8 The PSD: A Microdomain for Converging Molecular Abnormalities in Schizophrenia 125
A. Banerjee, K.E. Borgmann-Winter, R. Ray and C.-G. Hahn

Introduction...125
The Postsynaptic Density, the Hub of Postsynaptic Signaling 126
    Dendritic Spines..126
    The Postsynaptic Density 127

Signaling Pathways in the PSD 128
Proteomic Landscape of the PSD 131
Evidence Supporting the Role of the PSD in Schizophrenia 133
Genetic Evidence 133
Postmortem Evidence for PSD Dysregulation in Schizophrenia 134
Dysregulations in PSD Proteins and Their Signaling in Schizophrenia 134
PSD as a Microdomain for Converging Molecular Alterations in Schizophrenia 136
Summary and Future Directions 137
References 138

CHAPTER 9 Targeting Cognitive Deficits in Schizophrenia via $GABA_A$ Receptors: Focus on $\alpha 2$, $\alpha 3$, and $\alpha 5$ Receptor Subtypes 149

U. Rudolph and H. Möhler

Excitatory and Inhibitory Tuning 150
Multiplicity of $GABA_A$ Receptors 150
Maturation of $GABA_A$ Receptors 151
$GABA_A$ Receptor Subtypes in Cortical Period Plasticity 152
$GABA_A$ Receptor Genetics and Schizophrenia 152
Deficits of GABAergic Cortical Interneurons and Impaired Cortical
Oscillations in Schizophrenia 152
Physiological and Pharmacological Functions of $GABA_A$ Receptor
Subtypes 154
Attempts to Enhance Cognition via $\alpha 2/\alpha 3$ $GABA_A$ Receptor Modulation 155
Inhibition of Dopaminergic Neurons by $\alpha 3$-Containing $GABA_A$ Receptors 156
Indirect Inhibition of Dopaminergic Neurons by $\alpha 5$-Containing $GABA_A$
Receptors 157
Multispecific Modulation of $\alpha 2$-, $\alpha 3$-, and $\alpha 5$-Containing $GABA_A$ Receptors 158
Proof-of-Concept Clinical Study With the Partial Positive Allosteric Modulator
Bretazenil 159
$\alpha 5$-Negative Allosteric Modulators and Cognition Enhancement 159
References 160

PART IV THE BIOCHEMICAL BASIS OF SCHIZOPHRENIA

CHAPTER 10 Metabolomics of Schizophrenia 167

D. Rujescu and I. Giegling

Introduction 167
Lipidomics in Schizophrenia 169

Energy Metabolism in Schizophrenia 172
Outlook 174
References 174

CHAPTER 11 The Role of Inflammation and the Immune System in Schizophrenia 179
N. Müller, E. Weidinger, B. Leitner and M.J. Schwarz

Introduction 179
Inflammatory Mechanisms in the CNS 180
Microglia as an Important Cellular Basis of Inflammation in the CNS 180
Kindling and Sensitization of the Immune Response: The Basis for the
Stress-Induced Inflammatory Response in Psychiatric Disorders 181
The Vulnerability-Stress-Inflammation Model of Schizophrenia 181
The Immune Dysbalance in Schizophrenia Is Associated With Chronic
Inflammation 182
The Impact of Inflammation on Neurotransmitters in Schizophrenia 182
The Possible Role of Infection in Schizophrenia 183
CNS Volume Loss in Imaging Studies: A Consequence of an
Inflammatory Process? 184
Cyclooxygenase-2 Inhibition as an Anti-Inflammatory Therapeutic
Approach in Schizophrenia 184
Further Immune-Related Substances in the Therapy of Schizophrenia 185
Methodological Aspects of the Response to Immune-Based Therapy in
Schizophrenia 185
Conclusions 186
Acknowledgment 188
Statement of Conflict of Interest 188
References 188

CHAPTER 12 Proteomics of Schizophrenia 195
M.P. Coba

Biomarkers 196
Postmortem Studies 196
Protein Interaction Pathways and Signaling Networks 197
Embryonic Brain 202
Conclusions 204
References 205

PART V THE ELECTRO-PHYSIOLOGICAL BASIS OF SCHIZOPHRENIA

CHAPTER 13 EEG and MEG Probes of Schizophrenia Pathophysiology ... 213
E. Neustadter, K. Mathiak and B.I. Turetsky

Introduction ... 213
Measures of Inhibitory Failure ... 215
    P50 Auditory Sensory Gating ... 215
    Prepulse Inhibition of Startle ... 218
Measures of Aberrant Salience Detection ... 220
    Mismatch Negativity ... 220
    P300 ... 223
Measures of Abnormal Gamma Band Oscillations ... 225
    Measuring Neural Oscillations ... 226
    Neurobiology of Gamma Oscillations ... 226
    Gamma Oscillations and Schizophrenia ... 228
Conclusion ... 229
References ... 230

CHAPTER 14 Cellular and Circuit Models of Increased Resting State Network Gamma Activity in Schizophrenia ... 237
R.S. White and S.J. Siegel

Introduction ... 238
Studies in Schizophrenia ... 238
    Resting State Brain Activity ... 238
    Electroencephalography ... 238
    Magnetoencephalography ... 239
    Functional Magnetic Resonance Imaging ... 239
Findings in Disease ... 239
    Resting State Network and Gamma Oscillations in Schizophrenia ... 239
    Power and Coherence of Brain Activity ... 241
    Inherent Noise and Relationship to Symptoms in Schizophrenia ... 242
    Circuit Models and Evidence in Humans ... 242
    Interneuron Network Gamma Generation ... 243
    The Pyramidal and Interneuron Model of Generating Gamma Oscillations ... 244
Animal Models of a Noisy Brain ... 244
    Defining Baseline Gamma in Animal Models ... 244
    Animal Models of Schizophrenia With Relation to Resting Gamma Activity ... 245
    In Vitro Examination of Animal Models of Increased Resting Gamma ... 247
    Relating Behavior to EEG/LFP Validation of Models ... 248
Summary and Conclusions ... 250
References ... 251

PART VI THE STRUCTURE AND FUNCTION OF NEURAL CIRCUITS IN SCHIZOPHRENIA

CHAPTER 15 Computational Neuroanatomy of Schizophrenia 263
C. Davatzikos and N. Koutsouleris

Introduction 263
Methods for Computational Neuroanatomy 266
    Region of Interest Versus Voxel-Based Analysis 266
    Regional Tissue Volumetrics 267
    Optimally Discriminant Voxel-Based Analysis 269
    Multivariate Pattern Analysis (MVPA) and Machine Learning 269
Topography of Reduced Brain Volumes in Schizophrenia 270
    Single-Subject Classification: From Computational Neuroanatomy of Schizophrenia to Individualized Diagnostic Tests 274
    Single-Subject Differentiation Between the Psychosis Prodrome and the At-Risk for Mental Psychosis Without Subsequent Disease Transition 275
    Are SCZ-Like Neuroanatomical Patterns Possibly Endophenotypes of Disease? 276
References 278

CHAPTER 16 Brain Computations in Schizophrenia 283
R.A. Adams and K.J. Friston

The Bayesian Brain, Precision, and Hierarchical Models 283
Psychosis and Synaptic Gain 285
Computational Implications of Decreased High-Level Synaptic Gain 287
Computational Implications of Striatal Presynaptic Dopamine Elevation 288
    Tonic Dopamine Signaling 289
    Phasic Dopamine Signaling 289
    Incentive (and Aberrant) Salience 290
    Jumping to Conclusions: Overweighting Evidence versus Lowered Decision Threshold 290
Conclusions and Further Questions 291
Acknowledgments 292
References 292

CHAPTER 17 Schizophrenia and Functional Imaging 297
K. Pauly and C. Moessnang

Functional Magnetic Resonance Imaging: Challenges and Perspectives for the Study of Schizophrenia 297
Functional Imaging of Schizophrenia Symptom Clusters 298
    Cerebral Correlates of Delusions 298
    Cerebral Correlates of Hallucinations 306
Summary and Outlook 308
References 308

CHAPTER 18 Anatomical and Functional Brain Network Architecture in Schizophrenia 313
G. Collin and M.P. van den Heuvel

Introduction 313
    What Is the Brain Network? 314
    Exploring the Brain's Connectional Anatomy 314
    Chapter Structure 316
Part 1. Brain Connectivity in Schizophrenia 317
    Structural Connectivity 317
    Functional Connectivity 318
    Structural-Functional Coupling 319
Part 2. Connectome Topology in Schizophrenia 321
    Connectome Segregation 322
    Connectome Integration 324
    Relation to Clinical Symptoms and Outcome 327
Conclusion and Future Directions 328
    Specificity of Hub Pathology to Schizophrenia 328
    Relation Between Connectome Organization and Cognitive Deficits 329
    Potential of Connectomics in Informing Long-Term Outcome and Clinical Practice 329
References 330

CHAPTER 19 Statistical Learning of the Neurobiology of Schizophrenia: Neuroimaging Meta-Analyses and Machine-Learning in Schizophrenia Research 337
D. Bzdok and S.B. Eickhoff

Introduction 337
Quantitative Meta-Analysis 338
    Structural Meta-Analyses in Schizophrenia 339
    Functional Meta-Analyses of Schizophrenia 341
    Discussion 342
Machine Learning 343
    Structural ML Studies in Schizophrenia 344
    Hypothesis-Guided ML Studies in Schizophrenia 345
    More Advanced ML Studies in Schizophrenia 346
    Discussion 346
Conclusion 347
References 348

PART VII MODELING SCHIZOPHRENIA IN ANIMALS

CHAPTER 20 Modeling Schizophrenia in Animals: Old Challenges and New Opportunities 353
Y. Ayhan, C.E. Terrillion and M.V. Pletnikov

Introduction 353
Genetic Animal Models of Schizophrenia 355
    Pathophysiological Models 356
    Models Based on Etiological Findings 358
Challenges in Modeling Schizophrenia 366
Conclusions 367
Acknowledgments 370
References 370

CHAPTER 21 Behavioral Phenotypes of Genetic Mouse Models: Contributions to Understanding the Causes and Treatment of Schizophrenia 383
P.M. Moran

Introduction 383
How Do You Measure Psychotic Behavior in a Mouse? 384
Models Targeting Dopaminergic Transmission 385
Models Targeting Glutamatergic Transmission 388
Future Directions 389
Conclusion 392
References 392

CHAPTER 22 Genetic Mechanisms Emerging from Mouse Models of CNV-Associated Neuropsychiatric Disorders 397
A. Nishi and N. Hiroi

Copy Number Variants and Diverse Neuropsychiatric Disorders 398
Dimensional Behavioral Models of Neuropsychiatric Disorders 400
Mechanisms by Which CNV-Encoded Genes Cause Behavioral Dimensions of Neuropsychiatric Disorders 402
    Noncontiguous Effects 406
    Mass Action and Net Effects 407
    Pleiotropy 407
    Developmental Trajectories 407
    Phenotypic Variation 408
Acknowledgment 409
References 409

Author Index 419
Subject Index 433

List of Contributors

T. Abel
Department of Biology, University of Pennsylvania, Philadelphia, PA, United States; Institute for Translational Medicine and Therapeutics, University of Pennsylvania, Philadelphia, PA, United States

R.A. Adams
Institute of Cognitive Neuroscience, University College London, London, United Kingdom; Division of Psychiatry, University College London, Charles Bell House, London, United Kingdom

S. Akbarian
Department of Psychiatry, Friedman Brain Institute, Icahn School of Medicine at Mount Sinai, New York, NY, United States

Y. Ayhan
Department of Psychiatry, Faculty of Medicine, Hacettepe University, Ankara, Turkey; Department of Psychiatry and Behavioral Sciences, School of Medicine, Johns Hopkins University, Baltimore, MD, United States

A. Banerjee
Department of Psychiatry, University of Pennsylvania, Philadelphia, PA, United States

K.E. Borgmann-Winter
Department of Psychiatry, University of Pennsylvania, Philadelphia, PA, United States; The Children's Hospital of Philadelphia, University of Pennsylvania School of Medicine, Philadelphia, PA, United States

D. Bzdok
Institut für Neurowissenschaften und Medizin (INM-1), Forschungszentrum Jülich GmbH, Jülich, Germany; Institut für klinische Neurowissenschaften und Medizinische Psychologie, Heinrich-Heine Universität Düsseldorf, Düsseldorf, Germany; Parietal Team, INRIA, Neurospin, Gif-sur-Yvette, France

S. Cichon
Division of Medical Genetics, University Hospital Basel, Basel, Switzerland; Department of Biomedicine, University of Basel, Basel, Switzerland; Institute of Human Genetics, University of Bonn, Bonn, Germany; Department of Genomics, Life & Brain Center, University of Bonn, Bonn, Germany; Institute of Neuroscience and Medicine (INM-1), Research Center Jülich, Jülich, Germany

M.P. Coba
Department of Psychiatry and Behavioral Sciences, Zilkha Neurogenetic Institute, Keck School of Medicine, University of Southern California, Los Angeles, CA, United States

G. Collin
Department of Psychiatry, University Medical Center Utrecht, Utrecht, The Netherlands; Brain Center Rudolf Magnus, Utrecht, The Netherlands

P. Cumming
School of Psychology and Counselling, Queensland University of Technology, and QIMR Berghofer Medical Research Institute, Brisbane, QLD, Australia

C. Davatzikos
Center for Biomedical Image Computing and Analytics, University of Pennsylvania, Philadelphia, PA, United States

S.B. Eickhoff
Institut für Neurowissenschaften und Medizin (INM-1), Forschungszentrum Jülich GmbH, Jülich, Germany; Institut für klinische Neurowissenschaften und Medizinische Psychologie, Heinrich-Heine Universität Düsseldorf, Düsseldorf, Germany

K.J. Friston
The Wellcome Trust Centre for Neuroimaging, University College London, London, United Kingdom

I. Giegling
Department of Psychiatry, Psychotherapy, and Psychosomatics, Martin Luther University Halle-Wittenberg, Halle, Germany

G. Gründer
Department of Psychiatry, Psychotherapy and Psychosomatics, RWTH Aachen University, Aachen, Germany

C.-G. Hahn
Department of Psychiatry, University of Pennsylvania, Philadelphia, PA, United States

H. Hakonarson
Center for Applied Genomics, Children's Hospital of Philadelphia Research Institute, Philadelphia, PA, United States; Department of Pediatrics, Perelman School of Medicine, University of Pennsylvania, Philadelphia, PA, United States

M. Hawrylycz
Allen Institute for Brain Science, Seattle, WA, United States

N. Hiroi
Department of Psychiatry and Behavioral Sciences, Albert Einstein College of Medicine, Bronx, NY, United States; Dominick P. Purpura Department of Neuroscience, Albert Einstein College of Medicine, Bronx, NY, United States; Department of Genetics, Albert Einstein College of Medicine, Bronx, NY, United States

N. Koutsouleris
Department of Psychiatry and Psychotherapy, Ludwig-Maximilians-University, Munich, Germany

M. Kundakovic
Department of Psychiatry, Friedman Brain Institute, Icahn School of Medicine at Mount Sinai, New York, NY, United States; Department of Biological Sciences, Fordham University, Bronx, NY, United States

B. Leitner
Department of Psychiatry and Psychotherapy, Ludwig Maximilian University, Munich, Germany

K. Mathiak
Department of Psychiatry, Psychotherapy, and Psychosomatics, Medical School, RWTH Aachen University, Aachen, Germany; Jülich Aachen Research Alliance-Translational Brain Medicine, RWTH Aachen University, Aachen, Germany

G.A. Miller
Department of Psychology and Department of Psychiatry and Biobehavioral Sciences, University of California, Los Angeles, CA, United States

C. Moessnang
Department of Psychiatry and Psychotherapy, Central Institute of Mental Health, Medical Faculty Mannheim/Heidelberg University, Mannheim, Germany

H. Möhler
Institute of Pharmacology, University of Zurich, Zurich, Switzerland; Department of Chemistry and Applied Biosciences, Swiss Federal Institute of Technology (ETH), Zurich, Switzerland

P.M. Moran
School of Psychology, University of Nottingham, Nottingham, United Kingdom

N. Müller
Department of Psychiatry and Psychotherapy, Ludwig Maximilian University, Munich, Germany

E. Neustadter
Yale University School of Medicine, New Haven, CT, United States

T. Nickl-Jockschat
Department of Psychiatry, Psychotherapy and Psychosomatics, RWTH Aachen University, Aachen, Germany; Juelich-Aachen Research Alliance–Translational Brain Medicine, Juelich/Aachen, Germany

A. Nishi
Department of Psychiatry and Behavioral Sciences, Albert Einstein College of Medicine, Bronx, NY, United States; Department of Psychiatry, Course of Integrated Brain Sciences and Medical Informatics, Institute of Biomedical Sciences, Tokushima University Graduate School, Tokushima, Japan

K. Pauly
Department of Psychiatry, Psychotherapy and Psychosomatic Medicine, RWTH Aachen University, Aachen, Germany

C. Peter
Department of Psychiatry, Friedman Brain Institute, Icahn School of Medicine at Mount Sinai, New York, NY, United States

M.V. Pletnikov
Department of Psychiatry and Behavioral Sciences, School of Medicine, Johns Hopkins University, Baltimore, MD, United States; Department of Molecular and Comparative Pathobiology, School of Medicine, Johns Hopkins University, Baltimore, MD, United States; Solomon H. Snyder Department of Neuroscience, School of Medicine, Johns Hopkins University, Baltimore, MD, United States; Department of Molecular Microbiology and Immunology, Bloomberg School of Public Health, Johns Hopkins University, Baltimore, MD, United States

R. Ray
Department of Psychiatry, University of Pennsylvania, Philadelphia, PA, United States

S. Ripke
Analytic and Translational Genetics Unit, Department of Medicine, Massachusetts General Hospital and Harvard Medical School, Boston, MA, United States; Stanley Center for Psychiatric Research and Medical and Population Genetics Program, Broad Institute of MIT and Harvard, Cambridge, MA, United States; Department of Psychiatry and Psychotherapy, Charité – Universitätsmedizin Berlin, Campus Mitte, Berlin, Germany

B.S. Rockstroh
Department of Psychology, University of Konstanz, Konstanz, Germany

P. Roussos
Department of Psychiatry, Department of Genetics and Genomic Science, Institute for Multiscale Biology, Friedman Brain Institute, Icahn School of Medicine at Mount Sinai, New York, NY, United States; Mental Illness Research, Education, and Clinical Center (VISN 3), James J. Peters VA Medical Center, Bronx, NY, United States

U. Rudolph
Laboratory of Genetic Neuropharmacology, McLean Hospital, Belmont, MA, United States; Department of Psychiatry, Harvard Medical School, Boston, MA, United States

D. Rujescu
Department of Psychiatry, Psychotherapy, and Psychosomatics, Martin Luther University Halle-Wittenberg, Halle, Germany

M.J. Schwarz
Institute of Laboratory Medicine, Medical Center of Ludwig Maximilian University (LMU), Munich, Germany

S.J. Siegel
Department of Psychiatry, University of Pennsylvania, Philadelphia, PA, United States

P.M.A. Sleiman
Center for Applied Genomics, Children's Hospital of Philadelphia Research Institute, Philadelphia, PA, United States; Department of Pediatrics, Perelman School of Medicine, University of Pennsylvania, Philadelphia, PA, United States

S. Sunkin
Allen Institute for Brain Science, Seattle, WA, United States

C.E. Terrillion
Department of Psychiatry and Behavioral Sciences, School of Medicine, Johns Hopkins University, Baltimore, MD, United States

B.I. Turetsky
Neuropsychiatry Program, Department of Psychiatry, Perelman School of Medicine, University of Pennsylvania, Philadelphia, PA, United States

M.P. van den Heuvel
Department of Psychiatry, University Medical Center Utrecht, Utrecht, The Netherlands; Brain Center Rudolf Magnus, Utrecht, The Netherlands

E. Weidinger
Department of Psychiatry and Psychotherapy, Ludwig Maximilian University, Munich, Germany

R.S. White
Department of Psychiatry, University of Pennsylvania, Philadelphia, PA, United States

Acknowledgments

We acknowledge the collaborative research communities at the University of Pennsylvania and RWTH Aachen University. Our colleagues, students, and postdoctoral fellows at Penn and at Aachen have taught us much about how to understand complex neuropsychiatric and neurodevelopmental disorders. Work in the Abel laboratory has been generously supported for more than 15 years by the National Institute of Mental Health, and our collaborative work has been fostered by an International Research Training Grant from the Deutsche Forschungsgemeinschaft. Dr Abel is supported by the Brush Family Professorship at the University of Pennsylvania. The Deutsche Forschungsgemeinschaft has also generously supported research in the Nickl-Jockschat laboratory.

We thank all of the contributors to this volume who wrote creative pieces that brought together ideas ranging from genetics to imaging to mouse models. Their efforts and expertise enabled us to compile this volume. Working with them has been a great and enriching experience.

This book would not have come together without the dedicated efforts of the editorial team at Elsevier, including Kathy Padilla and Melanie Tucker.

PART I

INTRODUCTION

CHAPTER 1

HISTORICAL AND CLINICAL OVERVIEW: IMPLICATIONS FOR SCHIZOPHRENIA RESEARCH

T. Nickl-Jockschat[1,2] and T. Abel[3,4]

[1]*Department of Psychiatry, Psychotherapy and Psychosomatics, RWTH Aachen University, Aachen, Germany*
[2]*Juelich-Aachen Research Alliance–Translational Brain Medicine, Juelich/Aachen, Germany* [3]*Department of Biology, University of Pennsylvania, Philadelphia, PA, United States* [4]*Institute for Translational Medicine and Therapeutics, University of Pennsylvania, Philadelphia, PA, United States*

CHAPTER OUTLINE

Introduction 3
A Brief History of the Definition of Schizophrenia 5
Lessons From Therapeutic Approaches 9
The Neurobiology of Schizophrenia 10
References 12

INTRODUCTION

Schizophrenia is a severe neuropsychiatric disorder that not only causes a high burden of disease but also challenges our understanding of how the mind and brain work. Certainties for healthy subjects (eg, that our thoughts or our actions are controlled by ourselves) are shattered for the schizophrenia patient. Consequently, a better understanding of the neurobiology of this disorder not only might help to identify better strategies for early diagnosis, therapy, and personalized medicine but also may answer some open questions about the biological correlates of our mind and consciousness.

Schizophrenia is defined as a syndrome. The diagnosis is based on a constellation of clinical symptoms and not on a common pathomechanism, as is the case for ischemic stroke or cardiac infarction. The two most widely used diagnostic systems, ICD-10 (*International Statistical Classification of Diseases and Related Health Problems*, WHO, 2010) and DSM-V (*Diagnostic and Statistical Manual of Mental Disorders*, American Psychiatric Association, 2013), both provide a catalog of symptoms and demand that a certain number of this pool must to be present over a given period of time for a diagnosis to be made (Table 1.1). It should be noted that because of these two classification systems, two patients can be diagnosed with schizophrenia even though they do not share a single symptom at the time of examination. According to DSM-V, a patient presenting with delusions and hallucinations can be diagnosed with schizophrenia, as can a patient presenting with disorganized speech and negative symptoms. Although

Table 1.1 Diagnostic Criteria for Schizophrenia According to DSM-V and ICD-10	
DSM-V	**ICD-10**
Two (or more) of the following, each present for a significant portion of time during a 1-month period (or less if successfully treated) At least one of these should include 1–3	Either at least one of the syndromes, symptoms, and signs listed as 1–4 or at least two of the symptoms and signs listed as 5–8 should be present for most of the time during an episode of psychotic illness lasting for at least 1 month (or at some time during most of the days).
1. Delusions	1. Thought echo, thought insertion or withdrawal, or thought broadcasting
2. Hallucinations	2. Delusions of control, influence, or passivity, clearly referred to body or limb movements or specific thoughts, actions, or sensations; delusional perception
3. Disorganized speech	3. Hallucinatory voices giving a running commentary on the patient's behavior or discussing him between themselves, or other types of hallucinatory voices coming from some part of the body
4. Grossly disorganized or catatonic behavior	4. Persistent delusions of other kinds that are culturally inappropriate and completely impossible (eg, being able to control the weather or being in communication with aliens from another world)
5. Negative symptoms (ie, diminished emotional expression or avolition)	5. Persistent hallucinations in any modality occurring every day for at least 1 month, accompanied by delusions (which may be fleeting or half-formed), without clear affective content, or accompanied by persistent overvalued ideas
	6. Neologisms, breaks, or interpolations in the train of thought, resulting in incoherence or irrelevant speech
	7. Catatonic behavior, such as excitement, posturing or waxy flexibility, negativism, mutism, and stupor
	8. "Negative" symptoms such as marked apathy, paucity of speech, and blunting or incongruity of emotional responses (it must be clear that these are not due to depression or to neuroleptic medication)
It should be noted that both diagnostic systems require additional tests, mainly to exclude secondary symptom genesis due to a somatic disorder.	

both hypothetical patients arguably differ in their clinical presentations, they are both given the same diagnosis. In other words, schizophrenia can manifest with totally different clinical phenotypes.

It should be clear that this heterogeneity poses significant challenges for the identification of the neurobiological underpinnings of this disorder. If schizophrenia can present with strikingly distinct symptoms in different patients, then distinct pathophysiological mechanisms with specific etiological factors might be subsumed under the broad concept of "schizophrenia" (cf. Tandon et al., 2013). Consequently, the identification of causal factors might be severely hampered due to this heterogeneity.

To understand the current definition of schizophrenia, it seems inevitable that we must look back at its historical roots. The diagnosis criteria in both ICD-10 and DSM-V are the result of clinical and scientific developments spanning several decades and are founded on descriptions that were published in the first half of the twentieth century. Thus, when speaking of schizophrenia today, we do not refer to an etiologically defined disease but rather to a syndrome with a history of its own.

Here, we briefly summarize the developments that influenced the definitions of this disorder. We point out that we do not claim completeness with regard to medical history; rather, we aim to exemplify the importance—and the inherent challenges—that these definitions pose for current research into the neurobiological basis of schizophrenia.

A BRIEF HISTORY OF THE DEFINITION OF SCHIZOPHRENIA

Since the beginning of psychiatry as a medical discipline, the syndrome called "schizophrenia" has been a major challenge for clinicians and scientists. Although schizophrenia-like symptoms have been reported by physicians of ancient Greece, we start our brief history of schizophrenia with the work of the German psychiatrist *Wilhelm Griesinger* (1817–68) (Fig. 1.1). Griesinger's ideas were influential on a scientific level and a clinical level. In general, he demanded that clinical psychiatry should be based on empirical research rather than speculation, and he highlighted the need for neurobiological (specifically, neuroanatomical) studies. Given his predecessors who stemmed from psychiatry in the era of German Romanticism and, accordingly, emphasized the importance of individual passions and irrational impulses for psychopathology, this concept appeared revolutionary (Hoff and Theodoridou, 2008). The famous phrase attributed to Griesinger that "mental illnesses are illnesses of the brain ("*Geisteskrankheiten sind Gehirnkrankheiten*")" is certainly an abbreviation and simplification of his ideas, but this phrase underlines his passionate attempt to further neurobiological research in

FIGURE 1.1

Wilhelm Griesinger (1817–68).

FIGURE 1.2

Emil Kraepelin (1856–1926).

psychiatry. With regard to schizophrenia, Griesinger formulated the concept of the "unitary psychosis" (*Einheitspsychose*) that is based on the assumption of a continuum between all psychiatric disorders. Griesinger proposed a model that starts with an early affective phase, followed by a delusional stage that finally leads to irreversible cognitive deficits (Griesinger, 1861). Given recent genetic studies that suggest a continuum between variants mediating vulnerability for schizophrenia and bipolar disorder, this idea of a continuum between psychiatric disorders appears surprisingly modern.

Distinct from Griesinger's unitary psychosis, *Emil Kraepelin* (1856–1926) (Fig. 1.2) proclaimed a dichotomy of psychoses. He distinguished two major classes of psychotic disorders: *dementia praecox* (premature dementia)—corresponding to our current use of the term schizophrenia—and *manisch-depressives Irreseyn* (manic depression)—corresponding to the current range of affective disorders (Kraepelin, 1896). Remarkably, the long-term outcome and the course of the illness played a decisive role in Kraepelin's classification. Broadly speaking, *dementia praecox* was thought to be associated with a poor prognosis, whereas manic depression was considered a comparatively benign disease. Kraepelin also explicitly acknowledged that individual symptoms did not distinguish between the two disease entities, but rather the pattern of symptoms leads to a distinction between these disorders. This basic Kraepelian dichotomy laid the foundation for the idea of distinct psychiatric disease entities and can be regarded as the first step toward our modern classification systems.

Although the Kraepelian dichotomy allowed a valid distinction between two types of disorders based on their prognosis and their course, there was still a lively discussion about which symptoms were able to distinguish between schizophrenia and disorders with a better prognosis. The most important input at this time that still influences our concepts of schizophrenia today came from the Swiss psychiatrist *Eugen Bleuler* (1857–1939) and his German contemporary *Kurt Schneider* (1887–1967).

FIGURE 1.3

Eugen Bleuler (1857–1939).

Bleuler (Fig. 1.3) was the one to coin the term "schizophrenia" or, more exactly, the "group of schizophrenias" (Bleuler, 1911). Before this, the Kraepelian term of *dementia praecox* was widely used. One of the reasons for Bleuler to shift the terminology away from the term "dementia" to the phrase "group of schizophrenias" was that he disagreed with Kraepelin in two important ways. Kraepelin treated schizophrenia as a single condition that was invariably incurable. Blueler pointed out that some cases had a comparatively good prognosis and did not proceed into a dementia-like phenotype. Moreover, Kraepelin regards *dementia praecox* as a disorder that manifests in young people; therefore, he coined the term *premature* dementia. In contrast to this, Bleuler pointed out that the disease could also manifest at later ages (Fusar-Poli and Politi, 2008). This challenged the Kraepelian notion that distinguished *dementia praecox*, as its name insinuates, from other disorders based on its malignant progress of debilitating symptoms.

Moreover, Beuler introduced diagnostic criteria for schizophrenia focused on fundamental symptoms that are regarded as core symptoms of the disorder and accessory symptoms that are often seen but are not necessary for diagnosis. His fundamental symptoms consisted of the loosening of associations (today referred to as formal thought disorder or disorganized speech), disturbances of affectivity, ambivalence, and autism (the latter to describe the social withdrawal of the patients and the first mentioning of the term "autism" in the medical literature). These have become famous as the "four As" (for an overview of Bleuler's fundamental and accessory symptoms, please see Table 1.2). It is important to point out that Bleuler's fundamental symptoms, which are central to his definition of the disorder, largely reflect what are today called "negative symptoms," whereas positive symptoms, such as delusions or hallucinations, are missing from Bleuler's description.

Similar to Bleuler's fundamental and accessory symptoms, Kurt Schneider (Fig. 1.4) distinguished between first-rank (core) and second-rank (less important) symptoms; however, in contrast to Bleuler's approach, Schneider's first-rank symptoms are closely related to positive symptoms. In

Table 1.2 Diagnostic Criteria for Schizophrenia (the "Four As") According to Eugen Bleuler

Affect	Inappropriate or flattened affect
Autism	Social withdrawal
Ambivalence	Holding of conflicting attitudes and emotions toward others and self
Associations	Loosening of thought associations leading to disorganized speech/thought disorder

FIGURE 1.4

Kurt Schneider (1887–1967).

this classification system, auditory hallucinations per se—and especially imperative voices—were not first-rank symptoms. Only auditory hallucinations of a dialoguing or commenting quality were viewed as first-rank symptoms. Schneider characterized two or more voices that are talking about the patient like a dialogue as "dialoguing," whereas voices that are commenting on the actions of the patient were labeled as "commenting." Many of his first-rank symptoms are centered around his idea of *Ich-Störung* (in the English language usually only referred to as a subgroup of delusions): bizarre phenomena such as thought withdrawal (the idea that thoughts are taken away from the patient's mind), thought insertion (the intrusion of thoughts into the patient's mind, usually believed to be caused by another person), thought dispersion or broadcast (the idea that the thoughts of the patient can be read or heard by other people), and passivity experiences or delusions of control (over the mind or body of the patient). Moreover, Schneider highlighted the importance of delusional perceptions, which are normal sensory perceptions linked to a bizarre, delusional conclusion by the patient (eg, seeing a "STOP" traffic sign leads the patient to the believe that this signals his or her own near death) (Schneider, 1950) (for an overview of Schneider's first- and second-rank symptoms, please see Table 1.3).

Kurt Schneider's diagnostic concept has exerted its influence over the years. Three out of the first four diagnostic criteria of the ICD-10 directly refer to first-rank symptoms, and only the fourth criterion—bizarre delusional beliefs—is a second-rank symptom. Although the authors certainly realize that current diagnostic systems are the result of decade-long discussions rather than the simple continuation

Table 1.3 First- and Second-Rank Symptoms of Schizophrenia According to Kurt Schneider

First-Rank Symptoms	Second-Rank Symptoms
Hearing voices conversing with one another (dialoguing voices)	Other disorders of perception
Voices heard commenting on one's actions (commenting voices)	Sudden delusional ideas
Thought echo (the patient hears his/her thoughts spoken aloud)	Perplexity
Thought withdrawal (the delusional belief that thoughts have been "taken out" of the patient's mind)	Affective symptoms
Thought insertion (thoughts are believed to be intruding into the patient's mind from outside, typically due to the actions of another person)	Feeling of emotional impoverishment
Thought broadcasting (the belief that the thoughts of the patients can be heard or perceived by other persons, also called thought dispersion)	
Passivity experiences (in which the individual has the experience of the mind or body being under the influence or control of some kind of external force or agency)	
Delusional perception (linking a normal sensory perception to a bizarre conclusion, eg, seeing a stop sign indicates that the patient is going to die soon.)	

of old concepts, they think that Schneider's impact on the current concepts in ICD-10 might serve well as an example of the importance of historical definitions of schizophrenia in other contemporary classification systems. Despite decades of discussions between basic researchers and clinicians, one of the two most important diagnostic systems still is largely based on Schneider's first-rank criteria.

It is important in this context that the two major classification systems—ICD-10 and DSM-V—that are used in research and in clinics do not use the same criteria for the diagnosis of schizophrenia. One of the differences most easily recognizable is the time during which symptoms need to be present: 1 month according to ICD-10, but 6 months in most cases according to DSM-V (WHO, 2010; American Psychiatric Association, 2013). Consequently, these differences define at least the fringes of the patient pools included in the respective diagnosis and thus might lead to different conclusions with regard to the neurobiological underpinnings of this complex disorder.

LESSONS FROM THERAPEUTIC APPROACHES

On January 23, 1934, the Hungarian psychiatrist *Ladislas Meduna* (1896–1964) used a camphor solution to induce a generalized seizure in a schizophrenia patient. As reported in Meduna's autobiography, the patient, 33-year-old Zsoltan L., achieved full remission after 4 years of illness and was discharged from the hospital (Meduna, 1985). Although other sources questioned the therapeutic success, as described by Meduna (Baran et al., 2008), first chemoconvulsive therapy, and later its modified version of electroconvulsive therapy (Bini, 1937) became the first effective biological treatment strategy that is still used today, although admittedly rarely for schizophrenia (cf. Loh et al., 2013).

With the discovery of chlorpromazine as the first antipsychotic drug in 1952, treatment strategies were revolutionized. Chlorpromazine was initially used as a sedative agent to enhance anesthesia. However, due to its sedating potential, it was believed to have a potential role in the treatment of psychiatric patients. On January 19, 1952, it was first administered to a patient with acute mania. Remarkably, this patient was discharged after 3 weeks of treatment (López-Muñoz et al., 2005). Following this remarkable success, the first clinical trial followed the same year (Delay and Deniker, 1955). For the first time in medical history, drugs existed that were easy to administer and had the potential to cure a disease that was previously regarded as incurable (López-Muñoz et al., 2005). Although the advent of atypical antipsychotics helped to minimize some of the extrapyramidal motor side effects, the basic molecular mechanism was still antidopaminergic (Gründer et al., 2009). With regard to our topic of neurobiological schizophrenia research, the use of antipsychotics led to important conclusions and helped to point out the dopaminergic system as a key player in schizophrenia pathophysiology (Carlsson, 2006). However, if dopamingergic antagonism alone would explain the antipsychotic properties, then a given patient should be fully remitted minutes after intravenous application of haloperidole, when the receptor occupancy needed for the antipsychotic effect is reached. The fact that a therapeutic response takes much longer points to the idea that other mechanisms are potentially involved.

The use of antipsychotics remains the gold standard in schizophrenia treatment (American Psychiatric Association, 2013). However, despite their overall efficacy, some patients benefit earlier and more completely from antipsychotic treatment than others, and a subgroup does not respond sufficiently to pharmacological treatment (Hasan et al., 2012). Consequently, new therapeutic strategies are needed. However, only a better understanding of schizophrenia pathophysiology can help to discover new treatment approaches. Therefore, a better understanding of the underlying neurobiological pathomechanisms not only will help to shed more light on basic functions of our minds and brains but also will help to develop new treatment strategies for this severe mental disorder.

THE NEUROBIOLOGY OF SCHIZOPHRENIA

The idea that schizophrenia was a biological disorder is nearly as old as the first depiction of its symptoms. Hippocrates tried to find physical causes for psychiatric disorders, including those with schizophrenia-like symptoms, but he suspected that imbalances of body fluids, not pathologies of the brain, caused the symptoms.

More than two millennia later, the improved clinical characterization of psychiatric disorders that were previously labeled as "insanity" without any further distinction also sparked new interest and new approaches in neurobiological research. As depicted, Wilhelm Kraepelin was the first to distinguish schizophrenia (or, in his words, *dementia praecox*) from other psychiatric disorders. Remarkably, his clinical ideas were compatible with increased scientific efforts to unravel the pathophysiological mechanisms underlying these diseases. Kraepelin took the initiative to found the *Deutsche Forschungsanstalt für Psychiatrie* (German Research Institute of Psychiatry), a predecessor of the Max Planck Institute for Psychiatry. Under this framework, scientists including *Alois Alzheimer* (1864–1915), *Franz Nissl* (1860–1919), and *Robert Gaupp* (1870–1953) researched the genetics and neuropathology of schizophrenia.

The next important revolution in neurobiological research came with the introduction of antipsychotics as a treatment strategy. As depicted, antipsychotic medication has directed the attention toward distinct neurotransmitter systems early and has helped to unravel some of the molecular changes in schizophrenia patients.

Although neuroanatomical research on the brains of schizophrenia patients has already been systematically conducted in Kraepelin's research institute, several important obstacles prevented the researchers from a robust characterization of structural changes in schizophrenia. As we know today, brain structure changes are subtle and inter-individual variance among patients is high. Consequently, there is a need for large groups in such studies, but these are difficult to acquire for a postmortem study. The introduction of modern imaging techniques, especially magnetic resonance imaging (MRI), allowed the inclusion of cohorts with sufficient sample sizes to counter that problem (Shenton et al., 2001). They also yielded the great advantage of longitudinal study designs that helped to characterize the temporal dynamics of these brain structure changes. In addition to structural imaging, functional imaging techniques have led to important breakthroughs. They have provided the unique chance to study functional changes in the brains of schizophrenia patients. Currently, functional MRI (fMRI) is a widely used method to characterize neural correlates of schizophrenia symptoms. The acquisition of large cohorts, often by collaborative approaches, has significantly increased the number of individuals available for a given analysis and, therefore, helped to solidify the reliability of their results.

It has been known since the beginning of the twentieth century that schizophrenia is a disorder with high heritability. Consequently, the identification of susceptibility variants has been a major effort in psychiatric genetics. However, most common variants in schizophrenia have rather low odds ratios for the actual manifestation of the disease. It took the pooling of large cohorts to achieve the robust identification of genetic variants with genome-wide significance. Modern genetics has also shifted the focus to large deletions or duplications of the genome, so-called copy number variations (CNVs). In contrast to more common variants, some CNVs show rather high odds ratios for schizophrenia and other neuropsychiatric disorders. Given this, the biological characterization of CNVs seems to be a promising approach to understanding the molecular underpinnings of this disorder.

With the recent major advances in psychiatric genetics and the robust identification of schizophrenia susceptibility genes, genetic animal models have gained increasing importance. These models yield several great advantages. Besides a homogeneous genetic background, animals can also be raised under standardized conditions. Consequently, environmental factors can be expected to play only a minor role in shaping the behavioral phenotype. They are also perfect models to study molecular and electrophysiological changes due to the modeled genetic variants. As depicted, there is considerable clinical heterogeneity among schizophrenia cases. Due to DSM-V and ICD-10, these heterogeneous cases are summarized under one diagnostic label. Consequently, considerable heterogeneity of the underlying pathophysiological processes might obscure the identification of the neurobiological underpinnings. One potential option to address this dilemma is the use of newly developed classification schemes that are based on dimensions of observable behavior and neurobiological measures rather than an "artificial grouping of heterogeneous syndromes with different pathophysiological mechanisms into one disorder" (Wong et al., 2010; Ford et al., 2014). Because of these considerations, the National Institute of Mental Health (NIMH) has developed so-called research domain criteria (RDoCs) that treat psychopathological symptoms as situated within a continuum that reaches from normal functioning to pathological extremes. This is intended to minimize biological heterogeneity within the studied cases and might be a decisive step toward the identification of neurobiological foundations of schizophrenia.

Our aim with this textbook is to provide an overview of the current state of this lively field of research. Therefore, we focus on behavioral, cognitive, clinical, electrophysiological, molecular, and genetic levels.

As depicted, schizophrenia is a group of complex and potentially pathophysiologically heterogeneous disorders with high heritability involved with pronounced changes in brain structure and function. The symptoms are complex and sometimes bizarre, involving disturbances of a multitude of cognitive and affective domains. Despite significant progress in various areas of research, the etiopathogenesis of these complex disorders is still not fully understood. The intricate interplay between various susceptibility genes and structural alterations in neuronal connectivity along with changes at epigenetic and neurochemical levels make the neurobiology of schizophrenia a complex topic. Therefore, we have organized and focused discussions on critical areas of research.

Given the complex nature of schizophrenia neurobiology, various professions are involved in research, ranging from molecular biologists to physicians and psychologists. However, students of the respective disciplines are often confronted with—and confused by—the puzzling riddle of this disorder. We explicitly aimed for our textbook to be easily comprehensible for this heterogeneous audience. Therefore, we include chapters briefly describing basic methods and techniques reported in the section. We hope that this approach makes it easy for MD and PhD students to introduce themselves to this field of research and help to spark new scientific approaches for advanced researchers.

REFERENCES

American Psychiatric Association, 2013. Diagnostic and Statistical Manual of Mental Disorders, fifth ed. Author, Washington, DC.

Baran, B., Bitter, I., Ungvari, G.S., Nagy, Z., Gazdag, G., 2008. The beginnings of modern psychiatric treatment in Europe. Lessons from an early account of convulsive therapy. Eur. Arch. Psychiatry Clin. Neurosci. 25, 434–440.

Bini, L., 1937. Richerche sperimentali nell'accesso epilettico da corrente elettrica. In: Die Therapie der Schizophrenie: Insulinschock, Cardiazol, Dauerschlaf. Bericht über die wissenschaftlichen Verhandlungen auf der 89. Versammlung der Schweizerischen Gesellschaft für Psychiatrie in Münsingen b. Bern am 29.-31. Mai 1937. Schweiz Arch Neurol Psychiatr, 39 (Suppl.), pp. 121–122.

Bleuler, E., 1911. Dementia praecox oder Gruppe der Schizophrenien. In: Aschaffenburg, G. (Ed.), Handbuch der Psychiatrie. Spezieller Teil, 4. Abteilung, 1. Hälfte Deuticke, Leipzig.

Carlsson, A., 2006. The neurochemical circuitry of schizophrenia. Pharmacopsychiatry 39 (Suppl. 1), S10–S14.

Delay, J., Deniker, P., 1955. Neuroleptic effects of chlorpromazine in therapeutics of neuropsychiatry. J. Clin. Exp. Psychopathol. 16 (2), 104–112.

Ford, J.M., Morris, S.E., Hoffman, R.E., Sommer, I., Waters, F., McCarthy-Jones, S., et al., 2014. Studying hallucinations within the NIMH RDoC framework. Schizophr. Bull. 40 (Suppl. 4), S295–304.

Fusar-Poli, P., Politi, P., 2008. Paul Eugen Bleuler and the birth of schizophrenia (1908). Am. J. Psychiatry 165 (11), 1407.

Griesinger, W., 1861. Die Pathologie und Therapie der psychischen Krankheiten. *Krabbe, Stuttgart*.

Gründer, G., Hippius, H., Carlsson, A., 2009. The "atypicality" of antipsychotics: a concept re-examined and redefined. Nat. Rev. Drug Discov. 8 (3), 197–202.

Hasan, A., Falkai, P., Wobrock, T., Lieberman, J., Glenthoj, B., Gattaz, W.F., et al., 2012. World Federation of Societies of Biological Psychiatry (WFSBP) guidelines for biological treatment of schizophrenia, part 1: update 2012 on the acute treatment of schizophrenia and the management of treatment resistance. World J. Biol. Psychiatry 13 (5), 318–378.

Hoff, P., Theodoridou, A., 2008. Schizophrene Psychosen im Spannungsfeld von Kognition, Affekt und Volition – Die psychiatriehistorische Perspektive. In: Kircher, T., Gauggel, S. (Eds.), Neuropsychologie der Schizophrenie Springer, Heidelberg.

Kraepelin, E., 1896. Psychiatrie. Ein Lehrbuch für Studierende und Aerzte. Barth, Leipzig.

REFERENCES

Loh, N., Nickl-Jockschat, T., Sheldrick, A.J., Grözinger, M., 2013. Accessibility, standards and challenges of electroconvulsive therapy in Western industrialized countries: a German example. World J. Biol. Psychiatry 14 (6), 432–440.

López-Muñoz, F., Alamo, C., Cuenca, E., Shen, W.W., Clervoy, P., Rubio, G., 2005. History of the discovery and clinical introduction of chlorpromazine. Ann. Clin. Psychiatry 17 (3), 113–135.

Meduna, L., 1985. Autobiography. Convuls. Ther. 1 (43–57), 121–135.

Schneider, K., 1950. Klinische Psychopathologie. Thieme, Stuttgart.

Shenton, M.E., Dickey, C.C., Frumin, M., McCarley, R.W., 2001. A review of MRI findings in schizophrenia. Schizophr. Res. 49, 1–52.

Tandon, R., Gaebel, W., Barch, D.M., Bustillo, J., Gur, R.E., Heckers, S., et al., 2013. Definition and description of schizophrenia in the DSM-5. Schizophr. Res. 150 (1), 3–10.

Wong, E.H., Yocca, F., Smith, M.A., Lee, C.M., 2010. Challenges and opportunities for drug discovery in psychiatric disorders: the drug hunters' perspective. Int. J. Neuropsychopharmacol. 13 (9), 1269–1284.

World Health Organization, 2010. International Statistical Classification of Diseases and Related Health Problems. <http://www.who.int/classifications/icd/en/bluebook.pdf?ua=1> (accessed 07.03.16.).

PART II

THE GENETIC AND EPIGENETIC BASIS OF SCHIZOPHRENIA

CHAPTER 2

PROGRESS AND PROSPECTS FOR ENDOPHENOTYPES FOR SCHIZOPHRENIA IN THE TIME OF GENOMICS, EPIGENETICS, OSCILLATORY BRAIN DYNAMICS, AND THE RESEARCH DOMAIN CRITERIA

G.A. Miller[1] and B.S. Rockstroh[2]

[1]Department of Psychology and Department of Psychiatry and Biobehavioral Sciences, University of California, Los Angeles, CA, United States [2]Department of Psychology, University of Konstanz, Konstanz, Germany

CHAPTER OUTLINE

Endophenotypes in Schizophrenia: Current Status 22
An Emerging Frontier for Endophenotypes: Epigenetics 26
A New Frontier for Endophenotypes: Brain Connectivity Dynamics 28
Limitations and Future Directions 30
References 31

Endophenotypes are defined as biological or psychological phenomena of a disorder believed to be in the causal chain between genetic contributions to a disorder and diagnosable symptoms of psychopathology (Glahn et al., 2014; Gould and Gottesman, 2006; Gottesman and Shields, 1972, 1973; Gottesman and Gould, 2003; Iacono and Lykken, 1979; Lenzenweger, 2010, 2013; Miller and Rockstroh, 2013; Pearlson, 2015). Endophenotypes can be biological or psychological, as long as there is a foreseeable causal path from the gene through endophenotype(s) to phenotype (clinical expression). The endophenotype literature frames the causal paths to illness as beginning with genes (rather than with environments—social, economic, physical—or in terms of a larger temporal continuum along which individuals are manifestations of slower processes), which is an arguable starting point (eg, Schumann, 2014). But the identification of endophenotypes should facilitate the identification of genetic contributions to the psychiatric disorder and the causal mechanisms between them.

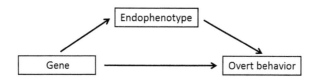

FIGURE 2.1

An endophenotype as an interim step in the causal chain between genes and observable clinical manifestation.

This chapter identifies and examines several conceptual issues concerning endophenotypes in schizophrenia research and in psychopathology research more generally. A selective review of recent endophenotype research in schizophrenia is enriched with some insights from genomics, epigenetics, and oscillatory brain dynamics to enhance the goal of understanding mechanisms of schizophrenia pathophysiology and psychopathology. The potential of adopting the Research Domain Criteria (RDoC) in pursuit of this goal is briefly discussed. Profound logical and conceptual issues remain to be resolved concerning the causal and other relationships between psychological and biological phenomena (Berenbaum, 2013; Lilienfeld, 2007, 2014; Miller, 1996, 2010; Miller and Keller, 2000; Miller and Rockstroh, 2013), but the empirical literature is nonetheless advancing.

If one assumes that there is a causal path between genes and manifest psychopathology and that that the path has multiple stages (whether or not there are multiple paths), then one believes that there are endophenotypes, which are the stages along that path. Fig. 2.1 illustrates the simplest possible example of such a model, which comprises essentially what is known in the statistical literature and is known as a mediation model. The concept is that at least some of the causal force of genetic input works via an intermediate step, the endophenotype. The figure conveys how the endophenotype may be closer, causally, to both genetic input and clinical manifestation.

The extent to which one should target endophenotypes is a strategic question for the psychopathology research literature. Much of the rationale in the endophenotype literature is that we can make progress by dissecting and understanding small spans along the causal path. One can focus on the mechanisms by which genes lead to endophenotypes or the mechanisms by which endophenotypes lead to a clinically significant disorder. One might treat some endophenotypes as milder versions of the disorder (such as was proposed for schizotypy regarding schizophrenia) or as fundamentally nonclinical features of one's personality (eg, low levels of behavioral inhibition), one's biochemistry (eg, aberrant DNA methylation), or other individual differences that in certain circumstances can foster the development or recurrence of disorder.

Nothing about the concept of endophenotype diminishes the role of environment, nor does it entail an assumption that genetic contributors are more important than environmental contributions. The assumptions are merely that there are genetic contributors and that it will prove very fruitful to identify their sequelae through the causal chain(s). As discussed, it is now widely understood that genes alone are likely to play only a limited role in most forms of psychopathology, that the genetic contributions are likely to be numerous and complex, and that the relationships of genetic contributions to environmental contributions are also likely to be complex. There is no doubt that environment will be a critical part of the schizophrenia story.

Table 2.1 Criteria for Endophenotypes
1. An endophenotype is associated with illness in the population
2. An endophenotype is heritable
3. An endophenotype is state-independent (manifests in an individual whether or not illness is active) but age-normed and might need to be elicited by a challenge (eg, glucose tolerance test in relatives of diabetics)
4. Within families, endophenotype and illness cosegregate
5. An endophenotype identified in probands is found in their unaffected relatives at a higher rate than in the general population
6. The endophenotype should be a trait that can be measured reliably, and ideally is more strongly associated with the disease of interest than with other psychiatric conditions (ie, specificity)
Source: Proposed by Gottesman and Shields (1972). See also Hasler (2006), Lenzenweger (2010), and Ritsner and Gottesman (2011).

The endophenotype concept was introduced to the psychopathology literature more than 40 years ago (Gottesman and Shields, 1972), which had little impact until an invitation to Gottesman 30 years later to revisit the concept (Gottesman and Gould, 2003). In the schizophrenia literature, a substantial body of evidence has evaluated potential endophenotypes following the criteria of Gottesman (discussed at length in Lenzenweger, 2010 and in Miller and Rockstroh, 2013). Table 2.1 lists the criteria proposed by Gottesman and colleagues. Table 2.2 cites some endophenotypes proposed for schizophrenia as meeting some, and in some cases all, of the Gottesman criteria.

In the past decade, the potential of endophenotypes has been an active subject of debate (eg, Baker, 2014; Braff, 2014; Ford, 2014; Goldman, 2014; Iacono, 2014; Iacono et al., 2014; Miller et al., 2014; Miller et al., 2016; Munafò and Flint, 2014; Patrick, 2014; Schumann, 2014; Wilhelmsen, 2014; Yee et al., 2015). A common misunderstanding is that endophenotypes must be biological phenomena. In fact, interest in psychological endophenotypes reflects the growing recognition that cognitive deficits are among the most important symptoms of schizophrenia in terms of impact on society and on individual quality of life. Accordingly, efforts are growing to develop and evaluate psychological interventions for specific phenomena associated with schizophrenia, especially the cognitive remediation of performance deficits. An important reason for this shift is that pharmacological treatment, emphasizing biological features of psychopathology, has not delivered on its promise to alleviate or substantially improve cognitive decline and negative symptoms, and, if anything, prospects have declined in recent years (Keefe and Harvey, 2012; Yee et al., 2015).

Intellectually, this literature is something of an antidote to the nearly exclusive focus on genetics and pharmacology that has dominated the schizophrenia treatment literature in recent decades. That tradition reflected a widely shared but naively reductionist view of mental illness that overemphasizes the clearly sizable contribution of genetic and other biological factors at the expense of psychological factors and gene × environment (G × E) interactions, correlations, and other joint relationships. Fig. 2.2 illustrates this much richer but much more challenging picture now taking shape. In addition to showing that an endophenotype may have far more complex inputs, it shows that both endophenotype and clinical manifestation may feed back into those inputs, altering their contribution over time. Such feedback loops over time are surely crucial to the prevention, development, remission, resurgence, and treatment of schizophrenia.

Table 2.2 Commonly Proposed Endophenotypes for Schizophrenia

System	Proposed Endophenotype	Task/Context	Example
Brain structure	Gray matter (reduction)		Glahn et al. (2015) Radulescu et al. (2014)
Brain structure	Hippocampus volume		Mathew et al. (2014) Seidman et al. (2002)
Brain structure	Subcortical and limbic shape		Roalf et al. (2015)
Brain structure	Regional white matter abnormalities		Bohlken et al. (2014) Lyu et al. (2014)
Brain motor	Neurological soft signs (motor coordination, sequencing, complex acts, sensory integration)	NES, CNI	Chan and Gottesman (2008) Neelam et al. (2011)
Connectivity	PFC–ACC–subcortical network	Executive functions	White and Gottesman (2012)
EEG oscillations	Resting alpha or gamma		Ivleva et al. (2014) Uhlhaas and Singer (2013)
EEG ERP	P50/M50/N100/M100 EEG gating suppression	Paired clicks	Adler et al. (1982) Bramon et al. (2004) Heinrichs (2004)
EEG ERP	MMN (reduction)	Acoustic deviance	Light et al. (2014) Turetsky et al. (2007)
EEG ERP	N100	Clicks	Ethridge et al. (2015) Turetsky et al. (2008)
EEG ERP	N200	Oddball	Ethridge et al. (2015)
EEG ERP	P300	Oddball	Malone et al. (2014) Turetsky et al. (2015)
EMG	Prepulse inhibition	Acoustic startle	Swerdlow et al. (2008)
Vision	Startle–blink reflex	Acoustic startle	Vaidyanathan et al. (2014)
Electrodermal	Nonresponding	Auditory habituation	Iacono (1985)
Vision	Smooth pursuit eye movements	Moving target	Calkins et al. (2008) Iacono (1985)
Vision	Saccade/antisaccade		Kathmann et al. (2014) Mazhari et al. (2011)
Cognitive	Verbal working memory	COGS battery cognitive tests	Gur et al. (2007) Pearlson and Calhoun (2009)
Cognitive	Episodic memory	COGS battery	Seidman et al. (2015)
Cognitive	Spatial WM	SWM test	Saperstein et al. (2006)
Cognitive	Visual (memory) span	Span tasks	Tuulio-Henriksson et al. (2003)
Cognitive	Attention	Degraded CPT	Gur et al. (2007)
Cognitive	Visual abstraction	COGS battery	Seidman et al. (2015)

Examples (not comprehensive) updated from review by Miller and Rockstroh (2013).

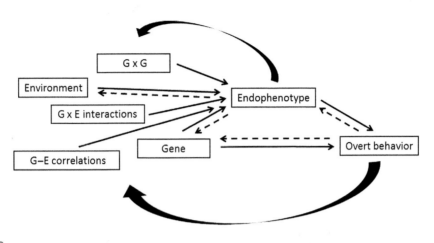

FIGURE 2.2

Numerous inputs to an endophenotype and feedback pathways over time. G, gene; E, environment.

The overemphasis on genetics and pharmacology in the research agenda for schizophrenia and other severe mental illness has been undermined by the lack of progress in pharmacological treatments of major mental illness in recent decades, with no major advances, few prospects, and no impact on some critical areas of clinical manifestation, specifically negative symptoms (such as flat affect and poverty of speech) and social and cognitive function. It has come to be appreciated that treating the positive symptoms (such as delusions and hallucinations), for which medications are often beneficial (although often with substantial side effects), is far from adequate if we are to help patients beyond reducing positive symptoms, including how well they perform at work and in social functions in daily life.

In response to the growing realization that heavy reliance on pharmacological methods has been unsuccessful, a variety of promising psychological interventions are now undergoing development, targeting sensory, cognitive, and social deficits in schizophrenia (and other mental disorders) (eg, Cuthbert and Instel, 2013; Crocker et al., 2013; Miller, 2010; Insel and Cuthbert, 2015). Furthermore, psychological symptoms have increasingly come to be appreciated as potentially valuable outcome variables for evaluating pharmacological interventions. We have come to realize that positive symptoms are not the primary barrier to effective function in the daily lives of people with schizophrenia. We need ways of improving their cognitive and social functions, which phenomenologically may appear as or contribute to negative symptoms, as critical factors in vocational performance and quality of life. Thus, there is some return of appreciation of the fundamentally psychological nature of schizophrenia. However, the field is not where it needs to be in terms of the appropriate technologies for measuring cognitive and social functions. Some of the sensory, cognitive, and social targets in the recent psychological treatment literature may qualify as endophenotypes. In fact, approaching endophenotypes as intervention targets, rather than limiting interventions to monolithic disorders, is appealing, especially when it may be beneficial to intervene before the full clinical picture develops.

ENDOPHENOTYPES IN SCHIZOPHRENIA: CURRENT STATUS

Miller and Rockstroh (2013) summarized research on endophenotypes in schizophrenia (see also Table 2.2), emphasizing a significant role for executive functions, several psychophysiological phenomena including functional brain (event-related and oscillatory) measures, and the overlap in phenomena meeting Gottesman's criteria for an endophenotype between schizophrenia and psychotic bipolar disorder. Some endophenotypes span traditional diagnostic categories, as conceived in the RDoC initiative (Kozak and Cuthbert, 2016; Miller et al., 2016).

A literature search in December 2014 (key words endophenotype, schizophrenia) confirmed that despite the overwhelming research on psychiatric genetics and some skepticism about the limited utility of endophenotypes (eg, Patrick, 2014), interest in endophenotypes in schizophrenia is unbowed. Endophenotype research is alive and prospering: for 2014, PubMed identified almost 80 publications. Moreover, results of consortia of researchers providing data from very large samples (eg, *B-SNIP, COGS-1, COGS-2*) offer an increasingly solid basis for optimism.

Many of the promising endophenotypes for schizophrenia are biological (in most cases psychophysiological), and several are psychological. Interestingly, overt performance measures appear to tap variance separate from those of individual biological measures. Recent factor analyses by Seidman et al. (2015) of data from 83 schizophrenia patients, 151 unaffected siblings, and 209 community comparison participants yielded five distinct cognitive factors (episodic memory, working memory, perceptual vigilance, inhibitory processing, and visual abstraction) with low association with neurophysiological measures. This divergence suggests that both classes of phenomena will be important for advancing explanatory stories about schizophrenia. Importantly, many such studies have evaluated a given potential endophenotype for schizophrenia with samples confined to a schizophrenia diagnosis (perhaps including patients' relatives), so conclusions about the diagnostic or dimensional specificity of most endophenotypes often cannot yet be drawn. The need for such specificity recedes if one pursues a transdiagnostic approach to psychopathology, as the United States National Institute of Mental Health now advocates in its RoDC initiative, but the need for evaluation across diagnostic categories becomes even clearer.

Ambitious recent studies have delivered on the analytic approach pursued by Glahn and colleagues (eg, Glahn et al., 2012, 2015), defining endophenotypes in very large samples randomly selected from extended pedigrees. Results are important for confirming cognitive and macrolevel cortical endophenotypes grounded in significant associations between these measures and genetic liability for schizophrenia: digit-symbol substitution, facial memory, emotion recognition, and temporal and prefrontal cortical surface abnormality.

In parallel with traditional endophenotype research, psychiatric genetics, represented by large-scale genome-wide association studies (GWAS), is beginning to identify replicated genetic patterns in schizophrenia. Regarding the vast and diverse literature, some major insights from GWAS are:

1. There is a growing consensus that genetic contributions to schizophrenia risk are carried by a large number of alleles, including common alleles with small effect sizes (eg, Munafò and Flint, 2014) together with rare mutations of fewer genetic loci of unknown heterogeneity (eg, Iacono et al., 2014). It may be that some very rare combinations of individually rare genes provide a potent genetic input to risk (Lykken et al., 1992; Miller et al., 2014; Vrieze et al., 2014). Although "no common [genetic] variant with odds ratio more than 1.2 has been identified" (Svrakic et al., 2013, p. 189; see also Hosak, 2013), large-scale GWAS indicate that a combination of common

genetic polymorphisms (eg, single nucleotide polymorphisms (SNPs)) explains up to 30% of the liability for schizophrenia (complemented by rare mutations of structural lesions of the genome, which account for an additional 2–4% of liability variance; International Schizophrenia Consortium, 2009; Sebat et al., 2009).

2. One or more polymorphisms appear to be shared by schizophrenia, bipolar disorder, and major depressive disorder (Cross-Disorder Group of the Psychiatric Genomics Consortium, 2013; Pearlson, 2015; Schulze et al., 2014). Thus, shared polymorphism cannot account for diagnosis-specific symptoms and features. A polygenetic risk score not only explained the most variance in phenotypes but also correlated significantly with treatment resistance (Frank et al., 2014).

3. GWAS approaches continue to generate controversy. "More than 70 genes have recently been suspected to be involved in the genetic background of schizophrenia based on GWAS's result. They are typically related to neurodevelopment/neuroplasticity, immunology and neuroendocrinology." (Hosak, 2013, p. 57). However, Schumann (2014) argued that "the sole reliance on canonical genome-wide significance thresholds is not sufficient to describe the complex relation of genotype and phenotype...[Rather] the development of novel phenotypical characterization and methods to describe the relation between genotype and phenotype are needed. This will require the development of innovative analytical strategies, as well as corroborative approaches linking association studies with functional characterization" (p. 1335).

4. No available genetic model sufficiently accounts for schizophrenia risk, many genetic findings are diagnostically nonspecific, and intensive, adequately powered consideration of additional factors (eg, epigenetic and brain oscillatory mechanisms) is overdue. Much larger sample sizes are likely to improve replicability of findings. However, effects that are visible only with very large samples must be very small, raising concerns about their importance and the efficiency of spending precious clinical research dollars in pursuit of apparently small contributions.

5. Of particular interest in endophenotype research, established associations with schizophrenia are high for loci with shared effects that are related to the regulation of neuronal development (Schizophrenia Psychiatric GWAS Consortium, 2011). Moreover, genetic risk variants (SNPs) associated with total brain volume and white matter volume (van Scheltinga et al., 2013) are located in genes with neuronal functions such as involvement in glutamatergic neurotransmission and synaptic plasticity (Schizophrenia Working Group of the Psychiatric Genomics Consortium, 2014).

6. Many different specific gene variants have been proposed over the years, generally with a poor record of replication. But the picture is improving. For example, healthy relatives of schizophrenia, bipolar, and major depressive disorder patients who carry a *CACNA1C* gene variant (rs1006737) that has consistently been associated with psychiatric disorders show less hippocampal and anterior cingulate activity during episodic memory recall and more depression and anxiety than individuals who are not carriers of this gene variant (Erk et al., 2014). This result suggests a specific contribution of a subtle genetic factor to information transfer in the brain that is potentially area-specific and task-specific, because *CACNA1C* encodes an alpha1 subunit of a voltage-dependent calcium channel, thereby affecting signal transfer between neurons.

Another example is the *NRGN* gene, coding neurogranin, a postsynaptic protein regulating the availability of Ca^{2+} in neurons exclusively expressed in the brain. An *NRGN* variant has been associated with schizophrenia, memory performance, and posterior cingulate activation in a memory retrieval task (Krug et al., 2013). Associations of GWAS-identified risk variants with functional brain measures have been shown for a variant of the *ZNF804A* (zinc finger) allele and prefrontal connectivity in a working memory task in a healthy sample (Esslinger et al., 2011), but they still need to be followed up in a patient sample. Other relationships will no doubt be identified.

Associations between genetic liability and cognitive dysfunction cut across diagnostic categories, with high associations between gene set enrichment and performance in spatial attention, verbal abilities, and Stroop interference tests in schizophrenia and bipolar patients (Fernandes et al., 2013). Gene set enrichment as a measure of genetic contribution to neurocognitive dysfunction in schizophrenia is determined as follows: (1) single SNPs are assigned to genes, and gene-based scores for the number of SNPs are ranked to produce candidate gene sets (more than one gene involved makes a set); (2) gene sets for selected test performance-based cognitive traits (such as those deficient in schizophrenia) are determined by correlation; (3) gene set enrichment analysis, such as cell type-specific expression analysis (Xu et al., 2014), assesses whether a predefined gene set shows significant enrichment of signal (an association in GWAS), for example, of those gene sets associated with cognitive deficits and with schizophrenia (and in bipolar patients) (Fernandes et al., 2013). If so, then this enrichment is considered evidence for a joint genetic contribution.

In summary, there is consensus that "multiple and variable genetic and environmental factors interact to influence the risk of schizophrenia. Both rare variants with large effect and common variant with small effect contribute to the genetic risk…with no indication for differential impact on its clinical features. Accumulating evidence supports a genetic architecture of schizophrenia with multiple scenarios, including additive polygenic, heterogeneity, and mixed polygenic heterogeneity" (Svrakic et al., 2013, p. 188). Although long suspected, the latest results from large consortia appear to have settled the general questions of whether single genetic contributions predominate or whether small, multifactor genetic contributions make diverse contributions in severe mental illness. The latter probably involves multiple causal paths with some endophenotypic or phenotypic convergence. Identifying the mechanisms of these effects remains the great challenge.

Several conclusions can be offered about the current status of endophenotypes for schizophrenia. First, as Kendler (2005) argued persuasively a decade ago, the hunt for genes "causing" schizophrenia as a diagnostic entity has failed so thoroughly, despite years of extraordinary efforts that we should stop looking for them in the sense of finding a simple genetic account of schizophrenia. We are beginning to accept that "genes do not code for complex disorders but for biological processes" (Diwadkar et al., 2014, p. 2), such as endophenotypes. "Schizophrenia is not a disease but a cluster of clinical symptoms. No biological marker is involved in schizophrenia diagnostics at present. This means that very few schizophrenia patients probably share identical genomic causation" (Hosak, 2013, p. 59). Read literally, this characterization of biomarkers is too pessimistic, but the claim that at present we lack biomarkers with sufficient sensitivity and specificity to be useful in explaining, detecting, predicting, or treating schizophrenia is valid.

Most promising is research on psychological and biological endophenotypes that appear to be causally close to clinical manifestations of disorders, either traditional psychological symptoms or some type of compromise in daily function. The promise of studying endophenotypes is affirmed, for example, by the recent evidence of small but replicable genetic effects driving psychological and biological phenomena associated with schizophrenia, as reviewed by Meyer-Lindenberg (2010) for the impact of *CACNA1C* on hippocampal activation and connectivity during episodic memory tasks or the functional polymorphism of *COMT* and other candidate genes for schizophrenia on prefrontal connectivity. Another example of a functional endophenotype may be gray matter reduction. Consistently reported brain structure changes, their physiological functions, and relation with disease duration (Nickl-Jockschat et al., 2011, p. S166) revealed significant clusters of gray matter reduction that were related to disease duration (left anterior insula), reward processing, and language functions.

Second, rather than hunting for genes, we should be hunting for endophenotypes, not only as a way forward to identify genetic contributions to schizophrenia in general or, for example, executive dysfunction in schizophrenia in particular but also as a prime target for assessment and intervention in their own right. A common assumption in gene-hunting is that discovering a (hopefully not too complex) gene story about schizophrenia is the most valuable means of preventing or treating it. But it is not apparent that genes will provide the best leverage. It may be easier to identify endophenotypes, identify the mechanisms by which they lead to disorder, and intervene in targeting them. Such interventions may be closer to the genes than to the clinical manifestation, or closer to the clinical manifestation. If the gene story we understand is more complex, then the best strategy along the causal continuum may change.

At least for the foreseeable future, we advocate dethroning genetics as the assumed key to understanding, preventing, and treating schizophrenia. The field should be agnostic about how tractable the genetic contributions to schizophrenia will prove to be and about what the most fruitful research investments will turn out to be in the foreseeable future. This is not at all an argument against continued research on genes. It is an argument in favor of a broader strategy than has been dominant in recent decades.

Third, the hunt for schizophrenia endophenotypes may be more successful if it emphasizes the understanding of mechanisms of genetic translation into function rather than identification of genes. The critical role of genes becomes clear not from enumerating genes, but rather from discovering how they work. Therefore, imaging genetics seems promising because these techniques "define neural systems that mediate heritable risk linked to candidate and genome-wide-supported common variants" (Meyer-Lindenberg, 2010, p. 194). This effort to characterize "the neural risk architecture" (p. 194), which includes characteristic oscillatory patterns and neural network communication, is less about targeting genetic origins of schizophrenia and more about understanding mechanisms of pathophysiology and psychopathology and their heterogeneity (Diwadkar et al., 2014), which seems more promising for diagnostic and therapeutic advances.

Finally, this systems emphasis is very much in line with the NIMH RDoC initiative. As discussed in Cuthbert (2014), Cuthbert and Insel (2010), Cuthbert and Instel (2013), Cuthbert and Kozak (2013), Insel and Cuthbert (2015), Insel et al. (2010), Kozak and Cuthbert (2016), Miller and Rockstroh (2013), Miller et al. (2016), and Yee et al. (2015), RDoC cultivates a balanced, comprehensive consideration of a wide range of units of analysis (genes, molecules, cells, circuits, central and peripheral physiology, overt behavior, and self-report extensible to social units of analysis and even to economic factors) and a wide range of psychological and biological constructs (such as acute threat, approach motivation, working memory, social communication, and biological rhythms). The RDoC matrix of units of analysis and domains of constructs (Fig. 2.3) is designed to facilitate the development constructs that are hybrids between psychological and biological phenomena. RDoC is careful not to assume a causal direction at present. For schizophrenia, for example, integration of different units of analysis in the domain of cognitive systems should identify relevant mechanisms and how their dysfunction contributes to the most salient feature of schizophrenia—cognitive impairment. It can also be noted that RDoC advocates, to the fullest extent possible, that human research be grounded in appropriate animal neuroscience, very much in accord with the endophenotype approach.

v. 5.1, 07/15/2012	RESEARCH DOMAIN CRITERIA MATRIX							
	‐ ‐ ‐ ‐ ‐ ‐ ‐ ‐ ‐ ‐ UNITS OF ANALYSIS ‐ ‐ ‐ ‐ ‐ ‐ ‐ ‐ ‐ ‐ ‐ ‐							
DOMAINS/CONSTRUCTS	Genes	Molecules	Cells	Circuits	Physiology	Behavior	Self-Reports	Paradigms
Negative Valence Systems Acute threat ("fear") Potential threat ("anxiety") Sustained threat Loss Frustrative nonreward								
Positive Valence Systems Approach motivation Initial responsiveness to reward Sustained responsiveness to reward Reward learning Habit								
Cognitive Systems Attention Perception Working memory Declarative memory Language behavior Cognitive (effortful) control								
Systems for Social Processes Affiliation/attachment Social Communication Perception/Understanding of Self Perception/Understanding of Others								
Arousal/Modulatory Systems Arousal Biological rhythms Sleep-wake								

FIGURE 2.3

The NIMH Research Domain Criteria matrix.

Reproduced from Kozak, M.J., Cuthbert, B.N., 2016. The NIMH Research Domain Criteria Initiative: background, issues, and pragmatics. Psychophysiology 53(3):286–297. See https://www.nimh.nih.gov/research-priorities/rdoc/research-domain-criteria-matrix.shtml for an expanded, interactive version.

Two major limitations must be acknowledged, however. First, we are very far away from being able to visualize the epigenetic machinery in schizophrenia. We have a list of psychological and biological findings that seem to matter, but we do not have a model that ties them together, from epigenetics to imminent risk or clinical expression. Second, RDoC does not presently spell out a view of schizophrenia. The onus is on the research community to select constructs in the matrix, or to add constructs to it, and to demonstrate how mechanisms in multiple units of analysis play a role in the development, clinical presentation, and treatment of the disorder. These are not limitations of RDoC itself, because they reflect the state of the literature.

AN EMERGING FRONTIER FOR ENDOPHENOTYPES: EPIGENETICS

Enormous efforts to identify genes contributing to schizophrenia have not proven sufficient to identify any key genetic, nongenetic, or interactive mechanisms that contribute to psychopathology (we have correlates but no mechanisms). There is recent progress in replicating some specific gene findings, and there is more longstanding evidence of correlations with environmental insults such as maternal infections, maternal psychological stress, hypoxia-inducing obstetric complications, cannabis use, climate, and factors such as migration and urbanicity (eg, Bradbury and Miller, 1985; Schmitt et al., 2014;

Oh and Petronis, 2008; Van Os, 2011). Recent thinking has increasingly emphasized the potential role of a class of G× E contributions in which environmental factors affect gene expression in specific ways that contribute to mental illness (Skinner et al., 2010).

The concept of epigenetics is used somewhat inconsistently in the psychopathology literature, especially regarding whether it encompasses nonheritable changes in gene expression, but two common, closely related meanings are heritable changes in gene expression in the absence of change in the DNA sequence and the biochemistry that mediates such changes in gene expression. Epigenetic effects are thought to play a crucial role in mechanisms of interaction between G and E (Diwadkar et al., 2014; Nishioka et al., 2012; Peedicayil, 2011; Svrakic et al., 2013), especially (perhaps) via a chain of deviant epigenetic events (Oh and Petronis, 2008). For example, preimputation is an epigenetic change that takes place during gametogenesis or embryogenesis, which increases the risk for psychosis by a process that unfolds depending on other prenatal and postnatal interactions of environmental effects with such preimputed epigenetic changes (Oh and Petronis, 2008; see also Diwadkar et al., 2014). Moreover, Mansuy and colleagues have demonstrated epigenetic transmission of early stress across generations (Franklin et al., 2010; Gapp et al., 2014) and its relation to cognitive functions (Bohacek et al., 2015; Franklin and Mansuy, 2010).

Not only popular media but also many scholars seem to not understand the potential malleability of genetic contributions to normal and abnormal behavior that epigenetics enables far from one's genetic blueprint being fixed for a lifetime from conception; it is not a blueprint. Gene expression is subject to modulation by other genes and by the environment on a timescale of minutes. These relationships undermine any notion of a nature–nurture split in accounts of psychopathology.

Epigenetic phenomena appear to be very promising candidates for endophenotypes. Altered epigenetic mechanisms in schizophrenia are of particular interest, because epigenetic mechanisms have been shown to play a role in normal neurodevelopment, neuronal plasticity, learning, and memory (Peedicayil, 2011), which are processes associated with the development of psychosis. Moreover, environmental factors have been shown to affect epigenetic mechanisms such as DNA methylation and histone modification (Nishioka et al., 2012; Svrakic et al., 2013) and microRNA changes (Zucchi et al., 2013). Considerable literatures confirm that epigenetic programming of the genome is particularly active prenatally, although epigenetic modulations of DNA methylation are also found across the life span (Fraga et al., 2005).

Although much work remains to be done, available evidence confirms differential DNA methylation in schizophrenia etiology (Wockner et al., 2014), suggesting the role for epigenetic abnormalities in schizophrenia, and indicates abnormal DNA methylation of genes associated with neurodevelopment (eg, neuronal migration) and neurotransmitters that are dysregulated in schizophrenia. For example, epigenetic studies of postmortem brain tissue have indicated prefrontal hypermethylation of *RELN* (which encodes extracellular matrix protein involved in neuronal migration) (Nishioka et al., 2012; Peedicayil, 2011) and *SOX10* (which encodes oligodendrocyte-specific transcription factors). In a methylome-wide association study of 759 schizophrenia and 738 comparison individuals, Aberg et al. (2014) found microRNA changes to be prominent in networks "that can be linked to neuronal differentiation and dopaminergic gene expression" (p. 255), in addition to methylation effects at *RELN*-related sites. From a genome-wide analysis of DNA methylation in monozygotic twins discordant for psychosis (schizophrenia or bipolar disorders), Dempster et al. (2011) reported epigenetic changes in brain networks and pathways related to neurodevelopment, particularly hypomethylation in the promoter of *ST6GALNAC1*, which is involved in protein glycosylation and thereby mediates cell–cell interaction. "Such massive involvement of neurodevelopmental genes suggests a nonspecific, global epigenetic effect and an early lesion...Global epigenetic defects are unlikely due to transcription errors in copying DNA methylation...but rather to epigenetic dysregulation resulting from pathological G×E interactions" (Svrakic et al., 2013, p.190).

Neurotransmitter-related epigenetic alterations involve hypermethylation of the *GAD1* gene, which encodes the enzyme catalyzing excitatory glutamic acid into inhibitory GABA (Peedicayil, 2011). Epigenetic dysfunction can start a cascade of presynaptic and postsynaptic hyperactivity and hypoactivity of GABA and glutamate, impacting the release of dopamine. Epigenetic effects on the gene encoding membrane-bound *COMT* have also been reported, although less consistently (Nishioka et al., 2012; Peedicayil, 2011). Postmortem findings are complemented by peripheral blood-derived epigenetic results, which also indicate altered methylation in genes related to dopamine (DRD2) and GABA (GABRB2) receptor encoding (Nishioka et al., 2012).

Although these results suggest that epigenetic alterations result in functional brain deficits in schizophrenia (Nishioka et al., 2012) and may help to understand the proximal biochemical origin of well-known neurotransmitter abnormalities in schizophrenia, the specific mechanisms of environmental impact on gene expression via epigenetics seem more important for understanding schizophrenia development and psychopathology. Several studies have documented transgenerational signatures of prenatal maternal psychological stress on epigenetic factors in offspring. Although evidence for environmental epigenetic alterations in schizophrenia is scarce, these general results may be relevant for schizophrenia as well. Zucchi et al. (2013) reported that stress in the dam during gestation was accompanied by upregulated microRNA profiles in offspring brains, including components that target putative marker genes for schizophrenia and bipolar disorders in humans. In the adolescent children of mothers exposed to intimate partner violence during pregnancy, Radtke et al. (2011) found increased methylation of the glucocorticoreceptor promoter gene, indicating transgenerational epigenetic programming of the HPA axis. In a methylation-wide association study, Aberg et al. (2014) reported hypoxia-related changes related to cigarette smoking and nicotine. As a result, smoking during pregnancy might contribute to pathology-relevant epigenetic effects.

Schizophrenia is "particularly well-suited for investigating how environmental conditions early in an organism's lifespan are translated into faulty epigenetic regulation that continues to shape behavior well into adulthood" (Puckett and Lubin, 2011, p.10). The authors considered epigenetics a crucial link for neurodevelopmental models of schizophrenia. They reviewed results showing hypermethylation of the *RELN* promoter in schizophrenia (involved in synaptic plasticity, specific to GABAergic cortical interneurons). As a consequence, diminished inhibitory GABAergic cortical activity might lead to hypofunctionality in frontal cortex, which is associated with poor executive function. In summary, epigenetic endophenotypes show promise in the sense that there is good reason to believe that they exist and can be discovered.

A NEW FRONTIER FOR ENDOPHENOTYPES: BRAIN CONNECTIVITY DYNAMICS

We anticipate that promising research on communication within and across neuronal networks will provide another growth area for endophenotypes. The concept of chronnectomics has emerged from the growing focus on oscillatory brain activity as a means of neuronal communication, with oscillatory dynamics called "a hallmark of neural signals" (Calhoun et al., 2014, p. 262), as a class of neural mechanisms enabling cognitive function and dysfunction. Beyond a description of the wiring network of the brain (eg, Sporns et al., 2005), recent discussions of chronnectomics emphasize temporal dynamics—relationships varying as a function of time, context, or modality (structure and function; even structure is not fixed over time): "the term chronnectome is used to describe a focus on identifying time-varying, but reoccurring, patterns of coupling among brain regions" (p. 262; see also Sporns, 2011). We suggest that the accumulating evidence for altered oscillatory dynamics, neuronal network communication, and network connectivity in schizophrenia warrants the investigation of features of the chronnectome as endophenotypes. Such phenomena have already become targets of treatment in schizophrenia (eg, Fisher et al., 2009, 2014; Popov et al., 2011, 2012, 2015; Popova et al., 2014).

Prasad and Keshavan (2008) reviewed the literature on anatomic alterations in schizophrenia for their potential as endophenotypes (for more recent, more focused reviews on gray matter loss and morphological alterations as endophenotypes, see Glahn et al., 2015 and Van der Velde et al., 2015). They saw candidates with reduced whole-brain volume, reduced gray matter (particularly in temporal lobe and hippocampus), morphometric alterations such as gray matter density reduction in medial temporal lobe, and superior temporal gyrus. White matter abnormalities in frontal regions, hippocampus, frontooccipital, and inferior longitudinal fascicula already meet criteria for endophenotypes. However, Prasad and Keshavan emphasized the role of structural endophenotypes, primarily in combination with functional (brain and performance) measures in neurocognitive "extended endophenotypes." As extended endophenotypes, they proposed, for example, prefrontally mediated cognitive functions (spanning structural abnormalities in DLPFC, hippocampus, etc., and abnormal working memory function). Structural connectivity analyses show changes in clustering and modularity (Van den Heuvel and Fornito, 2014) that have been characterized as a segregated pattern of network organization (Deco and Kringelbach, 2014). Developing tools to capture combinations of structural and functional phenomena involved in brain dynamics will be critically important to progress in this area.

Aspects of neuromagnetic oscillatory activity have also been evaluated as candidate endophenotypes, with complex and inconsistent results. Whereas Narayanan et al. (2014) found that augmented slow resting state beta activity met endophenotype criteria (also augmented in first-degree relatives of schizophrenia and bipolar patients), Ranlund et al. (2014) could not confirm abnormal resting state activity in first-degree relatives. Considerable research may be needed to determine which potentially endophenotypic features of the frequency/power spectrum are informative for those with a diagnosable disorder versus those at risk for it. For example, Narayanan et al. (2014) reported a wider frequency range of abnormalities in patients than in unaffected relatives.

Uhlhaas and Singer (2013) and Buzsaki and Watson (2012) reviewed evidence suggesting that "neural oscillations are ideally suited as a measure to establish links between genes, physiology, and behavior in schizophrenia, and eventually may contribute to the identification of pathophysiological mechanisms… [and that] high-frequency oscillations qualify as an endophenotype" (Uhlhaas & Singer, p. 308). Decreased power in localized gamma oscillations and perturbed interregional gamma coherence appear to meet endophenotype criteria because they are present in first-degree relatives of schizophrenia patients (Buzsaki and Watson, 2012). Moreover, animal studies show that variation of some putative schizophrenia susceptibility genes (*DISC1, Neuregulin 1*) influences parameters of the excitation–inhibition balance that determine phase synchrony (eg, Carlson et al., 2011; Sigurdsson et al., 2010) and alter long-range gamma coherence in schizophrenia and bipolar patients (Ozerdem et al., 2010). Evaluating task-induced oscillatory activity, Ethridge et al. (2015) reported auditory oddball-induced time-frequency alterations in schizophrenia and bipolar patients that they viewed as a functional endophenotype.

Large-scale functional neuronal networks coherently active during the resting state within or between anatomical regions have been defined as resting state networks (Nakagawa et al., 2014) and have been shown to be abnormal in schizophrenia (Bullmore, 2012; see also Camchong et al., 2011; Meyer-Lindenberg, 2010). Complementing the traditional focus on lesions, abnormal connectivity is increasingly seen as a source of "primary pathogenesis" (White and Gottesman, 2012, p. 2396). Results are obtained by various approaches to the analysis of network connectivity. When determining time-varying spatial maps by clustering algorithms such as ICA, seed-voxel correlational analyses, or graph metrics (Bullmore, 2012; Calhoun et al., 2014; Fornito and Bullmore, 2012; Lynall et al., 2010; White and Gottesman, 2012), the results converge to suggest that functional networks are disrupted, fragmented, less flexible, or less dynamic in schizophrenia than in control groups, thereby showing decreased small-world properties (Deco and Kringelbach, 2014). Patients with schizophrenia exhibit a higher probability of transition of specific network microstates and spend more time in relatively less connected dynamic states (Miller et al., 2014). Both global communication efficiency and communication between local segregated networks seem reduced (Deco and Kringelbach, 2014). In particular, brain regions subserving cognitive functions seem less functionally connected over time in schizophrenia than in control groups. This phenomenon is evident in the resting state (eg, Meda et al., 2012) and in task-related activities (eg, Popov et al., 2013). Such findings are in line with anatomical (Camchong et al., 2011) and functional (Lynall et al., 2010; Wotruba et al., 2014) measures. Such functional connectivity measures appear to meet criteria for an endophenotype (Fornito and Bullmore, 2012; Meda et al., 2014; White and Gottesman, 2012). Similarities in the resting state abnormalities for schizophrenia and psychotic bipolar patients as well as diagnosis-specific network characteristics point to transdiagnostic dysfunction (Calhoun et al., 2014; Meda et al., 2012).

A variety of features of brain chronnectomics show promise as endophenotypes, but in a young and relatively small literature. It is easy to imagine that most of the connectivity phenomena that will prove informative have not even been identified or tied to schizophrenia. However, the combinatorial explosion of potential risk factors, diagnostic features, and analysis methods make the forecast for a successful research timeline quite difficult.

LIMITATIONS AND FUTURE DIRECTIONS

The growing schizophrenia endophenotype literature indicates that a number of phenomena already qualify as endophenotypes but that considerable work lies ahead to identify and validate more of them, and then they must be stitched together to form an explanatory framework. More than 40 years since Gottesman and Shields (1972) proposed that the psychopathology literature should pursue endophenotypes have, like so many directions in the psychopathology literature, produced a disappointing yield. However, the field is moving beyond the expectation that the genetic story in schizophrenia will be simple, which is a premise that has surely held back progress. The field is also beginning to move beyond the naïve biological reduction fostered by the US-declared Decade of the Brain, with psychological phenomena such as cognition and social function now seen as crucial to the story. The outlook for epigenetic and oscillatory phenomena that support critical psychological functions is very good in terms of grounds for betting on important relationships to be discovered, but is exceptionally challenging pragmatically. Finally, it can be noted that the endophenotype literature to date does not encourage faith in traditional diagnostic boundaries. Commonalities frequently reported between schizophrenia and psychotic bipolar patients support the transdiagnostic perspective of the RDoC initiative.

Limitations of all approaches in understanding the origins of schizophrenia—by endophenotypes, genomics, epigenetics, and chronnectomics—have been noted. We do not share the skepticism of some about the utility of the endophenotype concept. On the contrary, the striking failure of a simplistic genetic approach to accounting for psychopathology and of the naïve biological reduction manifested in much that was written during and about the Decade of the Brain is compelling evidence that more sophisticated strategies are needed that, almost paradoxically, tackle a large set of individually smaller questions. As discussed, the endophenotype concept suits current trends in the literature quite well. Importantly, the RDoC initiative is restoring the centrality of psychological phenomena, methods, and intervention targets to the psychopathology literature, not in competition with the recent biological emphasis but in explicit integration with it. This can only be a good thing for our science and for those we hope to help clinically.

REFERENCES

Aberg, K.A., McClay, J.L., Nerella, S., Clark, S., Kumar, G., Wenan, C., et al., 2014. Methylome-wide association study of schizophrenia identifying blood biomarker signatures of environmental insults. JAMA Psychiatry 71, 255–264. http://dx.doi.org/10.1001/jamapsychiatry.2013.3730.

Adler, L.E., Pachtman, E., Franks, R., Pecevich, M., Waldo, M.C., Freedman, R., 1982. Neurophysiological evidence for a deficit in inhibitor mechanisms involved in sensory gating in schizophrenia. Biol. Psychiatry 17, 639–654.

Baker, L.A., 2014. Do our "big data" in genetic analysis need to get bigger? Psychophysiology 51, 1321–1322.

Berenbaum, H., 2013. Classification and psychopathology research. J. Abnorm. Psychol. 122, 894–901. http://dx.doi.org/10.1037/a0033096.

Bohacek, J., Farinelli, M., Mirante, O., Steiner, G., Gapp, K., Coiret, G., et al., 2015. Pathological brain plasticity and cognition in the offspring of males subjected to postnatal traumatic stress. Mol. Psychiatry 20, 621–631. http://dx.doi.org/10.1038/mp.2014.80

Bohlken, M.M., Mandl, R.C., Brouwer, R.M., van den Heuvel, M.P., Hedman, A.M., Kahn, R.S., et al., 2014. Heritability of structural brain network topology: a DTI study of 156 twins. Hum. Brain Mapp. 35, 5295–5305. http://dx.doi.org/10.1002/hbm.22550.

Bradbury, T.N., Miller, G.A., 1985. Season of birth in schizophrenia: a review of evidence, methodology, and etiology. Psychol. Bull. 98, 569–594.

Braff, D.L., 2014. Genomic substrates of neurophysiological endophenotypes: where we've been and where we're going. Psychophysiology 51, 1323–1324.

Bramon, E., Rabe-Hesketh, S., Sham, P., Murray, R.M., Frangou, S., 2004. Meta-analysis of the P300 and P50 waveforms in schizophrenia. Schizophr. Res. 70, 315–329.

Bullmore, E.T., 2012. Functional network endophenotypes of psychotic disorders. Biol. Psychiatry 71, 844–845. http://dx.doi.org/10.1016/j.biopsych.2012.03.019.

Buzsaki, G., Watson, B.O., 2012. Brain rhythms and neural syntax: implications for efficient coding of cognitive content and neuropsychiatric disease. Dialogues Clin. Neurosci. 14, 345–367.

Calhoun, V.D., Miller, R., Pearlson, G., Adali, T., 2014. The chronnectome: time-varying connectivity networks as the next frontier in fMRI data discovery. Neuron 84, 262–274. http://dx.doi.org/10.1016/j.neuron.2014.10.015.

Calkins, M.E., Iacono, W.G., Ones, D.S., 2008. Eye movement dysfunction in first-degree relatives of patients with schizophrenia: a meta-analytic evaluation of candidate endophenotypes. Brain Cogn. 68, 436–461.

Camchong, J., MacDonald III, A.W., Müller, B.A., Lim, K.O., 2011. Altered functional and anatomical connectivity in schizophrenia. Schizophr. Bull. 37 (3), 640–656. http://dx.doi.org/10.1093/schbul/spb131.

Carlson, G.C., Talbot, K., Halene, T.B., et al., 2011. Dysbindin-1 mutant mice implicate reduced fast-phasic inhibition as a final common disease mechanism in schizophrenia. Proc. Natl. Acad. Sci. U.S.A. 108, E962–E970.

Chan, R.C., Gottesman, I.I., 2008. Neurological soft signs as candidate endophenotypes for schizophrenia: a shooting star or a Northern star? Neurosci. Biobehav. Rev. 32, 957–971.

Crocker, L.D., Heller, W., Warren, S.L., O'Hare, A.J., Infantolino, Z.P., Miller, G.A., 2013. Relationships among cognition, emotion, and motivation: implications for intervention and neuroplasticity in psychopathology. Front. Hum. Neurosci. 7, 1–19. article 261.

Cross-Disorder Group of the Psychiatric Genomics Consortium, 2013. Genetic relationship between five psychiatric disorders estimated from genome-wide SNPs. Nat. Genet. 9, 984–995. http://dx.doi.org/10.1038/ng.2711.

Cuthbert, B.N., 2014. Translating intermediate phenotypes to psychopathology: the NIMH Research Domain Criteria. Psychophysiology 51, 1205–1206.

Cuthbert, B.N., Insel, T.R., 2010. Toward new approaches to psychotic disorders: the NIMH Research Domain Criteria project. Schizophr. Bull. 36, 1061–1062.

Cuthbert, B.N., Insel, T.R., 2013. Toward precision medicine in psychiatry: the NIMH Research Domain Criteria project. BMC Med. 11, 126.

Cuthbert, B.N., Kozak, M.J., 2013. Constructing constructs for psychopathology: the NIMH Research Domain Criteria. J. Abnorm. Psychol. 122, 928–937.

Deco, G., Kringelbach, M.L., 2014. Great expectations: using whole-brain computational connectomics for understanding neuropsychiatric disorders. Neuron 84, 892–905.

Dempster, E.L., Pidsley, R., Schalkwyk, L.C., Owens, S., Georgiades, A., Kane, F., et al., 2011. Disease-associated epigenetic changes in monozygotic twins discordant for schizophrenia and bipolar disorders. Hum. Mol. Genet. 20 4786-479. http://dx.doi.org/10.1093/hmg/ddr416.

Diwadkar, V.A., Bustamante, A., Rai, H., Uddin, M., 2014. Epigenetic, stress, and their potential impact on brain network function: a focus on the schizophrenia diatheses. Front. Psychiatry 5, 71. http://dx.doi.org/10.3389/fpsyt.2014.00071.

Erk, S., Meyer-Lindenberg, A., Schmierer, P., Mohnke, S., Grimm, O., Garbusow, M., et al., 2014. Hippocampal and frontolimbic function as intermediate phenotype for psychosis: evidence from healthy relatives and a common risk variant for CACNA1A. Biol. Psychiatry 76, 466–475. http://dx.doi.org/10.1016/j.biopsych.2013.11.025.

Esslinger, C., Kirsch, P., Haddad, L., Mier, D., Sauer, C., Erk, S., et al., 2011. Cognitive state and connectivity effects of the genome-wide significant psychosis variant in ZNF804A. Neuroimage 54 (3), 2514–2523. http://dx.doi.org/10.1016/j.neuroimage.2010.10.012.

Ethridge, L.E., Hamm, J.P., Pearlson, G.D., Tamminga, C.A., Sweeney, J.A., Keshavan, M.S., et al., 2015. Event-related potential and time-frequency endophenotypes for schizophrenia and psychotic bipolar disorder. Biol. Psychiatry 7, 127–136. http://dx.doi.org/10.1016/j.biopsych.2014.03.032.

Fernandes, C.P., Christoforou, A., Giddaluru, S., Ersland, K.M., Djurovic, S., Mattheisen, M., et al., 2013. A genetic deconstruction of neurocognitive traits in schizophrenia and bipolar disorders. PLoS One 8, e81052. http://dx.doi.org/10.1371/journal.pone.0081052.

Fisher, M., Holland, C., Merzenich, M.M., Vinogradov, S., 2009. Using neuroplasticity-based auditory training to improve verbal memory in schizophrenia. Am. J. Psychiatry 166, 472–485.

Fisher, M., Loewy, R., Carter, C., Lee, A., Ragland, J.D., Niendam, T., et al., 2014. Neuroplasticity-based auditory training via laptop computer improves cognition in young individuals with recent onset schizophrenia. Schizophr. Bull. 41, 250–258.

Ford, J.M., 2014. Decomposing P300 to identify its genetic basis. Psychophysiology 51, 1325–1326.

Fornito, A., Bullmore, E.T., 2012. Connectomic intermediate phenotypes for psychiatric disorders. Front. Psychiatry 3, 32. http://dx.doi.org/10.3389/fpsyt.2012.00032.

Fraga, M.F., Bassestar, E., Paz, M.F., Ropero, S., Setien, F., et al., 2005. Epigenetic differences arise during lifespan in monozygotic twins. Proc. Natl. Acad. Sci. U.S.A. 102, 10604–10609.

Frank, J., Lang, M., Witt, S.H., Strohmaier, J., Rujescu, D., Cichon, S., et al., 2014. Identification of increased genetic risk scores for schizophrenia in treatment-resistant patients. Mol. Psychiatry, 1–2. http://dx.doi.org/10.10138/mp.2014.56.

Franklin, T.B., Mansuy, I.M., 2010. The prevalence of epigenetic mechanisms in the regulation of cognitive functions and behaviour. Curr. Opin. Neurobiol. 20, 441–449. http://dx.doi.org/10.1016/j-conb.2010.04.007.

Franklin, T.B., Russig, H., Weiss, I.C., Gräff, J., Linder, N., Michalin, A., et al., 2010. Epigenetic transmission of the impact of early stress across generations. Biol. Psychiatry 68, 408–415. http://dx.doi.org/10.1016/j.biopsych.2010.05.036.

Gapp, K., Soldado-Magraner, S., Alvarez-Sanchez, M., Bohacek, J., Vernaz, G., Shu, H., et al., 2014. Early life stress in fathers improves behavioral flexibility in their offspring. Nat. Commun. 18 (5), 5466. http://dx.doi.org/10.1038/ncomms6466.

Glahn, D.C., Curran, J.E., Winkler, A.M., Carless, M.A., Kent, C.W., et al., 2012. High dimensional endophenotype ranking in the search for major depression risk genes. Biol. Psychiatry 71, 6–14.

Glahn, D.C., Knowles, E.E., McKay, D.R., Sprooten, E., Raventós, H., Blangero, J., et al., 2014. Arguments for the sake of endophenotypes: examining common misconceptions about the use of endophenotypes in psychiatric genetics. Am. J. Med. Genet. B Neuropsychiatr. Genet. 165, 122–130.

Glahn, D.C., Williams, J.T., McKay, D.R., Knowles, E.E., Sprooten, E., Mathias, S.R., et al., 2015. Discovering schizophrenia endophenotypes in randomly ascertained pedigrees. Biol. Psychiatry 77, 75–83. http://dx.doi.org/10.1016/j.biopsych.2014.06.027.

Goldman, D., 2014. The missing heritability of behavior: the search continues. Psychophysiology 51, 1327–1328.

Gottesman, I.I., Gould, T.D., 2003. The endophenotype concept in psychiatry: etymology and strategic intentions. Am. J. Psychiatry 160, 636–645.

Gottesman, I.I., Shields, J., 1972. Schizophrenia and Genetics: A Twin Study Vantage Point. Academic Press, New York, NY.

Gottesman, I.I., Shields, J., 1973. Genetic theorizing and schizophrenia. Br. J. Psychiatry 122, 15–30.

Gould, T.D., Gottesman, I.I., 2006. Psychiatric endophenotypes and the development of valid animal models. Genes, Brain Behav. 5, 113–119.

Gur, R.E., Calkins, M.E., Gur, R.C., Horan, W.P., Nuechterlein, K.H., Seidman, L.J., et al., 2007. l. The Consortium on the Genetics of Schizophrenia: neurocognitive endophenotypes. Schizophr. Bull. 33, 49–68.

Hasler, G., 2006. Evaluating endophenotypes for psychiatric disorders. Rev. Bras. Psiquiatr. 28, 91–92.

Heinrichs, R.W., 2004. Meta-analysis and the science of schizophrenia: variant evidence or evidence of variants? Neurosci. Biobehav. Rev. 28, 379–394.

Hosak, L., 2013. New findings in the genetics of schizophrenia. World J. Psychiatry 22 (3), 57–61. http://dx.doi.org/10.5498/wjp.v3.i3.57.

Iacono, W.G., 1985. Psychophysiological markers of psychopathology: a review. Canadian Psychology 26, 96–112.

Iacono, W.G., 2014. Genome-wide scans of genetic variants for psychophysiological endophenotypes: introduction to this special issue of *Psychophysiology*. Psychophysiology 51, 1201–1202.

Iacono, W.G., Lykken, D.T., 1979. Eye tracking and psychopathology: new procedures applied to a sample of normal monozygotic twins. Arch. Gen. Psychiatry 36, 1361–1369.

Iacono, W.G., Vaidyanathan, U., Vrieze, S.I., Malone, S.M., 2014. Knowns and unknowns for psychophysiological endophenotypes: integration and response to commentaries. Psychophysiology 51, 1339–1347.

Insel, T.R., Cuthbert, B.N., 2015. Brain disorders? Precisely: precision medicine comes to psychiatry. Science 348, 499–500.

Insel, T.R., Cuthbert, B.N., Garvey, M.B., Heinssen, R., Pine, D.S., Quinn, K., et al., 2010. Research Domain Criteria (RDoC): toward a new classification framework for research on mental disorders. Am. J. Psychiatry 167, 748–750.

International Schizophrenia Consortium, 2009. Common variants conferring risk of schizophrenia. Nature 460, 744–747.

Ivleva, E.I., Moates, A.F., Hamm, J.P., Bernstein, I.H., O'Neill, H.B., Cole, D., et al., 2014. Smooth pursuit eye movement, prepulse inhibition, and auditory paired stimuli processing endophenotypes across the schizophrenia-bipolar disorder psychosis dimension. Schizophr. Bull, 40, 642–652.

Kathmann, N., Hochrein, A., Uwer, R., Bondy, B., 2014. Deficits in gain of smooth pursuit eye movements in schizophrenia and affective disorder patients and their unaffected relatives. Am. J. Psychiatry 160, 696–702.

Keefe, R.S., Harvey, P.D., 2012. Cognitive impairment in schizophrenia. Handbook of Experimental Pharmacology 213, 11–37. http://dx.doi.org/10.1007/978-3-642-25758-2.

Kendler, K.S., 2005. "A gene for…": the nature of gene action in psychiatric disorders. Am. J. Psychiatry 162, 1243–1252.

Kozak, M.J., Cuthbert, B.N., 2016. The NIMH Research Domain Criteria Initiative: background, issues, and pragmatics. Psychophysiology 53 (3), 286–297.

Krug, A., Krach, S., Jansen, A., Nieratschker, V., Witt, S.H., Shah, N.J., et al., 2013. The effect of neurogranin on neural correlates of episodic memory encoding and retrieval. Schizophr. Bull. 39, 141–150. http://dx.doi.org/10.1093/schbul/sbr076.

Lenzenweger, M.F., 2010. Schizotypy and Schizophrenia: The View From Experimental Psychopathology. Guilford, New York, NY.

Lenzenweger, M.F., 2013. Thinking clearly about the endophenotype–intermediate phenotype–biomarker distinctions in developmental psychopathology research. Dev. Psychopathol. 25 (4 Pt. 2), 1347–1357.

Light, G.A., Swerdlow, N.R., Thomas, M.L., Calkins, M.E., Green, M.F., Greenwood, T.A., et al., 2014. Validation of mismatch negativity and P3a for use in multi-site studies of schizophrenia: characterization of demographic, clinical, cognitive, and functional correlates in COGS-2. Schizophr. Res. 163, 63–72.

Lilienfeld, S.O., 2007. Cognitive neuroscience and depression: legitimate versus illegitimate reductionism and five challenges. Cognit. Ther. Res. 31, 263–272.

Lilienfeld, S.O., 2014. The Research Domain Criteria (RDoC): an analysis of methodological and conceptual challenges. Behav. Res. Ther. 62, 129–139.

Lykken, D.T., McGue, M., Tellegen, A., Bouchard Jr., T.J., 1992. Emergenesis: genetic traits that may not run in families. Am. Psychol. 47, 1565–1977.

Lynall, M.E., Bassett, D.S., Kerwin, R., McKenna, P.J., Kitzbichler, M., Muller, U., et al., 2010. Functional connectivity and brain networks in schizophrenia. J. Neurosci. 30, 9477–9487. http://dx.doi.org/10.1523/JNEUROSCI.0333-10.2010.

Lyu, H., Hu, M., Eyler, L.T., Jin, H., Wang, J., Ou, J., et al., 2015. Regional white matter abnormalities in drug-naive, first-episode schizophrenia patients and their healthy unaffected siblings. Aust. NZ J. Psychiatry 49 (3), 246–254. 0004867414554268.

Malone, S.M., Vaidyanathan, U., Basu, S., Miller, M.B., McGue, M., Iacono, W.G., 2014. Heritability and molecular-genetic basis of the P3 event-related brain potential: a genome-wide association study. Psychophysiology 51, 1246–1258. http://dx.doi.org/10.1111/psyp.12345.

Mathew, I., Gardin, T.M., Tandon, N., Eacks, S., Francis, A.N., Seidman, L.J., et al., 2014. Medial temporal lobe structures and hippocampal subfields in psychotic disorders: findings from the Bipolar-Schizophrenia Network of Intermediate Phenotypes (SB-SNIP) study. JAMA Psychiatry 71, 769–777. http://dx.doi.org/10.1001/jamapsychiatry.2014.453.

Mazhari, S., Price, G., Dragovic, M., Waters, F.A., Clissa, P., Jablensky, A., 2011. Revisiting the suitability of antisaccade performance as an endophenotype in schizophrenia. Brain Cogn. 77, 223–230.

Meda, S.A., Gill, A., Stevens, M.C., Lorenzoni, R.P., Glahn, D.C., Calhoun, V.D., et al., 2012. Differences in resting-state fMRI functional network connectivity between schizophrenia and psychotic bipolar probands and their unaffected first-degree relatives. Biol. Psychiatry 71, 881–889. http://dx.doi.org/10.1016/j.biopsych.2012.01.025.

Meda, S.A., Ruano, G., Windemuth, A., O'Neil, K., Berwise, C., et al., 2014. Multivariate analysis reveals genetic associations of the resting default mode network in psychotic bipolar disorder and schizophrenia. Proc. Natl. Acad. Sci. U.S.A. 111, E2066–E2075. http://dx.doi.org/10.1073/pnas.1313093111.

Meyer-Lindenberg, A., 2010. From maps to mechanisms through neuroimaging of schizophrenia. Nature 468, 194–202. http://dx.doi.org/10.1038/nature09569.

Miller, G.A., 1996. How we think about cognition, emotion, and biology in psychopathology. Psychophysiology 33, 615–628.

Miller, G.A., 2010. Mistreating psychology in the decades of the brain. Perspect. Psychol. Sci. 5, 716–743.

Miller, G.A., Keller, J., 2000. Psychology and neuroscience: making peace. Curr. Dir. Psychol. Sci. 9, 212–215.

Miller, G.A., Rockstroh, B., 2013. Endophenotypes in psychopathology research: where do we stand? Annu. Rev. Clin. Psychol. 9, 177–213. http://dx.doi.org/10.1146/annurev-clinpsy-05212-185540.

Miller, G.A., Clayson, P.E., Yee, C.M., 2014. Hunting genes, hunting endophenotypes. Psychophysiology 51, 1329–1330. NIHMS627013.

Miller, G.A., Rockstroh, B., Hamilton, H.K., Yee, C.M., 2016. Psychophysiology as a core strategy in RDoC. Psychophysiology 53 (3), 410–414.

Miller, R.L., Yaesoubi, M., &Calhoun, V.D. (2014, August). Higher dimensional analysis shows reduced dynamism of time-varying network connectivity in schizophrenia patients. In: Engineering in Medicine and Biology Society (EMBC) 36th Annual International Conference of the IEEE, 2014. pp. 3837–3840.

Munafò, M.R., Flint, J., 2014. The genetic architecture of psychophysiological phenotypes. Psychophysiology 51, 1331–1332.

Nakagawa, T.T., Woolrich, M., Luckhoo, H., Joensson, M., Mohseni, H., Kringelbach, M.L., et al., 2014. How delays matter in an oscillatory whole-brain spiking-neuron network model for MEG alpha-rhythms at rest. NeuroImage 87, 383–394.

Narayanan, B., O'Neill, K., Berwise, C., Stevens, M.C., Calhoun, V.D., Clementz, B.A., et al., 2014. Resting state electroencephalogram oscillatory abnormalities in schizophrenia and psychotic bipolar patients and their relatives from the bipolar and schizophrenia network on intermediate phenotype study. Biol. Psychiatry 76 (6), 456–465. http://dx.doi.org/10.1016/j.biopsych.2013.12.008.

Neelam, K., Garg, D., Marshall, M., 2011. A systematic review and meta-analysis of neurological soft signs in relatives of people with schizophrenia. BMC Psychiatry 11, 139.

Nickl-Jockschat, T., Schneider, F., Pagl, A.D., Laird, A.R., Fox, P.T., Eickhoff, S.B., 2011. Progressive pathology is functionally linked to the domains of language and emotion: meta-analysis of brain structure changes in schizophrenia patients. Eur. Arch. Psychiatry Clin. Neurosci. 261 (Suppl. 2), S166–S171. http://dx.doi.org/10.1007/s00406-011-0249-8.

Nishioka, M., Bundo, M., Kasai, K., Iwamoto, K., 2012. DNA methylation in schizophrenia: progress and challenges of epigenetic studies. Genome Med. 4, 96. http://dx.doi.org/10.1186/gm397.

Oh, G., Petronis, A., 2008. Environmental studies of schizophrenia through the prism of epigenetics. Schizophr. Bull. 34, 1122–1129. http://dx.doi.org/10.1093/schbul/sbn105.

Ozerdem, A., Guntekin, B., Saatci, E., Tunca, Z., Basar, E., 2010. Disturbances in long-distance gamma coherence in bipolar disorder. Progr. Neuro-Psychopharmacol. Biol. Psychiatry 34, 861–865.

Patrick, C.J., 2014. Genetics, neuroscience, and psychopathology: clothing the emperor. Psychophysiology 51, 1333–1334.

Pearlson, G.D., 2015. Etiologic, phenomenologic, and endophenotypic overlap of schizophrenia and bipolar disorder. Annu. Rev. Clin. Psychol. 11, 251–281. http://dx.doi.org/10.1146/annurev-clinpsy-032814-112915.

Pearlson, G.D., Calhoun, V.D., 2009. Convergent approaches for defining functional imaging endophenotypes in schizophrenia. Front. Hum. Neurosci. 3, 1–11.

Peedicayil, J., 2011. Epigenetic management of major psychosis. Clin. Epigenet. 2, 249–256. http://dx.doi.org/10.1007/s13148-011-0038-2.

Popov, T., Jordanov, T., Rockstroh, B., Elbert, T., Merzenich, M.M., Miller, G.A., 2011. Specific cognitive training normalizes auditory sensory gating in schizophrenia: a randomized trial. Biol. Psychiatry 69, 465–471. http://dx.doi.org/10.1016/j.biopsych.2010.09.028.

Popov, T., Rockstroh, B., Weisz, N., Elbert, T., Miller, G.A., 2012. Adjusting brain dynamics in schizophrenia by means of perceptual and cognitive training. PLoS ONE 7 (7), e39051. http://dx.doi.org/10.1371/journal.pone.0039051.

Popov, T., Miller, G.A., Rockstroh, B., Weisz, N., 2013. Neuromagnetic oscillatory activity in somatosensory cortex indexes recognition of facial affect expression. J. Neurosci. 33, 6018–6026.

Popov, T.G., Carolus, A., Schubring, D., Popova, P., Miller, G.A., Rockstroh, B.S., 2015. Targeted training modifies oscillatory brain activity in schizophrenia patients. NeuroImage Clin. 7, 807–814.

Popova, P., Popov, T., Wienbruch, C., Carolus, A., Miller, G.A., Rockstroh, B., 2014. Changing facial affect recognition in schizophrenia: effects of training on brain dynamics. NeuroImage Clin. 6, 156–165. http://dx.doi.org/10.1016/j.nicl.2014.08.026.

Prasad, K.M., Keshavan, M.S., 2008. Structural cerebral variations as useful endophenotypes in schizophrenia: do they help construct "extended endophenotypes"? Schizophr. Bull. 34 (4), 774–790. http://dx.doi.org/10.1093/schbul/sbn017.

Puckett, R.E., Lubin, F.D., 2011. Epigenetic mechanisms in experience-driven memory formation and behavior. Epigenomics 3, 649–664. http://dx.doi.org/10.2217/epi.11.86.

Radtke, K.M., Ruf, M., Gunter, H.M., Dorhmann, K., Schauer, M., Meyer, A., et al., 2011. Transgenerational impact of intimate partner violence on methylation of the promoter of the glucocorticoid receptor. Transl. Psychiatry 19, 1:e21. http://dx.doi.org/10.1038/tp.2011.21.

Radulescu, E., Ganeshan, B., Shergill, S.S., Medford, N., Chatwin, C., Young, R.C., et al., 2014. Grey-matter texture abnormalities and reduced hippocampal volume are distinguishing features of schizophrenia. Psychiatry Res. 223 (3), 179–186. http://dx.doi.org/10.1016/j.psychresns.2014.05.014.

Ranlund, S., Nottage, J., Shaikh, M., Dutt, A., Constante, M., Walshe, M., et al., 2014. Resting EEG in psychosis and at-risk populations—a possible endophenotype? Schizophr. Res. 153 (1-3), 96–102. http://dx.doi.org/10.1016/j.schres.2013.12.017.

Roalf, D.R., Vandekar, S.N., Almasy, L., Ruparel., K., Satterthwaite, T.D., Elliott, M.A., et al., 2015. Heritability of subcortical and limbic brain volume and shape in multiplex-multigenerational families with schizophrenia. Biol. Psychiatry 77, 137–146. http://dx.doi.org/10.1016/j.biopsych.20140.05.009.

Saperstein, A.M., Fuller, R.L., Avila, M.T., Adami, H., McMahon, R.P., Thaker, G.L., et al., 2006. Spatial working memory as a cognitive endophenotype of schizophrenia: assessing risk for pathophysiologicyl dysfunction. Schizophr. Bull. 32, 498–506.

Schizophrenia Psychiatric GWAS Consortium, 2011. Genome-wide association study identifies five new schizophrenia loci. Nat. Genet. 43, 969–976. http://dx.doi.org/10.1038/ng.940.

Schizophrenia Working Group of the Psychiatric Genomics Consortium, 2014. Biological insights from 108 schizophrenia-associated genetic loci. Nature 511, 421–427. http://dx.doi.org/10.1038/nature13595.

Schmitt, A., Malchow, B., Hasa, A., Falkai, P., 2014. The impact of environmental factors in severe psychiatric disorders. Front. Neurosci. 8, 19. http://dx.doi.org/10.3389/fnins.2014.00019.

Schulze, T.G., Akula, N., Breuer, R., Steele, J., Nalls, M.A., Singleton, A.B., et al., 2014. Molecular genetic overlap in bipolar disorder, schizophrenia, and major depressive disorder. World J. Biol. Psychiatry 15, 200–208. http://dx.doi.org/10.3109/15622975.2012.

Schumann, G., 2014. Are we doing enough to extract genomic information from our data? Psychophysiology 51, 1335–1336.

Sebat, J., Levy, D.L., McCarthy, S.E., 2009. Rare structural variants in schizophrenia: one disorder, multiple mutations; one mutation, multiple disorders. Trends Genet. 25, 528–535.

Seidman, L.J., Faraone, S.V., Goldstein, J.M., Kremen, W.S., Horton, N.J., Makris, N., et al., 2002. Left hippocampal volume as a vulnerability indicator for schizophrenia. Arch. Gen. Psychiatry 59, 839–849.

Seidman, L.J., Hellemann, G., Nuechterlein, K.H., Greenwood, T.A., Braff, D.L., Cadenhead, K.S., et al., 2015. Factor structure and heritability of endophenotypes in schizophrenia: findings from the Consortium on the Genetics of Schizophrenia (COGS-1). Schizophr. Res. 163, 73–79.

Sigurdsson, T., Stark, K.L., Karayiorgou, M., Gogos, J.A., Gordon, J.A., 2010. Impaired hippocampal-prefrontal synchrony in a genetic mouse model of schizophrenia. Nature 464, 763–767.

Skinner, M.K., Mannikam, M., Guerrero-Bosagna, C., 2010. Epigenetic transgenerational actions of environmental factors in disease etiology. Trends Endocrinol. Metab. 21, 214–222. http://dx.doi.org/10.1016/j.tem.2009.12.007.

Sporns, O., 2011. The human connectome: a complex network. Ann. N.Y. Acad. Sci. 1224, 109–125.

Sporns, O., Tononi, G., Kokls, R., 2005. The human connectome: A structural description of the human brain. PLoS Comut. Biol. 1, e42.

Svrakic, D.M., Zorumski, C.F., Svrakic, N.M., Zwir, I., Cloninger, C.R., 2013. Risk architecture of schizophrenia: the role of epigenetics. Curr. Opin. Psychiatry 26, 188–195. http://dx.doi.org/10.1097/YCO.0b013e2835d8329.

Swerdlow, N.R., Weber, M., Qu, Y., Light, G., Braff, D.L., 2008. Realistic expectations of prepulse inibition in translational models for schizophrenia research. Psychopharmacology 199, 331–388.

Turetsky, B.I., Calkins, M.E., Light, G.A., Olincy, A., Radant, A.D., Swerdlow, N.R., 2007. Neurophysiological endophenotypes of schizophrenia: The viability of selected candidate measures. Schizoophr. Bull 33, 69–94.

Turetsky, B.I., Greenwood, T.A., Olincy, A., Radant, A.D., Braff, K.S., Cadenhead, D.J., et al., 2008. Abnormal auditory N100 amplitude: a heritable endophenotype in first-degree relatives of schizophrenia probands. Biol. Psychiatry 64, 1051–1059. http://dx.doi.org/10.1016/j.biopsych.2008.06.018.

Turetsky, B.I., Dress, E.M., Braff, D.L., Calkins, M.E., Green, M.F., Greenwood, T.A., et al., 2015. The utility of P300 as a schizophrenia endophenotype and predictive biomarker: clinical and socio-demographic modulators in COGS-2. Schizophr. Res. 163, 53–62. http://dx.doi.org/10.1016/j.schres.2014.09.024.

Tuulio-Henriksson, A., Arajärvi, R., Partonen, T., Hauka, J., Variolo, T., Schreck, M., et al., 2003. Familial loading associates with impairment in visual span among healthy siblings of schizophrenia patients. Biol. Psychiatry 54, 623–628.

Uhlhaas, P.J., Singer, W., 2013. High-frequency oscillations and the neurobiology of schizophrenia. Dialogues Clin. Neurosci. 15, 301–313.

Vaidyanathan, U., Malone, S.M., Miller, M.B., McGue, M., Iacono, W.G., 2014. Heritability and molecular genetic basis of acoustic startle eye blink and affectively modulated startle response: a genome-wide association study. Psychophysiology 51, 1285–1299. http://dx.doi.org/10.1111/psyp.12348.

Van den Heuvel, M.P., Fornito, A., 2014. Brain networks in schizophrenia. Neuropsychol. Rev. 24, 32–48. http://dx.doi.org/10.1007/s11065-014-9248-7.

Van der Velde, J., Gromann, P.M., Swart, M., de Haan, L., Wiersma, D., Bruggeman, R., et al., 2015. Grey matter, an endophenotype for schizophrenia? A voxel-based morphometry study in siblings of patients with schizophrenia. J. Psychiatry Neurosci. 40, 207–213. http://dx.doi.org/10.1503/jpn.140064.

Van Os, J., 2011. From schizophrenia metafacts to non-schizophrenia facts. Schizophr. Res. 127, 16–27. http://dx.doi.org/10.1016/j.schres.2011.026.

van Scheltinga, A.F.T., Bakker, S.C., van Haren, N.E., Derks, E.M., Buizer-Voskamp, J.E., Boos, H.B.M., et al., 2013. Genetic schizophrenia risk variants jointly modulate total brain and white matter volume. Biol. Psychiatry 73, 525–531. http://dx.doi.org/10.1016/j.biopsych.2012.08.017.

Vrieze, S.I., Malone, S.M., Vaidyanathan, U., Kwong, A., Kang, H.M., Zhan, X., et al., 2014. In search of rare variants: preliminary results from whole genome sequencing of 1,325 individuals with psychophysiological endophenotypes. Psychophysiology 51, 1309–1320.

White, T., Gottesman, I., 2012. Brain connectivity and gyrification as endophenotypes schizophrenia: weight and evidence. Curr. Topics Med. Chem. 12, 2393–2403.

Wilhelmsen, K.C., 2014. The feasibility of genetic dissection of endophenotypes. Psychophysiology 51, 1337–1338.

Wockner, L.F., Noble, E.P., Lawford, B.R., Young, R.M., Morris, C.P., Whitehall, V.L., et al., 2014. Genome-wide DNA methylation analysis of human brain tissue for schizophrenia patients. Transl. Psychiatry 4, e339. http://dx.doi.org/10.1038/tp.2013.111.

Wotruba, D., Michels, L., Buechler, R., Metzler, S., Theodoridou, A., Gerstenberg, M., et al., 2014. Aberrant coupling within and across default mode, task-positive, and salience network in subjects at risk for psychosis. Schizophr. Bull. 40, 1095–1104. http://dx.doi.org/10.1093/schbul/sbt161.

Xu, X., Wells, A.B., O'Brien, D.R., Nehorai, A., Dougherty, J.D., 2014. Cell type-specific expression analysis to identify putative cellular mechanisms for neurogenetic disorders. J. Neurosci. 34, 1420–1431.

Yee, C.M., Javitt, D.C., Miller, G.A., 2015. Replacing categorical with dimensional analyses in psychiatry research: the RDoC initiative. JAMA Psychiatry 72 (12), 1159–1160.

Zucchi, F.C., Yao, Y., Ward, I.D., Ilniytskyy, Y., Olson, D.M., Nemzies, K., et al., 2013. Maternal stress induces epigenetic signatures of psychiatric and neurological diseases in the offspring. PLoS One 8, e56967. http://dx.doi.org/10.1371/journal.pone.0056967.

CHAPTER 3

INSIGHTS FROM GENOME-WIDE ASSOCIATION STUDIES (GWAS)

S. Cichon[1,2,3,4,5] and S. Ripke[6,7,8]

[1]*Division of Medical Genetics, University Hospital Basel, Basel, Switzerland* [2]*Department of Biomedicine, University of Basel, Basel, Switzerland* [3]*Institute of Human Genetics, University of Bonn, Bonn, Germany* [4]*Department of Genomics, Life & Brain Center, University of Bonn, Bonn, Germany* [5]*Institute of Neuroscience and Medicine (INM-1), Research Center Jülich, Jülich, Germany* [6]*Analytic and Translational Genetics Unit, Department of Medicine, Massachusetts General Hospital and Harvard Medical School, Boston, MA, United States* [7]*Stanley Center for Psychiatric Research and Medical and Population Genetics Program, Broad Institute of MIT and Harvard, Cambridge, MA, United States* [8]*Department of Psychiatry and Psychotherapy, Charité – Universitätsmedizin Berlin, Campus Mitte, Berlin, Germany*

CHAPTER OUTLINE

The Pre-GWAS Era: Linkage and Candidate Gene Association Studies	40
The Developments That Made GWAS Possible	41
The Principle of GWAS	42
GWAS Findings in Schizophrenia	44
Genetic Relationship Between Schizophrenia and Other Psychiatric Disorders	46
Were GWAS of Schizophrenia Successful So Far and Should They Continue?	48
Future Directions	49
References	49

There is compelling evidence that schizophrenia is highly heritable, suggesting that genetic factors are substantially involved in disease development. This knowledge has led to important endeavors to discover the responsible genes since the 1980s. The genetic strategies were heavily influenced by theoretical considerations regarding the nature and effect size of the genetic factors involved in schizophrenia as well as by the knowledge about genetic variation in humans and the laboratory technologies to analyze genetic variation.

THE PRE-GWAS ERA: LINKAGE AND CANDIDATE GENE ASSOCIATION STUDIES

In the late 1980s and the 1990s, the hope was that monogenic forms of schizophrenia exist, particularly in some multiaffected multigeneration families. Based on the formal genetic findings, it was clear that this would only apply to a small fraction of the disease cases in the population. The majority of schizophrenia families were thought to be influenced by a small number of disease genes with relatively high penetrance (so-called major disease genes). It was therefore reasonable to assume that similar family-based positional cloning strategies used to successfully identify thousands of genes for Mendelian diseases should also be powerful to pinpoint these genes. The central concept of this approach was to find genetic linkage (ie, to identify regions of the genome that are cotransmitted with the disease over generations) in families with two or more affected individuals. The principle is very straightforward: using a few hundred evenly spaced genetic markers across the whole genome (usually microsatellite markers) that allow tracing back the inheritance of each homologous chromosome pair in each family, linkage has the potential to locate disease genes by virtue of chromosomal position alone, without any prior knowledge of disease etiology. In hindsight, hopes of Mendelian forms of the illness have not materialized, and most linkage studies have failed to achieve stringent "genome-wide" levels of significance or to replicate preexisting findings. A common explanation used to be that the effect size (penetrance) of the disease genes was smaller than previously assumed. Most studies were therefore heavily underpowered. Another problem was the insufficient genomic coverage of the marker sets used. Even the largest metaanalysis of genetic linkage studies performed at that time (Lewis et al., 2003) did not come up with statistically convincing evidence for genomic loci harboring highly penetrant major disease genes that play a role in the majority of schizophrenia patients.

A second major research strategy applied during the 1990s and early 2000s was candidate gene association studies involving common genetic variants. The principle of these studies is to detect allelic variants that are more (or less) common in patients with schizophrenia than they are within the general population. These studies are more powerful than linkage studies in detecting genes of small effect (ie, reduced penetrance), assuming that a high proportion of the disease alleles are relatively common in the population and that they are attributable to a common founder. Most of the candidate genes were selected because they are either functional candidates (ie, they encode a protein implicated by an etiological hypothesis) or positional candidates located in chromosomal regions implicated by previous linkage studies. A major disadvantage of candidate gene association studies is that they leave a large proportion of the genome uninvestigated. A systematic, comprehensive investigation of the genetic variability present in the candidate genes could not be performed due to the limited knowledge of their variability in the population. Apart from these significant shortcomings, most of the early candidate gene studies did not comply sample sizes necessary to detect smaller effects. This led to a confusing picture of some candidate gene association findings that could not be robustly replicated.

Although the results of early linkage and candidate gene association were disappointing, they contributed to advance the field by showing that early genetic model assumptions of schizophrenia were obviously wrong. Still, heritability estimations and formal genetic studies left no doubt that genetic factors substantially contribute to schizophrenia. The questions were in regard to what these genetic factors realistically look like and how they can be identified. An extensive debate started over the genetic contribution to individual susceptibility, not only to schizophrenia but also to common complex diseases in general (eg, Schork et al., 2009). One hypothesis, the "Common Disease, Common Variant (CDCV)" hypothesis, argues that many common genetic variants with relatively low penetrance are the major contributors to genetic susceptibility to common diseases. A second hypothesis, the "Common Disease, Rare Variant (CDRV)" hypothesis, however, proposes that multiple rare genetic variants, each with relatively high penetrance, are the major contributors. To comprehensively test these hypotheses, our knowledge about genetic variation in the human genome as well as the technological tools to analyze it in many individuals were not yet sufficiently developed at the turn of the millennium. Since then, however, considerable progress has been made and, today, both hypotheses have their place in current research efforts to identify the genetic factors involved in the development of schizophrenia.

THE DEVELOPMENTS THAT MADE GWAS POSSIBLE

The past 15 years have seen major advances in the field of human disease genetics. These developments have paved the way for novel research strategies such as genome-wide association studies (GWAS) to identify the genetic factors involved in schizophrenia and other genetically complex disorders.

The beginning of a new era in human genetics was marked by the completion of the Human Genome Project in 2003. For the first time since the description of the molecular structure of DNA by Francis Crick and James D. Watson in 1953, there was a reference sequence available for the 3 billion bases that comprise the human genome. There was a strong focus on systematically describing the extent of interindividual genetic variation, particularly biallelic single nucleotide polymorphisms (SNPs), which are the most abundant type of variation. It became evident that even though the majority of bases do not differ, or are invariant, between any two individuals in the world, there are more than 9 million unique SNPs in the genome where healthy European individuals differ in their DNA sequence (The 1000 Genomes Project Consortium, 2012) and where the less frequent allele in the population (the minor allele) has an allele frequency of at least 1% (ie, minor allele frequency (MAF) is >1%). Today, comprehensive catalogs of SNPs are deposited in publicly available databases and can be used without restrictions. Another important feature of the genome that was discovered when SNPs were investigated in many individuals from different populations was the haplotype block structure. The human genome comprises recombination hot spots, where most of the recombinations take place. In between these recombination hot spots, there are stretches of DNA (so-called blocks) where recombination events are extremely rare, to the effect that SNPs that are located in these blocks are often correlated to one another. This phenomenon is also called linkage disequilibrium (LD) and has important consequences for the selection of SNPs for systematic genetic association studies; because many of the SNPs in a block are in LD, only a few—the so-called haplotype tagging SNPs—need to be investigated in the course of systematic association studies. Soon after the completion of the Human Genome Project, the International HapMap Project was initiated (The International HapMap Project, 2005); its goal was to establish a catalog of common genetic variation in humans and describe the haplotype block structure in different populations. Much of this information was used for the development of SNP arrays that play a key role in GWAS. Today, the International HapMap Consortium has outgrown into the 1000 Genomes Project Consortium with a stronger focus on rare variation in different populations.

Another crucial step to make GWAS feasible was the development of DNA microarrays. Microarrays are a technology in which thousands to millions of nucleic acids are bound to a surface (microarray) and are used as a fishing rod to measure the relative concentration of nucleic acid sequences in a mixture via hybridization and subsequent detection of the hybridization events (Bumgarner, 2013). Companies started to develop microarrays as SNP genotyping platforms to allow the simultaneous, rapid genotyping of hundreds of thousands of SNPs in an individual at affordable costs. The most commonly used approaches are allele discrimination by hybridization, as used by Affymetrix and the Infinium Assay of Illumina. Both the Affymetrix and the Illumina methods for SNP genotyping have been highly successful and are highly used. The microarrays have constantly become more informative and denser. Today, SNP arrays capable of detecting >1 M different human SNPs are available from both vendors. Call rates (the fraction of the SNPs on the array that can be reliably called) and technical reproducibility of SNP calls exceed 99.5%. In addition, the same arrays or variations thereof can also be used to detect copy number variants (CNVs).

THE PRINCIPLE OF GWAS

In a GWAS, many SNPs are genotyped in a large number of individuals using commercially available high-throughput genotyping platforms (SNP microarrays). For all individuals, phenotype data are available. The phenotype can be either continuous, quantitative (such as height, blood pressure), or dichotomous, resulting in two distinct groups, usually individuals with disease (cases) and without disease (controls). All individuals are genotyped using genotyping platforms that measure between 300,000 and 1,000,000 unique SNPs in each sample. Because many SNPs located in the same haplotype block are correlated via LD, it is computationally feasible to estimate the genotypes of correlated SNPs (not genotyped by the microarray) across the genome using a reference data set, such as that provided by the 1000 Genomes Project (The 1000 Genomes Project Consortium, 2012). This method of estimating missing genotypes is called imputation and enables a more comprehensive analysis of the SNP variation present in the human genome as well as metaanalyses across multiple data sets that have used different genotyping platforms (and thus genotyped different sets of SNPs). This is a crucial step for collaboration and combining GWAS data from multiple sources, thereby increasing sample size and boosting the power to detect novel genetic associations.

In a case-control GWAS, the genotype data from patients and unaffected, population-matched controls are compared. Other than the correct identification of individuals with the disease, there is no requirement for prior biological knowledge of the trait undergoing investigation. A GWAS is an unbiased, genome-wide search for SNPs that are involved in the development of a disease. Such SNPs will show statistically significant allele frequency differences between the two compared groups, patients and controls, and point to regions of the genome that harbor potential causal variants. After the often inconsistent and frustrating results of the linkage and candidate gene era, the GWAS approach has led to a breakthrough and has discovered several thousands of genetic loci reliably associated with a range of complex phenotypes and diseases (Fig. 3.1). These findings not only provide novel biological insight into diseases but also have the potential to guide therapeutic development.

GWAS have been used as a method for approximately 10 years and have been a very valuable tool for gaining more fundamental insight into the etiology of complex diseases. The costs of SNP microarrays have decreased substantially over time, so that it is now feasible to analyze tens of thousands of individuals. This, in combination with the formation of large international collaborations that jointly

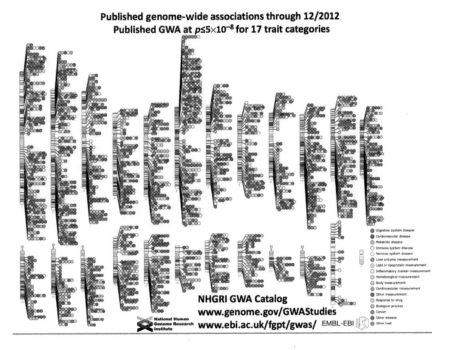

FIGURE 3.1

SNPs associated with traits per chromosome. Nervous system disease traits are represented in light yellow.

analyze GWAS data from multiple sites, has been a crucial requirement to achieve the statistical power to detect the small genetic effects mediated by common risk variants.

One limitation of GWAS is that it cannot identify the functionally causal variant(s) in an associated region. As a consequence of the LD between many variants within a haplotype block, the association signal typically covers a broad chromosomal region with several SNPs showing similar statistical significance, although these regions are usually much smaller than the typical size of a linkage peak in the pre-GWAS era. The associated regions may span 500 kb or more, sometimes containing numerous genes. In practice, however, at least half of the associated regions encompass only one or two genes. In other regions, pathway or gene set analyses or functional annotations may help prioritize candidate genes, but the identification of the responsible gene/functionally relevant variant remains a challenge.

Another limitation of GWAS is the large number of tested SNPs, posing a massive multiple testing problem. This is addressed by correction methods, and a SNP allele is assumed to be associated with the disease when the allele frequency difference between patients and controls produces $p < 5 \times 10^{-08}$. This significance threshold, also named genome-wide significance threshold, accounts for testing 1 million independent SNPs, assuming a type 1 error rate of 5%. It should be noted that for real effects a sample size increase will eventually overcome the multiple testing burden. An extensive increase of sample size is usually only to be reached with extensive international collaboration. Successful collaborations achieved this with centralized analysis structures and centralized data storage accessible to all contributing centers.

GWAS FINDINGS IN SCHIZOPHRENIA

The first GWAS of schizophrenia was published in 2006 (Mah et al., 2006). It comprised 320 patients and 325 controls and genotyped approximately 25,000 gene-based SNPs. Unsurprisingly, in retrospect, no SNP was identified that was even close to significantly meeting today's statistical requirements for GWAS. The next two published GWAS (Lencz et al., 2007; Sullivan et al., 2008) used the Affymetrix 500K genotyping platform and already had much more comprehensive coverage of the genome. By today's standards, sample sizes were still small (178 patients/144 controls in Lencz et al., 2007; 738 patients/733 controls in Sullivan et al., 2008), and only suggestive association findings could be reported; none of them exceeded the threshold for genome-wide significance. With the fourth published GWAS by O'Donovan et al. (2008), the first robust association finding could be reported. Although the initial GWAS by these authors, comprising 479 patients and 2937 controls, did not yield a genome-wide significant p value, there were a few findings that were just shy of meeting this significance threshold. The authors took these SNPs and performed follow-up in a number of schizophrenia case–control samples that they put together through international collaborations. Overall, this replication sample had an impressive size of 6666 patients and 9897 controls. In these follow-up samples, a SNP located in the transcription factor *ZNF804A* showed a consistent association pattern with the disease, and this gene is considered by many in the field as the first success of GWAS in schizophrenia. These early days were very informative and moved the field forward once again. It became clear that the effect size of common risk genes for schizophrenia is generally less than genotype relative risks (GRR) of 1.2, much lower than expected and that several hundreds of patients and controls did not suffice to detect genome-wide significant associations when appropriately correcting for multiple testing. It was during that time when researchers began to intensify collaborations and jointly analyzed GWAS data sets to substantially increase samples sizes and therefore boost the power of their analyses to systematically detect small genetic effects contributing to schizophrenia across the genome. Consortia like ISC (International Schizophrenia Consortium), SGENE, MGS (Molecular Genetics of Schizophrenia), the BoMa (Bonn–Mannheim) were founded, subsequently supported the PGC (Psychiatric Genomics Consortium) and substantially contributed to the success of GWAS in schizophrenia and other neuropsychiatric disorders in recent years. A collaboration formed by the ISC, SGENE, and MGS came up with three further genome-wide significant GWAS findings in the genes encoding transcription factor 4 (*TCF4*) and neurogranin (*NRGN*) and in the major histocompatibility complex (MHC) region on chromosome 6 (Purcell et al., 2009; Shi et al., 2009; Stefansson et al., 2009). TCF4 plays a role in nervous system development, and NRGN is a protein kinase substrate that binds calmodulin and is involved in neuronal calcium signaling. None of these genes were in the focus of historical candidate gene studies, and this illustrates the power of the GWAS approach: it leads to novel disease-relevant, biological processes. A highly interesting finding was that of the MHC region. Genes in this genetically highly complex region are involved in immune response, and the region has long been postulated to harbor variants conferring a risk for schizophrenia because there is evidence for linkage in this region (Lewis et al., 2003) and research has suggested the involvement of infection in disease development (Brown and Derkits, 2010).

Apart from the fact that the aforementioned studies were capable of cracking the genome-wide significance threshold and adding new biological insights to schizophrenia, the study by Purcell et al. (2009) introduced a new statistical method and analysis strategy that, since then, has become very important: the polygenic risk score. The assumption is that each GWAS of schizophrenia potentially contains thousands of risk SNPs that are not strong enough to reach genome-wide significance. Collectively, however, they could account for a substantial proportion of variation in risk. In practice, for instance, all nominally associated (ie, $p < 0.05$) SNPs of a GWAS can be used to summarize an individual score (in principle a sum of the risk alleles that an individual carries at all these putative schizophrenia-related SNPs) in another independent data set. It has been shown in many different data sets that the polygenic risk score derived with a SNP set from an initial GWAS is capable of distinguishing schizophrenia patients from controls in independent data sets, albeit with low sensitivity and specificity. Notably, the predictive power of the polygenic test improves (in terms of phenotypic variance explained) when the p value threshold is lowered and SNPs with p values between 0.05 and 0.5 are also included. This means that even though the number of false-positive SNPs increases when lowering the p value threshold, many schizophrenia-associated SNPs with small genetic effects (eg, GRR = 1.05) still must be contained in this p value range. Thus, the most important implication of that work was that further increasing the sample size would push many more yet unknown schizophrenia-related SNPs and genes beyond the genome-wide significance threshold.

This again fostered the collaborative approach to reach substantially larger sample sizes, and most international research groups working on GWAS joined the PGC that performed megaanalyses of almost all available GWAS data. The first schizophrenia analysis of the Schizophrenia Working Group of the PGC already comprised 21,856 individuals (patients and controls) in a GWAS step, as well as 29,939 individuals in a follow-up replication step (The Schizophrenia GWAS Consortium, 2011). It yielded five novel gene loci: *MIR137*, *PCGEM1*, *CSMD1*, *MMP16*, and *CNNM2/NT5C2*. The strongest finding (*MIR137*) was in an intron of a primary transcript for a microRNA137, a known regulator of neuronal development. Some of the other genome-wide significant loci contained predicted targets of *MIR137*, suggesting *MIR137*-mediated dysregulation as a previously unknown etiologic mechanism in schizophrenia.

In the following years, the PGC continued to involve more and more groups and samples in this worldwide collaboration and recently published a second megaanalysis that represents the largest GWAS of schizophrenia so far. It included contributions of almost 37,000 individuals with schizophrenia, 302 investigators, 35 countries, and 4 continents (Schizophrenia Working Group of the Psychiatric Genomics Consortium, 2014). The group reported 128 statistically independent associations, implicating at least 108 schizophrenia-associated loci (Fig. 3.2).

Whereas 83 of these loci were novel, 25 had previously been reported by other studies, thus confirming that the use of large samples results in reproducible findings. Preliminary biological pathway analyses performed by the PGC schizophrenia group using these impressive data did not identify any preannotated biological pathways that were enriched for associations. This may be a problem of the limited pathway information that these analyses are based upon. Maybe the relevant biological pathways are not yet depicted in the databases. Another explanation might be that there really is such a high degree of polygenicity that even larger samples would be necessary to pinpoint the relevant biological pathways with high statistical confidence in the preannotated pathways data.

However, even the known functions of many of the genes identified in the current GWAS data provide valuable insights in the biological processes obviously involved. It is notable that within the schizophrenia-associated loci there are a number of genes that are involved in the following functional categories: synaptic function and plasticity, glutamatergic (and possibly dopaminergic) neurotransmission, neuronal calcium signaling, neurodevelopment, and immune processes. Interestingly, the validity of some of these genes is also supported by the presence of rare variants. These findings are consistent with leading pathophysiological hypotheses. There are also findings that are not yet understood, such

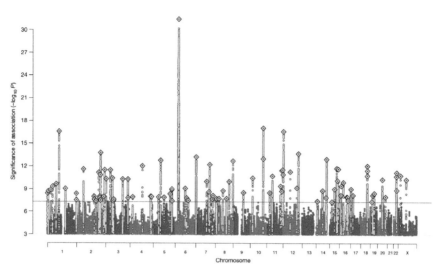

FIGURE 3.2

Manhattan plot showing the results of the latest megaanalysis of the Schizophrenia Working Group of the PGC, implicating 108 independent schizophrenia-associated loci (Schizophrenia Working Group of the Psychiatric Genomics Consortium, 2014).

as an enrichment of association signals in brain-expressed genes that both interact with the Fragile X mental retardation protein (FMRP). Research in this direction will hopefully help bridge the knowledge of genetic association patterns and pathophysiology.

GENETIC RELATIONSHIP BETWEEN SCHIZOPHRENIA AND OTHER PSYCHIATRIC DISORDERS

The comprehensive GWAS data that are now available have made it possible to address some long-standing unanswered questions in psychiatry. A new method, "genomic-restricted maximum likelihood estimation" (GREML; Lee et al., 2011), implemented in the computational tool "Genome-wide Complex Trait Analysis" (GCTA; Yang et al., 2011) is based on the intuition that under a polygenic model, cases should, on average, be more similar to other cases than to controls. This method is widely used to estimate the variance explained by all the autosomal SNPs as well as to estimate the genetic correlation between two diseases using SNP data.

This method has also been proven to be extremely helpful in analyzing what proportion of the SNPs involved in the development of schizophrenia also plays a role in the development of other common psychiatric disorders, such as bipolar disorder (BPD), major depressive disorder (MDD), attention-deficit hyperactivity disorder (ADHD), and autism spectrum disorders (ASDs). The Cross-Disorder Working Group of the PGC has recently published the results of the comparison of the aforementioned five common psychiatric disorders (Cross-Disorder Group of the Psychiatric Genomics Consortium,

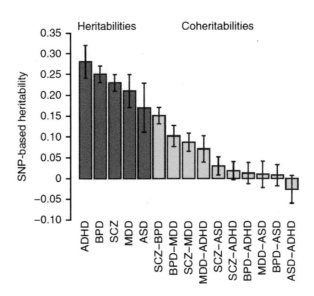

FIGURE 3.3

SNP-based heritabilities of five common psychiatric disorders (schizophrenia (SCZ), bipolar disorder (BPD), major depressive disorder (MDD), attention-deficit hyperactivity disorder (ADHD), and autism spectrum disorder (ASD)) and SNP-based correlations (coheritabilities) between any pair of two of the five disorders (Cross-Disorder Working Group of the Psychiatric Genomics Consortium, 2013).

2013). In a first step, the SNP-based heritabilities for each of these disorders were estimated from the large GWAS data sets available in the PGC. This value measures the variance between patient–control status explained by SNPs (Fig. 3.3). In schizophrenia, SNPs explain 23% of the variance, indicating that common variants play a substantial role in the development of the disease. In a second step, pairwise comparisons were performed between the five disorders to estimate the proportion of genetic information that overlaps between any two of these disorders. A surprisingly high correlation of 0.68 was found between schizophrenia and BPD, a moderate correlation (0.43) was found between schizophrenia and MDD, and low correlation (0.16) was found between schizophrenia and ASD. No significant correlation was found between schizophrenia and ADHD.

The phenomenon that a gene or a genetic variant contributes to multiple phenotypes is called "pleiotropy." Not only GWAS but also other genetic studies (such as whole-exome or whole-genome sequencing studies) as well as epidemiological studies provide increasing evidence that pleiotropy is a very common phenomenon for genetic risk factors for neuropsychiatric disorders. The dissection of the considerable pleiotropy between schizophrenia and BPD and MDD, for instance, not only will require very large cohorts but also will require this work to be undertaken across current diagnostic boundaries (Tansey et al., 2015).

The above mentioned methods require access to raw genotypes. Recently new methods were developed that only require summary statistics, eg, LDScore and the Brainstorm Project (Bulik-Sullivan et al., 2015).

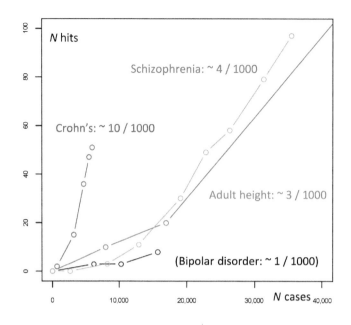

FIGURE 3.4

Number of genome-wide significant findings as a function of sample size, reflected as number of cases. Red: Crohn's disease; yellow: schizophrenia; black: bipolar disorder; green: adult height.

WERE GWAS OF SCHIZOPHRENIA SUCCESSFUL SO FAR AND SHOULD THEY CONTINUE?

There has been an extensive debate in the field regarding to what extent GWAS will be successful in psychiatric disorders compared to GWAS in other genetically complex diseases/phenotypes, such as diabetes mellitus type 2, heart infarction, or human height. In part, this related with the long period of time when psychiatric genetics produced inconsistent or negative results. However, looking at the plethora of new schizophrenia-associated loci from the latest GWAS of schizophrenia (Schizophrenia Working Group of the Psychiatric Genomics Consortium, 2014), an important conclusion of that work must be to continue along this successful path. It is of note that the >100 identified SNPs explain only 3% of the variance, whereas the total contribution of SNPs has been estimated at 23%. Further increasing the sample size will deliver more genome-wide significant loci and thus more comprehensive biological insights.

When we look at the genome-wide significant SNP association findings as a function of sample size in schizophrenia and compare them to two other genetically complex diseases/phenotypes that are generally considered as phenotypes for which GWAS were a successful strategy, namely, Crohn's disease and human height, findings regarding schizophrenia are comparable to those for human height (Fig. 3.4). For schizophrenia, it appears that there was a crucial inflection point in approximately 15,000 cases. Past this point, there is an almost linear relationship between sample size and the number of novel discoveries (approximately 4 new genome-wide significant SNPs per 1000 additional cases; Fig. 3.4). It is therefore reasonable to aim at further increasing the samples sizes to get a more complete picture of the genetic loci involved in schizophrenia.

FUTURE DIRECTIONS

Without doubt, schizophrenia is a success story in terms of identification of common risk variants using GWAS. Despite diagnostic difficulties and presumed high heterogeneity, the field succeeded in identifying far more than 100 genetic loci that are involved in the development of schizophrenia with unprecedented statistical confidence. No other psychiatric disorder (eg, BPD and MDD, ADHD, and ASD) has produced a comparable list of genome-wide significant findings of common risk variants. The reasons for this are not well understood at this point. However, this may be due to the fact that sample numbers that were put together in large collaborative efforts are currently larger for schizophrenia than for other psychiatric disorders. Also, the genetic architecture of schizophrenia might be more amenable to GWAS than that of other psychiatric disorders. Despite these encouraging findings, it will be important to continue to increase sample sizes (and power) and discover more risk genes to get a more complete picture of the biological processes involved in the disorder. Improved bioinformatics methods, such as refined biological pathway analyses, and more comprehensive databases with preannotated pathways will be necessary to better exploit GWAS results in this direction. It should also be emphasized that schizophrenia is characterized by a broad (and probably continuous) allelic spectrum of disease alleles, ranging from rare variants with relatively high penetrance to common variants with relatively low penetrance.

Comprehensive individual information on polygenic risk scores will likely be very useful for improved clinical (sub)classifications of patients and for guiding options for medication and therapy.

In terms of investigating the pathophysiological consequences of the many risk variants identified so far, there is a lot of work ahead. At this point, for the risk SNPs identified through GWAS, it is unclear whether they are the functionally relevant SNPs or just "proxies" in LD. This cannot be fully resolved by statistical or bioinformatics methods. Eventually, experimental approaches (molecular and cellular investigations) will be key to understanding the functional consequences of risk alleles and, thus, pathophysiological processes.

Some of the identified disease loci may already be interesting therapeutic targets. However, the hope is that with deeper insights into the biology of schizophrenia and the underlying pathophysiological processes, yet undiscovered treatments will become possible.

REFERENCES

Brown, A.S., Derkits, E.J., 2010. Prenatal infection and schizophrenia: a review of epidemiologic and translational studies. Am. J. Psychiatry 167 (3), 261–280.

Bulik-Sullivan, B.K., Loh, P.R., Finucane, H.K., Ripke, S., Yang, J., Schizophrenia Working Group of the Psychiatric Genomics Consortium. In: Patterson, N., et al., 2015. LD Score regression distinguishes confounding from polygenicity in genome-wide association studies. Nat. Genet. 47 (3), 291–295.

Bumgarner, R., 2013. DNA microarrays: types, applications and their future. Curr. Protoc. Mol. Biol. 22 Unit-22.1.

Cross-Disorder Group of the Psychiatric Genomics Consortium, 2013. Genetic relationship between five psychiatric disorders estimated from genome-wide SNPs. Nat. Genet. 45, 984–994.

Lee, S.H., Wray, N.R., Goddard, M.E., Visscher, P.M., 2011. Estimating missing heritability for disease from genome-wide association studies. Am. J. Hum. Genet. 88, 294–305.

Lencz, T., Morgan, T.V., Athanasiou, M., Dain, B., Reed, C.R., Kane, J.M., et al., 2007. Converging evidence for a pseudoautosomal cytokine receptor gene locus in schizophrenia. Mol Psychiatry 12 (6), 572–580.

Lewis, C.M., Levinson, D.F., Wise, L.H., DeLisi, L.E., Straub, R.E., Hovatta, I., et al., 2003. Genome scan meta-analysis of schizophrenia and bipolar disorder, part II: schizophrenia. Am. J. Hum. Genet. 73, 51–59.

Mah, S., Nelson, M.R., DeLisi, L.E., Reneland, R.H., Markward, N., James, M.R., et al., 2006. Identification of the semaphorin receptor PLXNA2 as a candidate for susceptibility to schizophrenia. Mol Psychiatry 11 (5), 471–478.

O'Donovan, M.C., Craddock, N., Norton, N., Williams, H., Peirce, T., Moskvina, V., et al., 2008. Identification of loci associated with schizophrenia by genome-wide association and follow-up. Nat. Genet. 40 (9), 1053–1055.

Purcell, S.M., Wray, N.R., Stone, J.L., Visscher, P.M., O'Donovan, M.C., Sullivan, P.F., et al., 2009. Common polygenic variation contributes to risk of schizophrenia and bipolar disorder. Nature 460 (7256), 748–752.

Schizophrenia Working Group of the Psychiatric Genomics Consortium, 2014. Biological insights from 108 schizophrenia-associated genetic loci. Nature 511, 421–427.

Schork, N.J., Murray, S.S., Frazer, K.A., Topol, E.J., 2009. Common vs. rare allele hypotheses for complex diseases. Curr. Opin. Genet. Dev. 19 (3), 212–219.

Shi, J., Levinson, D.F., Duan, J., Sanders, A.R., Zheng, Y., Pe'er, I., et al., 2009. Common variants on chromosome 6p22.1 are associated with schizophrenia. Nature 460 (7256), 753–757.

Stefansson, H., Ophoff, R.A., Steinberg, S., Andreassen, O.A., Cichon, S., Rujescu, D., et al., 2009. Common variants conferring risk of schizophrenia. Nature 460 (7256), 744–747.

Sullivan, P.F., Lin, D., Tzeng, J.Y., van den Oord, E., Perkins, D., Stroup, T.S., et al., 2008. Genomewide association for schizophrenia in the CATIE study: results of stage 1. Mol Psychiatry 13 (6), 570–584.

Tansey, K.E., Owen, M.J., O'Donovan, M.C.O., 2015. Schizophrenia genetics: building the foundations of the future. Schizophr. Bull. 41, 15–19.

The 1000 Genomes Project Consortium, 2012. An integrated map of genetic variation from 1,092 human genomes. Nature 491, 56–65.

The International HapMap Consortium, 2005. A haplotype map of the human genome. Nature 437, 1299–1320.

The Schizophrenia Psychiatric Genome-Wide Association Study (GWAS) Consortium, 2011. Genome-wide association study identifies five new schizophrenia loci. Nat. Genet. 43, 969–976.

Yang, J., Lee, S.H., Goddard, M.E., Visscher, P.M., 2011. GCTA: a tool for genome-wide complex trait analysis. Am. J. Hum. Genet. 88, 76–82.

CHAPTER 4

SEQUENCING APPROACHES TO MAP GENES LINKED TO SCHIZOPHRENIA

P.M.A. Sleiman[1,2] and H. Hakonarson[1,2]
[1]Center for Applied Genomics, Children's Hospital of Philadelphia Research Institute, Philadelphia, PA, United States [2]Department of Pediatrics, Perelman School of Medicine, University of Pennsylvania, Philadelphia, PA, United States

CHAPTER OUTLINE

Introduction and Background..51
Next-Generation Sequencing 52
Exome Versus Whole Genome..53
Sequencing Studies to Date..55
Conclusions..57
Future Directions 57
References..58

INTRODUCTION AND BACKGROUND

Schizophrenia is a debilitating psychiatric disorder that poses a worldwide public health burden. The disorder is characterized by a heterogeneous clinical presentation that complicates diagnosis of the disorder and the studies attempting to identify its genetic etiology.

Evidence from family, twin, and epidemiological studies indicates that schizophrenia is a complex disorder that arises as a result of interplay between genetic and environmental factors. The genetic component of the disorder has been shown to be significant: heritability (h^2) estimates range between approximately 0.6 and 0.8 (Gejman et al., 2011). Multiple environmental factors have been associated with schizophrenia, including early childhood and maternal stress (Jablensky, 2000). Recent studies have also implicated viral infection of the central nervous system as a potential environmental causative agent (Borglum et al., 2014).

As a complex disorder, schizophrenia does not display classical Mendelian inheritance. Only one single gene mutation has been reported, a translocation in a Scottish pedigree affecting the DISC1 gene, the causality of which remains uncertain (Sullivan, 2013; Porteous et al., 2014).

The first definitive schizophrenia-associated mutations to be identified were copy-number variations (CNV). Deletions at chromosome 22q11.2 were shown to significantly increase the risk of schizophrenia (Karayiorgou et al., 1995) as well as other psychiatric and neurodevelopmental phenotypes (Malhotra and Sebat, 2012). Since that initial discovery, improvements in genotyping technology have enabled a more comprehensive analysis of the role of CNVs in schizophrenia, leading to the discovery of several additional well-replicated schizophrenia-associated CNVs (Malhotra and Sebat, 2012; Rees et al., 2014; Szatkiewicz et al., 2014). The most well-established loci include NRXN1 (Rujescu et al., 2009), VIPR2 (Vacic et al., 2011), 1q21 (International Schizophrenia Consortium, 2008), 3q29 (Itsara et al., 2009), 7q11 (Kirov et al., 2012), 15q11 (Kirov et al., 2009), 15q13 (International Schizophrenia Consortium, 2008), 16p13(McCarthy et al., 2009), 16p11 (McCarthy et al., 2009), and 17q12 (Moreno-De-Luca et al., 2010). Additionally, schizophrenia patients have been shown to have an increased burden of large, rare CNVs compared with controls, as well as increased frequency of de novo CNVs (Malhotra and Sebat, 2012).

Although schizophrenia-associated CNVs are individually rare, accounting for a fraction of cases, they have yielded important insights into biological processes underlying the disease. CNVs have been shown to cluster around genes involved in two postsynaptic density pathways, the N-methyl-D-aspartate receptor (NMDAR) complex and the activity-regulated cytoskeleton-associated (ARC) protein complex (Szatkiewicz et al., 2014; Kirov et al., 2012).

Genome-wide association studies (GWAS) for schizophrenia have shown that unlike CNVs, which have large individual effect sizes but are extremely rare in the population, common variants that associate with the disorder have small individual effect sizes. The largest GWAS to date included 36,989 cases and identified 108 loci that surpassed genome-wide significance, all with odds ratios ≤1.2 (Schizophrenia Working Group of the Psychiatric Genomics Consortium, 2014). Despite the low effect sizes, the GWAS data offer important biological insights into the disorder. Associations were shown to cluster around genes known to function in glutamatergic synaptic and calcium channel pathways. Combined GWAS of schizophrenia, bipolar disorder, autism spectrum disorders, major depressive disorder, and attention-deficit hyperactivity disorder (ADHD) showed evidence for shared risk (Cross-Disorder Group of the Psychiatric Genomics Consortium, 2013) with substantial overlap between the three adult-onset disorders, schizophrenia, bipolar disorder, and major depressive disorder, and a reduced yet still significant overlap between schizophrenia and autism spectrum disorder (Cross-Disorder Group of the Psychiatric Genomics Consortium et al., 2013).

Finally, significant overlap was also detected between the schizophrenia GWAS loci and genes found to carry de novo nonsynonymous mutations in sequencing studies of schizophrenia (Schizophrenia Working Group of the Psychiatric Genomics Consortium, 2014), which are discussed further in this chapter.

NEXT-GENERATION SEQUENCING

Current sequencing technologies such as Illuminia's HiSeq were developed to decrease the cost of sequencing compared with traditional Sanger sequencing approaches. The Human Genome Project took more than 10 years at a cost of nearly $3 billion to generate one complete human sequence; using current NGS technology, an entire human genome can be sequenced in a matter of days and at a cost of less than $5000. To achieve these savings, the entire process was reimagined by using the Illumina solution as an example rather than randomly terminating nascent DNA strands from a single target and separating them by size as with Sanger sequencing. Targets are fixed to a substrate, and bases are detected as they are incorporated into the growing DNA strands, allowing for massively parallel interrogation of targets. The resultant sequences are then mapped to a reference or assembled de novo using a suite of bioinformatic techniques. Variant bases can then be called from the aligned sequence. The technology can be applied to a variety of input templates for DNA that include small targeted panels of genes, the entire coding sequence (known as an exome), or whole genomes.

EXOME VERSUS WHOLE GENOME

A small fraction of the human genome, estimated at 2%, encodes the transcribed sequences that ultimately form the regulatory RNAs and protein products that are the building blocks of our cells. Given their known functional roles, it is thought that these sequences will be enriched with the disease-causing variants that geneticists are attempting to identify. As such, library preparation techniques have been devised to enrich these exome sequences, allowing them to be effectively separated from the remainder of the genome. Selectively sequencing the coding fraction of the genome has been termed whole-exome sequencing (WES) and currently accounts for the majority of the efforts underway in the human genetics community. Whole-exome sequence capture methods involve DNA fragmentation, end repair, and A-tailing of the fragmented DNA, followed by hybridization to RNA (in the case of the Agilent SureSelect system) or DNA baits (in the case of the NimbleGen or Illumina systems) that are designed to physically capture specific DNA sequences. The captured DNA is then eluted from the baits, purified using biotin-based precipitation, and then amplified by PCR to yield enough material for sequencing.

Whole-genome sequencing (WGS) Exome sequencing has been successful in identifying mutations that cause highly penetrant monogenic diseases; however, by restricting focus to the coding regions, it becomes more difficult to identify structural variants and information on potentially important regulatory regions is overlooked. Although an exome comprises approximately 2% of the human genome, recent publications by the ENCODE project linked more than 80% of the human genome to a biological function. This clearly indicates that WGS will ultimately be needed to fully catalog all deleterious changes in the human genome. Although WGS provides a more comprehensive view of the genome and the variation within, the cost of WGS remains prohibitively expensive for the study of complex disease, for which large numbers of samples are required to achieve statistical significance.

Overview bioinformatics NGS outputs short reads of between 150 and 200 bp that have to be aligned, most commonly against a reference genome using an aligner such as the Burrows-Wheeler transform (BWA) (Li and Durbin, 2009). Variant sites, SNPs, short insertions or deletions (INDELs), and CNVs can be called following alignment for downstream analysis. Prior to analysis, the called variants have to be annotated and their effects have to be predicted. Several programs have been developed, such as ANNOVAR (Wang et al., 2010), the Variant Effect Predictor (VEP), and SnpEff (Cingolani et al., 2012). These programs predict the coding effects of genetic variation with respect to protein function such as synonymous or nonsynonymous missenses, stop codon gains or losses, and splice site gains or losses. For noncoding variants, genomic locations within genes are returned, such as intronic, 5′ UTR, or 3′ UTR. In addition, variants can be annotated with a broad array of corollary information such as allele frequency in public databases, functional importance scores from SIFT and PolyPhen, noncoding functional elements predicted by ENCODE (The ENCODE Project Consortium et al., 2007), and previously reported pathogenic mutations from OMIM and HGMD.

Annotation and functional prediction are essential before variants can be assessed for association with disease. Rare variant analysis can be broadly separated into family-based analysis typically in search of high-penetrance variance that is clearly transmitted through pedigree and case control analyses, similar to GWAS, for which lower-penetrance alleles with smaller effect sizes are the target. Both approaches are applicable to schizophrenia and are described further in the following sections.

Familial Family-based analysis can be used to detect high-penetrance recessive, dominant, and de novo mutations of single nucleotide and copy number variants. Workflows typically begin with a filtration step to exclude variants that have a population frequency that exceeds the prevalence of the disease and variants that are predicted as benign. Frequency data have been generated for tens of thousands of samples, mainly exomes, and made publically available in multiple resources, including dbSNP, the 1000 Genomes Project (1 kGP) (Durbin et al., 2010), NHLBI GO Exome Sequencing Project (ESP), and the Exome Aggregation Consortium (ExAC), as well as national efforts such as the Genome of the Netherlands Consortium (Go-NL) (Genome of the Netherlands Consortium, 2014) and UK10K (Muddyman et al., 2013). The ExAC is currently the largest resource, with a collation of sequence data from 63,358 unrelated individuals that were sequenced as part of various disease-specific and population genetic studies.

Following filtration, the remaining rare and potentially deleterious variants are fit to an appropriate genetic model that is recessive, dominant, X-linked, or de novo, based on the transmission of the trait through the family pedigree. The remaining candidate genes can be subsequently followed up with further bioinformatic analyses including expression data, pathway analysis, and animal model comparisons or validated functionally using transfected cell cultures or rapid animal knockdown systems such as the zebrafish.

Complex GWAS and the common variant common disease hypothesis (Reich and Lander, 2001) underlying them have proven to be successful at identifying loci but only partially successful at explaining the sum of the phenotypic variance. Numerous common variant loci, 108 in the case of schizophrenia, have been associated with a wide array of phenotypes; however, these loci predominantly confer modest effect sizes that account for a small proportion of the phenotypic variance. To account for the missing variability, attention has now turned to assessing the role of rare variants in conferring risk to complex disease, known as the common disease rare variant hypothesis (Iyengar and Elston, 2007). Because rare variants are not adequately tagged on genotyping arrays, their contribution to complex disease remains understudied. Recent advances in next-generation sequencing, coupled with a reduction in the cost of sequencing, have enabled such studies in sufficient numbers to identify rare variants underlying complex disease. For a rare variant to be causal of a complex disease, it must display reduced penetrance; as such, statistical tests of association are more powerful to detect such variants compared with the familial approached described. We briefly outline two statistical approaches to the analysis of rare variants in a complex disease such as schizophrenia under a case-control paradigm, single point tests, and agglomerative tests (Zhou and Stephens, 2012). The simplest approach is to test each individual variant for association by comparing the allele counts in the cases and controls using a Fisher exact test. However, single variant tests are not very powerful unless a single variant accounts for a substantial number of cases, which can occur due to bottlenecks (Gudbjartsson et al., 2015) in population isolates but are generally quite rare in more heterogeneous populations. It is more likely that multiple variants will exist within a gene; by testing the overall mutation burden in the gene as a whole, these can be identified. Multiple agglomerative gene-based association tests have been developed, such as the Sequence Kernel Association Test (SKAT), the Variable Threshold Test (Price et al., 2010), and the Combined Multivariate and Collapsing (CMC) test (Li and Leal, 2008). Gene-based methods for analysis of sequencing data can be broadly divided into collapsing tests such as CMC, whereby the total number of rare alleles across a gene is directly correlated with the phenotype and burden tests such as VT and SKAT that, instead of comparing the total number of variants per individual, examine the number of variants with nonzero effect sizes (whether positive or negative) to see if they exceed chance expectations. Collapsing tests assume all rare variants in the target region have effects on the phenotype in the same direction and of similar magnitude, which makes filtering and frequency cutoffs important considerations because neutral variation will reduce the power of the test (Neale et al., in press). Burden tests are more robust in the presence of protective and deleterious variants and null variants but are less powerful than collapsing tests when a large number of variants in a region are causal and in the same direction.

SEQUENCING STUDIES TO DATE

Several sequencing studies regarding schizophrenia have been published. Studies published prior to 2014 were of smaller scale and aimed primarily toward providing support for the hypothesis that, just as for de novo CNVs, the rates of de novo single-nucleotide variants (SNVs) and small INDELs mutations were increased in schizophrenia (Girard et al., 2011; Gulsuner et al., 2013; Xu et al., 2012). Although increased de novo rates have been reported in autism (Iossifov et al., 2012; Neale et al., 2012; O'Roak et al., 2012; Sanders et al., 2012) and intellectual disability (ID) (de Ligt et al., 2012; Rauch et al., 2012), the largest study of de novo mutations in schizophrenia to date reports no evidence of increased nonsynonymous or loss-of-function (LOF) de novo mutations (Fromer et al., 2014), suggesting that this class of mutation may not be as important in schizophrenia as had been previously suggested (Girard et al., 2011; Gulsuner et al., 2013; Xu et al., 2012) or observed in other disorders.

However, although de novo mutations do not appear to be associated with the risk of schizophrenia, LOF mutations did occur more frequently than expected in people with schizophrenia who also had lower premorbid educational attainment, suggesting that de novo LOF mutations may have a role in neurodevelopmental impairment across diagnostic boundaries (Fromer et al., 2014). These findings are consistent with the results from the Psychiatric Genomics Consortium (PGC2) GWAS.

The schizophrenia de novo mutations were also predicted to be less damaging than those found in autism or ID (Fromer et al., 2014), which is consistent with schizophrenia being on the milder end of the neurodevelopmental impairment scale compared with autism or ID. Finally, de novo mutations were shown to be enriched in glutamatergic postsynaptic proteins comprising the ARC and NMDAR complexes (Kirov et al., 2012).

Moving beyond de novo mutations, a large WES study investigating the role of recessively inherited alleles in general and LOF mutations in particular in a Swedish sample was recently published. The authors focused on homozygous or compound heterozygous LOF that would result in complete gene-product knockouts and in regions of homozygosity (ROH), which arise due to recent inbreeding (Ruderfer et al., 2015). Although elevated rates of LOF have been identified in autism (Lim et al., 2013), and although elevated rates of ROH have been identified in schizophrenia using genotyping arrays (Keller et al., 2012), neither was shown to be significantly associated with schizophrenia following WES (Ruderfer et al., 2015).

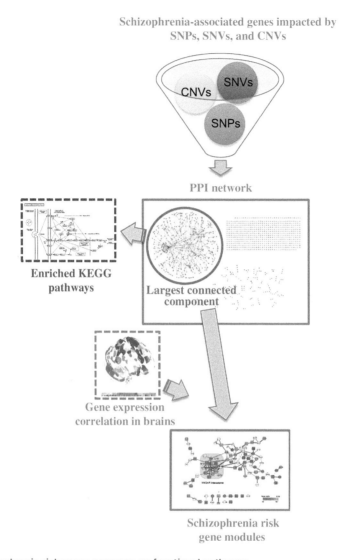

FIGURE 4.1 Schizophrenia risk genes converge on functional pathways.

Beginning with the individual genes that have been implicated in the risk of developing schizophrenia in disparate genetic studies, including genome-wide association studies (GWAS), copy number variation (CNV) studies, and sequencing studies, protein–protein interaction networks can be built that link these risk genes to their known interactors. The resulting connected component can then be assessed for enrichment of known functional pathways. Additional data sources such as differential gene expression data from brain tissue of schizophrenia cases and controls can be included to weight the edges of the graph. Such approaches allow for the identification of pathways that are disrupted in schizophrenia, thus shedding light on the underlying biology of the disorder and also possibly highlighting suitable pathways for therapeutic intervention.

Finally, the largest study to date examining rare variants predicted to be damaging under a case-control study design has identified some putatively associated variants in a set of candidate genes (Purcell et al., 2014). The study, which included approximately 2500 cases and an equal number of controls, used a hypothesis-free model and did not identify any genes with significant genome-wide excess of rare mutations in cases. However, when the authors examined a set of candidate genes enriched for potential schizophrenia susceptibility genes (ie, genes with de novo mutations, in GWAS regions, mapping to CNVs, members of the ARC, NMDAR, and postsynaptic density pathways), they were able to demonstrate an increased burden of rare nonsense and disruptive variants in cases compared with controls (Purcell et al., 2014). The mutation burden in cases was spread across a large number of genes, suggesting that similar to what has been observed in GWAS of common variants, the mutational target of schizophrenia encompasses many hundreds of genes (Fig. 4.1).

CONCLUSIONS

To date, next-generation sequencing technologies have only been applied to schizophrenia in a few studies. Although these studies lacked power to implicate specific genes and rare mutations in schizophrenia, they have been useful in allowing us to conclude that rare and also common single-nucleotide variations play roles in schizophrenia, although the relative contribution of the two classes of variant remains uncertain pending much larger sequencing studies, as does the contribution of rare variation outside the exome. Both GWAS and WES studies suggest that genetic variation contributes substantially to the phenotypic variance and is highly polygenic. In contrast to some diseases such as Alzheimer disease or autoimmune disorders in which a single gene or locus accounts for a significant proportion of risk, the presence of a gene or locus harboring common or rare variants with large effect sizes can be ruled out in schizophrenia. Although variants in the small effect size may not be useful diagnostically, they have immense potential to assist in unraveling the underlying biology of the disorder and may highlight suitable pathways for therapeutic intervention. The cross-disorder analyses that have been conducted for both rare and common variants have revealed substantial genetic overlap between the adult-onset disorders and more modest overlap between schizophrenia and autism. This shared risk highlights the importance and potential utility of genetic findings for the etiology and pathophysiology of these related disorders.

FUTURE DIRECTIONS

Even more so than for GWAS that preceded sequencing approaches, large sample sizes will be required to fully map out the spectrum of rare predisposition alleles.

Current WES studies are only able to interrogate the coding fraction of the genome, which accounts for approximately 3% of the entire sequence. Although coding sequence is a natural starting place to begin the search for schizophrenia predisposing variants, the noncoding fraction of the genome codes for functionally important elements such as gene promoters, transcription factor binding elements, and nontranslated transcribed sequences such as antisense RNAs, all of which may harbor disease predisposition variants. Whole-genome studies, which are currently uneconomical at large scales, will be required to ultimately determine the ensemble of genetic variants associated with schizophrenia. With rapidly evolving technology and associated decreases in sequencing costs, we are rapidly approaching a time when such studies will become feasible.

REFERENCES

Borglum, A.D., et al., 2014. Genome-wide study of association and interaction with maternal cytomegalovirus infection suggests new schizophrenia loci. Mol. Psychiatry. 19, 325–333.

Cingolani, P., et al., 2012. A program for annotating and predicting the effects of single nucleotide polymorphisms, SnpEff: SNPs in the genome of *Drosophila melanogaster* strain w1118; iso-2; iso-3. Fly 6, 80–92.

Cross-Disorder Group of the Psychiatric Genomics Consortium, 2013. Identification of risk loci with shared effects on five major psychiatric disorders: a genome-wide analysis. Lancet 381 (9875), 1371–1379.

Cross-Disorder Group of the Psychiatric Genomics Consortium, 2013. Genetic relationship between five psychiatric disorders estimated from genome-wide SNPs. Nat. Genet. 45, 984–994.

de Ligt, J., et al., 2012. Diagnostic exome sequencing in persons with severe intellectual disability. N. Engl. J. Med. 367, 1921–1929.

Durbin, R.M., et al., 2010. A map of human genome variation from population-scale sequencing. Nature 467, 1061–1073.

Fromer, M., et al., 2014. De novo mutations in schizophrenia implicate synaptic networks. Nature 506, 179–184.

Gejman, P.V., Sanders, A.R., Kendler, K.S., 2011. Genetics of schizophrenia: new findings and challenges. Annu. Rev. Genomics. Hum. Genet. 12, 121–144.

Genome of the Netherlands Consortium, 2014. Whole-genome sequence variation, population structure and demographic history of the Dutch population. Nat. Genet. 46, 818–825.

Girard, S.L., et al., 2011. Increased exonic de novo mutation rate in individuals with schizophrenia. Nat. Genet. 43, 860–863.

Gudbjartsson, D.F., et al., 2015. Large-scale whole-genome sequencing of the Icelandic population. Nat. Genet. 47, 435–444.

Gulsuner, S., et al., 2013. Spatial and temporal mapping of de novo mutations in schizophrenia to a fetal prefrontal cortical network. Cell 154, 518–529.

International Schizophrenia Consortium, 2008. Rare chromosomal deletions and duplications increase risk of schizophrenia. Nature 455, 237–241.

Iossifov, I., et al., 2012. De novo gene disruptions in children on the autistic spectrum. Neuron 74, 285–299.

Itsara, A., et al., 2009. Population analysis of large copy number variants and hotspots of human genetic disease. Am. J. Hum. Genet. 84, 148–161.

Iyengar, S.K., Elston, R.C., 2007. The genetic basis of complex traits: rare variants or "common gene, common disease"? Methods. Mol. Biol. 376, 71–84.

Jablensky, A., 2000. Epidemiology of schizophrenia: the global burden of disease and disability. Eur. Arch. Psychiatry. Clin. Neurosci. 250, 274–285.

Karayiorgou, M., et al., 1995. Schizophrenia susceptibility associated with interstitial deletions of chromosome 22q11. Proc. Natl. Acad. Sci. U. S. A. 92, 7612–7616.

Keller, M.C., et al., 2012. Runs of homozygosity implicate autozygosity as a schizophrenia risk factor. PLoS. Genet. 8, e1002656.

Kirov, G., et al., 2012. De novo CNV analysis implicates specific abnormalities of postsynaptic signalling complexes in the pathogenesis of schizophrenia. Mol. Psychiatry. 17, 142–153.

Kirov, G., et al., 2009. Support for the involvement of large copy number variants in the pathogenesis of schizophrenia. Hum. Mol. Genet. 18, 1497–1503.

Li, B., Leal, S., 2008. Methods for detecting associations with rare variants for common diseases: application to analysis of sequence data. Am. J. Hum. Genet. 83, 311–321.

Li, H., Durbin, R., 2009. Fast and accurate short read alignment with Burrows-Wheeler transform. Bioinformatics 25, 1754–1760.

Lim, E.T., et al., 2013. Rare complete knockouts in humans: population distribution and significant role in autism spectrum disorders. Neuron 77, 235–242.

Malhotra, D., Sebat, J., 2012. CNVs: harbingers of a rare variant revolution in psychiatric genetics. Cell 148, 1223–1241.

McCarthy, S.E., et al., 2009. Microduplications of 16p11.2 are associated with schizophrenia. Nat. Genet. 41, 1223–1227.

Moreno-De-Luca, D., et al., 2010. Deletion 17q12 is a recurrent copy number variant that confers high risk of autism and schizophrenia. Am. J. Hum. Genet. 87, 618–630.

Muddyman, D., Smee, C., Griffin, H., Kaye, J., 2013. Implementing a successful data-management framework: the UK10K managed access model. Genome Med. 5, 100.

Neale, B.M. et al., 2011. Testing for an unusual distribution of rare variants. PLoS. Genet. 7(3):e1001322.

Neale, B.M., et al., 2012. Patterns and rates of exonic de novo mutations in autism spectrum disorders. Nature 485, 242–245.

O'Roak, B.J., et al., 2012. Sporadic autism exomes reveal a highly interconnected protein network of de novo mutations. Nature 485, 246–250.

Porteous, D.J., et al., 2014. DISC1 as a genetic risk factor for schizophrenia and related major mental illness: response to Sullivan. Mol. Psychiatry. 19, 141–143.

Price, A.L., et al., 2010. Pooled association tests for rare variants in exon-resequencing studies. Am. J. Hum. Genet. 86, 832–838.

Purcell, S.M., et al., 2014. A polygenic burden of rare disruptive mutations in schizophrenia. Nature 506, 185–190.

Rauch, A., et al., 2012. Range of genetic mutations associated with severe non-syndromic sporadic intellectual disability: an exome sequencing study. Lancet 380, 1674–1682.

Rees, E., et al., 2014. Analysis of copy number variations at 15 schizophrenia-associated loci. Br. J. Psychiatry 204, 108–114.

Reich, D.E., Lander, E.S., 2001. On the allelic spectrum of human disease. Trends. Genet. 17, 502–510.

Ruderfer, D.M., et al., 2015. No evidence for rare recessive and compound heterozygous disruptive variants in schizophrenia. Eur. J. Hum. Genet. 23, 555–557.

Rujescu, D., et al., 2009. Disruption of the neurexin 1 gene is associated with schizophrenia. Hum. Mol. Genet. 18, 988–996.

Sanders, S.J., et al., 2012. De novo mutations revealed by whole-exome sequencing are strongly associated with autism. Nature 485, 237–241.

Schizophrenia Working Group of the Psychiatric Genomics Consortium, 2014. Biological insights from 108 schizophrenia-associated genetic loci. Nature 511, 421–427.

Sullivan, P.F., 2013. Questions about DISC1 as a genetic risk factor for schizophrenia. Mol. Psychiatry. 18, 1050–1052.

Szatkiewicz, J.P., et al., 2014. Copy number variation in schizophrenia in Sweden. Mol. Psychiatry. 19, 762–773.

The ENCODE Project Consortium, 2007. Identification and analysis of functional elements in 1% of the human genome by the ENCODE pilot project. Nature 447, 799–816.

Vacic, V., et al., 2011. Duplications of the neuropeptide receptor gene VIPR2 confer significant risk for schizophrenia. Nature 471, 499–503.

Wang, K., Li, M., Hakonarson, H., 2010. ANNOVAR: functional annotation of genetic variants from high-throughput sequencing data. Nucleic. Acids. Res. 38, e164.

Xu, B., et al., 2012. De novo gene mutations highlight patterns of genetic and neural complexity in schizophrenia. Nat. Genet. 44, 1365–1369.

Zhou, X., Stephens, M., 2012. Genome-wide efficient mixed-model analysis for association studies. Nat. Genet. 44, 821–824.

CHAPTER 5

EPIGENETIC APPROACHES TO DEFINE THE MOLECULAR AND GENETIC RISK ARCHITECTURES OF SCHIZOPHRENIA

M. Kundakovic[1,2], C. Peter[1], P. Roussos[1] and S. Akbarian[1]

[1]*Department of Psychiatry, Friedman Brain Institute, Icahn School of Medicine at Mount Sinai, New York, NY, United States* [2]*Department of Biological Sciences, Fordham University, Bronx, NY, United States*

CHAPTER OUTLINE

Introduction	62
An Epigenetic Link Between Environmental Risk Factors and Schizophrenia	64
Evidence From Human Studies	64
Evidence From Animal Studies	65
DNA Methylation in Schizophrenia	65
DNA Methylation and Gene Regulation in the Brain	65
DNA Methylation Changes in Schizophrenia	66
Studies in the Peripheral Tissue and Possible Epigenetic Biomarkers in Schizophrenia	66
Epigenetic Approaches to the Molecular and Genetic Risk Architectures of Schizophrenia	67
Do Epigenetic Mechanisms Contribute to Long-Lasting Alterations in Gene Expression in Schizophrenia Brain?	67
Functional Neuroepigenomics Could Inform About Disease-Associated Genetic Risk Polymorphisms	69
Higher-Order Chromatin and the Genetic Risk Architecture of Schizophrenia	69
Epigenetic Mechanisms and the Treatment of Schizophrenia	71
Future Directions	72
Chromatin-Modifying Drugs: A Future Treatment for Schizophrenia?	72
Comprehensive, Region-Specific, and Cell Type–Specific Epigenome Mappings in the Brain	73
Larger and More Ambitious Clinical Studies	73
Acknowledgment	74
References	74

INTRODUCTION

Schizophrenia (SCZ) is a major psychiatric disorder, often with onset during adolescence or young adulthood, with positive symptoms such as delusions and hallucinations and negative symptoms such as social withdrawal and apathy. The life span of subjects diagnosed with SCZ is reduced, on average, by 15 years in comparison with the general population, and cardiovascular disease and suicide account for a significant share of this increase in mortality (Hennekens et al., 2005; Laursen et al., 2012; Saha et al., 2007). Despite the widespread prescription of antipsychotic medications, which are mostly aimed at dopaminergic and serotonergic receptor systems (Taly, 2013; Kim and Stahl, 2010), many patients still experience chronic and debilitating symptoms (Lieberman et al., 2005; Swartz et al., 2007). Rationale drug development in SCZ is extremely challenging, given the lack of a unifying neuropathology (Catts et al., 2013; Dorph-Petersen and Lewis, 2011) in conjunction with a highly heterogeneous genetic risk architecture (Andreassen et al., 2014; Rodriguez-Murillo et al., 2012).

Here, we outline how neuroepigenetic approaches (Sweatt, 2013)—broadly defined as the study of chromatin structure and function in the developing and adult nervous systems, including their role in neuronal and behavioral plasticity—could advance our knowledge of the neurobiology of SCZ. Epigenetic markings, such as DNA cytosine methylation and histone modifications and variants (Fig. 5.1), could

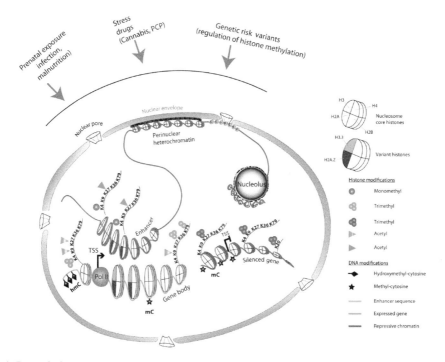

FIGURE 5.1 The building blocks of the epigenome.

The epigenome comprises DNA modifications, including (but not limited to) cytosine methylation and hydroxymethylation, and a large number of site-specific and residue-specific histone modifications and histone variants (only some of which are shown in this figure as representative examples). These molecular building

be viewed as a "molecular bridge" used by myriads of external ("environmental") or internal factors mold and shape the nascent genetic material throughout the entire life span of a brain cell (Akbarian and Nestler, 2013). Although a comprehensive discussion on epigenetic (epi is Greek for over, above) regulation in the nervous system is beyond the scope of this book chapter, we refer the reader to recent handbooks and special journal volumes in this field (Akbarian and Nestler, 2013; Sweatt, 2013; Petronis and Mill, 2011; Appasani, 2012).

We discuss the potential contributions of epigenetic approaches toward advancing a better understanding of the underlying neurobiology and the potential for improved treatment of schizophrenia. This includes brief discussions of epigenetic mechanisms linked to environmental and genetic risk factors of the disease, chromatin-associated alterations in SCZ brain examined postmortem, and emerging findings from animal and other preclinical model systems that could pave the way for novel treatment avenues for psychosis.

◀ **FIGURE 5.1** (Continued) blocks largely define the epigenetic landscapes that organize genomic DNA, often in a locus-specific fashion, into active transcriptional units (green) including promoter and enhancer sequences (blue) and condensed chromatin including silenced genes (red). These epigenetic signatures are thought to distinguish between various cell types and developmental stages sharing the same genome (Rodriguez-Paredes and Esteller, 2011; Li and Reinberg, 2011). Many heterochromatic sequences are tethered to the nuclear envelope and pore complex and also are enriched at the periphery of the nucleolus (an intranuclear compartment for ribosomal biogenesis). This figure shows only a small representative subset of histone variants and site-specific lysine (K) residues at histone H3 and H4 N-terminal tails that are subject to posttranslational modification by methylation and/or acetylation. The regulation of chemical histone modifications is extremely complex, and there could exist far more than 100 amino acid residue-specific posttranslational modifications (PTMs) in a typical vertebrate cell (Tan et al., 2011), including mono- (me1), di- (me2), and tri- (me3) methylation, acetylation, and crotonylation, polyADP-ribosylation, and small protein (ubiquitin, SUMO) modification of specific lysine residues, as well as arginine (R) methylation and "citrullination," serine (S) phosphorylation, tyrosine (T) hydroxylation, and several others (Tan et al., 2011; Kouzarides, 2007; Taverna et al., 2007). Multiple combinatorial sets of histone PTMs contribute to functional chromatin states that differentially define gene proximal promoters and gene bodies as opposed to enhancer and other regulatory sequences, condensed heterochromatin, and the "insulator" sequences that compartmentalize and provide boundaries for these various domains of chromatin (Zhou et al., 2011). The spectrum of DNA modifications is much less complex when compared to histone PTM, and the bulk of DNA modifications exist as cytosine methylation (m) and hydroxymethylation (hm) (Kriaucionis and Heintz, 2009). The mC and hmC markings show a differential (but not mutually exclusive) pattern of genomic occupancy. The hmC5 mark broadly correlates with local gene expression levels (Jin et al., 2011; Song et al., 2011), whereas methyl-cytosine (mC5) markings, particularly when positioned around the 5′ end of genes, are thought to function primarily as a negative regulator of transcription (Maunakea et al., 2010; Sharma et al., 2005).

As discussed in this chapter, there is ample evidence that brain epigenomes are sensitive to a number of risk factors associated with SCZ, including prenatal infection or malnutrition and a variety of types of drug abuse, including but not limited to phenylcyclidine (PCP) and cannabis (Paparelli et al., 2011). Furthermore, there is increasing evidence that polymorphisms and mutations affecting the chromatin remodeling and histone modification machinery in the nucleus, particularly of genes regulating histone lysine methylation, play an important role in the cause of disease in at least a subset of cases diagnosed with SCZ (Network and Pathway Analysis Subgroup of Psychiatric Genomics Consortium, 2015; Takata et al., 2014).

AN EPIGENETIC LINK BETWEEN ENVIRONMENTAL RISK FACTORS AND SCHIZOPHRENIA

SCZ develops as a result of complex interactions of genetic and environmental risk factors. In light of the developmental origin of SCZ, several in utero exposures have been associated with an increased risk of this disorder, including severe dietary restriction, exposure to viral infections, and maternal stress, as well as exposure to hypoxia due to gestational and birth complications (Khashan et al., 2008; Oh and Petronis, 2008). In addition, postnatal risk factors for SCZ include childhood trauma, growing up in an urban environment, cannabis use, and minority group position, among others (van Os et al., 2010). Because the brain epigenome is responsive to environmental cues throughout life and is particularly vulnerable to disruption during development, it has emerged as a plausible substrate through which adverse environments can exert lasting effects on the brain and contribute to mental disorders, including SCZ (Bale et al., 2010; Kundakovic, 2013; Kundakovic and Champagne, 2014; Tsankova et al., 2007). Moreover, as discussed further, many DNA sequence variants associated with hereditary risk for schizophrenia, including the majority of single-nucleotide polymorphisms (SNPs) identified in genome-wide association studies (GWAS), are positioned within noncoding portions of the genome, potentially affecting structure and function of the surrounding chromatin in a complex interplay with environmental factors.

EVIDENCE FROM HUMAN STUDIES

The Dutch Famine Study was the first to infer a possible epigenetic link between prenatal adverse environment and adult SCZ (Kirkbride et al., 2012). This seminal epidemiological study established associations between maternal exposure to severe famine during the Nazi blockade in western Holland during World War II (October 1944–May 1945) and the offspring's long-term health outcomes (Stein et al., 1972). In particular, maternal exposure during the famine's peak at the approximately time of conception was associated with increased risk of schizoid diagnoses at age 18 years and hospitalization for SCZ in adulthood (approximately twofold increase) (Susser and Lin, 1992; Brown and Susser, 2008). Although maternal stress may have also played an important role in initiating these effects, it is believed that maternal nutritional deficiency due to the famine is more strongly associated with impaired neurodevelopment and increased SCZ risk in offspring (Kirkbride et al., 2012). Deficiency of many micronutrients and macronutrients may contribute to the adverse effects of famine on neurodevelopment and, among those, folic acid emerged as one of the main candidates (Kirkbride et al., 2012). Folic acid plays an important role in brain development, and maternal folate supplementation during the periconceptional period is associated with decreased risk of neurodevelopmental disorders in children (Kirkbride et al., 2012; Roth et al., 2011; Suren et al., 2013), some of which could be antecedents of SCZ and related disease (Kirkbride et al., 2012). Folate is also important for the production of methyl donors and DNA methylation. This implies that nutritional (particularly folate) deficiency could affect neurodevelopment and later risk of schizophrenia, at least in part, via DNA methylation-dependent mechanisms. Within the Dutch Famine cohort, DNA methylation was examined in peripheral blood and compared between the individuals who were prenatally exposed to famine and their unexposed same-sex siblings at approximately 60 years of age (Heijmans et al., 2008). Periconceptional exposure to famine was shown to be associated with DNA methylation changes in multiple genes implicated in growth and metabolic pathways, suggesting long-term and widespread epigenetic dysregulation induced by prenatal famine exposure (Heijmans et al., 2008; Tobi et al., 2014, 2009).

EVIDENCE FROM ANIMAL STUDIES

Animal studies have shown that environmental risk factors associated with SCZ, such as prenatal stressors (Dong et al., 2014; Mueller and Bale, 2008), early life adversity (Roth et al., 2009; Weaver et al., 2004; Murgatroyd et al., 2009), as well as drug abuse and social defeat in adolescence/adulthood (Nestler, 2014; Walker et al., 2015), could induce lasting epigenetic changes, resulting in behavioral phenotypes relevant to psychiatric disorders. Modeling SCZ in animals, discussed in more detail in some of the other chapters in this book, is particularly challenging; this is one of the major limitations in establishing the causality between the effects of environmental factors on the epigenome and SCZ-relevant phenotypes (Kundakovic, 2014). However, a few recent studies focused on certain SCZ-relevant risk factors and endophenotypes that are feasible to model in rodents. Dong et al. used restraint stress during pregnancy as a paradigm of prenatal environmental risk factors and established its effects on locomotor activity and social interaction (Dong et al., 2014). They showed that prenatal stress induces lasting epigenetic dysregulation of SCZ-relevant genes and that this is associated with hyperactivity and impaired social interaction, drawing a possible link between epigenetic effects of prenatal stress and SCZ. Furthermore, some of the behavioral phenotypes relevant to SCZ, including hyperactivity and defective prepulse inhibition of startle, are induced by mild isolation stress during adolescence, when this environmental stressor is combined with a genetic risk factor (Niwa et al., 2013). Although these and other behavioral phenotypes may turn out to be primarily driven by changes in neuronal connectivity or some other adaptive mechanisms outside of the neuronal nucleus, it is noteworthy that in the aforementioned study, the synergistic effect of social isolation and *DISC1* (disrupted in schizophrenia 1) genetic variant converged on the mesocortical projection of dopaminergic neurons and was mediated, in part, through increased DNA methylation of the tyrosine hydroxylase gene involved in dopamine synthesis. This study supports the hypothesis that genetically predisposed individuals may be more sensitive to environmental risk factors and reveals DNA methylation as one of the epigenetic mechanisms through which gene–environment interaction could lead to SCZ.

DNA METHYLATION IN SCHIZOPHRENIA
DNA METHYLATION AND GENE REGULATION IN THE BRAIN

DNA methylation is one of the key epigenetic mechanisms involved in regulation of gene expression (Klose and Bird, 2006). Methylation occurs at position 5 of cytosine, primarily in the context of cytosine–guanine (CpG) dinucleotides. Cytosine methylation located in gene promoters is often implicated in gene repression acting via two possible mechanisms: (1) by directly impeding the binding of transcription factors or (2) by locally inducing repressive chromatin structure that is nonpermissive to transcription. Importantly, DNA methylation and regulation of gene expression are complex, often defying overly simplistic models, including the notion that increased CpG methylation near the 5′ end of genes is equal to "gene repression." First, in addition to CpG methylation, a significant portion of methylation is positioned at non-CpG sites, particularly in embryonic tissues as well as in neuronal cells (Lister et al., 2013; Guo et al., 2014). Second, the interrelation between cytosine methylation and gene expression is perhaps more complex than previously thought, with many other genomic regions in addition to promoters subject to DNA methylation, including gene bodies, resulting in repression and activation of genes, depending on context (Klose and Bird, 2006; Wu et al., 2010). Third, additional methylation-related cytosine modifications have been described with growing interest in hydroxymethylation, particularly in the brain (Lister et al., 2013; Guo et al., 2011; Kaas et al., 2013; Rudenko et al., 2013). DNA methylation is critical for gene regulation during neural cell differentiation and brain development (Golebiewska et al., 2009; Liu and Casaccia, 2010; Miller and Gauthier, 2007; Fan et al., 2005) and is actively involved in the regulation of memory formation and learning in adulthood (Day and Sweatt, 2011). Therefore, aberrant DNA methylation may be an important player in the pathophysiology of SCZ, potentially contributing to the initiation of SCZ pathology during brain development and to the maintenance of SCZ symptoms, including cognitive impairments and others. It is not surprising, then, that many epigenetic studies of SCZ focused on DNA methylation.

DNA METHYLATION CHANGES IN SCHIZOPHRENIA

Early studies examined DNA methylation status of several candidate genes with dysregulated expression in brains of SCZ patients. Several gene regulatory regions showed differential DNA methylation profiles in postmortem SCZ brains compared to controls, including those of reelin (*RELN*) (Abdolmaleky et al., 2005; Grayson et al., 2005), catechol-*O*-methyltransferase (*COMT*) (Abdolmaleky et al., 2006), serotonin receptor type-2 (*HTR2A*) (Abdolmaleky et al., 2011), and sex-determining region Y-box containing gene 10 (*SOX10*) (Iwamoto et al., 2005). In the majority of studies, DNA methylation changes were inversely correlated with changes in gene expression, supporting the hypothesis that epigenetic alterations could contribute to brain dysfunction in schizophrenia by changing mRNA and protein abundance in particular neuronal and/or glial cell types. However, there is considerable heterogeneity, and not all candidate gene DNA methylation studies reported positive findings in SCZ postmortem brain (Siegmund et al., 2007). Studies on candidate SCZ genes were followed by assessment of DNA methylation on a genome-wide scale. One study (Mill et al., 2008) examined approximately 12,000 regulatory regions in frontal cortices of psychotic patients and control subjects. They found differential DNA methylation patterns at numerous loci, including several genes involved in GABAergic and glutamatergic neurotransmission and brain development. Interestingly, this study included both SCZ and bipolar disorder patients, implying that some epigenetic signatures may not be specific to schizophrenia but rather may be associated with psychosis. A second study interrogating 450,000 (2%) CpG sites genome-wide reported that CpGs associated with close to 3000 genes were differentially methylated in schizophrenia (Wockner et al., 2014). However, these early genome-scale DNA methylomics studies, like the aforementioned candidate genes studies, had only allowed for limited conclusions because of small sample size, and the field eagerly awaits larger-scale studies with many hundreds of clinical cases and controls to allow for more firm assessment of whether there is a tractable epigenetic risk architecture in the brain for such a heterogeneous patient population such as schizophrenics.

STUDIES IN THE PERIPHERAL TISSUE AND POSSIBLE EPIGENETIC BIOMARKERS IN SCHIZOPHRENIA

Because brain tissue is not accessible in living patients, it would be of interest to find possible epigenetic biomarkers in accessible tissues such as blood, saliva, or buccal cells. Although it is clear that between-tissue variation in DNA methylation greatly exceeds interindividual differences within any tissue, recent studies have shown that some interindividual variation remains tractable across brain and blood (Davies et al., 2012; Kundakovic et al., 2015). Consistent with this, several studies have shown DNA methylation changes in peripheral tissues of SCZ patients; those studies included genes that have also been shown to be differentially methylated in schizophrenic brains such as *HTR2A* and brain-derived neurotrophic factor (*BDNF*) (Carrard et al., 2011; Ghadirivasfi et al., 2011; Ikegame et al., 2013; Nohesara et al., 2011). Comparison between two genome-wide studies using Illumina platforms in postmortem brains and blood of schizophrenic patients showed that there was an overlap in 100 differentially methylated regions (Wockner et al., 2014). Finally, a recent study included almost 1500 schizophrenia cases and controls and explored DNA methylation profiles (although not a single base resolution) (Aberg et al., 2014). The major differentially methylated regions in this study are linked to differentiation and dopaminergic gene expression and to hypoxia and infection. These early studies, if independently confirmed, are promising because of the implication that peripheral markers possibly linked to early environmental insults associated with SCZ may provide a "mirror" for the DNA methylation changes occurring in the diseased brain. Of note, the same blood-based study implicated differential methylation of *RELN* (encoding a glycoprotein critical for neuronal migration and cortical connectivity) as one of the major findings consistent with the early postmortem studies (Abdolmaleky et al., 2005; Grayson et al., 2005; Aberg et al., 2014).

Furthermore, it should be mentioned that the aforementioned blood DNA methylation study (Aberg et al., 2014) drew, at least indirectly, a link between microRNA networks targeting neuronal differentiation and dopaminergic signaling pathways. MicroRNAs are small noncoding RNAs implicated in posttranscriptional regulation of gene expression via a variety of mechanisms ranging from mRNA degradation to translational inhibition (O'Carroll and Schaefer, 2013). There is increasing evidence that a subset of microRNAs could contribute to defective neuroplasticity and synaptic function in schizophrenia (Mellios and Sur, 2012). Finally, a small set of microRNAs, including microRNAs miR-130b and miR-193a-3p, were in a very recent study involving approximately 1000 subjects identified as state-independent biomarkers of schizophrenia based on plasma microRNA measurements (Wei et al., 2015). Although the tissue and cellular origins of these circulating microRNAs remain to be identified, these highly provocative findings, if further confirmed by independent studies, could provide an extremely valuable molecular toolbox for clinical studies aimed at predicting important parameters such as, for example, course of illness or response to treatment.

EPIGENETIC APPROACHES TO THE MOLECULAR AND GENETIC RISK ARCHITECTURES OF SCHIZOPHRENIA

DO EPIGENETIC MECHANISMS CONTRIBUTE TO LONG-LASTING ALTERATIONS IN GENE EXPRESSION IN SCHIZOPHRENIA BRAIN?

There can be little doubt that despite the lack of a unifying neuropathology, many cases of SCZ are affected by gene expression alterations in the cerebral cortex and other brain regions, often including transcripts important for oligodendrocyte function and myelination (Martins-de-Souza et al., 2009; Regenold et al., 2007; Katsel et al., 2005; Aston et al., 2004; Hakak et al., 2001; Tkachev et al., 2003) or inhibitory and excitatory neurotransmission (Duncan et al., 2010; Charych et al., 2009; Woo et al., 2008; Akbarian and Huang, 2006; Guidotti et al., 2005; Dracheva et al., 2004; Hashimoto et al., 2008;

Benes, 2010; Beneyto et al., 2007; Meador-Woodruff and Healy, 2000; Hemby et al., 2002; Mirnics et al., 2000; Middleton et al., 2002). The rationale to explore certain types of epigenetic modifications in postmortem brains of subjects diagnosed with psychiatric disease is often based on the hypotheses that changes in RNA expression are associated with altered epigenetic decoration at the site of the corresponding gene promoter and related regulatory sequences. Quite often the accompanying abnormalities in DNA methylation and histone modifications are then discussed in terms of a stable and long-lasting epigenetic "lesion" in response to an environmental insult or some other pathogenic effect operating in early life, many years before adulthood and disease onset. Animal studies have successfully shown that some of the epigenetically dysregulated genes in the SCZ postmortem brain, such as *GAD1* (encoding GABA synthesis enzyme) (Huang et al., 2007), are sensitive to different grades of maternal care during the early postnatal period, resulting in differential regulation of *Gad1* promoter-associated DNA methylation and histone acetylation in the hippocampus of adult rats (Zhang et al., 2010). Furthermore, postmortem studies involving adult suicide victims provided evidence that abuse in early childhood years is associated with a lasting DNA methylation imprint on the stress-regulated glucocorticoid receptor *NR3C1* promoter (McGowan et al., 2009) in the hippocampus and on DNA repeats encoding ribosomal RNAs (McGowan et al., 2008).

However, it is fair to admit that little is known about the stability and dynamic turnover of epigenetic markings in the human brain; therefore, it remains unclear whether any of the aforementioned epigenetic alterations in the SCZ brain reflect a "trait" stably maintained for years or, alternatively, whether disease-associated chromatin changes merely reflect the functional state of the brain at the time of death. Obviously, postmortem brain studies are cross-sectional, and it will be impossible to resolve this critical question, at least by not studying the diseased tissue—brain—of subjects afflicted with the illness. Given that most, or perhaps all, epigenetic markings studied to date are subject to bidirectional regulation in cell culture systems and animal models, it is reasonable to assume that the epigenetic decoration of human brain genomes is subject to similar types of dynamic regulation. For example, DNA methylation at specific promoter sequences is subject to rapid upregulation or downregulation on a scale of minutes to hours (Kundakovic et al., 2007; Levenson et al., 2006). Hippocampal DNA methylation signatures are highly sensitive to acute depolarization (Martinowich et al., 2003; Nelson et al., 2008) and electroconvulsive seizures, affecting regulatory sequences regulating NMDA and GABA-A receptor genes, Notch signaling pathways, and other systems with a key regulatory role in synaptic signal and plasticity (Guo et al., 2011). Changes in neuronal activity result in robust changes in expression and activity of multiple DNA methylation–associated proteins, including MeCP2 and Gadd45b (Ma et al., 2009; Cohen et al., 2011; Li et al., 2011). Furthermore, physiological activation of hippocampal circuitry during learning and memory is sufficient to elicit highly dynamic DNA methylation changes at *PP1*, *RELN*, and other gene promoters regulating synaptic plasticity (Day and Sweatt, 2010; Miller et al., 2010). These findings, taken together, would suggest that some of the epigenetic alterations reported in the diseased postmortem brain are not necessarily stable for very long periods of time; instead, they are regulated by mechanisms that operate on a much shorter timescale, perhaps lasting only a few weeks or days, or even less. There is an ever-increasing list of conditions that reportedly affect chromatin structure and function in the brain, including but by far not limited to, ischemia (Endres et al., 2000), environmental toxins (Bollati et al., 2007; Desaulniers et al., 2005; Salnikow and Zhitkovich, 2008), nicotine (Satta et al., 2007, 2008), alcohol (Marutha Ravindran and Ticku, 2004), psychostimulants (Numachi et al., 2004, 2007; LaPlant et al., 2010), and many of the drugs that are prescribed for the treatment of SCZ, including antipsychotic and mood-stabilizer drugs (Mill et al., 2008; Cheng et al., 2008; Shimabukuro et al., 2006; Li et al., 2004; Dong et al., 2008;

Kwon and Houpt, 2010; Bredy et al., 2007). Therefore, a plethora of factors that have the potential for long-term and short-term "epigenetic impact" affect genome organization and function in brain across all phases of development and aging.

FUNCTIONAL NEUROEPIGENOMICS COULD INFORM ABOUT DISEASE-ASSOCIATED GENETIC RISK POLYMORPHISMS

Over the past decade, multiple GWAS have produced strongly significant evidence that specific common DNA genetic variants among people influence their genetic susceptibility to a number of complex neuropsychiatric illnesses, including schizophrenia (Ripke et al., 2013; Purcell et al., 2014) and bipolar disorder (Sklar et al., 2011). The majority of common variant loci associated with genetic risk for these complex diseases reside within a noncoding sequence of unknown function, and many are far from discovered genes. For example, the major histocompatibility complex (MHC) locus—long implicated in psychiatric disease (Purcell et al., 2009; Shi et al., 2009; Stefansson et al., 2009)—harbors approximately 50 disease-associated SNPs in the 26- to 33-megabase region of the MHC locus on chromosome 6; however, strikingly, 50% of these SNPs were not located near genes, but rather were separated by many thousands of base pairs from the closest transcription start site or RNA. To be able to understand these associations mechanistically, it is critical to develop strategies for honing in on regions and genetic variants more likely to have functional effects. Thus, the elucidation of the function of noncoding sequences through neuroepigenomics is an important next step toward the development of testable hypotheses regarding biological processes involved in the pathogenesis of neuropsychiatric disorders (Fig. 5.2). A recent study discovered that on a genome-wide scale, a significant portion of SCZ-associated genetic polymorphisms match to noncoding sequences with active epigenetic regulation in human brain tissue, as evidenced, for example, by enrichment for open chromatin-associated histone methylation and acetylation marks that define many of the promoters, enhancers, and other DNA elements involved in transcriptional regulation (Roussos et al., 2014).

HIGHER-ORDER CHROMATIN AND THE GENETIC RISK ARCHITECTURE OF SCHIZOPHRENIA

In the context of the aforementioned risk-associated noncoding variants of SCZ, it is important to emphasize that DNA methylation, epigenetic histone modifications, and other types of epigenetic regulation would fall short of adequately describing the epigenome and localized chromatin architectures at any given genomic locus inside the nucleus of brain cells. This is because the chromosomal arrangements in the interphase nucleus are not random. Specifically, loci at sites of active gene expression are more likely to be clustered together and positioned toward a central position within the nucleus, whereas heterochromatin and silenced loci move more toward the nuclear periphery (Cremer and Cremer, 2001; Duan et al., 2010). Chromosomal loopings, in particular, are among the most highly regulated supranucleosomal structures and are associated with transcriptional regulation by positioning distal regulatory enhancer or silencer elements positioned on the linear genome far apart from their target gene to interact directly with that specific promoter (Wood et al., 2010; Gaszner and Felsenfeld, 2006). Proper regulation of such types of higher-order chromatin is critical. For example, Cornelia de Lange syndrome (CdLS), with an estimated incidence of 1:10–30,000 live births, is among the more frequent genetic disorders (source http://ghr.nlm.nih.gov) and is associated with severe developmental delay and a range

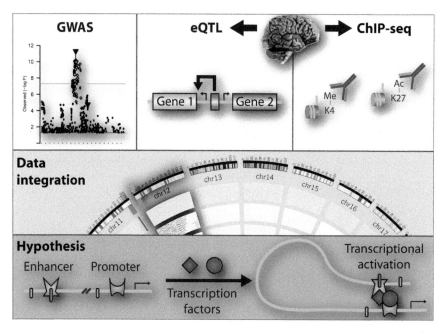

FIGURE 5.2 Exploring the genetic architecture of schizophrenia by using functional neuroepigenomic approaches.

Work flow. First, noncoding risk variants for schizophrenia emerge from genome-wide association studies (GWAS) such as the ones pursued by the Psychiatric Genomics Consortium (PGC) (Purcell et al., 2014). These sequences are then annotated using functional datasets including expression quantitative trait loci (Zhong et al., 2010; Cookson et al., 2009; Degner et al., 2012; Nicolae et al., 2010) and epigenome regulatory sequences, including histone methylation and acetylation profiling (ChIP-seq) from human brain tissue. Together with proper data integration and bioinformatical approaches, according to a recent study, schizophrenia risk variants are enriched for alleles that affect gene expression and lie within promoters and enhancers (Roussos et al., 2014). This led to the hypothesis that a great proportion of risk loci lie within enhancer sequences that physically interact, via chromosomal loop formation, with the promoter regions and are associated with the transcriptional regulation in the human brain. Mechanistic follow-up could include measuring transcription factor occupancies at promoter and enhancer sequences, chromosomal loop mappings, and other epigenomic assays.

of neuropsychiatric symptoms (Moss et al., 2008). CdLS (including Online Mendelian Inheritance of Man (OMIM) 122470 and 300590) has been linked to mutations in the cohesin complex (Deardorff et al., 2012; Gervasini et al., 2013). Cohesins and their associated proteins form ring-like structures that, in a donut-like shape, "bundle" together DNA segments from different locations; by interaction with transcriptional coactivators, the complex could promote the physical interaction of promoters with enhancers separated by thousands of base pairs on the linear genome, thereby regulating gene expression (Kagey et al., 2010). Interestingly, there is early evidence that abnormal weakening, or some other alterations affecting chromosomal loop structures, in conjunction with loop-bound genetic risk variants could contribute to dysregulated expression of GABA synthesis genes (Bharadwaj et al., 2013) and NMDA glutamate receptor subunits (Bharadwaj et al., 2014) and the calcium channel gene *CACNA1C* (Roussos et al., 2014) in cerebral cortex and other brain regions of subjects diagnosed with SCZ.

EPIGENETIC MECHANISMS AND THE TREATMENT OF SCHIZOPHRENIA

Antipsychotic drugs, mainly targeting dopamine D_2, and serotonergic and adrenergic receptor systems have been the mainstay of treatment for SCZ for more than half a century, but they still leave many patients with an incomplete therapeutic response and a poor outcome (Thornicroft et al., 2004; an der Heiden and Häfner, 2011). Whether or not the knowledge gained by neuroepigenetic approaches will eventually lead to improved treatment options in the future is unclear at this point. Of note, both typical antipsychotics acting as dopamine D_2 receptor antagonists and atypical drugs with more of mixed receptor profile affect DNA methylation and histone modification levels in cerebral cortex and striatum, which are two key nodes in the neural circuits of psychosis (Li et al., 2004; Dong et al., 2008; Akbarian, 2010; Kurita et al., 2012). However, it is not clear whether this indicates a critical role of chromatin-regulatory mechanisms for antipsychotic drug action. Here, we briefly discuss, as one potential example of future epigenetic therapy, the inhibition of histone deacetylase activity to exemplify the significant challenges that need to be overcome before such a type of epigenetic therapy would become acceptable as a clinical trial.

Histone acetylation is associated with a more flexible and "open" chromatin state, thereby facilitating gene expression and enabling enhancer and other regulatory sequences separated from a gene target site by thousands of base pairs to engage with distant promoters (Schwarzenbacher et al., 2006; Fig. 5.1). Histone acetylation is regulated by the opposing effects of histone acetyltansferases (HATs) and deacetylases (HDACs). There are at least 18 different HDACs encoded in the human genome, and they are commonly divided into four classes based on their equivalents in yeast (Dokmanovic et al., 2007). Class I includes HDAC1, 2, 3, and 8, class II/IIa includes HDAC 4, 5, 6, 7, 9, and 10, and HDAC11 is the sole representative of class IV (Dokmanovic et al., 2007). These HDACs are defined by a zinc ion site in the catalytic binding pocket, which also explains why many classical HDAC inhibitor (HDACi) drugs, including short-chain fatty acids (eg, butyrates), related compounds (eg, sodium valproate), and trichostatin A (Dokmanovic et al., 2007), have a broad profile and act on multiple HDACs. On a side note, clinically effective doses of valproate—a widely prescribed mood stabilizer and anticonvulsant—are probably too low to induce histone hyperacetylaton in brain (Hasan et al., 2013). Class III HDACs, which are also known as sirtuins, are defined by a different catalytic site, without the zinc ion but with nicotineamide dinucleotide (NAD+) as an essential cofactor (Dokmanovic et al., 2007). Most, or perhaps all, of the HDACs are thought to target various nuclear and cytoplasmic nonhistone proteins for deacetylation (Dokmanovic et al., 2007). Interestingly, expression of the class I histone deacetylase, *HDAC1*, was increased (on average 30–50%) in the prefrontal cortex and hippocampus of multiple SCZ postmortem brain cohorts (Sharma et al., 2008; Narayan et al., 2008; Benes et al., 2007; Jakovcevski et al., 2013). Therefore, abnormal *HDAC1* expression in corticolimbic circuitry is a type of molecular pathology representative of a significant portion of subjects with SCZ. Furthermore, overexpression of *Hdac1* in young adult mouse prefrontal cortex resulted in robust impairments in working memory, increased repetitive behaviors, and abnormal locomotor response profiles in novel environments in conjunction with dysregulated expression of more than 300 transcripts, including several that are located in the MHC risk locus on chromosome 6p21.3-22.1 (Jakovcevski et al., 2013). Interestingly, *Hdac1* expression becomes successively downregulated during the course of postnatal development, which could point to a neurodevelopmental cause of the observed excessive *HDAC1* expression in adult SCZ (Jakovcevski et al., 2013). Interestingly, *Hdac2*, which is a class I HDAC like Hdac1, has also recently been implicated in SCZ. Specifically, overexpression of *Hdac2* in a mouse prefrontal cortex resulted in SCZ-like phenotypes, including diminished prepulse inhibition (Kurita et al., 2012). However, overexpression of *Hdac2* in cerebral cortex, striatum, and other areas of the adult brain resulted in improved attentional set-shifting in the adult (Morris et al., 2013), which would suggest that alterations in expression or activity of HDAC2 result in very complex brain phenotypes dependent on cell type and developmental stage. These findings would suggest that drug-induced inhibition of neuronal and/or glial *HDACs* could result in a therapeutic effect for SCZ. However, there are significant challenges to explore this hypothesis in a clinical context. As discussed in a recent review (Hasan et al., 2013), there are newly developed potent HDAC inhibitor drugs (HDACi) either approved or undergoing clinical trials, such as the benzamide-based MS-275 (trade name Entinostat), that when orally administered exert a therapeutic effect in preclinical models of traumatic brain injury and neurodegeneration (Cao et al., 2013; Zhang and Schluesener, 2013). However, these drugs, which are mostly used as anticancer agents, broadly inhibit multiple HDAC isoforms, lack CNS specificity, and, although not directly cytotoxic, exhibit a safety profile that would mandate additional investigations prior to any experimental use in psychiatric patients (Hasan et al., 2013). In addition, animal studies suggest that HDACi potentially augments therapeutic effects of atypical antipsychotic drugs (Kurita et al., 2012; Grayson et al., 2010) and antidepressants (Schroeder et al., 2007; Zhu et al., 2009; Covington et al., 2011; Lin et al., 2012). However, combination therapies obviously would require even stricter safety criteria as compared to single drug regimens. We have argued that it may be premature to initiate trials with HDACi or other epigenetic drug targets in SCZ, but given that this is a rapidly evolving field, pending the availability of HDACi with favorable safety profiles, such trials should be given serious consideration (Hasan et al., 2013). Interestingly, the human genome also encodes 50 proteins containing "bromodomains," which essentially recognize and bind acetylated histone lysine residues (Filippakopoulos et al., 2012) and are thought to provide an important scaffold to recruit transcriptional proteins (Helin and Dhanak, 2013). The interaction of the acetyl-lysine binding pocket of these proteins is considered "druggable" and already identified as a potential therapeutic target in some cancers and inflammatory disease (Helin and Dhanak, 2013). Some of the bromodomain-containing proteins, including *BRD1*, are differentially regulated in cerebral cortex and hippocampus after electronconvulsive seizures (Fryland et al., 2012) and could provide important drug targets to treat major psychiatric disorders, including SCZ (Severinsen et al., 2006).

FUTURE DIRECTIONS
CHROMATIN-MODIFYING DRUGS: A FUTURE TREATMENT FOR SCHIZOPHRENIA?

The introduction of dopamine D_2 receptor antagonists revolutionized the treatment of schizophrenia in the 1950s, and the consecutive development of new antipsychotics increased the quality of life of schizophrenia patients. Nevertheless, available treatment options are still limited and chromatin-modifying drugs are certainly one avenue of research that needs to be further explored in preclinical research, including animal model systems. Once a class of chromatin-modifying drugs shows some promise in preclinical studies, as in the case of the HDAC inhibitors (Schroeder et al., 2007), it will be important to pinpoint the critical mechanism of action. For example, are the beneficial effects of HDAC inhibitors in various acute and chronic neurodegenerative and cognitive disorders due to their broad effects on histone modifications, or are they due to changes in the acetylation of nonhistone proteins such as the presynaptic molecules and regulators of vesicle release, or a combination thereof? (Jakovcevski and Akbarian, 2012). Of note, the fields of oncology and general medicine are currently pursuing hundreds of clinical trials with epigenetic drug targets. It is possible that clinical trials for neuropsychiatric disease will soon be enriched by an array of chromatin compounds that could emerge from ongoing preclinical and translational research.

COMPREHENSIVE, REGION-SPECIFIC, AND CELL TYPE-SPECIFIC EPIGENOME MAPPINGS IN THE BRAIN

Recent multiinstitutional consortia, such as the Encyclopedia of DNA Elements (ENCODE), using next-generation sequencing–based technologies (Zhao and Grant, 2011; Park, 2009) mapped DNA and histone modification landscapes and transcription factor binding patterns on comprehensive genome-scale maps, often at base pair resolution, in a variety of peripheral cell lines and tissues (Thurman et al., 2012; Bernstein et al., 2012). In contrast, only very few epigenetic markings, including DNA methylation and hydroxymethylation (Lister et al., 2013), and a small number of histone methylation and acetylation markings that differentiate between active and inactive/repressed promoter and enhancer sequences (Zhu et al., 2013) have been charted with next-generation sequencing technology in human brain (Maze et al., 2014), and there are no comparable studies using SCZ postmortem brain on a larger scale. In the near future, government-sponsored programs, such as PsychENCODE, and/or private or industry-sponsored efforts are expected to catalyze the generation of a much larger brain epigenomic dataset than the one currently available, with the expectation of a much deeper and mechanistic understanding about epigenetic regulation in the nervous system, including the interrelation with the genetic risk architectures of major psychiatric disease.

LARGER AND MORE AMBITIOUS CLINICAL STUDIES

The evidence summarized here strongly supports the hypothesis that dysregulation of DNA methylation may play an important role in SCZ pathophysiology. To advance this field, future studies need to: (1) include larger cohorts of patients and controls to increase the power to detect differentially methylated regions in schizophrenia; (2) explore DNA methylation profiles of neuronal and nonneuronal cell populations separately, given that a significant portion of epigenetic regulation is cell type-specific (Lister et al., 2013; Siegmund et al., 2007; Cheung et al., 2010; Iwamoto et al., 2011); (3) include more comprehensive DNA methylation methods, such as whole-genome bisulfite sequencing, that provide single base resolution and whole-genome coverage; and (4) include more comprehensive comparisons between DNA methylation profiles in the brain and peripheral tissues of SCZ patients. The final goal should be to integrate DNA methylation data with genetic information, gene expression data, and histone modification profiles to understand the functional role of DNA methylation changes and to use that data in the wider context of gene (dys)regulation in SCZ. More broadly, the epigenetic link between environmental risk factors and SCZ, although plausible, is still very scarce. We need more direct evidence that an environmental risk factor induces functional epigenetic changes that can contribute to SCZ in humans. One of the ways to address this question is to follow large, longitudinal birth cohorts and include tissue collection for epigenetic analyses (Mill and Heijmans, 2013; Pishva et al., 2014). This approach would allow for the analyses of epigenetic alterations over time. It would also provide the opportunity to relate epigenetic changes to specific environmental exposures and to the development of the disease. Because studies involving humans are mainly correlational, complementary animal models that could dissect the effects of different environmental exposures on the brain/blood epigenome and link this to SCZ-related phenotypes will also be needed. Although it is obvious that this field is still very young, these studies hold great promise for the development of novel preventive approaches and early interventions in SCZ.

ACKNOWLEDGMENT

Work in the authors' laboratory is supported by grants from the National Institutes of Mental Health and the Brain Behavior Research Foundation (BBRF/NARSAD). The authors report no financial conflicts of interest.

REFERENCES

Abdolmaleky, H.M., Cheng, K.H., Russo, A., et al., 2005. Hypermethylation of the reelin (RELN) promoter in the brain of schizophrenic patients: a preliminary report. Am. J. Med. Genet. B. Neuropsychiatr. Genet. 134B (1), 60–66.

Abdolmaleky, H.M., Cheng, K.H., Faraone, S.V., et al., 2006. Hypomethylation of MB-COMT promoter is a major risk factor for schizophrenia and bipolar disorder. Hum. Mol. Genet. 15 (21), 3132–3145.

Abdolmaleky, H.M., Yaqubi, S., Papageorgis, P., et al., 2011. Epigenetic dysregulation of HTR2A in the brain of patients with schizophrenia and bipolar disorder. Schizophr. Res. 129 (2–3), 183–190.

Aberg, K.A., McClay, J.L., Nerella, S., et al., 2014. Methylome-wide association study of schizophrenia: identifying blood biomarker signatures of environmental insults. JAMA Psychiatry 71 (3), 255–264.

Akbarian, S., 2010. Epigenetics of schizophrenia. Curr. Top. Behav. Neurosci. 4, 611–628.

Akbarian, S., Huang, H.S., 2006. Molecular and cellular mechanisms of altered GAD1/GAD67 expression in schizophrenia and related disorders. Brain Res. Rev. 52 (2), 293–304.

Akbarian, S., Nestler, E.J., 2013. Epigenetic mechanisms in psychiatry. Neuropsychopharmacology. 38 (1), 1–2.

an der Heiden, W., Häfner, H., 2011. Course and outcome, third ed. In: Weinberger, D. Harrison, P. (Eds.), Schizophrenia, vol. 3 Wiley-Blackwell, Oxford.

Andreassen, O.A., Thompson, W.K., Dale, A.M., 2014. Boosting the power of schizophrenia genetics by leveraging new statistical tools. Schizophr. Bull. 40 (1), 13–17.

Appasani, K., 2012. Epigenomics, From Chromatin Biology to Therapeutics. Cambridge University Press, Cambridge, MA; New York, NY.

Aston, C., Jiang, L., Sokolov, B.P., 2004. Microarray analysis of postmortem temporal cortex from patients with schizophrenia. J. Neurosci. Res. 77 (6), 858–866.

Bale, T.L., Baram, T.Z., Brown, A.S., et al., 2010. Early life programming and neurodevelopmental disorders. Biol. Psychiatry. 68 (4), 314–319.

Benes, F.M., 2010. Amygdalocortical circuitry in schizophrenia: from circuits to molecules. Neuropsychopharmacology. 35 (1), 239–257.

Benes, F.M., Lim, B., Matzilevich, D., Walsh, J.P., Subburaju, S., Minns, M., 2007. Regulation of the GABA cell phenotype in hippocampus of schizophrenics and bipolars. Proc. Natl. Acad. Sci. U.S.A. 104 (24), 10164–10169.

Beneyto, M., Kristiansen, L.V., Oni-Orisan, A., McCullumsmith, R.E., Meador-Woodruff, J.H., 2007. Abnormal glutamate receptor expression in the medial temporal lobe in schizophrenia and mood disorders. Neuropsychopharmacology. 32 (9), 1888–1902.

Bernstein, B.E., Birney, E., Dunham, I., Green, E.D., Gunter, C., Snyder, M., 2012. An integrated encyclopedia of DNA elements in the human genome. Nature. 489 (7414), 57–74.

Bharadwaj, R., Jiang, Y., Mao, W., et al., 2013. Conserved chromosome 2q31 conformations are associated with transcriptional regulation of GAD1 GABA synthesis enzyme and altered in prefrontal cortex of subjects with schizophrenia. J. Neurosci. 33 (29), 11839–11851.

Bharadwaj, R., Peter, C.J., Jiang, Y., et al., 2014. Conserved higher-order chromatin regulates NMDA receptor gene expression and cognition. Neuron. 84 (5), 997–1008.

Bollati, V., Baccarelli, A., Hou, L., et al., 2007. Changes in DNA methylation patterns in subjects exposed to low-dose benzene. Cancer. Res. 67 (3), 876–880.

Bredy, T.W., Wu, H., Crego, C., Zellhoefer, J., Sun, Y.E., Barad, M., 2007. Histone modifications around individual BDNF gene promoters in prefrontal cortex are associated with extinction of conditioned fear. Learn. Mem. 14 (4), 268–276.

Brown, A.S., Susser, E.S., 2008. Prenatal nutritional deficiency and risk of adult schizophrenia. Schizophr. Bull. 34 (6), 1054–1063.

Cao, P., Liang, Y., Gao, X., Zhao, M.G., Liang, G.B., 2013. Administration of MS-275 improves cognitive performance and reduces cell death following traumatic brain injury in rats. CNS Neurosci. Ther. 19 (5), 337–345.

Carrard, A., Salzmann, A., Malafosse, A., Karege, F., 2011. Increased DNA methylation status of the serotonin receptor 5HTR1A gene promoter in schizophrenia and bipolar disorder. J. Affect. Disord. 132 (3), 450–453.

Catts, V.S., Fung, S.J., Long, L.E., et al., 2013. Rethinking schizophrenia in the context of normal neurodevelopment. Front. Cell. Neurosci. 7, 60.

Charych, E.I., Liu, F., Moss, S.J., Brandon, N.J., 2009. GABA(A) receptors and their associated proteins: implications in the etiology and treatment of schizophrenia and related disorders. Neuropharmacology. 57 (5-6), 481–495.

Cheng, M.C., Liao, D.L., Hsiung, C.A., Chen, C.Y., Liao, Y.C., Chen, C.H., 2008. Chronic treatment with aripiprazole induces differential gene expression in the rat frontal cortex. Int. J. Neuropsychopharmacol. 11 (2), 207–216.

Cheung, I., Shulha, H.P., Jiang, Y., et al., 2010. Developmental regulation and individual differences of neuronal H3K4me3 epigenomes in the prefrontal cortex. Proc. Natl. Acad. Sci. U.S.A. 107 (19), 8824–8829.

Cohen, S., Gabel, H.W., Hemberg, M., et al., 2011. Genome-wide activity-dependent MeCP2 phosphorylation regulates nervous system development and function. Neuron. 72 (1), 72–85.

Cookson, W., Liang, L., Abecasis, G., Moffatt, M., Lathrop, M., 2009. Mapping complex disease traits with global gene expression. Nat. Rev. Genet. 10 (3), 184–194.

Covington III, H.E., Vialou, V.F., LaPlant, Q., Ohnishi, Y.N., Nestler, E.J., 2011. Hippocampal-dependent antidepressant-like activity of histone deacetylase inhibition. Neurosci. Lett. 493 (3), 122–126.

Cremer, T., Cremer, C., 2001. Chromosome territories, nuclear architecture and gene regulation in mammalian cells. Nat. Rev. Genet. 2 (4), 292–301.

Davies, M.N., Volta, M., Pidsley, R., et al., 2012. Functional annotation of the human brain methylome identifies tissue-specific epigenetic variation across brain and blood. Genome. Biol. 13 (6), R43.

Day, J.J., Sweatt, J.D., 2010. DNA methylation and memory formation. Nat. Neurosci. 13 (11), 1319–1323.

Day, J.J., Sweatt, J.D., 2011. Epigenetic mechanisms in cognition. Neuron. 70 (5), 813–829.

Deardorff, M.A., Bando, M., Nakato, R., et al., 2012. HDAC8 mutations in Cornelia de Lange syndrome affect the cohesin acetylation cycle. Nature. 489 (7415), 313–317.

Degner, J.F., Pai, A.A., Pique-Regi, R., et al., 2012. DNase I sensitivity QTLs are a major determinant of human expression variation. Nature. 482 (7385), 390–394.

Desaulniers, D., Xiao, G.H., Leingartner, K., Chu, I., Musicki, B., Tsang, B.K., 2005. Comparisons of brain, uterus, and liver mRNA expression for cytochrome p450s, DNA methyltransferase-1, and catechol-*O*-methyltransferase in prepubertal female Sprague-Dawley rats exposed to a mixture of aryl hydrocarbon receptor agonists. Toxicol. Sci. 86 (1), 175–184.

Dokmanovic, M., Clarke, C., Marks, P.A., 2007. Histone deacetylase inhibitors: overview and perspectives. Mol. Cancer. Res. 5 (10), 981–989.

Dong, E., Nelson, M., Grayson, D.R., Costa, E., Guidotti, A., 2008. Clozapine and sulpiride but not haloperidol or olanzapine activate brain DNA demethylation. Proc. Natl. Acad. Sci. U.S.A. 105 (36), 13614–13619.

Dong, E., Dzitoyeva, S.G., Matrisciano, F., Tueting, P., Grayson, D.R., Guidotti, A., 2014. Brain-derived neurotrophic factor epigenetic modifications associated with schizophrenia-like phenotype induced by prenatal stress in mice. Biol. Psychiatry.

Dorph-Petersen, K.A., Lewis, D.A., 2011. Stereological approaches to identifying neuropathology in psychosis. Biol. Psychiatry. 69 (2), 113–126.

Dracheva, S., Elhakem, S.L., McGurk, S.R., Davis, K.L., Haroutunian, V., 2004. GAD67 and GAD65 mRNA and protein expression in cerebrocortical regions of elderly patients with schizophrenia. J. Neurosci. Res. 76 (4), 581–592.

Duan, Z., Andronescu, M., Schutz, K., et al., 2010. A three-dimensional model of the yeast genome. Nature. 465 (7296), 363–367.

Duncan, C.E., Webster, M.J., Rothmond, D.A., Bahn, S., Elashoff, M., Shannon Weickert, C., 2010. Prefrontal GABA(A) receptor alpha-subunit expression in normal postnatal human development and schizophrenia. J. Psychiatr. Res.

Endres, M., Meisel, A., Biniszkiewicz, D., et al., 2000. DNA methyltransferase contributes to delayed ischemic brain injury. J. Neurosci. 20 (9), 3175–3181.

Fan, G., Martinowich, K., Chin, M.H., et al., 2005. DNA methylation controls the timing of astrogliogenesis through regulation of JAK-STAT signaling. Development 132 (15), 3345–3356.

Filippakopoulos, P., Picaud, S., Mangos, M., et al., 2012. Histone recognition and large-scale structural analysis of the human bromodomain family. Cell. 149 (1), 214–231.

Fryland, T., Elfving, B., Christensen, J.H., Mors, O., Wegener, G., Borglum, A.D., 2012. Electroconvulsive seizures regulates the Brd1 gene in the frontal cortex and hippocampus of the adult rat. Neurosci. Lett. 516 (1), 110–113.

Gaszner, M., Felsenfeld, G., 2006. Insulators: exploiting transcriptional and epigenetic mechanisms. Nat. Rev. Genet. 7 (9), 703–713.

Gervasini, C., Parenti, I., Picinelli, C., et al., 2013. Molecular characterization of a mosaic NIPBL deletion in a Cornelia de Lange patient with severe phenotype. Eur. J. Med. Genet. 56 (3), 138–143.

Ghadirivasfi, M., Nohesara, S., Ahmadkhaniha, H.R., et al., 2011. Hypomethylation of the serotonin receptor type-2A Gene (HTR2A) at T102C polymorphic site in DNA derived from the saliva of patients with schizophrenia and bipolar disorder. Am. J. Med. Genet. B. Neuropsychiatr. Genet. 156B (5), 536–545.

Golebiewska, A., Atkinson, S.P., Lako, M., Armstrong, L., 2009. Epigenetic landscaping during hESC differentiation to neural cells. Stem Cells 27 (6), 1298–1308.

Grayson, D.R., Jia, X., Chen, Y., et al., 2005. Reelin promoter hypermethylation in schizophrenia. Proc. Natl. Acad. Sci. U.S.A. 102 (26), 9341–9346.

Grayson, D.R., Kundakovic, M., Sharma, R.P., 2010. Is there a future for histone deacetylase inhibitors in the pharmacotherapy of psychiatric disorders? Mol. Pharmacol. 77 (2), 126–135.

Guidotti, A., Auta, J., Davis, J.M., et al., 2005. GABAergic dysfunction in schizophrenia: new treatment strategies on the horizon. Psychopharmacology. 180 (2), 191–205.

Guo, J.U., Ma, D.K., Mo, H., et al., 2011. Neuronal activity modifies the DNA methylation landscape in the adult brain. Nat. Neurosci. 14 (10), 1345–1351.

Guo, J.U., Su, Y., Zhong, C., Ming, G.L., Song, H., 2011. Hydroxylation of 5-methylcytosine by TET1 promotes active DNA demethylation in the adult brain. Cell. 145 (3), 423–434.

Guo, J.U., Su, Y., Shin, J.H., et al., 2014. Distribution, recognition and regulation of non-CpG methylation in the adult mammalian brain. Nat. Neurosci. 17 (2), 215–222.

Hakak, Y., Walker, J.R., Li, C., et al., 2001. Genome-wide expression analysis reveals dysregulation of myelination-related genes in chronic schizophrenia. Proc. Natl. Acad. Sci. U.S.A. 98 (8), 4746–4751.

Hasan, A., Mitchell, A., Schneider, A., Halene, T., Akbarian, S., 2013. Epigenetic dysregulation in schizophrenia: molecular and clinical aspects of histone deacetylase inhibitors. Eur. Arch. Psychiatry Clin. Neurosci. 263 (4), 273–284.

Hashimoto, T., Bazmi, H.H., Mirnics, K., Wu, Q., Sampson, A.R., Lewis, D.A., 2008. Conserved regional patterns of GABA-related transcript expression in the neocortex of subjects with schizophrenia. Am. J. Psychiatry 165 (4), 479–489.

Heijmans, B.T., Tobi, E.W., Stein, A.D., et al., 2008. Persistent epigenetic differences associated with prenatal exposure to famine in humans. Proc. Natl. Acad. Sci. U.S.A. 105 (44), 17046–17049.

Helin, K., Dhanak, D., 2013. Chromatin proteins and modifications as drug targets. Nature. 502 (7472), 480–488.

Hemby, S.E., Ginsberg, S.D., Brunk, B., Arnold, S.E., Trojanowski, J.Q., Eberwine, J.H., 2002. Gene expression profile for schizophrenia: discrete neuron transcription patterns in the entorhinal cortex. Arch. Gen. Psychiatry. 59 (7), 631–640.

Hennekens, C.H., Hennekens, A.R., Hollar, D., Casey, D.E., 2005. Schizophrenia and increased risks of cardiovascular disease. Am. Heart J. 150 (6), 1115–1121.

Huang, H.S., Matevossian, A., Whittle, C., et al., 2007. Prefrontal dysfunction in schizophrenia involves mixed-lineage leukemia 1-regulated histone methylation at GABAergic gene promoters. J. Neurosci. 27 (42), 11254–11262.

Ikegame, T., Bundo, M., Sunaga, F., et al., 2013. DNA methylation analysis of BDNF gene promoters in peripheral blood cells of schizophrenia patients. Neurosci. Res. 77 (4), 208–214.

Iwamoto, K., Bundo, M., Yamada, K., et al., 2005. DNA methylation status of SOX10 correlates with its downregulation and oligodendrocyte dysfunction in schizophrenia. J. Neurosci. 25 (22), 5376–5381.

Iwamoto, K., Bundo, M., Ueda, J., et al., 2011. Neurons show distinctive DNA methylation profile and higher interindividual variations compared with non-neurons. Genome Res. 21 (5), 688–696.

Jakovcevski, M., Akbarian, S., 2012. Epigenetic mechanisms in neurological disease. Nat. Med. 18 (8), 1194–1204.

Jakovcevski, M., Bharadwaj, R., Straubhaar, J., et al., 2013. Prefrontal cortical dysfunction after overexpression of histone deacetylase 1. Biol. Psychiatry 74 (9), 696–705.

Jin, S.G., Wu, X., Li, A.X., Pfeifer, G.P., 2011. Genomic mapping of 5-hydroxymethylcytosine in the human brain. Nucleic. Acids. Res. 39 (12), 5015–5024.

Kaas, G.A., Zhong, C., Eason, D.E., et al., 2013. TET1 controls CNS 5-methylcytosine hydroxylation, active DNA demethylation, gene transcription, and memory formation. Neuron. 79 (6), 1086–1093.

Kagey, M.H., Newman, J.J., Bilodeau, S., et al., 2010. Mediator and cohesin connect gene expression and chromatin architecture. Nature. 467 (7314), 430–435.

Katsel, P., Davis, K.L., Haroutunian, V., 2005. Variations in myelin and oligodendrocyte-related gene expression across multiple brain regions in schizophrenia: a gene ontology study. Schizophr. Res. 79 (2–3), 157–173.

Khashan, A.S., Abel, K.M., McNamee, R., et al., 2008. Higher risk of offspring schizophrenia following antenatal maternal exposure to severe adverse life events. Arch. Gen. Psychiatry 65 (2), 146–152.

Kim, D.H., Stahl, S.M., 2010. Antipsychotic drug development. Curr. Topics Behav. Neurosci. 4, 123–139.

Kirkbride, J.B., Susser, E., Kundakovic, M., Kresovich, J.K., Davey Smith, G., Relton, C.L., 2012. Prenatal nutrition, epigenetics and schizophrenia risk: can we test causal effects? Epigenomics.

Klose, R.J., Bird, A.P., 2006. Genomic DNA methylation: the mark and its mediators. Trends Biochem. Sci. 31 (2), 89–97.

Kouzarides, T., 2007. Chromatin modifications and their function. Cell. 128 (4), 693–705.

Kriaucionis, S., Heintz, N., 2009. The nuclear DNA base 5-hydroxymethylcytosine is present in Purkinje neurons and the brain. Science 324 (5929), 929–930.

Kundakovic, M., 2013. Prenatal programming of psychopathology: the role of epigenetic mechanisms. J. Med. Biochem 32 (4), 313–324.

Kundakovic, M., 2014. Postnatal risk environments, epigenetics, and psychosis: putting the pieces together. Soc. Psychiatry Psychiatr. Epidemiol. 49 (10), 1535–1536.

Kundakovic, M., Champagne, F.A., 2014. Early-life experience, epigenetics, and the developing brain. Neuropsychopharmacology 40, 141–153.

Kundakovic, M., Chen, Y., Costa, E., Grayson, D.R., 2007. DNA methyltransferase inhibitors coordinately induce expression of the human reelin and glutamic acid decarboxylase 67 genes. Mol. Pharmacol. 71 (3), 644–653.

Kundakovic, M., Gudsnuk, K., Herbstman, J.B., Tang, D., Perera, F.P., Champagne, F.A., 2015. DNA methylation of BDNF as a biomarker of early-life adversity. Proc. Natl. Acad. Sci. U.S.A. 112 (22), 6807–6813.

Kurita, M., Holloway, T., Garcia-Bea, A., et al., 2012. HDAC2 regulates atypical antipsychotic responses through the modulation of mGlu2 promoter activity. Nat. Neurosci. 15 (9), 1245–1254.

Kwon, B., Houpt, T.A., 2010. Phospho-acetylation of histone H3 in the amygdala after acute lithium chloride. Brain. Res. 1333, 36–47.

LaPlant, Q., Vialou, V., Covington III, H.E., et al., 2010. Dnmt3a regulates emotional behavior and spine plasticity in the nucleus accumbens. Nat. Neurosci. 13 (9), 1137–1143.

Laursen, T.M., Munk-Olsen, T., Vestergaard, M., 2012. Life expectancy and cardiovascular mortality in persons with schizophrenia. Curr. Opin. Psychiatry 25 (2), 83–88.

Levenson, J.M., Roth, T.L., Lubin, F.D., et al., 2006. Evidence that DNA (cytosine-5) methyltransferase regulates synaptic plasticity in the hippocampus. J. Biol. Chem. 281 (23), 15763–15773.

Li, G., Reinberg, D., 2011. Chromatin higher-order structures and gene regulation. Curr. Opin. Genet. Dev. 21 (2), 175–186.

Li, H., Zhong, X., Chau, K.F., Williams, E.C., Chang, Q., 2011. Loss of activity-induced phosphorylation of MeCP2 enhances synaptogenesis, LTP and spatial memory. Nat. Neurosci. 14 (8), 1001–1008.

Li, J., Guo, Y., Schroeder, F.A., et al., 2004. Dopamine D2-like antagonists induce chromatin remodeling in striatal neurons through cyclic AMP-protein kinase A and NMDA receptor signaling. J. Neurochem. 90 (5), 1117–1131.

Lieberman, J.A., Stroup, T.S., McEvoy, J.P., et al., 2005. Effectiveness of antipsychotic drugs in patients with chronic schizophrenia. N. Engl. J. Med. 353 (12), 1209–1223.

Lin, H., Geng, X., Dang, W., et al., 2012. Molecular mechanisms associated with the antidepressant effects of the class I histone deacetylase inhibitor MS-275 in the rat ventrolateral orbital cortex. Brain Res. 1447, 119–125.

Lister, R., Mukamel, E.A., Nery, J.R., et al., 2013. Global epigenomic reconfiguration during mammalian brain development. Science 341 (6146), 1237905.

Liu, J., Casaccia, P., 2010. Epigenetic regulation of oligodendrocyte identity. Trends Neurosci. 33 (4), 193–201.

Ma, D.K., Jang, M.H., Guo, J.U., et al., 2009. Neuronal activity-induced Gadd45b promotes epigenetic DNA demethylation and adult neurogenesis. Science 323 (5917), 1074–1077.

Martinowich, K., Hattori, D., Wu, H., et al., 2003. DNA methylation-related chromatin remodeling in activity-dependent BDNF gene regulation. Science 302 (5646), 890–893.

Martins-de-Souza, D., Gattaz, W.F., Schmitt, A., et al., 2009. Alterations in oligodendrocyte proteins, calcium homeostasis and new potential markers in schizophrenia anterior temporal lobe are revealed by shotgun proteome analysis. J. Neural. Transm. 116 (3), 275–289.

Marutha Ravindran, C.R., Ticku, M.K., 2004. Changes in methylation pattern of NMDA receptor NR2B gene in cortical neurons after chronic ethanol treatment in mice. Brain. Res. Mol. Brain. Res. 121 (1–2), 19–27.

Maunakea, A.K., Nagarajan, R.P., Bilenky, M., et al., 2010. Conserved role of intragenic DNA methylation in regulating alternative promoters. Nature. 466 (7303), 253–257.

Maze, I., Shen, L., Zhang, B., et al., 2014. Analytical tools and current challenges in the modern era of neuroepigenomics. Nat. Neurosci. 17 (11), 1476–1490.

McGowan, P.O., Sasaki, A., Huang, T.C., et al., 2008. Promoter-wide hypermethylation of the ribosomal RNA gene promoter in the suicide brain. PLoS. One 3 (5), e2085.

McGowan, P.O., Sasaki, A., D'Alessio, A.C., et al., 2009. Epigenetic regulation of the glucocorticoid receptor in human brain associates with childhood abuse. Nat. Neurosci. 12 (3), 342–348.

Meador-Woodruff, J.H., Healy, D.J., 2000. Glutamate receptor expression in schizophrenic brain. Brain. Res. Brain. Res. Rev. 31 (2–3), 288–294.

Mellios, N., Sur, M., 2012. The emerging role of micrornas in schizophrenia and autism spectrum disorders. Front. Psychiatry 3, 39.

Middleton, F.A., Mirnics, K., Pierri, J.N., Lewis, D.A., Levitt, P., 2002. Gene expression profiling reveals alterations of specific metabolic pathways in schizophrenia. J. Neurosci. 22 (7), 2718–2729.

Mill, J., Heijmans, B.T., 2013. From promises to practical strategies in epigenetic epidemiology. Nat. Rev. Genet. 14 (8), 585–594.

Mill, J., Tang, T., Kaminsky, Z., et al., 2008. Epigenomic profiling reveals DNA-methylation changes associated with major psychosis. Am. J. Hum. Genet. 82 (3), 696–711.

Miller, C.A., Gavin, C.F., White, J.A., et al., 2010. Cortical DNA methylation maintains remote memory. Nat. Neurosci. 13 (6), 664–666.

Miller, F.D., Gauthier, A.S., 2007. Timing is everything: making neurons versus glia in the developing cortex. Neuron. 54 (3), 357–369.

Mirnics, K., Middleton, F.A., Marquez, A., Lewis, D.A., Levitt, P., 2000. Molecular characterization of schizophrenia viewed by microarray analysis of gene expression in prefrontal cortex. Neuron. 28 (1), 53–67.

Morris, M.J., Mahgoub, M., Na, E.S., Pranav, H., Monteggia, L.M., 2013. Loss of histone deacetylase 2 improves working memory and accelerates extinction learning. J. Neurosci. 33 (15), 6401–6411.

Moss, J.F., Oliver, C., Berg, K., Kaur, G., Jephcott, L., Cornish, K., 2008. Prevalence of autism spectrum phenomenology in Cornelia de Lange and Cri du Chat syndromes. Am. J. Ment. Retard. 113 (4), 278–291.

Mueller, B.R., Bale, T.L., 2008. Sex-specific programming of offspring emotionality after stress early in pregnancy. J. Neurosci. 28 (36), 9055–9065.

Murgatroyd, C., Patchev, A.V., Wu, Y., et al., 2009. Dynamic DNA methylation programs persistent adverse effects of early-life stress. Nat. Neurosci. 12 (12), 1559–1566.

Narayan, S., Tang, B., Head, S.R., et al., 2008. Molecular profiles of schizophrenia in the CNS at different stages of illness. Brain. Res. 1239, 235–248.

Nelson, E.D., Kavalali, E.T., Monteggia, L.M., 2008. Activity-dependent suppression of miniature neurotransmission through the regulation of DNA methylation. J. Neurosci. 28 (2), 395–406.

Nestler, E.J., 2014. Epigenetic mechanisms of depression. JAMA Psychiatry 71 (4), 454–456.

Network and Pathway Analysis Subgroup of Psychiatric Genomics Consortium, 2015. Psychiatric genome-wide association study analyses implicate neuronal, immune and histone pathways. Nat. Neurosci. 18 (2), 199–209.

Nicolae, D.L., Gamazon, E., Zhang, W., Duan, S., Dolan, M.E., Cox, N.J., 2010. Trait-associated SNPs are more likely to be eQTLs: annotation to enhance discovery from GWAS. PLoS. Genet. 6 (4), e1000888.

Niwa, M., Jaaro-Peled, H., Tankou, S., et al., 2013. Adolescent stress-induced epigenetic control of dopaminergic neurons via glucocorticoids. Science 339 (6117), 335–339.

Nohesara, S., Ghadirivasfi, M., Mostafavi, S., et al., 2011. DNA hypomethylation of MB-COMT promoter in the DNA derived from saliva in schizophrenia and bipolar disorder. J. Psychiatr. Res. 45 (11), 1432–1438.

Numachi, Y., Yoshida, S., Yamashita, M., et al., 2004. Psychostimulant alters expression of DNA methyltransferase mRNA in the rat brain. Ann. N. Y. Acad. Sci. 1025, 102–109.

Numachi, Y., Shen, H., Yoshida, S., et al., 2007. Methamphetamine alters expression of DNA methyltransferase 1 mRNA in rat brain. Neurosci. Lett. 414 (3), 213–217.

O'Carroll, D., Schaefer, A., 2013. General principals of miRNA biogenesis and regulation in the brain. Neuropsychopharmacology. 38 (1), 39–54.

Oh, G., Petronis, A., 2008. Environmental studies of schizophrenia through the prism of epigenetics. Schizophr. Bull. 34 (6), 1122–1129.

Paparelli, A., Di Forti, M., Morrison, P.D., Murray, R.M., 2011. Drug-induced psychosis: how to avoid star gazing in schizophrenia research by looking at more obvious sources of light. Front. Behav. Neurosci. 5, 1.

Park, P.J., 2009. ChIP-seq: advantages and challenges of a maturing technology. Nat. Rev. Genet. 10 (10), 669–680.

Petronis, A., Mill, J., 2011. Brain, Behavior, and Epigenetics. Springer-Verlag Berlin. Heidelberg. http://dx.doi.org/10.1007/978-3-642-17426-1.

Pishva, E., Kenis, G., van den Hove, D., et al., 2014. The epigenome and postnatal environmental influences in psychotic disorders. Soc. Psychiatry Psychiatr. Epidemiol. 49 (3), 337–348.

Purcell, S.M., Wray, N.R., Stone, J.L., et al., 2009. Common polygenic variation contributes to risk of schizophrenia and bipolar disorder. Nature. 460 (7256), 748–752.

Purcell, S.M., Moran, J.L., Fromer, M., et al., 2014. A polygenic burden of rare disruptive mutations in schizophrenia. Nature. 506 (7487), 185–190.

Regenold, W.T., Phatak, P., Marano, C.M., Gearhart, L., Viens, C.H., Hisley, K.C., 2007. Myelin staining of deep white matter in the dorsolateral prefrontal cortex in schizophrenia, bipolar disorder, and unipolar major depression. Psychiatry. Res. 151 (3), 179–188.

Ripke, S., O'Dushlaine, C., Chambert, K., et al., 2013. Genome-wide association analysis identifies 13 new risk loci for schizophrenia. Nat. Genet.

Rodriguez-Murillo, L., Gogos, J.A., Karayiorgou, M., 2012. The genetic architecture of schizophrenia: new mutations and emerging paradigms. Annu. Rev. Med. 63, 63–80.

Rodriguez-Paredes, M., Esteller, M., 2011. Cancer epigenetics reaches mainstream oncology. Nat. Med. 17 (3), 330–339.

Roth, C., Magnus, P., Schjolberg, S., et al., 2011. Folic acid supplements in pregnancy and severe language delay in children. JAMA. 306 (14), 1566–1573.

Roth, T.L., Lubin, F.D., Funk, A.J., Sweatt, J.D., 2009. Lasting epigenetic influence of early-life adversity on the BDNF gene. Biol. Psychiatry. 65 (9), 760–769.

Roussos, P., Mitchell, A.C., Voloudakis, G., et al., 2014. A role for noncoding variation in schizophrenia. Cell Rep. 9 (4), 1417–1429.

Rudenko, A., Dawlaty, M.M., Seo, J., et al., 2013. Tet1 is critical for neuronal activity-regulated gene expression and memory extinction. Neuron. 79 (6), 1109–1122.

Saha, S., Chant, D., McGrath, J., 2007. A systematic review of mortality in schizophrenia: is the differential mortality gap worsening over time? Arch. Gen. Psychiatry 64 (10), 1123–1131.

Salnikow, K., Zhitkovich, A., 2008. Genetic and epigenetic mechanisms in metal carcinogenesis and cocarcinogenesis: nickel, arsenic, and chromium. Chem. Res. Toxicol. 21 (1), 28–44.

Satta, R., Maloku, E., Costa, E., Guidotti, A., 2007. Stimulation of brain nicotinic acetylcholine receptors (nAChRs) decreases DNA methyltransferase 1 (DNMT1) expression in cortical and hippocampal GABAergic neurons of Swiss albino mice. Soc. Neurosci. Abs.

Satta, R., Maloku, E., Zhubi, A., et al., 2008. Nicotine decreases DNA methyltransferase 1 expression and glutamic acid decarboxylase 67 promoter methylation in GABAergic interneurons. Proc. Natl. Acad. Sci. U.S.A. 105 (42), 16356–16361.

Schroeder, F.A., Lin, C.L., Crusio, W.E., Akbarian, S., 2007. Antidepressant-like effects of the histone deacetylase inhibitor, sodium butyrate, in the mouse. Biol. Psychiatry. 62 (1), 55–64.

Schwarzenbacher, R., McMullan, D., Krishna, S.S., et al., 2006. Crystal structure of a glycerate kinase (TM1585) from *Thermotoga maritima* at 2.70 A resolution reveals a new fold. Proteins. 65 (1), 243–248.

Severinsen, J.E., Bjarkam, C.R., Kiaer-Larsen, S., et al., 2006. Evidence implicating BRD1 with brain development and susceptibility to both schizophrenia and bipolar affective disorder. Mol. Psychiatry. 11 (12), 1126–1138.

Sharma, R.P., Grayson, D.R., Guidotti, A., Costa, E., 2005. Chromatin, DNA methylation and neuron gene regulation--the purpose of the package. J. Psychiatry. Neurosci. 30 (4), 257–263.

Sharma, R.P., Grayson, D.R., Gavin, D.P., 2008. Histone deactylase 1 expression is increased in the prefrontal cortex of schizophrenia subjects: analysis of the National Brain Databank microarray collection. Schizophr. Res. 98 (1–3), 111–117.

Shi, J., Levinson, D.F., Duan, J., et al., 2009. Common variants on chromosome 6p22.1 are associated with schizophrenia. Nature. 460 (7256), 753–757.

Shimabukuro, M., Jinno, Y., Fuke, C., Okazaki, Y., 2006. Haloperidol treatment induces tissue- and sex-specific changes in DNA methylation: a control study using rats. Behav. Brain. Funct. 2, 37.

Siegmund, K.D., Connor, C.M., Campan, M., et al., 2007. DNA methylation in the human cerebral cortex is dynamically regulated throughout the life span and involves differentiated neurons. PLoS One 2 (9), e895.

Sklar, P., Ripke, S., Scott, L.J., et al., 2011. Large-scale genome-wide association analysis of bipolar disorder identifies a new susceptibility locus near ODZ4. Nat. Genet. 43 (10), 977–983.

Song, C.X., Szulwach, K.E., Fu, Y., et al., 2011. Selective chemical labeling reveals the genome-wide distribution of 5-hydroxymethylcytosine. Nat. Biotechnol. 29 (1), 68–72.

Stefansson, H., Ophoff, R.A., Steinberg, S., et al., 2009. Common variants conferring risk of schizophrenia. Nature. 460 (7256), 744–747.

Stein, Z., Susser, M., Saenger, G., Marolla, F., 1972. Nutrition and mental performance. Science 178 (4062), 708–713.

Suren, P., Roth, C., Bresnahan, M., et al., 2013. Association between maternal use of folic acid supplements and risk of autism spectrum disorders in children. JAMA. 309 (6), 570–577.

Susser, E.S., Lin, S.P., 1992. Schizophrenia after prenatal exposure to the Dutch Hunger Winter of 1944-1945. Arch. Gen. Psychiatry. 49 (12), 983–988.

Swartz, M.S., Perkins, D.O., Stroup, T.S., et al., 2007. Effects of antipsychotic medications on psychosocial functioning in patients with chronic schizophrenia: findings from the NIMH CATIE study. Am. J. Psychiatry. 164 (3), 428–436.

Sweatt, J.D., 2013. Epigenetic Regulation in the Nervous System: Basic Mechanisms and Clinical Impact. Academic Press, London; Waltham, MA.

Sweatt, J.D., 2013. The emerging field of neuroepigenetics. Neuron. 80 (3), 624–632.

Takata, A., Xu, B., Ionita-Laza, I., Roos, J.L., Gogos, J.A., Karayiorgou, M., 2014. Loss-of-function variants in schizophrenia risk and SETD1A as a candidate susceptibility gene. Neuron. 82 (4), 773–780.

Taly, A., 2013. Novel approaches to drug design for the treatment of schizophrenia. Exp. Opin. Drug Discov. 8 (10), 1285–1296.

Tan, M., Luo, H., Lee, S., et al., 2011. Identification of 67 histone marks and histone lysine crotonylation as a new type of histone modification. Cell. 146 (6), 1016–1028.

Taverna, S.D., Li, H., Ruthenburg, A.J., Allis, C.D., Patel, D.J., 2007. How chromatin-binding modules interpret histone modifications: lessons from professional pocket pickers. Nat. Struct. Mol. Biol. 14 (11), 1025–1040.

Thornicroft, G., Tansella, M., Becker, T., et al., 2004. The personal impact of schizophrenia in Europe. Schizophr. Res. 69 (2–3), 125–132.

Thurman, R.E., Rynes, E., Humbert, R., et al., 2012. The accessible chromatin landscape of the human genome. Nature. 489 (7414), 75–82.

Tkachev, D., Mimmack, M.L., Ryan, M.M., et al., 2003. Oligodendrocyte dysfunction in schizophrenia and bipolar disorder. Lancet. 362 (9386), 798–805.

Tobi, E.W., Lumey, L.H., Talens, R.P., et al., 2009. DNA methylation differences after exposure to prenatal famine are common and timing- and sex-specific. Hum. Mol. Genet. 18 (21), 4046–4053.

Tobi, E.W., Goeman, J.J., Monajemi, R., et al., 2014. DNA methylation signatures link prenatal famine exposure to growth and metabolism. Nat. Commun. 5, 5592.

Tsankova, N., Renthal, W., Kumar, A., Nestler, E.J., 2007. Epigenetic regulation in psychiatric disorders. Nat. Rev. Neurosci. 8 (5), 355–367.

van Os, J., Kenis, G., Rutten, B.P., 2010. The environment and schizophrenia. Nature. 468 (7321), 203–212.

Walker, D.M., Cates, H.M., Heller, E.A., Nestler, E.J., 2015. Regulation of chromatin states by drugs of abuse. Curr. Opin. Neurobiol. 30C, 112–121.

Weaver, I.C., Cervoni, N., Champagne, F.A., et al., 2004. Epigenetic programming by maternal behavior. Nat. Neurosci. 7 (8), 847–854.

Wei, H., Yuan, Y., Liu, S., et al., 2015. Detection of circulating miRNA levels in schizophrenia. Am. J. Psychiatry 172, 1141–1147. appiajp201514030273.

Wockner, L.F., Noble, E.P., Lawford, B.R., et al., 2014. Genome-wide DNA methylation analysis of human brain tissue from schizophrenia patients. Transl. Psychiatry 4, e339.

Woo, T.U., Kim, A.M., Viscidi, E., 2008. Disease-specific alterations in glutamatergic neurotransmission on inhibitory interneurons in the prefrontal cortex in schizophrenia. Brain. Res. 1218, 267–277.

Wood, A.J., Severson, A.F., Meyer, B.J., 2010. Condensin and cohesin complexity: the expanding repertoire of functions. Nat. Rev. Genet. 11 (6), 391–404.

Wu, H., Coskun, V., Tao, J., et al., 2010. Dnmt3a-dependent nonpromoter DNA methylation facilitates transcription of neurogenic genes. Science 329 (5990), 444–448.

Zhang, T.Y., Hellstrom, I.C., Bagot, R.C., Wen, X., Diorio, J., Meaney, M.J., 2010. Maternal care and DNA methylation of a glutamic acid decarboxylase 1 promoter in rat hippocampus. J. Neurosci. 30 (39), 13130–13137.

Zhang, Z.Y., Schluesener, H.J., 2013. Oral administration of histone deacetylase inhibitor MS-275 ameliorates neuroinflammation and cerebral amyloidosis and improves behavior in a mouse model. J. Neuropathol. Exp. Neurol. 72 (3), 178–185.

Zhao, J., Grant, S.F., 2011. Advances in whole genome sequencing technology. Curr. Pharm. Biotechnol. 12 (2), 293–305.

Zhong, H., Beaulaurier, J., Lum, P.Y., et al., 2010. Liver and adipose expression associated SNPs are enriched for association to type 2 diabetes. PLoS. Genet. 6, e1000932.

Zhou, V.W., Goren, A., Bernstein, B.E., 2011. Charting histone modifications and the functional organization of mammalian genomes. Nat. Rev. Genet. 12 (1), 7–18.

Zhu, H., Huang, Q., Xu, H., Niu, L., Zhou, J.N., 2009. Antidepressant-like effects of sodium butyrate in combination with estrogen in rat forced swimming test: involvement of 5-HT(1A) receptors. Behav. Brain Res. 196 (2), 200–206.

Zhu, J., Adli, M., Zou, J.Y., et al., 2013. Genome-wide chromatin state transitions associated with developmental and environmental cues. Cell 152 (3), 642–654.

CHAPTER

EXPLORING NEUROGENOMICS OF SCHIZOPHRENIA WITH *ALLEN INSTITUTE FOR BRAIN SCIENCE* RESOURCES

6

M. Hawrylycz[1], T. Nickl-Jockschat[2,3] and S. Sunkin[1]

[1]*Allen Institute for Brain Science, Seattle, WA, United States* [2]*Department of Psychiatry, Psychotherapy and Psychosomatics, RWTH Aachen University, Aachen, Germany* [3]*Juelich-Aachen Research Alliance–Translational Brain Medicine, Juelich/Aachen, Germany*

CHAPTER OUTLINE

Introduction	83
Introduction to the Portal	84
Mouse Atlas Resources	84
Common Reference Space in the Mouse Brain	86
Integrated Search and Visualization in the Allen Brain Atlas	87
The Mouse Developmental Atlas	91
Connectivity in the Mouse Brain	93
Human Atlas Resources	97
The Allen Human Brain Atlas	97
BrainSpan Atlas of the Developing Human Brain	99
NIH Blueprint Non-Human Primate Atlas	101
Beyond the Atlases	102
Applications of the Allen Institute for Brain Science Resources in Schizophrenia Research	102
Acknowledgments	104
References	104

INTRODUCTION

Schizophrenia is highly heritable, and genetic risk for vulnerability is conferred by a large number of alleles, including common alleles of small effect that have been detected through genome-wide association studies (GWAS) (Lencz and Malhotra, 2015; Ripke et al., 2013; Roussos et al., 2015; Wang et al., 2015). Because the public health impact of schizophrenia is so substantial, data and tools to further our study of this complex disease are needed. The Allen Brain Atlas (www.brain-map.org) (Sunkin et al., 2012) is an online, publicly available resource that integrates gene expression and connectivity data with neuroanatomical information for mouse, human, and nonhuman primate, and it may be of interest to researchers working in these areas. Launched in 2004 by the Allen Institute for Brain Science, the portal currently has approximately 45,000 unique users each month and has more than a petabyte of gene expression data points generated to date. As one of the most comprehensive gene expression resources for the nervous system, scientists regularly use these data to learn about the expression profile of genes in various neuroanatomical regions. Additional uses include searching for biomarkers, correlating gene expression to anatomy, and other large-scale correlative data analysis. This chapter reviews the resources available and describes how certain ones may be of use to those studying schizophrenia. In addition to substantial raw data, visualization and search tools are available to analyze the massive amount of data generated. Examples are provided regarding how these tools may be leveraged for scientific discovery.

INTRODUCTION TO THE PORTAL

The Allen Brain Atlas portal (www.brain-map.org) contains a growing collection of data and software suites that integrates gene expression and connectivity data with neuroanatomy and other modalities. There are currently nine major gene expression and connectivity atlases that span across species and developmental timepoints. In addition, there are three data sets focused on gene expression in sleep, diversity of gene expression across mouse strains, and gene expression profiles of glioblastoma in humans. Also, there is the newly released Allen Cell Types Database that focuses on electrophysiology and morphology of genetically modified cells from Cre lines, along with abstract point models and more biophysically detailed models generated from the same profiled cells. Although the portal serves as the entry point for these data resources, it also provides an integrated environment that provides information to the scientific community across all portal resources. The portal is updated a few times each year, and accompanying each release is an announcement that highlights the newly released data and new application features, some of which may be useful to those studying schizophrenia.

MOUSE ATLAS RESOURCES

There are four large-scale atlases associated with the mouse: The Allen Mouse Brain Atlas, the Allen Developing Mouse Brain Atlas, the Allen Mouse Brain Connectivity Atlas, and the Allen Spinal Cord Atlas. The recently released Allen Cell Types Database is not discussed in this chapter. The inaugural project of the Allen Institute for Brain Science, the Allen Mouse Brain Atlas (Lein et al., 2007) (see Fig. 6.1), is a genome-wide 3D atlas of gene expression in the adult mouse brain. The atlas contains 600 terabytes of high-resolution in situ hybridization (ISH) images of the P56 mouse brain. Building on the foundations of the adult mouse data pipeline (Ng et al., 2007; Ng et al., 2012), the Allen Developing Mouse Brain Atlas (Thompson et al., 2014) profiles changes of gene expression during development from E11 through P56 during key stages in the Theiler developmental framework (www.emouseatlas.org). Effectively a 4D atlas, the data set, comprises gene expression profiles of more than 2000 genes in seven different developmental stages ranging from early embryonic to adult. The atlas serves as a framework to explore when and where genes are activated or deactivated as the mouse brain develops.

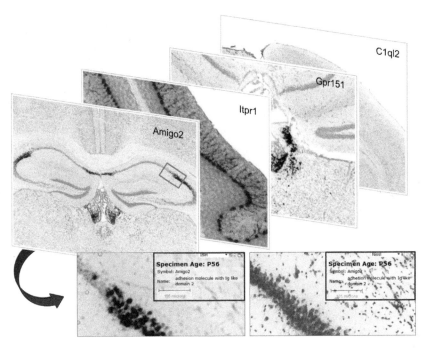

FIGURE 6.1 In situ hybridization images from the Allen Mouse Brain Atlas.

Genes *Amigo2*, *Itpr1*, *Gpr151*, and *C1ql2* are markers for specific anatomic regions such as subregions of the hippocampus (*Amigo2*) or cerebellum (*Itpr1*). Images can be launched in an independent viewer together with Nissl histological stains and expression-labeled masks, as shown in the inset. *Itpr1* has a key role in determining synaptic plasticity and is often disrupted in schizophrenia (Tsuboi et al., 2015).

A central principle of modern neuroscience research is that the nervous system is a network of diverse types of neurons and supporting cells communicating with each other, mainly through synaptic connections. To further build on previous gene expression studies, the Allen Mouse Brain Connectivity Atlas examines the neural connections in the adult mouse brain (Oh et al., 2014). The Allen Mouse Brain Connectivity Atlas has the desired features summarized in a recent mesoscale connectome position essay (Bohland et al., 2009): brain-wide coverage, validated and extensible experimental techniques, a single standardized data format, a quantifiable and integrated neuroinformatics resource, and an open-access public online database. Thus, the Allen Mouse Brain Connectivity Atlas may provide a foundational resource for structural and functional investigations into brain circuits that underlie behavioral and cognitive processes and into diseases that affect these networks. The atlas contains axonal projections mapped from approximately 300 regions of the adult mouse brain and diverse neuronal populations defined by approximately 100 transgenic mice genetically engineered to target specific cell types (Madisen et al., 2015). Complementing these atlases associated with the mouse brain, the Allen Spinal Cord Atlas (http://mousespinal.brain-map.org/) is a genome-wide gene expression map of the juvenile P4 and adult mouse P56 spinal cord. Similar to other mouse gene expression atlases, the Allen Spinal Cord Atlas uses ISH, which covers all anatomic segments of the spinal cord.

COMMON REFERENCE SPACE IN THE MOUSE BRAIN

Common coordinate frameworks are important for meaningful access and interpretation of data. The Allen Reference Atlas (Dong, 2008) was originally created for the Allen Mouse Brain Atlas. It was designed to serve as a reference resource for gene expression pattern comparisons to neuroanatomical structures and to provide standard neuroanatomical ontology structural labels. It was also intended to serve as a template for the development of 3D computer graphic models of the mouse brain and automated informatics annotation tools. Since its initial construction, it has been refined and leveraged in other Allen Institute for Brain Science resources, particularly the Allen Mouse Brain Connectivity Atlas, and will be used in future work involving adult mouse systems.

The center of the common reference space and the automated data processing informatics pipeline is an annotated 3D reference space based on the C57Bl/6J mouse brain, which is the same brain specimen used for the coronal Allen Reference Atlas (Dong, 2008). A reference brain was sectioned to span a nearly complete specimen, resulting in 528 Nissl-stained sections that were each 25 µm thick. A brain volume was reconstructed from the images using a combination of high-frequency section-to-section histology registration with low-frequency histology and (ex cranio) magnetic resonance imaging (MRI) registration (Yushkevich et al., 2006). This first stage of reconstruction was then aligned with a reconstructed sagittally sectioned specimen to obtain the correct mid-sagittal alignment. Using the Allen Reference Atlas as a guide, more than 800 structures were extracted from the 2D coronal reference plates (132 plates, 100 µm apart) and interpolated to create symmetric 3D annotations (see Fig. 6.5 for an example). This atlas provides a 3D extension of the Allen Reference Atlas and can also be used to interface with existing atlases such as Paxinos (Paxinos and Franklin, 2012) and Waxholm space (Johnson et al., 2010).

The reference atlases and 3D models provide a framework for which gene expression data and axonal projections from the Allen Brain Atlas resources containing the adult P56 mouse can be automatically annotated and mapped into the same coordinate space. In the Allen Mouse Brain Atlas, an automated informatics pipeline (Ng et al., 2007) was developed to process the images containing the expression of approximately 20,000 genes, thereby allowing for online anatomic structural search, visualization, and data mining of ISH data. Using the same architectural pipeline, new informatics modules were developed to detect and map axonal projections. Unlike the brightfield images of the Allen Mouse Brain Atlas and Allen Developing Mouse Brain Atlas, data from the Allen Mouse Brain Connectivity Atlas contain fluorescent images of axonal projections targeting different anatomic regions or various cell types. This pipeline consists of modules similar to each of the other pipelines, including routines for image registration, signal segmentation and quantification, and data presentation. Although the modules are different, the output of the pipeline is similar to that of the Allen Mouse Brain Atlas in that quantified signal values are obtained at a grid voxel level and annotated according to the Allen Reference Atlas ontology.

In 2012, the Allen Institute launched a 10-year plan to investigate how the brain works. The plan includes cataloguing different kinds of individual cells in the brain, understanding the relationships between those different kinds of cells, comprehending how information is encoded and decoded by the brain, and modeling how the entire brain processes and computes information. To facilitate this effort, a next-generation Common Coordinate Framework (CCF) was needed to support integration of data generated as part of this plan. Version 3 (v3) of the CCF is based on a 3D, 10-µm, isotropic, and highly detailed population average of 1675 specimens. Currently, CCF v3 consists of 185 newly drawn structures in 3D: 123 subcortical structures, 41 fiber tracts (plus ventricular systems), and 21 cortical regions including primary visual and higher visual areas. The final product of CCF will consist of approximately 300 gray matter structures, cortical layers, approximately 80 fiber tracts, and ventricle structures in 3D. The Allen Reference Atlas ontology (Dong, 2008) was utilized in the annotation and updated to include higher visual areas.

INTEGRATED SEARCH AND VISUALIZATION IN THE ALLEN BRAIN ATLAS

All data associated with Allen Brain Atlas resources have been indexed and can be searched. The data span several different modalities, including colorimetric ISH, dual fluorescence ISH, histological staining, microarray, RNA sequencing, and microarray data. The list of indexed items includes:

- Genes
- Image series/experiments
- Anatomic regions/structures
- Reference atlases
- Content
- Documentation keywords

Images from ISH are grouped into "image series" that comprise an experiment. Experiments consist of slides from the same specimen and that receive the same treatment, whether it is Nissl staining or ISH with a riboprobe for a particular gene. Some ISH data sets include anatomic reference atlases to provide a structural framework for the location of gene expression.

Basic visualization of one or more experiments is standardized across all mouse resources. At the core of this is the Experiment Image Viewer (see Fig. 6.2), which allows for zooming and panning of high-resolution images. A large number of experiments along with the corresponding reference atlases used in the data set can be selected for simultaneous viewing. A drag-and-drop feature allows for reordering of experiment windows. The number of columns of experiments displayed and window size can also be customized. When working with such large data sets, it is critical to be able to simultaneously view similar neuroanatomical regions across many experiments. A cross-plane and cross-timepoint–based "synchronize" feature in this viewer allows for all experiments displayed in the window to be zoomed to the same approximate zoom level and position in the brain based on linear alignment of images to the corresponding reference atlases. These features allow users to efficiently compare and analyze many experiments simultaneously.

As a search example, consider the dopamine receptor *Drd2*. This G-protein-coupled receptor is a target for dopamine agonists, is a major neurotransmitter in the human brain, and is associated with schizophrenia and other diseases. In particular, schizophrenia, for example, is often treated by blocking the type 2 dopamine receptors, and this gene has been a target for pharmaceuticals (Watanabe et al., 2015). Search for *Drd2* in the Allen Mouse Brain Atlas and launch the detailed viewer for the coronal image series; this produces the screenshot shown in Fig. 6.2. Searching by structure or fine annotation category for striatum will return this gene among the highest returns as well, due to its specificity for expression in the striatum.

In addition to different visualization mechanisms available in the atlases, another important component of a large data set is the ability to search and mine data. Because of inherent differences in data modalities between gene expression and axonal projection data, search constructs may vary by atlas

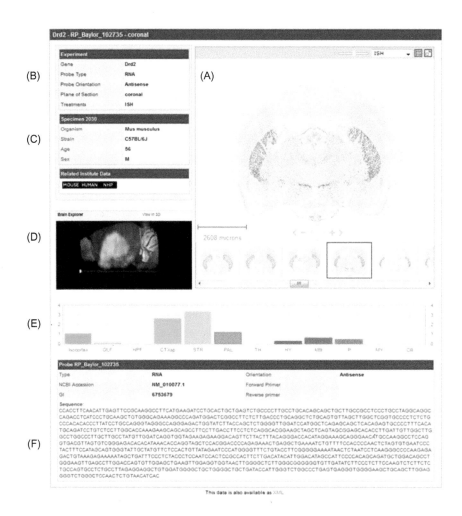

FIGURE 6.2 Detailed image viewer in the Allen Mouse Brain Atlas.

Coronal images series for *Drd2* in the Allen Mouse Brain Atlas showing (A) an individual image plate chosen from an image series. Gene expression is apparent in the putamen of the striatum. (B) Probe and section detail, (C) strain and age information, (D) small thumbnail of 3D gene expression viewer (this image can be rotated), (E) general expression profile for 12 major structures of the brain, and (F) detailed probe information are also available in the image viewer.

type. In the gene expression data sets of the Allen Mouse Brain Atlas and Allen Developing Mouse Brain Atlas, the core search is the Gene Search. This search allows for input of one or more genes when the results displayed are based on experiments available for those genes. The Differential Search feature allows users to search based on enriched gene expression in particular structures of the brain in contrast to other structure(s) entered. The Differential Search is a real-time search service whereby

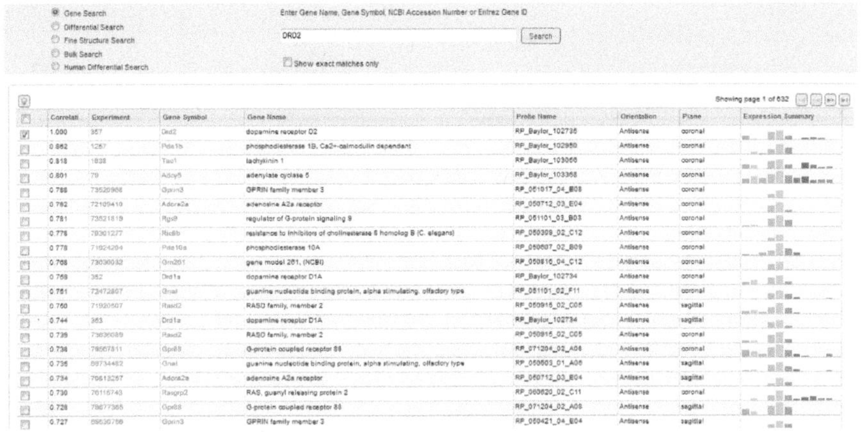

FIGURE 6.3 Detailed correlative search return.

Image series that are highly correlated with a seed image for *Drd2* are displayed. The correlation is given in the left-most numerical column, followed by experiment number, gene symbol, gene name, and, finally, section plane (coronal or sagittal) and expression summary for major structures.

gene expression calculations of fold change over all experiments for the particular atlas are computed. The result of the search displays experiments sorted by highest to lowest fold change between the target structure(s) and contrast structure(s).

When using gene expression databases, searching by structure or other keywords is sometimes not as powerful as "search by example." This feature is also highly desirable because genes with similar expression patterns may be related in function. The Correlative Search utility will accomplish this function, and this type of search by example facility is also available in the Allen Human Brain Atlas, Allen Developing Mouse Brain Atlas, and the BrainSpan Atlas of the Developing Human Brain. For example, once a gene of interest has been identified, to find other genes with spatial expression profiles similar to the gene of interest, first select the experiment by clicking on the gene name in the search results list. Applying this correlative search option to *Drd2* and setting the domain to be brain-wide will show a list of returns as in Fig. 6.3. The user sees that the experiment loads in the panel on the right side of the results list. Next, the brain structure(s) for which the user would like to see a similar expression pattern is selected from the drop-down menu. All data or just the coronal data can be searched by selecting the "Search Coronal Data Only" box prior to selecting "Search." This action will return experiments with a similar expression profile as brain region(s) of interest. The top returns for correlative search with *Drd2* show remarkable anatomic similarity to the seed image pattern. In Fig. 6.4, panels B–E show high spatial correlates of the image in panel A; for example, tachykinn, precursor 1 (*Tac1*) has $\rho = 0.818$. Gprin family member 3 (*Gprin3*, 0.788), phosphodiesterase 1B, calmodulin-dependent (*Pde1b*, 0.862), and adenylate cyclase 5 (*Adcy5*, 0.801) are the top correlates shown in the left column of Fig. 6.3. The closest section from the Allen Reference Atlas in the coronal plane is also shown. Of the top 20 returns, 10 have citations in the literature related to schizophrenia.

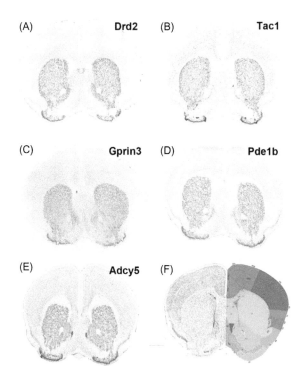

FIGURE 6.4 Individual genes from correlative search with *Drd2*.

Each of the genes displayed has high correlation with the seed *Drd2* shown in (A). Although all of these genes have a search pattern highly correlated with *Drd2*, only two genes, *Tac1* and *Gprin3*, have known associations with schizophrenia in the literature. The gene expression patterns are strikingly similar with their high spatial correlations. (F) Individual plate from the Allen Reference Atlas from a nearby section.

Similar to standardized features of the 2D visualization, basic navigation and 3D visualization in the downloadable Brain Explorer (Lau et al., 2008) are also standardized. In this desktop application, one can view spatially registered data from the adult mouse, developing mouse, and connectivity. There are three panes in the Brain Explorer window. The main pane shows the 3D brain anatomy and reconstructed experimental data at the voxel level (Fig. 6.5). Informatics values, location of voxels, and corresponding pointers to the original 2D images from the web application are given when selecting a voxel for a particular experiment. The upper right pane shows the structures of the ontology in hierarchical or alphabetical order of the corresponding reference atlas. Because these image panes are interconnected, users can control what structure(s) and gene expression voxel data for experiments are displayed in the main window. The third and final pane in the lower right shows which experiment(s) is currently displayed in the main pane. Within this pane, basic zoom, pan, and rotate functionalities can be found. For the Allen Developing Mouse Brain Atlas, multiple experiments at different timepoints can also be viewed simultaneously. Fig. 6.5 shows *Drd2* for the adult C57Bl/6J mouse in the Brain Explorer application.

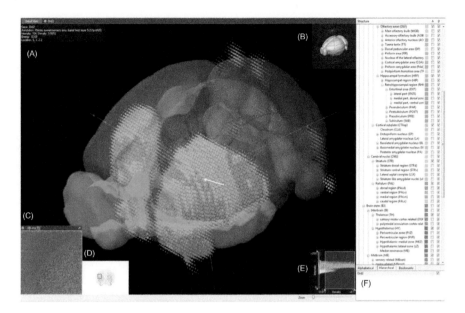

FIGURE 6.5 3D viewing of gene expression in the Brain Explorer.

The downloadable Brain Explorer application is tightly integrated with the online application, and gene expression profiles in 3D can be easily viewed and navigated. *Drd2* is shown with prominent striatal expression and the cursor is centered on the green dot. (A) Experimental details similar to Fig. 6.2B and 6.2C. (B) 3D common coordinate framework used for rotation, translation, and geometric navigation. (C) Gene expression heatmap shows expression profile near the white arrow. (D) Closest 2D section with original ISH image is displayed. (E) Expression threshold controls for changing resolution. (F) Anatomic structures from the Allen Reference Atlas are displayed in a hierarchical organization.

THE MOUSE DEVELOPMENTAL ATLAS

The Allen Developing Mouse Brain Atlas served to provide characterization of gene expression in the brain beginning from mid-gestation through juvenile/young adult (Thompson et al., 2014). Building on the foundation established by the Allen Mouse Brain Atlas, the Allen Developing Mouse Brain Atlas provides a framework to explore temporal and spatial regulation of gene expression, effectively a 4D atlas, with a highly accessible and easily navigable database. The atlas is an extensive data set and resource that provides spatial and temporal profiling of approximately 2100 genes across mouse C57Bl/6J embryonic and postnatal development with cellular-level resolution. Genes were surveyed by high-throughput ISH across seven embryonic and postnatal ages (E11.5, E13.5, E15.5, E18.5, P4, P14, and P28), in addition to P56 data available from the Allen Mouse Brain Atlas (see Fig. 6.6). This developmental survey comprises 18,358 sagittal and 1913 coronal ISH experiments. From a neuroanatomical perspective, the Allen Developing Mouse Brain Atlas defines a number of CNS subdivisions (described in 2D atlas plates and 3D structural models) based on an updated version of the prosomeric model of the vertebrate brain (Puelles and Ferran, 2012; Puelles and Rubenstein, 2003). A novel

CHAPTER 6 EXPLORING NEUROGENOMICS OF SCHIZOPHRENIA

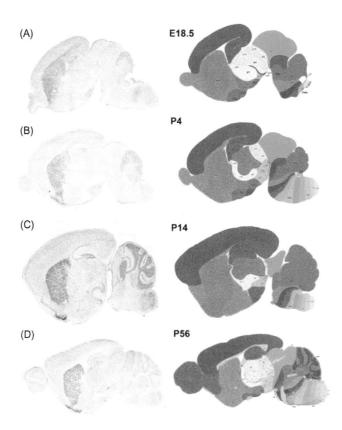

FIGURE 6.6 Gene expression of *Drd2* across developmental stages of the mouse.

(A–D) ISH images shown with yellow counterstain for *Drd2* for four stages: E18.5, P4, P14, and P56. Corresponding plates from the ontogenetic reference atlases are shown on the right. A Lucifer yellow counterstain is applied to the ISH data from P14 and younger. Emergent gene expression in the striatum is seen starting at E18.5.

informatics framework enables navigation of expression data within and across timepoints. In addition to stage-specific novel reference atlases, the resource provides innovative ontogenetic ontology of the full brain with more than 2500 hierarchically organized names and definitions, offering rapid access and a range of visualization and analysis tools.

The stages chosen in the developmental atlas were intended to survey diverse mechanisms, including regional specification, proliferation, neurogenesis, gliogenesis, migration, axon pathfinding, synaptogenesis, cortical plasticity, and puberty. Due to the requirement of surveying seven distinct timepoints, the number of genes profiled was necessarily limited. These include: (1) approximately 800 transcription factors representing 40% of total transcription factors, with nearly complete coverage of homeobox, basic helix-loop-helix, forkhead, nuclear receptor, high-mobility group, and POU domain genes; (2) neurotransmitters and their receptors, with extensive coverage of genes related to dopaminergic, serotonergic, glutamatergic, and GABA-ergic signaling, as well as neuropeptides and their receptors; (3) neuroanatomical marker genes delineating regions or cell types throughout development; (4) genes associated with signaling pathways relevant to brain development, including axon guidance (approximately 80% coverage), receptor tyrosine kinases, and their ligands, and Wnt and Notch signaling pathways; and (5) a category of highly studied genes coding for common drug targets, ion channels (approximately 37% coverage), G-protein-coupled receptors (GPCRs; approximately 7% coverage), cell adhesion genes (approximately 32% coverage), and genes involved in schizophrenia and other neurodevelopmental diseases, which were expected to be expressed in the adult brain or during development.

A new series of reference atlases were also created. The Allen Developing Mouse Brain Reference Atlas was created to provide a novel neuroanatomical framework based on genoarchitectonic data. The reference atlas is effectively seven atlases of different stages of the developing mouse brain with a common developmental ontology. Although the ontologies and structural delineations are different between the adult mouse of the Allen Reference Atlas and the developing stages of the Allen Developing Mouse Brain Reference Atlas, the two atlases share a common 3D volume for P56. This common 3D volume allows for ongoing development to spatially map and integrate the two atlases, thereby allowing cross-query of ontologies and comparisons of anatomical structure delineations. The informatics pipeline, similar to that of the Allen Mouse Brain Atlas, contains modules to preprocess the images, segment, quantitate, and create grid voxel annotation. Both spatial and temporal searches and visualization are possible for data across timepoints. The atlas is also integrated with the 3D Brain Explorer application as shown in Fig. 6.7.

CONNECTIVITY IN THE MOUSE BRAIN

The Allen Mouse Brain Connectivity Atlas is a 3D, high-resolution map of connections in the adult mouse. Axonal projections are traced from defined regions throughout the brain using enhanced green fluorescent protein (EGFP)-expressing adeno-associated viral vectors to trace axonal projections from defined regions and cell types (Fig. 6.8). These are mapped into a common 3D space, described previously, using a standardized platform to generate a comprehensive and quantitative database of inter-areal and cell type-specific projections. By adopting viral and genetic tracing approaches, this atlas offers a reproducible framework to facilitate future efforts in probing circuit functions. Thus, this atlas provides a foundational resource for structural and functional investigations into the brain circuits that underlie behavioral and cognitive processes and into the diseases that affect these networks (Fig. 6.9).

Captured images were processed in an informatics pipeline consisting of a series of modules for preprocessing, alignment, and registration to reference space, signal detection, gridding, and quantification (Oh et al., 2014). Subsequently, each image series was registered into a 3D Allen Reference Atlas model that contains more than 800 annotated brain structures derived from the 2D Allen Reference Atlas coronal plates. After segmentation and registration of all 424 image sets into the common reference 3D space, quantitative values were derived for the segmented signals in each voxel contained within each brain. These values can be used for a variety of analyses of brain circuits and network structure. The standardized projection data set generated and the informatics framework built around it provide a brain-wide, detailed, and quantitative connectivity map that is the most comprehensive to date for any vertebrate species.

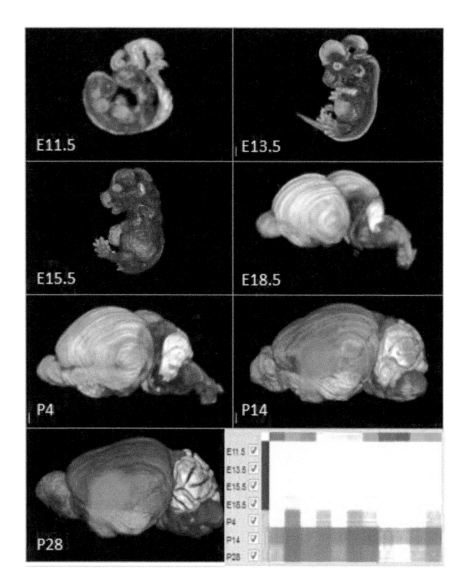

FIGURE 6.7 Three-dimensional (3D) expression summaries and heatmap for the developmental mouse.

Seven developmental stages from E11.5 to P28 are shown for gene expression of *Drd2*, together with the summary heatmap of expression values. The emergence of striatal expression of *Drd2* is apparent. Search and navigation based on these profiles are available online, as are the 3D expression summaries.

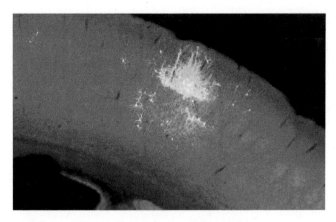

FIGURE 6.8 Local connectivity in the Allen Mouse Brain Connectivity Atlas.

Local connectivity in the somatosensory cortex barrel field of C57Bl/6J mouse is shown with enhanced green fluorescent protein (EGFP)-expressing adeno-associated viral vectors that trace axonal projections from defined regions and cell types. High-throughput serial two-photon tomography was utilized to image the EGFP-labeled axons throughout the brain.

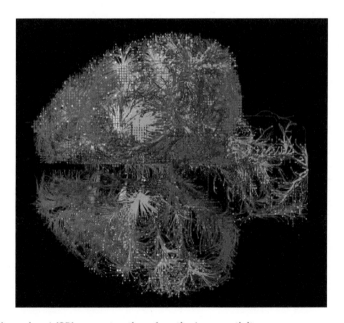

FIGURE 6.9 Three-dimensional (3D) reconstruction of cortical connectivity.

The entire connectivity data set preserves the 3D spatial relationship of different domains, pathways, and topography. This lays the groundwork for large-scale analyses of global neural networks, as well as networks within and between different neural systems, and can also help to further refine the reference 3D framework by adding connectivity information to improve anatomical delineations defined solely by cytoarchitecture and chemoarchitecture.

In addition to being able to view the experimental data within their respective environment in the Brain Explorer, the registration of the Allen Mouse Brain Atlas and the Allen Mouse Brain Connectivity Atlas into the same Allen Reference Atlas allows for visualization within the same environment. This feature is an important component when searching for potential correlations between gene expression and axonal projections. In the Allen Mouse Brain Connectivity Atlas, the basic search is searching for one or more injection sites. One can also browse the experiments based on anatomic locations of the injection sites. The search can be further refined by specifying the afferent structure(s), which limits the search result to experiments that include structures for which projection signals are detected. By default, the search returns experiments that have a projection signal from either hemisphere. This could further be refined by limiting results to a particular hemisphere. A correlative search is also available and allows for search of experiments with projection patterns similar to those of the selected experiment of interest (Fig. 6.10).

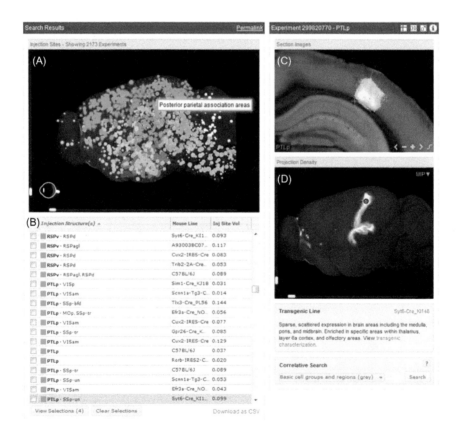

FIGURE 6.10 Allen Mouse Brain Connectivity Atlas.

(A) Search facility showing mapped experiments color-coded by injection site in major structures of the brain. The particular injection site chosen is in the posterior parietal association areas in the cortex (green). (B) List of specific injection sites, associated genetically modified Cre lines, and site volume. (C) Two-photon image of the quantified image injection site in fluorescent green. (D) Three-dimensional reconstruction of projection mapping from injection site into hippocampus and other subcortical areas. Correlative search by projection pattern is also available.

HUMAN ATLAS RESOURCES

In addition to the mouse data sets, there are three large human and nonhuman primate datasets within the Allen Brain Atlas portal. The Allen Human Brain Atlas is a multimodal atlas that maps gene expression in the adult human brain (Hawrylycz et al., 2015; Hawrylycz et al., 2012). The atlas contains genome-wide microarray gene expression sampled from 300 to 900 brain regions in each hemisphere and mapped to the MRI from the same donor brain. High-resolution ISH data are available for selected gene sets of specific brain regions. Complementing the adult human dataset, the BrainSpan Atlas of the Developing Human Brain is a foundational resource for studying transcriptional mechanisms in human brain development. The atlas profiles gene expression in 16 cortical and subcortical structures across the full course of human development using RNA sequencing. More detailed structural sampling is performed for four mid-gestational prenatal male and female specimens using microarrays. Similar to the strategy for ISH in the Allen Human Brain Atlas, the high-resolution ISH for this project is performed for selected sets of genes in developing and adult human brains. Finally, the National Institutes of Health (NIH) Blueprint Non-Human Primate (NHP) Atlas is a data set containing gene expression in the developing rhesus macaque brain consisting of microarray data across prenatal and postnatal development for fine subdivisions of five brain regions; ISH data sets in postnatal development in five major brain regions, whole-brain prenatal ISH for two genes, and serial analysis of selected genes across the entire adult brain; and reference data sets that include developmental stage-specific MRI and Nissl histology. All of these data sets are designed to complement one another when studying human brain development.

THE ALLEN HUMAN BRAIN ATLAS

The Allen Human Brain Atlas is a genome-wide, large-scale, anatomic profiling of six clinically neurotypical adult brains consisting of three Caucasian males, two African American males, and one Caucasian woman (Hawrylycz et al., 2015). For each brain, 500–900 samples spanning one ($n = 4$) or both ($n = 2$) hemispheres were analyzed using whole-genome microarrays. Unlike the fully automated data pipeline developed for the mouse atlases, the Allen Human Brain Atlas data pipeline is semi-automated and is composed of two main parts—the normalization of microarray data samples and the MR processing and registration of microarray data. A custom Agilent microarray containing 60,000 probes was designed and used for this project. This array contains 44,000 Agilent Whole Human Genome probes supplemented with an additional 16,000 probes with at least two different probes (and, when possible, on different exons) available for 93% of genes with Entrez Gene IDs. MRI was performed for each brain used in the data set. All T1-weighted volumes were anatomically segmented, and the entire volume was mapped into the MNI space. To enable unified navigation and visualization of the histology sections, microarray data, and MRI, manual registration was performed for each brain. Results from the registration processes allowed construction of the 2D and 3D MR-based navigation maps where all histology data, gene expression data for a particular gene, and MRI are interconnected. These views are presented within the web application and the Brain Explorer application.

The Allen Human Brain Atlas supports many search and access features common to other atlas resources. In particular, Fig. 6.11 shows the results of a correlation-based search for the human gene *DRD2*. Structures and patient information are given in the bar along the top panel (Fig. 6.11C) where expression values are presented in heatmap form. The genes listed show moderate correlation with the seed microarray series of *DRD2*. The striatum is color-coded purple in this representation and

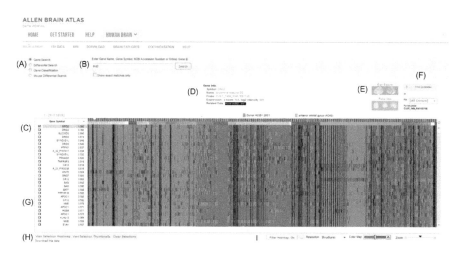

FIGURE 6.11 Search in the Allen Human Brain Atlas.

Example of correlative search in the Allen Human Brain Atlas for *DRD2*. (A) Basic types of gene search available in the atlas, including cross-comparison with mouse atlas data, (B) entry search point, (C) color-coded patient and anatomic structure bars for the six brains and entry seed gene expression, (D) experimental details, (E) 3D visualization control, (F) correlation search entry point, (G) heatmap representation of return results, (H) data download options, and (I) visualization and color controls.

the thalamus is light green. These genes represent the same basic expression structure as their mouse homologs and, in particular, the genes shown in Fig. 6.4.

The genomic scope of the data enables users to perform a variety of quantitative comparisons, for example, in a case study focused on comparing expression patterns for a smaller set of 96 brain regions that were sampled at least twice in at least five brains, pooling across hemispheres. Fig. 6.12 shows the number of genes expressed in all six brains that are differentially expressed between pairs of these regions in at least five of six specimens. This reveals a type of genetic topography of the human brain in which large transcriptional differences are seen between major structures. Heterogeneity within subdivisions is also evident, such as the distinctive patterning of primary visual cortex compared to other cortical regions and complex differentiation of nuclei in the brainstem. In contrast, cortex, cerebellum, and amygdala are notably homogenous across their constituent subdivisions. This representation highlights the magnitude of mesoscale (fine but not cellular resolution) similarities and differences between brain regions. This map, its associated genes, metadata, and anatomic guide are available for dynamic online browsing (http://casestudies.brain-map.org/ggb).

In addition to extensive microarray data, the atlas contains a variety of ISH studies, as summarized in Table 6.1, including large-scale cortical and subcortical profiling and specialized studies of both autism and schizophrenia.

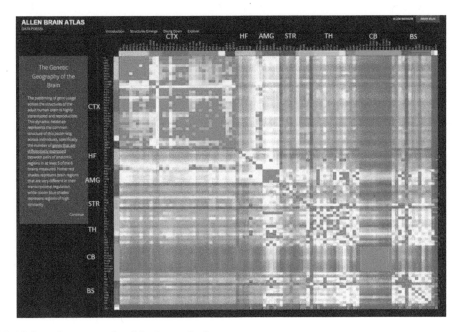

FIGURE 6.12 Genetic topography of the human brain.

Hotter red shades represent brain regions that are very different in their transcriptional regulation, whereas cooler blue shades represent regions of high similarity. Each entry in the consensus map represents the number of differentially expressing genes between a pair of regions. A given gene must be differential between a pair of structures in five out of six sample brains to be counted, and the counts displayed are in logarithmic scale. In constructing the map, samples were first pooled to each anatomic region. Independent sample t-tests were performed to test for the significance of fold change between each pair of regions. These tests were subsequently corrected for multiple hypothesis testing using the Benjamini-Hochberg method.

BRAINSPAN ATLAS OF THE DEVELOPING HUMAN BRAIN

The BrainSpan Atlas of the Developing Human Brain is a foundational resource for studying transcriptional mechanisms involved in human brain development and was formed as a consortium from several institutions (http://www.brainspan.org/static/home) with strong support from the National Institute for Mental Health (National Institutes of Health). A major component of the BrainSpan atlas is genome-wide transcriptional profiling aimed at identification of transcriptional programs differentially active at different stages of brain maturation throughout development. The data set profiles gene expression across 13 developmental stages in 8–16 brain structures with modalities of:

- *Developmental Transcriptome*: RNA sequencing and exon microarray data profiling up to 16 cortical and subcortical structures across the full course of human brain development.

Table 6.1 ISH Gene Expression Studies in the Human Brain

Study	Purpose	Tissue	Gene Selection
1000 gene survey in cortex (Zeng et al., 2012)	Characterize expression of genes from multiple gene classes in two cortical regions	Visual cortex and middle temporal cortex from multiple adult control cases (n = 2–6 per gene)	Multiple genes (1000) from various functional and marker gene classes, including genes related to neuropsychiatric or neurological disease and genes of interest in comparative genomics
Subcortex study	Characterize expression of neurotransmitter system genes in subcortical brain regions	Subcortical regions from front of caudate to just posterior to substantia nigra (left hemisphere); hypothalamus from right hemisphere (n = 2)	Neurotransmitter, receptor, and catabolic and metabolic enzyme genes (55 genes) for GABA and glutamate neurotransmitter system in subcortical regions; anatomic marker genes (10 genes) in hypothalamic tissue
Autism study (Stoner et al., 2014)	Compare cortical microstructure of control and autism cases	Frontal, temporal, and occipital cortical regions from young postmortem control and autism cases (n = 11 per condition)	Lamina-specific and cell type-specific molecular markers, including markers for neurons, glial, and risk genes for autism (25 genes selected from an initial panel of 64 genes)
Schizophrenia study (Guillozet-Bongaarts et al., 2014)	Compare controls and schizophrenics for differences in expression level or pattern in a brain region postulated to be affected in schizophrenia	Dorsolateral prefrontal cortex from control and schizophrenia cases (n = 33 and 19, respectively)	Schizophrenia candidate genes culled from the literature, and cell type and cortical layer markers (60 genes)

Several ISH experimental studies focusing on cortical and subcortical regions, autism genes, and schizophrenia genes were completed. An additional study of the neurotransmitter class of genes is also available.

- *Prenatal Laser Microdissection (LMD) Microarray*: High-resolution neuroanatomical transcriptional profiles of approximately 300 distinct structures spanning the entire brain for four mid-gestational prenatal specimens (Miller et al., 2014).
- *ISH*: High-resolution ISH image data covering selected genes and brain regions in developing and adult human brain.
- *Reference Atlas*: Full-color, high-resolution, anatomic reference atlases of prenatal and adult human brain. Full-color high-resolution web-based, digital reference atlases have been created for 15 pcw (early mid-prenatal) whole brain, 21 pcw (mid-prenatal) cerebrum, 21 pcw brainstem, and adult whole brain.

Most features for search and access are available in the BrainSpan atlas, except for those dependent on detailed spatial mapping of data. Fig. 6.13 shows *DRD2* expressed in the ventral striatum in a 29-year-old man. The reference atlases allow users to directly compare gene expression patterns to an annotated atlas, shown in Fig. 6.13, for the striatum region in a 34-year-old woman.

HUMAN ATLAS RESOURCES | 101

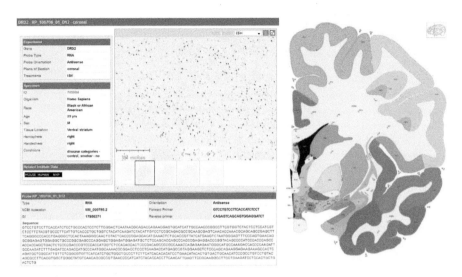

FIGURE 6.13 BrainSpan atlas expression viewer.

(Left) Similar to other atlases, the BrainSpan atlas offers an image viewer for detailed experimental browsing. *DRD2* is shown in the ventral striatum of a 29-year-old man. Detailed image viewers and expression detection maps are also available. (Right) Reference atlas plate from 34-year-old woman showing detail in the striatum region (olive).

NIH BLUEPRINT NON-HUMAN PRIMATE ATLAS

Although brain development is largely conserved across mammals (Workman et al., 2013), specific features are found only in primates and not in mouse. Therefore, an important and complementary resource to the Allen Institute mouse atlases is the NIH Blueprint Non-Human Primate Atlas, which consists of a suite of anatomical and molecular data sets intended to provide detailed characterization of nonhuman primate brain development and its underlying transcriptional mechanisms. The principal modalities include:

- Fine anatomical resolution transcriptome time series data generated with a combination of LMD and DNA microarrays. Whole-genome expression data were measured in forebrain regions associated with neurological and neuropsychiatric disorders, including the neocortex (medial prefrontal and visual cortices), hippocampus, striatum, and amygdala.
- Gross structure transcriptional profiling across postnatal development for the same structures.
- Cellular resolution gene expression data generated with a high-throughput ISH platform.
- Anatomical reference data sets consisting of MRI and corresponding densely sampled Nissl stains.

This resource captures transcriptional dynamics across the full span of brain development in discrete progenitor and postmitotic cell populations as they divide, differentiate, migrate, and mature. Developmental timepoints were chosen to correspond to peak periods of neurogenesis for neurons destined for different layers and glial cell types of primary visual cortex (Rakic, 1974) and major developmental epochs after birth (neonate, infant, juvenile, and young adult) (Bakken et al., 2015). The NHP Atlas has many of the same features as the Allen Human Brain Atlas, including heatmaps for individual genes across brain regions and development, correlative search to find genes with similar expression dynamics as the gene of interest, and differential search to identify genes enriched in particular brain regions or ages.

BEYOND THE ATLASES

We have only briefly surveyed the basic resources of the Allen Brain Atlas portal. Although each resource within the Allen Brain Atlas portal is a standalone application, when combined, they provide many opportunities to mine and analyze data in the mouse, human, and nonhuman primate. Additional resources provide capabilities to access data across projects as well as programmatic access to most of the data generated. Several advanced search mechanisms include text-based search and query that can be further refined in the user interface by selecting or deselecting the search facets. Another convenient data discovery mechanism is the "Related Data" present within each of the atlases. As with most large-scale databases, the ability to programmatically access the data is critical for large-scale data mining and analysis. The application programming interface (API) of the Allen Brain Atlas enables all data within the portal to be accessed and downloaded. Implemented as a series of REST-based web services (http://en.wikipedia.org/wiki/Representational_state_transfer), the API contains mechanisms to retrieve raw and normalized expression data, experiment metadata, computed grid-level statistics extracted from images, reference 3D models, and reference atlas ontologies. Web services are also available for all text and gene expression differential and correlative searches that are available throughout the atlases.

APPLICATIONS OF THE ALLEN INSTITUTE FOR BRAIN SCIENCE RESOURCES IN SCHIZOPHRENIA RESEARCH

Schizophrenia is highly heritable, but genes do not explain the risk of manifestation alone. A variety of environmental factors, such as prenatal environmental influences, an upbringing in an urban environment, cannabis use, and other factors, have been identified to influence the risk of disease manifestation. Consequently, epigenetic mechanisms as a potential point of convergence for both genetic and environmental noxae have been hypothesized to play a major role in the pathogenesis and pathophysiology of schizophrenia (van Os et al., 2010). Brain mapping tools that include spatial information about neural gene expression patterns are of high interest for the field of schizophrenia research and can be used in multiple applications.

Recent genome-wide association studies enrolling large cohorts have identified more than 100 SNPs associated with schizophrenia (Schizophrenia Working Group of the Psychiatric Genomics Consortium, 2014) (see also Chapter 3). Also, an increasing number of copy number variants (CNVs) that contribute to vulnerability have been recently identified (Gratten et al., 2014). However, little is known about the biological function and the expression patterns of some of the identified genes. The Allen Brain Atlas can help to characterize gene expression patterns of interest throughout the brain and, thus, can help to understand their relation to brain regions involved in schizophrenia.

The Allen Human Brain Atlas, offering genome wide broad spatial profiling of transcription in the brain, has made feasible exciting new studies that incorporate imaging data and gene expression. Because the coordinates of all microarray probes in the atlas are available in standard MNI space, the resources of the Allen Human Brain Atlas can be utilized in the analysis of imaging data sets. A recent analysis of gene expression patterns correlated with resting state networks proved that such an approach was feasible and potentially fruitful (Richiardi et al., 2015). In this study, coordinates of the microarray samples were mapped to several resting state networks. Transcriptional similarity of the resulting gene expression profiles was used to identify genes with correlated expression within the neural networks. In this way, a set of 136 genes was identified with a significant enrichment of ion channels. In a subsequent analysis using the Allen Mouse Brain Connectivity Atlas, the authors demonstrated that brain regions, in which 57 mouse orthologues of the previously identified genes were enriched, showed significantly higher axonal connectivity than expected by chance.

Although the study certainly does not deal with schizophrenia itself, it illustrates the enormous potential of the Allen Brain Atlas. Regarding schizophrenia patients, an enormous wealth of imaging studies exists in the canonical literature. Consequently, these resources can be used to identify gene expression patterns in brain regions identified by functional or structural imaging studies. Although the tissue in the Allen Human Brain Atlas was derived from neuropsychiatrically healthy donors, observer-independent hypotheses of potentially affected molecular pathways can thus be generated. As also demonstrated by Richiardi and colleagues, the Allen Mouse Brain Atlas can be used to check and corroborate the findings in humans.

Schizophrenia is largely perceived as a neurodevelopmental disorder, with changes in brain structure and function beginning long before the actual onset of symptoms (Insel, 2010). In this context of schizophrenia as a life-long disease, gene expression patterns over the course of brain development become relevant. The BrainSpan Atlas of the Developing Human Brain provides valuable insight into the trajectories of gene expression patterns throughout brain maturation. A better characterization of the temporal dynamics and the spatial expression profiles of genes involved in the pathogenesis in schizophrenia will help to further our understanding of the underlying molecular pathologies.

Although the aforementioned atlases rely on the brains of neuropsychiatrically healthy donors, the Allen Institute for Brain Science also had generated data from schizophrenia patients. Gene expression patterns of the dorsolateral prefrontal cortices of schizophrenia patients are also available within the data sets that can be used to explore gene expression patterns within brain tissues of actual patients (Guillozet-Bongaarts et al., 2014).

With an increasing characterization of the genetic architecture of schizophrenia, animal models have become an essential tool to identify behavioral phenotypes and molecular pathways involved in the pathogenesis. Mouse models are still among the most used animal models in this field of research (see Chapter 20). The Allen Mouse Brain Atlas resources provide a valuable tool to identify brain regions with high expression of a given gene at high spatial resolution. Consequently, experiments can be designed accordingly. Brain regions with high expression patterns of a given gene can be chosen for dissection and subsequent molecular analyses. Moreover, hypotheses based on these brain regions can be formulated.

The idea of schizophrenia as a "disorder of dysconnectivity" is an influential hypothesis with regard to neural systems level pathophysiology (Stephan et al., 2009). This hypothesis postulates that schizophrenia symptoms are the result of an abnormal (both hyper- and hypo-) connectivity throughout the brain. With the Allen Mouse Brain Connectivity Atlas, a detailed characterization of axonal connectivity in the mouse brain is available for the researcher. As demonstrated by Richiardi et al. (2015), this data set can be used to explore axonal connectivity of brain regions with high expression of schizophrenia susceptibility genes.

In addition to rodents, nonhuman primates are emerging as a fit model organism for psychiatric research in general and specifically for schizophrenia (Nelson and Winslow, 2009). Although it should be noted that this field is still in its infancy, the NIH Blueprint Non-Human Primate Atlas provides anatomical, cellular, and transcriptomic data.

In their entirety, the freely accessible resources of the Allen Institute provide an enormous richness of data. An increased utilization of these data sets will help to further our understanding of schizophrenia and increase the explanatory power of the single original study.

ACKNOWLEDGMENTS

The authors thank P.G. Allen and J. Allen for their vision, encouragement, and support throughout the development of this work. In addition, we acknowledge contributions from many at the Allen Institute for Brain Science as well as from our collaborators and Advisory Council members who were instrumental in the success of the discussed projects. The BrainSpan atlas project was supported by Award Number RC2MH089921 (PIs: Ed Lein and Michael Hawrylycz, Allen Institute for Brain Science) from the National Institute of Mental Health. The National Institutes of Health Blueprint project described was supported by contract HHSN-271-2008-0047 from the National Institute of Mental Health. Its contents are solely the responsibility of the authors and do not necessarily represent the official views of the National Institutes of Health or the National Institute of Mental Health.

REFERENCES

Bakken, T.E., Miller, J.A., Luo, R., et al., 2015. Spatiotemporal dynamics of the postnatal developing primate brain transcriptome. Hum Mol Genet. 24, 4327–4339. http://dx.doi.org/10.1093/hmg/ddv4166. Epub 2015 May 4327.

Bohland, J.W., Wu, C., Barbas, H., et al., 2009. A proposal for a coordinated effort for the determination of brainwide neuroanatomical connectivity in model organisms at a mesoscopic scale. PLoS Comput Biol 5, e1000334.

Dong, H.W., 2008. The Allen Reference Atlas: A Digital Color Brain Atlas of the C57BL/6J Male Mouse, first ed John Wiley & Sons, Inc, Hoboken, NJ.

Gratten, J., Wray, N.R., Keller, M.C., et al., 2014. Large-scale genomics unveils the genetic architecture of psychiatric disorders. Nat Neurosci. 17, 782–790. http://dx.doi.org/10.1038/nn.3708. Epub 2014 May 1027.

Guillozet-Bongaarts, A.L., Hyde, T.M., Dalley, R.A., et al., 2014. Altered gene expression in the dorsolateral prefrontal cortex of individuals with schizophrenia. Mol Psychiatry. 19, 478–485. http://dx.doi.org/10.1038/mp.2013.1030. Epub 2013 Mar 1026.

Hawrylycz, M., Miller, J.A., Menon, V., et al., 2015. Canonical genetic signatures of the adult human brain. Nat Neurosci. 18, 1832–1844. http://dx.doi.org/10.1038/nn.4171 Epub 2015 Nov 1816.

Hawrylycz, M.J., Lein, E.S., Guillozet-Bongaarts, A.L., et al., 2012. An anatomically comprehensive atlas of the adult human brain transcriptome. Nature 489, 391–399.

Insel, T.R., 2010. Rethinking schizophrenia. Nature. 468, 187–193. http://dx.doi.org/10.1038/nature09552.

Johnson, G.A., Badea, A., Brandenburg, J., et al., 2010. Waxholm space: an image-based reference for coordinating mouse brain research. NeuroImage 53, 365–372.

Lau, C., Ng, L.L., Thompson, C.L., et al., 2008. Exploration and visualization of gene expression with neuroanatomy in the adult mouse brain. BMC Bioinformatics 9, 153–163.

Lein, E., Hawrylycz, M., Ao, N., et al., 2007. Genome-wide atlas of gene expression in the adult mouse brain. Nature 445, 168–176.

Lencz, T., Malhotra, A.K., 2015. Targeting the schizophrenia genome: a fast track strategy from GWAS to clinic. Mol Psychiatry. 20, 820–826. http://dx.doi.org/10.1038/mp.2015.1028. Epub 2015 Apr 1014.

Madisen, L., Garner, A.R., Shimaoka, D., et al., 2015. Transgenic mice for intersectional targeting of neural sensors and effectors with high specificity and performance. Neuron. 85, 942–958. http://dx.doi.org/10.1016/j.neuron.2015.1002.1022.

Miller, J.A., Ding, S.L., Sunkin, S.M., et al., 2014. Transcriptional landscape of the prenatal human brain. Nature 508, 199–206. http://dx.doi.org/10.1038/nature13185. Epub 12014 Apr 13182.

Nelson, E.E., Winslow, J.T., 2009. Non-human primates: model animals for developmental psychopathology. Neuropsychopharmacology. 34, 90–105. http://dx.doi.org/10.1038/npp.2008.1150. Epub 2008 Sep 1017.

Ng, L.L., Pathak, S.D., Kuan, C.L., et al., 2007. Neuroinformatics for genome-wide 3D gene expression mapping in the mouse brain. IEEE Trans Comput Biol Bioinform 4, 382–393.

Ng, L.L., Sunkin, S.M., Feng, D., et al., 2012. Large-scale neuroinformatics for in situ hybridization data in the mouse brain. Int Rev Neurobiol 104, 159–182.

Oh, S.W., Harris, J.A., Ng, L., et al., 2014. A mesoscale connectome of the mouse brain. Nature 508, 207–214. http://dx.doi.org/10.1038/nature13186. Epub 12014 Apr 13182.

Paxinos, G., Franklin, K.B.J., 2012. The Mouse Brain in Stereotaxic Coordinates, fourth ed. Academic Press, New York.

Puelles, L., Ferran, J.L., 2012. Concept of neural genoarchitecture and its genomic fundament. Front Neuroanat 6, 47. http://dx.doi.org/10.3389/fnana.2012.00047. eCollection 02012.

Puelles, L., Rubenstein, J.L., 2003. Forebrain gene expression domains and the evolving prosomeric model. Trends Neurosci 26, 469–476.

Rakic, P., 1974. Neurons in rhesus monkey visual cortex: systematic relation between time of origin and eventual disposition. Science. 183, 425–427.

Richiardi, J., Altmann, A., Milazzo, A.C., et al., 2015. Brain networks. Correlated gene expression supports synchronous activity in brain networks. Science 348, 1241–1244. http://dx.doi.org/10.1126/science.1255905. Epub 1252015 Jun 1255911.

Ripke, S., O'Dushlaine, C., Chambert, K., et al., 2013. Genome-wide association analysis identifies 13 new risk loci for schizophrenia. Nat Genet. 45, 1150–1159. http://dx.doi.org/10.1038/ng.2742. Epub 2013 Aug 1125.

Roussos, P., Giakoumaki, S.G., Zouraraki, C., et al., 2015. The relationship of common risk variants and polygenic risk for schizophrenia to sensorimotor gating. Biol Psychiatry 27 00526-00520.

Stephan, K.E., Friston, K.J., Frith, C.D., 2009. Dysconnection in schizophrenia: from abnormal synaptic plasticity to failures of self-monitoring. Schizophr Bull. 35, 509–527. http://dx.doi.org/10.1093/schbul/sbn1176. Epub 2009 Jan 1020.

Stoner, R., Chow, M.L., Boyle, M.P., et al., 2014. Patches of disorganization in the neocortex of children with autism. N Engl J Med. 370, 1209–1219. http://dx.doi.org/10.1056/NEJMoa1307491.

Sunkin, S.M., Ng, L., Lau, C., et al., 2012. Allen Brain Atlas: an integrated spatio-temporal portal for exploring the central nervous system. Nucleic Acids Res 41, D996–D1008.

Thompson, C.L., Ng, L., Menon, V., et al., 2014. A high-resolution spatiotemporal atlas of gene expression of the developing mouse brain. Neuron. 83, 309–323. http://dx.doi.org/10.1016/j.neuron.2014.1005.1033. Epub 2014 Jun 1019.

Tsuboi, D., Kuroda, K., Tanaka, M., et al., 2015. Disrupted-in-schizophrenia 1 regulates transport of ITPR1 mRNA for synaptic plasticity. Nat Neurosci. 18, 698–707. http://dx.doi.org/10.1038/nn.3984. Epub 2015 Mar 1030.

van Os, J., Kenis, G., Rutten, B.P., 2010. The environment and schizophrenia. Nature 468, 203–212. http://dx.doi.org/10.1038/nature09563.

Wang, Z., Yang, B., Liu, Y., et al., 2015. Further evidence supporting the association of NKAPL with schizophrenia. Neurosci Lett. 605, 49–52. http://dx.doi.org/10.1016/j.neulet.2015.1008.1023.

Watanabe, Y., Shibuya, M., Someya, T., 2015. DRD2 Ser311Cys polymorphism and risk of schizophrenia. Am J Med Genet B Neuropsychiatr Genet. 168B, 224–228. http://dx.doi.org/10.1002/ajmg.b.32303. Epub 32015 Feb 32325.

Workman, A.D., Charvet, C.J., Clancy, B., et al., 2013. Modeling transformations of neurodevelopmental sequences across mammalian species. J Neurosci. 33, 7368–7383. http://dx.doi.org/10.1523/JNEUROSCI.5746-7312.2013.

Yushkevich, P., Avants, B., Ng, L., et al., 2006. 3D mouse brain reconstruction from histology using a coarse-to-fine approach. In: Third International Workshop on Biomedical Image Registration (WBIR), pp. 230–237.

Zeng, H., Shen, E.H., Hohmann, J.G., et al., 2012. Large-scale cellular-resolution gene profiling in human neocortex reveals species-specific molecular signatures. Cell. 149, 483–496. http://dx.doi.org/10.1016/j.cell.2012.1002.1052.

PART III

THE NEUROCHEMICAL BASIS OF SCHIZOPHRENIA

CHAPTER 7

THE DOPAMINE HYPOTHESIS OF SCHIZOPHRENIA: CURRENT STATUS

G. Gründer[1] and P. Cumming[2]

[1]*Department of Psychiatry, Psychotherapy and Psychosomatics, RWTH Aachen University, Aachen, Germany*
[2]*School of Psychology and Counselling, Queensland University of Technology, and QIMR Berghofer Medical Research Institute, Brisbane, QLD, Australia*

CHAPTER OUTLINE

Introduction: History of the Dopamine Hypothesis 109
Dopamine Receptor Occupancy Studies 110
Molecular Imaging Studies of Dopaminergic Neurotransmission in Schizophrenia 112
Prefrontal–Subcortical Dopamine Dysregulation 116
Ketamine Psychosis 118
Conclusion: Beyond the Dopamine Hypothesis 118
References 119

INTRODUCTION: HISTORY OF THE DOPAMINE HYPOTHESIS

The "dopamine hypothesis" of schizophrenia arose from the serendipitous discovery by Jean Delay and Pierre Deniker in 1952 of the antipsychotic effects of chlorpromazine, first developed as a presurgical sedative (Delay et al., 1952). Their findings in a psychiatric population were soon reproduced in at least 10 clinical studies conducted in subsequent years (Cowden et al., 1955). Carlsson and Lindqvist later found that chlorpromazine and the newly developed antipsychotic haloperidol increase the concentration of dopamine metabolites in mouse brain, without altering the dopamine concentration, leading to their proposal that blockade of dopamine neurotransmission relieves psychotic symptoms (Carlsson and Lindqvist, 1963). This was the basis for the formulation of the hypothesis that excessive dopamine transmission represents a core feature of schizophrenia (Van Rossum, 1966). Administration of high doses of amphetamine can produce in normal individuals an acute psychosis that is indistinguishable from the paranoid subtype of schizophrenia and that is rapidly ameliorated by antipsychotic treatment (Angrist et al., 1974); given that amphetamine is a powerful releaser of catecholamines in brain, this clinical observation lent further support to the dopamine hypothesis, which soon became the most popular theory for the integration of diverse biological and clinical findings in schizophrenia research (Meltzer and Stahl, 1976).

The dopamine hypothesis in its original formulation was implicitly confined to dopaminergic transmission in the basal ganglia, but a more current conceptualization links the positive symptoms of schizophrenia to excess dopamine transmission in subcortical brain structures, especially in the striatum, while associating negative symptoms and cognitive deficits with a dopaminergic impairment in the prefrontal cortex (Davis et al., 1991; Weinberger, 1987). However, substantiation of dopaminergic abnormalities in schizophrenia was initially hindered by a paucity of neurochemical data. Postmortem analyses did not reveal any significant differences in cerebral concentrations of dopamine or its metabolites between controls and patients dying with schizophrenia (Winblad et al., 1979), although others found increased dopamine levels in parts of striatum (Crow et al., 1979). Whereas some homogenate binding studies have suggested altered dopamine receptor or transporter expression, anatomically defined autoradiographic investigations did not reveal any such changes in brains from patients (Knable et al., 1994). Well-known caveats and confounds in such studies include treatment history and substantial postmortem delays, resulting in excessive variability in the neurochemical endpoints. Ex vivo studies in cerebrospinal fluid may provide a more reliable marker of central dopamine turnover, with the caveat that the main dopamine metabolite homovanillic acid (HVA) in the lumbar cerebrospinal fluid arises from cerebral cortex and the basal ganglia; nonetheless, a concordance of several such studies showed a positive correlation between HVA levels and severity of positive symptoms in the absence of a significant group difference relative to healthy controls (Maas et al., 1997). Neuroendocrine challenges present another channel for probing brain dopamine in schizophrenia. In one such study, the prolactin response to the dopamine agonist apomorphine was blunted in a group of untreated schizophrenia patients (Duval et al., 2003); others note that findings of abnormalities in the HPA axis are confounded by nonspecific aspects of psychosocial stress and agitation (Belvederi Murri et al., 2012). In general, these findings emphasize the need for more direct methods of assessing dopaminergic status in brains of patients living with schizophrenia, and they highlight the confounding effects of treatment on dopaminergic markers.

DOPAMINE RECEPTOR OCCUPANCY STUDIES

It has been known since the 1970s (Seeman et al., 1976) that the efficacy of antipsychotic compounds against positive symptoms of schizophrenia is linked to antagonism of dopamine D_2-like receptors (review in Gründer et al., 2011). However, this link is made inductively, based on observations of the affinity of antipsychotic compounds for receptors expressed in vitro. The advent of molecular imaging with single photo emission computer tomography (SPECT) and positron emission tomography (PET) made possible the assessment of dopamine receptor availability in living brain and the testing of hypotheses related to dopamine neurotransmission in schizophrenia. An early PET study with the dopamine antagonist [^{11}C]NMSP showed elevation of D_2-like receptors in striatum of drug-naïve schizophrenia patients (Wong et al., 1986), and a meta-analysis of molecular imaging studies showed a small but significant 13% increase in the receptor availability in striatum (Weinberger and Laruelle, 2002). A PET study with the D_3-preferring agonist [^{11}C]-(+)-PHNO failed to show any abnormality in striatal or extrastriatal binding in schizophrenia patients (Graff-Guerrero et al., 2009).

In an elegant series of PET studies, Farde and colleagues from the Karolinska Institute in Stockholm measured $D_{2/3}$ receptor occupancy by antipsychotic medications in relation to their clinical effects in schizophrenia; they found that therapeutic doses of first-generation antipsychotics such as haloperidol occupy 65–90% of dopamine $D_{2/3}$ receptors in the human striatum (Farde et al., 1992). Patients with acute extrapyramidal side (EPS) effects were found to have higher mean dopamine D_2 receptor occupancies (82%) than those without such side effects (74%). The studies also suggested a lower threshold of approximately 60–65% occupancy for sufficient treatment response (Nordström et al., 1993), although clinical improvement is not assured for every patient attaining this occupancy threshold. The favorable "therapeutic window" of 65–80% striatal $D_{2/3}$ receptor occupancy also applies for most of the newer antipsychotics. When their doses are raised above a certain threshold, striatal (and potentially extrastriatal) $D_{2/3}$ dopamine occupancy increases to levels that are associated with a higher incidence of EPS (Kapur et al., 1998; Nyberg et al., 1997, 1999). However, some of the second-generation "atypical" antipsychotics have therapeutic occupancies falling outside the therapeutic window. In the case of the partial agonist aripiprazole, 100% occupancy can be obtained in striatum without evoking EPS, whereas clozapine and quetiapine can be effective at lower occupancies, suggesting an action outside the striatum or at non-D_2 binding sites (Gründer et al., 2003, 2006, 2008, 2009; Vernaleken et al., 2010). In a multitracer extension of the occupancy paradigm, the atypical antipsychotic aripiprazole showed 90% occupancy at striatal $D_{2/3}$ sites, less than 60% occupancy at cortical serotonin $5HT_{2A}$ sites, and only 16% occupancy at $5HT_{1A}$ receptors (Mamo et al., 2007). Few drugs are entirely selective, so it must be considered that the balance between clinical benefits and side effects can be obtained by actions at multiple binding sites (Kapur et al., 2000).

Whereas most PET studies of dopamine receptor occupancy have utilized either butyrophenone ($D_{2,3,4}$, ie, D_2-like) or benzamide ligands ($D_{2,3}$), the D_3-preferring agonist [^{11}C]-(+)-PHNO now allows investigation of antipsychotic occupancy at the D_3 sites, which predominate in extrastriatal regions (Girgis et al., 2015). Nonetheless, treatment of patients with schizophrenia with haloperidol, olanzapine, or risperidone leads to markedly reduced [^{11}C]PHNO binding in the caudate/putamen, which must mainly represent D_2 receptors. Even more perplexingly, the same antipsychotic treatment leads to *increased* [^{11}C]PHNO binding in D_3-rich regions, notably globus pallidus (Mizrahi et al., 2011). Although the binding properties of [^{11}C]PHNO are not fully understood, these results suggest that paradoxical upregulation of extrastriatal D_3 sites may be a hitherto unappreciated aspect of antipsychotic treatment.

A dopamine hypothesis of schizophrenia might predict that failure of antipsychotic medication should be attributable to inadequate occupancy in vivo, for example, as a function of dose-dependent pharmacodynamics. However, in a PET study of D_2 receptor occupancy in patients before and after treatment with haloperidol, Wolkin et al. (1989) found comparable occupancy in responders and nonresponders, indicating that substantial occupancy at D_2 sites is not always sufficient for treatment response. This in itself is evidence for pharmacological heterogeneity among schizophrenia patients, of whom approximately 25% fail to respond to high-dose antipsychotic treatment (Agid et al., 2013). We draw an analogy to treatments of high blood pressure in that the mechanistically distinct modes of treatment may not be consistent with a unitary cause for the symptoms.

Despite the exceptions to the therapeutic window for typical antipsychotic medications, the use of dopamine $D_{2/3}$ receptor imaging during treatment with antipsychotics has become a widely accepted standard for aiding drug-dosing decisions. The relationship between doses of antipsychotic drugs and their (striatal) $D_{2/3}$ dopamine receptor occupancy has almost attained the status of a biomarker, an endpoint of proven utility in decision-making in the development of new antipsychotic drugs (Wong et al., 2009), albeit at the risk of misassignment of doses for atypical drugs. However, as noted, $D_{2/3}$ receptor occupancy is not unambiguously related to clinical outcome. Positive symptoms such as hallucinations or delusions, which are effectively controlled by antipsychotics in 75% of patients, represent only one dimension of the schizophrenia phenomenology, albeit an important dimension. Long-term functional outcome in schizophrenia, however, is more determined by improvement in cognitive function rather than control of positive symptoms (Bowie et al., 2006). Because the available antipsychotic medications are of limited efficacy in improving cognitive symptoms (Keefe et al., 2007), $D_{2/3}$ receptor occupancy in striatum is an incomplete marker of symptom control or prediction of disease progression. As such, the demonstrable importance of $D_{2/3}$ receptor occupancy for the control of positive symptoms does lend support for a dopamine hypothesis of schizophrenia, but it also draws attention to the inadequacy of a simple dopamine hypothesis to account for all cases.

MOLECULAR IMAGING STUDIES OF DOPAMINERGIC NEUROTRANSMISSION IN SCHIZOPHRENIA

The molecular imaging occupancy studies described typically entail competition between a radiopharmaceutical and a medication for a common binding site. However, the binding of dopamine, the endogenous agonist ligand, exerts competition against certain structural classes of dopamine D_2 ligands in living brain. Notably, dopamine competes against the benzamide class of ligands for its binding to D_2 receptors present in dorsal striatum and to D_3 receptors, which are relatively abundant in ventral striatum and predominating in certain extrastriatal structures. Since the early 1990s, this paradigm has been applied in PET and SPECT studies to try to reveal the dynamics of synaptic dopamine concentrations (review in Laruelle, 2011). In this model, the availability of radioligand binding sites decreases with increasing levels of dopamine, manifesting in a reduction in the binding potential (BP), which is proportional to the ratio of the abundance of $D_{2/3}$ receptors in the tissue (B_{max}) to the apparent affinity of the radioligand (K_D^{app}); it is this latter microparameter that is influenced by the endogenous neurotransmitter concentration prevailing at the time of the imaging study. It follows that the specific binding measured in vivo is reduced in proportion to the endogenous tonus of dopamine, and that pharmacological depletion of dopamine will unmask the proportion of binding sites normally occupied by dopamine. Conversely, a challenge with a psychostimulant such as amphetamine decreases the observed BP as an index of the increased dopamine release. The in vivo competition paradigm has proven very useful in the elucidation of perturbed dopamine transmission at $D_{2/3}$ receptors in schizophrenia and other neuropsychiatric disorders. Disappointingly, the endogenous competition paradigm fails in the case of dopamine D_1 receptors, and it has not yet been clearly generalized to neurotransmitter systems other than dopamine (ie, in PET studies of $5HT_2$ antagonists) (Quednow et al., 2012; Talbot et al., 2012).

In early reports of the psychostimulant-evoked dopamine release in the brain of healthy human subjects, administration of methylphenidate reduced striatal [^{11}C]raclopride on PET by 23% (Volkow et al., 1994) and reduced [^{123}I]IBZM binding on SPECT by 15% (Laruelle et al., 1995). In both studies, the magnitude of dopamine release inferred from these binding changes correlated with the subjective pleasurable experience of the subjects following psychostimulant administration. Conversely, acute dopamine depletion by treatment with the dopamine synthesis inhibitor α-methyl-para-tyrosine (AMPT) increased the binding of [^{11}C]raclopride or [^{123}I]IBZM in human striatum by an average of 18% and 28%, respectively (Laruelle et al., 1997; Verhoeff et al., 2001). However, the true tonic occupancy of $D_{2/3}$ sites by dopamine is likely higher because complete dopamine depletion would only be obtained with intolerable pharmacological treatments such as reserpine. Nonetheless, the individual increase in $D_{2/3}$ radioligand binding (%ΔBP) after AMPT-induced dopamine depletion correlated with the transiently impaired cognitive performance in healthy volunteers (Verhoeff et al., 2001).

As noted, the in vivo competition paradigm has generally pitted the endogenous agonist dopamine against $D_{2/3}$ antagonist radiotracers, all based on the benzamide structure. However, the general properties of G-protein-coupled receptors predict that agonist ligands may more sensitively detect fluctuations in endogenous dopamine at some subset of receptors in a high-affinity state for dopamine; this phenomenon was first demonstrated in a preclinical study of mice receiving dual injections of the benzamide antagonist [^{11}C]raclopride and the $D_{2/3}$ agonist [^{3}H]-N-propylnorapomorphine under a variety of pharmacological conditions (Cumming et al., 2002). Substantially higher sensitivity of agonist ligands to amphetamine challenge was subsequently confirmed in PET studies with [^{11}C]NPA, [^{11}C]MNPA, and [^{11}C]-(+)-PHNO (Narendran et al., 2010; Seneca et al., 2006; Willeit et al., 2008). Nonetheless, results of other competition studies (Finnema et al., 2009) and saturation binding studies in vitro (Cumming, 2011) challenge the notion that agonist binding sites represent a privileged subset of dopamine receptors.

Pharmacological challenge paradigms were soon utilized in groundbreaking [^{123}I]IBZM SPECT studies of patients with schizophrenia challenged with amphetamine to provoke an acute exacerbation of schizophrenia symptoms. Compared to healthy volunteers, this group of patients showed a significantly greater reduction in $D_{2/3}$ availability as compared to a healthy control group (Laruelle et al., 1996; Abi-Dargham et al., 1998). Breier et al. (1997) confirmed the SPECT finding in a [^{11}C]raclopride PET study of 11 patients with schizophrenia in comparison to 12 healthy subjects. These consistent results of the competition paradigm could be interpreted in several ways, such as an increase in the amphetamine-releasable dopamine pool in nerve terminals, desensitization of presynaptic autoreceptors, or sensitization of the postsynaptic $D_{2/3}$ receptors in patients with schizophrenia. Although these studies were conducted in drug-free patients, the apparent amphetamine-evoked dopamine release normalized in a group of patients with a stable response to antipsychotic medication (Laruelle et al., 1999). As such, greater "lability" of striatal $D_{2/3}$ receptor availability to altered dopamine tonus seems more of a state marker for psychosis.

Using the AMPT dopamine depletion paradigm, Abi-Dargham et al. (2000) also demonstrated a larger increase in radioligand binding in patients with schizophrenia (+19%) compared to matched controls (+9%). Here, the most parsimonious explanation is that tonic synaptic dopamine concentrations are higher in the schizophrenia group, because AMPT seems unlikely to have predominant effects on the affinity state of $D_{2/3}$ receptors. Here, higher magnitude of the individual AMPT-induced increase in [^{123}I]IBZM tracer binding (suggestive of higher basal dopamine occupancy) predicted better subsequent response to antipsychotic treatment (Abi-Dargham et al., 2000). Among those participants who underwent both challenge paradigms (ie, amphetamine and AMPT), there was a positive correlation between the individual amphetamine-induced decrease and the AMPT-induced increase in receptor availability in the patients, but not in the control group (Abi-Dargham et al., 2009). This finding suggests that there is a concurrent state of increased tonic dopamine release and elevated capacity for amphetamine-evoked release in untreated schizophrenia patients. However, the possibility of a type II error (false negative) in the control group cannot be excluded because the lower magnitudes of changes might disfavor the detection of a correlation between the amphetamine and AMPT conditions. Furthermore, the group differences in results of challenge studies, although highly significant, are not pathognomonic of schizophrenia because there is considerable overlap between patients and controls in the individual binding decrease after amphetamine (%ΔBP). A scatter plot of the results suggests the

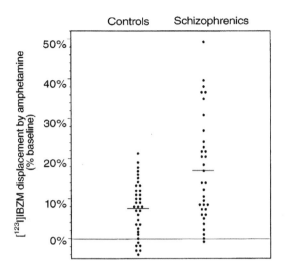

FIGURE 7.1

Amphetamine-induced dopamine release as assessed by reduction of dopamine $D_{2/3}$ receptor availability in 34 patients with schizophrenia compared to 36 healthy controls. Reduction in $D_{2/3}$ receptor availability was measured with single photon emission computed tomography and the $D_{2/3}$ receptor radiotracer $[^{123}I]IBZM$.

From Laruelle, M., Abi-Dargham, A., Gil, R., Kegeles, L., Innis R., 1999. Increased dopamine transmission in schizophrenia: relationship to illness phases. Biol Psychiatry 46:56–72.

presence of two groups of patients: those with amphetamine response in the normal range and those with a super-high response (Fig. 7.1).

Thus, the competition paradigm suggests the presence of neurochemical heterogeneity within the population of patients with schizophrenia, which may have important implications for the dopamine hypothesis. In particular, it might be predicted that the "high displacers" are those most likely to respond to antipsychotic medication, and it might be suggested that individual molecular imaging studies, typically undertaken in groups of 10–20 patients, are underpowered to detect a putative bimodal distribution of the response to pharmacological challenge.

Whereas the spatial resolution of $[^{123}I]IBZM$ SPECT does not allow resolution of anatomic divisions of the striatum, corresponding $[^{11}C]$raclopride PET studies can be used to test whether regional effects of AMPT are specific to certain subregions of the striatum (Kegeles et al., 2010). Although AMPT administration led to a significantly larger increase in $D_{2/3}$ receptor availability in patients (+15%) compared to controls (+10%), this difference was restricted to the associative striatum. No between-group differences were observed in the limbic and sensorimotor striatum. These authors concluded from their study that schizophrenia is associated with tonically elevated dopamine function in the caudate nucleus but not in the limbic striatum, which may question the therapeutic relevance of the mesolimbic selectivity of second-generation antipsychotic drugs (Kinon and Lieberman, 1996). Because the between-group difference was most pronounced in the precommissural dorsal caudate, a structure that processes signals arising from the dorsolateral prefrontal cortex, their observation might also suggest that a limited elevation in subcortical dopamine function in schizophrenia could adversely affect performance of cognitive tasks mediated by the dorsolateral prefrontal cortex (Kegeles et al., 2010).

Despite the evidence for greater vulnerability of dopamine agonist ligands to competition from endogenous dopamine, there have been no such PET studies of the amphetamine challenge paradigm in patients with schizophrenia. One imaging study revealed no regional differences in binding of the D_3-preferring agonist [^{11}C]-(+)-PNHO in untreated patients performing a cognitive task in comparison to healthy controls, but dopamine release could not be inferred from ΔBP due to the lack of a baseline scan in that study (Suridjan et al., 2013). Amphetamine-evoked decreases in [^{11}C]-(+)-PNHO binding were substantially higher in dorsal striatum of pathological gambling patients (Boileau et al., 2014), as was the stress-evoked [^{11}C]-(+)-PNHO binding in patients at risk of schizophrenia (Mizrahi et al., 2014). These studies taken together suggest that perturbed dopamine signaling might not be specific to psychosis, but rather may be a biomarker of aberrant salience attribution.

The competition paradigm reveals the vulnerability of postsynaptic $D_{2/3}$ receptors to task-evoked or pharmacologically evoked changes in occupancy by dopamine, which reflects a composite of presynaptic and postsynaptic factors. The functional state of dopamine neurons could be more specifically assessed using markers for strictly presynaptic elements. A meta-analysis of molecular imaging studies in schizophrenia did not reveal any abnormality in plasma membrane dopamine transporter availability (Fusar-Poli and Meyer-Lindenberg, 2013a), nor was there any evidence of altered levels of vesicular monoamine transporters to [^{11}C]DTBZ PET (Taylor et al., 2000).

The integrity of the biochemical pathway for dopamine synthesis can be assessed from PET studies of the trapping of [^{18}F]FDOPA and other substrates for DOPA decarboxylase, the ultimate enzyme in the pathway. Although the kinetics of [^{18}F]FDOPA metabolism and cerebral trapping is complicated, there is a general agreement that [^{18}F]FDOPA PET gives an index of the dopamine synthesis capacity in brain regions with abundant dopamine innervation (Kumakura and Cumming, 2009). An early [^{18}F]FDOPA PET study showed increased tracer uptake in striatum in a small group of patients with schizophrenia and in epilepsy patients with intraictal psychosis (Reith et al., 1994). The finding in schizophrenia has since been replicated a number of times; meta-analyses have shown a highly significant increase ($P < 0.001$) in the striatal utilization of [^{18}F]FDOPA and other DOPA decarboxylase tracers in a composite group of several hundreds of patients (Fusar-Poli and Meyer-Lindenberg, 2013b; Kambeitz et al., 2014; Fig. 7.2). Unfortunately, the meta-analyses were not fit to test the patient group for bimodality, as suggested by the distribution of %ΔBP in the challenge paradigm. It might be expected that individuals with particularly high %ΔBP in the amphetamine or AMPT challenge paradigm would also have high [^{18}F]FDOPA trapping, but this experiment has yet to be performed.

Although the meta-analyses of [^{18i}F]FDOPA utilization in schizophrenia patients showed a highly significant increase with a moderate to high effect size (Hedges $g = 0.8$), increased dopamine synthesis capacity cannot be considered pathognomonic for schizophrenia because PET scanning would provide false-negative or false-positive results in approximately one-third of cases. Similarly, a meta-analysis of serotonin transporter imaging studies involving patients with major depression revealed a highly significant reduction, which was of a magnitude far too low to be considered diagnostic (Gryglewski et al., 2014). These results indicate that the molecular imaging abnormalities, although real, cannot be considered either necessary or sufficient drivers for the pathophysiology of these psychiatric disorders. Nonetheless, a PET study involving individuals at risk of psychosis (ie, in a putative prodromal state) already had slightly increased [^{18}F]FDOPA uptake, which seems consistent with a neurochemical state imparting vulnerability (Fusar-Poli et al., 2010).

Although [^{18}F]FDOPA PET is of limited sensitivity for cortical measurements, studies using [^{11}C]DOPA have shown parallel elevation in tracer uptake in medial prefrontal cortex of a group of mostly drug-naïve schizophrenia patients (Lindström et al., 1999). This isolated finding was not supported by

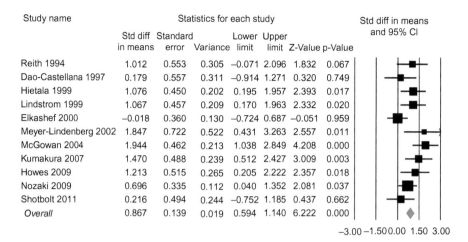

FIGURE 7.2

Meta-analysis of striatal dopamine synthesis capacity in schizophrenia using random effect models. Positive values of Hedges *g* indicate greater dopamine synthesis capacity in patients as compared with controls.

From Fusar-Poli, P., Meyer-Lindenberg, A., 2013b. Striatal presynaptic dopamine in schizophrenia, part II: meta-analysis of [(18)F/ (11)C]-DOPA PET studies. Schizophr Bull 39, 33–42. Referenced studies in the original publication.

a meta-analysis of extrastriatal studies (Kambeitz et al., 2014), and it seems at odds with the extended dopamine hypothesis, which indicates that dopamine transmission should be impaired in frontal cortex. Another isolated PET result involved an extended model of [^{18}F]FDOPA kinetics, which does not assume irreversible trapping of the decarboxylation product [^{18}F]fluorodopamine during PET scans lasting several hours. This approach revealed elevated [^{18}F]fluorodopamine turnover in brains of patients with schizophrenia compared to healthy controls (Kumakura et al., 2007), which suggested what might be termed "churning" of dopamine metabolism (ie, rapid synthesis and rapid catabolism). [^{18}F]FDOPA PET revealed potentiation of striatal tracer uptake in healthy volunteers following an acute challenge with haloperidol, consistent with autoreceptor-mediated regulation of the pathway (Vernaleken et al., 2006). However, treatment of schizophrenia patients with haloperidol for 4–6 weeks decreased striatal dopamine synthesis capacity (Gründer et al., 2003), which we interpreted to provide evidence for depolarization block of mesencephalic dopamine neurons, a phenomenon described in rats, and proposed as a mechanism for the delayed onset of antipsychotic response in humans (Grace, 1992).

PREFRONTAL–SUBCORTICAL DOPAMINE DYSREGULATION

Although the subcortical hyperdopaminergia in psychosis is a well-documented observation, the prefrontal cortical deficit in dopamine transmission is less well-established (Kambeitz et al., 2014). In one PET study, elevated cortical D_1 receptor availability correlated negatively with (impaired) working memory performance (Abi-Dargham et al., 2002). This was first interpreted as a compensatory (but ineffective) mechanism for sustained deficient prefrontal cortical function, and it could later only be reproduced in drug-naïve, but not drug-free, patients (Abi-Dargham et al., 2012). These authors speculated that antipsychotic exposure leads to a normalization of cortical D_1 receptors, and thus an elevation can only be detected in drug-naïve patients. However, Okubo et al. (1997) used a different radiotracer and found a *decrease* in prefrontal D_1 receptor availability in drug-naïve and drug-free patients with schizophrenia, the magnitude of which correlated with severity of negative symptoms and impairment in working memory performance. Clinical evidence and animal studies, however, clearly suggest that the cognitive deficits associated with schizophrenia are related to impaired prefrontal cortical function, and that this cortical dysfunction can be improved by dopamine agonists (Barch and Carter, 2005; Barch and Ceaser, 2012). A meta-analysis of functional imaging studies clearly suggests that activation of the prefrontal cortex is abnormally reduced during working memory tasks in patients with schizophrenia and in subjects with at-risk mental states (Minzenberg et al., 2009). In an important recent [^{11}C] FLB457 PET study using the competition paradigm, Slifstein et al. (2015) demonstrated for the first time that dopamine release following amphetamine challenge is blunted in prefrontal cortex and other extrastriatal brain regions in patients with schizophrenia to an extent directly related to the attenuated BOLD activation during an fMRI working memory task.

A number of in vivo multimodal imaging studies suggest that the impaired prefrontal function observed in schizophrenia is directly related to excessive subcortical dopamine function. Activation of prefrontal cerebral blood flow (as measured with [^{15}O]H$_2$O-PET) during the Wisconsin Card Sorting Test (a test of prefrontal function) was significantly reduced in patients with schizophrenia compared to controls, and in patients the reduction was inversely related to striatal dopamine synthesis capacity (as measured with [^{18}F]FDOPA-PET), that is, those patients with the most pronounced deficit in prefrontal function presented with the largest increase in striatal dopamine synthesis capacity (Meyer-Lindenberg et al., 2002). A similar multimodal study assessed prefrontal function with fMRI in patients at risk for schizophrenia. Here, again, the magnitude of prefrontal dysfunction during a working memory task was negatively correlated with striatal dopamine synthesis capacity, which suggests that those patients and at-risk subjects with the most pronounced prefrontal cortical function present with the most distinct subcortical hyperdopaminergia (Fusar-Poli et al., 2010).

Current hypotheses suggest that excessive subcortical dopamine function is secondary to a primary deficit in cortical N-methyl-D-aspartate (NMDA) receptor-mediated excitatory transmission in schizophrenia (review in Poels et al., 2014). In striatum and frontal cortex, presynaptic dopamine release is under inhibitory control of GABAergic neurons, which are activated by NMDA receptors. In particular, striatal dopamine release is modulated by $GABA_B$ receptors located on presynaptic dopamine terminals. NMDA receptor activation in cerebral cortex leads to GABA release from striatal GABAergic interneurons (Javitt et al., 2005). Given this neurochemical anatomy of the cortico-striatal pathway, impaired NMDA receptor signaling in cortex leads to excessive striatal dopamine function. This reciprocal pathway is supported by animal studies in which the amphetamine-induced dopamine release was attenuated by administration of glycine, a co-transmitter needed for proper NMDA receptor function, or glycine transporter inhibitors (Balla et al., 2012).

In patients with schizophrenia, due to a lack of suitable radioligands for PET or SPECT, glutamatergic function has been mainly studied with magnetic resonance spectroscopy (MRS). In one of the earliest such studies, the concentration of N-acetylaspartate (NAA; a biomarker of neuronal integrity) in the dorsolateral prefrontal cortex correlated negatively with the amphetamine-induced striatal dopamine release as measured with the amphetamine challenge PET paradigm (Bertolino et al., 2000). Another multimodal study showed an inverse relationship between the glutamate signal in prefrontal cortex and the individual

[^{18}F]FDOPA trapping in ventral striatum of medicated schizophrenia patients (Gleich et al., 2015). MR spectroscopy has also given the promising result that treatment of schizophrenia patients with a glycine transporter inhibitor produced the expected reduction in the (hippocampal) cortical glutamate complex signal (Strzelecki et al., 2015). These findings are grounded in animal studies showing that neonatal cortical lesions in nonhuman primates result in augmentation of amphetamine-evoked dopamine release in the adults (Saunders et al., 1998). In a similar rat model of schizophrenia, early ventral hippocampal lesions produced reduction of NAA in prefrontal cortex similar to those observed in schizophrenia, in conjunction with amphetamine-evoked hyperlocomotion (Bertolino et al., 2002). As such, the NMDA–dopamine link in schizophrenia seems consistent with a primary developmental disturbance in cortical function, leading to disinhibition of subcortical dopamine, which itself is responsible for the positive symptoms.

KETAMINE PSYCHOSIS

Administration of the NMDA antagonist ketamine to healthy subjects induces a transient psychotic state, which is characterized by negative symptoms and cognitive deficits. In this respect, NMDA blockade may be superior to the psychostimulant model of schizophrenia, which is marked mainly by positive symptoms. Better still is an extension of the simple dopamine hypothesis to a "dopamine/glutamate hypothesis" of schizophrenia, which has been tested in a number of SPECT and PET studies with administration of ketamine at high subanesthetic doses. Although some such studies have shown increased dopamine release in the striatum of healthy volunteers (Breier et al., 1998; Smith et al., 1998; Vollenweider et al., 2000), this result could not be confirmed in other studies (Aalto et al., 2002; Kegeles et al., 2002; Vernaleken et al., 2013). This inconsistent finding may be an instance of heterogeneity of vulnerability within a healthy population; in particular, we found that psychotogenic effects of ketamine correlated positively with baseline striatal $D_{2/3}$ receptor availability (Vernaleken et al., 2013), suggesting that high baseline $D_{2/3}$ receptor availability (resulting from low occupancy by dopamine) may impart benefits with regard to cognitive flexibility, but may increase the risk of maladaptive information processing in the face of environmental stresses and challenges. Returning to the amphetamine challenge [^{11}C]raclopride PET paradigm, it is notable that ΔBP in healthy volunteers additionally challenged with ketamine attained the magnitude seen in schizophrenia patients (Kegeles et al., 2000). Thus, NMDA blockade with ketamine may mimic primary cortical dysfunction, resulting in functional hyperdopaminergia in the basal ganglia. An fMRI study supports a top-down model of impaired control of reward processing in schizophrenia, manifesting in reduced coupling between activity in prefrontal cortex and ventral striatum (Richter et al., 2015), which is a pattern that is also implicated in a probabilistic decision-making paradigm (Rausch et al., 2014). In general, these results fit with a model in which hyperdopaminergia results in inappropriate attribution of salience and reward-based learning.

CONCLUSION: BEYOND THE DOPAMINE HYPOTHESIS

New molecular imaging studies lend partial support for a dopamine model of schizophrenia while avoiding some of the experimental confounds arising from postmortem studies. However, the PET/SPECT literature is plagued by small sample sizes, which may not capture heterogeneity of the disorder. This may call for multicenter studies using standard endpoints in a large population of patients. Looking ahead, we can expect that dopaminergic abnormalities will emerge as a key feature in psychotic disorders, but that in some proportion of patients, perhaps 25%, available markers may be completely normal. In addition, the relevant abnormalities in dopamine markers may be restricted to particular anatomic divisions of the basal ganglia or in cerebral cortex. If schizophrenia is only partly defined by subcortical hyperdopaminergia, then cortical glutamate transmission likely presents an important additional dimension. Thus, impaired NMDA transmission in cortical regions communicating with striatum may present a primary disturbance, perhaps arising as a developmental phenomenon imparting vulnerability to emergence of subcortical dopamine-mediated positive symptoms in young adulthood. We feel that multimodal imaging, for example, with hybrid PET-MR instruments, holds great promise for elucidating the causal sequence leading to schizophrenia and for revealing abnormal coupling between cortex and the basal ganglia, with the caveat that future studies must be powered to accommodate heterogeneity in populations of patients meeting the diagnostic criteria for schizophrenia.

REFERENCES

Aalto, S., Hirvonen, J., Kajander, J., Scheinin, H., Någren, K., Vilkman, H., et al., 2002. Ketamine does not decrease striatal dopamine D2 receptor binding in man. Psychopharmacology (Berl) 164, 401–406.

Abi-Dargham, A., Gil, R., Krystal, J., et al., 1998. Increased striatal dopamine transmission in schizophrenia: confirmation in a second cohort. Am J Psychiatry 155, 761–767.

Abi-Dargham, A., Rodenhiser, J., Printz, D., et al., 2000. Increased baseline occupancy of D2 receptors by dopamine in schizophrenia. Proc Natl Acad Sci USA 97, 8104–8109.

Abi-Dargham, A., Mawlawi, O., Lombardo, I., Gil, R., Martinez, D., Huang, Y., et al., 2002. Prefrontal dopamine D1 receptors and working memory in schizophrenia. J Neurosci 22, 3708–3719.

Abi-Dargham, A., van de Giessen, E., Slifstein, M., Kegeles, L.S., Laruelle, M., 2009. Baseline and amphetamine-stimulated dopamine activity are related in drug-naïve schizophrenic subjects. Biol Psychiatry 65, 1091–1093.

Abi-Dargham, A., Xu, X., Thompson, J.L., Gil, R., Kegeles, L.S., Urban, N., et al., 2012. Increased prefrontal cortical D1 receptors in drug naive patients with schizophrenia: a PET study with [^{11}C]NNC112. J Psychopharmacol 26, 794–805.

Agid, O., Schulze, L., Arenovich, T., Sajeev, G., McDonald, K., Foussias, G., et al., 2013. Antipsychotic response in first-episode schizophrenia: efficacy of high doses and switching. Eur Neuropsychopharmacol 23, 1017–1022.

Angrist, B., Lee, H.K., Gershon, S., 1974. The antagonism of amphetamine-induced symptomatology by a neuroleptic. Am J Psychiatry 131, 817–819.

Balla, A., Schneider, S., Sershen, H., Javitt, D.C., 2012. Effects of novel, high affinity glycine transport inhibitors on frontostriatal dopamine release in a rodent model of schizophrenia. Eur Neuropsychopharmacol 22, 902–910.

Barch, D.M., Carter, C.S., 2005. Amphetamine improves cognitive function in medicated individuals with schizophrenia and in healthy volunteers. Schizophr Res 77, 43–58.

Barch, D.M., Ceaser, A., 2012. Cognition in schizophrenia: core psychological and neural mechanisms. Trends Cogn Sci 16, 27–34.

Belvederi Murri, M., Pariante, C.M., Dazzan, P., Hepgul, N., Papadopoulos, A.S., Zunszain, P., et al., 2012. Hypothalamic-pituitary-adrenal axis and clinical symptoms in first-episode psychosis. Psychoneuroendocrinology 37, 629–644.

Bertolino, A., Breier, A., Callicott, J.H., Adler, C., Mattay, V.S., Shapiro, M., et al., 2000. The relationship between dorsolateral prefrontal neuronal N-acetylaspartate and evoked release of striatal dopamine in schizophrenia. Neuropsychopharmacology 22, 125–132.

Bertolino, A., Roffman, J.L., Lipska, B.K., van Gelderen, P., Olson, A., Weinberger, D.R., 2002. Reduced N-acetylaspartate in prefrontal cortex of adult rats with neonatal hippocampal damage. Cereb Cortex 12, 983–990.

Boileau, I., Payer, D., Chugani, B., Lobo, D.S., Houle, S., Wilson, A.A., et al., 2014. In vivo evidence for greater amphetamine-induced dopamine release in pathological gambling: a positron emission tomography study with [(11)C]-(+)-PHNO. Mol Psychiatry 19, 1305–1313.

Bowie, C.R., Reichenberg, A., Patterson, T.L., Heaton, R.K., Harvey, P.D., 2006. Determinants of real-world functional performance in schizophrenia subjects: correlations with cognition, functional capacity, and symptoms. Am J Psychiatry 163, 418–425.

Breier, A., Su, T.P., Saunders, R., Carson, R.E., Kolachana, B.S., de Bartolomeis, A., et al., 1997. Schizophrenia is associated with elevated amphetamine-induced synaptic dopamine concentrations: evidence from a novel positron emission tomography method. Proc Natl Acad Sci USA 94, 2569–2574.

Breier, A., Adler, C.M., Weisenfeld, N., Su, T.P., Elman, I., Picken, L., et al., 1998. Effects of NMDA antagonism on striatal dopamine release in healthy subjects: application of a novel PET approach. Synapse 29, 142–147.

Carlsson, A., Lindqvist, M., 1963. Effect of chlorpromazine or haloperidol on formation of 3-methoxytyramine and normetanephrine in mouse brain. Acta Pharmacol Toxicol (Copenh) 20, 140–144.

Cowden, R.C., Zax, M., Sproles, J.A., 1955. Reserpine alone and as an adjunct to psychotherapy in the treatment of schizophrenia. AMA Arch Neurol Psychiatry 74, 518–522.

Crow, T.J., Baker, H.F., Cross, A.J., Joseph, M.H., Lofthouse, R., Longden, A., et al., 1979. Monoamine mechanisms in chronic schizophrenia: post-mortem neurochemical findings. Br J Psychiatry 134, 249–256.

Cumming, P., 2011. Absolute abundances and affinity states of dopamine receptors in mammalian brain: a review. Synapse 65, 892–909.

Cumming, P., Wong, D.F., Gillings, N., Hilton, J., Scheffel, U., Gjedde, A., 2002. Specific binding of [(11)C] raclopride and N-[(3)H]propyl-norapomorphine to dopamine receptors in living mouse striatum: occupancy by endogenous dopamine and guanosine triphosphate-free G protein. J Cereb Blood Flow Metab 22, 596–604.

Davis, K.L., Kahn, R.S., Ko, G., Davidson, M., 1991. Dopamine in schizophrenia: a review and reconceptualization. Am J Psychiatry 148, 1474–1486.

Delay, J., Deniker, P., Harl, J.M., 1952. Therapeutic use in psychiatry of phenothiazine of central elective action (4560 RP). Ann Med Psychol (Paris) 110, 112–117.

Duval, F., Mokrani, M.C., Monreal, J., Bailey, P., Valdebenito, M., Crocq, M.A., et al., 2003. Dopamine and serotonin function in untreated schizophrenia: clinical correlates of the apomorphine and d-fenfluramine tests. Psychoneuroendocrinology 28, 627–642.

Farde, L., Nordström, A.L., Wiesel, F.A., Pauli, S., Halldin, C., Sedvall, G., 1992. Positron emission tomographic analysis of central D1 and D2 dopamine receptor occupancy in patients treated with classical neuroleptics and clozapine. Relation to extrapyramidal side effects. Arch Gen Psychiatry 49, 538–544.

Finnema, S.J., Halldin, C., Bang-Andersen, B., Gulyás, B., Bundgaard, C., Wikström, H.V., et al., 2009. Dopamine D(2/3) receptor occupancy of apomorphine in the nonhuman primate brain – a comparative PET study with [11C]raclopride and [11C]MNPA. Synapse 63, 378–389.

Fusar-Poli, P., Meyer-Lindenberg, A., 2013a. Striatal presynaptic dopamine in schizophrenia, part I: meta-analysis of dopamine active transporter (DAT) density. Schizophr Bull 39, 22–32.

Fusar-Poli, P., Meyer-Lindenberg, A., 2013b. Striatal presynaptic dopamine in schizophrenia, part II: meta-analysis of [(18)F/(11)C]-DOPA PET studies. Schizophr Bull 39, 33–42.

Fusar-Poli, P., Howes, O.D., Allen, P., Broome, M., Valli, I., Asselin, M.C., et al., 2010. Abnormal frontostriatal interactions in people with prodromal signs of psychosis: a multimodal imaging study. Arch Gen Psychiatry 67, 683–691.

Girgis, R.R., Xu, X., Gil, R.B., Hackett, E., Ojeil, N., Lieberman, J.A., et al., 2015. Antipsychotic binding to the dopamine-3 receptor in humans: A PET study with [(11)C]-(+)-PHNO. Schizophr Res 168, 373–376.

Gleich, T., Deserno, L., Lorenz, R.C., Boehme, R., Pankow, A., Buchert, R., et al., 2015. Prefrontal and striatal glutamate differently relate to striatal dopamine: potential regulatory mechanisms of striatal presynaptic dopamine function? J Neurosci 35, 9615–9621.

Grace, A.A., 1992. The depolarization block hypothesis of neuroleptic action: implications for the etiology and treatment of schizophrenia. J Neural Transm Suppl 36, 91–131.

Graff-Guerrero, A., Mizrahi, R., Agid, O., Marcon, H., Barsoum, P., Rusjan, P., et al., 2009. The dopamine D2 receptors in high-affinity state and D3 receptors in schizophrenia: a clinical [11C]-(+)-PHNO PET study. Neuropsychopharmacology 34, 1078–1086.

Gründer, G., Carlsson, A., Wong, D.F., 2003. Mechanism of new antipsychotic medications: occupancy is not just antagonism. Arch Gen Psychiatry 60, 974–977.

Gründer, G., Landvogt, C., Vernaleken, I., Buchholz, H.G., Ondracek, J., Siessmeier, T., et al., 2006. The striatal and extrastriatal D2/D3 receptor-binding profile of clozapine in patients with schizophrenia. Neuropsychopharmacology 31, 1027–1035.

Gründer, G., Fellows, C., Janouschek, H., Veselinovic, T., Boy, C., Bröcheler, A., et al., 2008. Brain and plasma pharmacokinetics of aripiprazole in patients with schizophrenia: an [18F]fallypride PET study. Am J Psychiatry 9165, 988–995.

Gründer, G., Hippius, H., Carlsson, A., 2009. The 'atypicality' of antipsychotics: a concept re-examined and re-defined. Nat Rev Drug Discov 8, 197–202.

Gründer, G., Hiemke, C., Paulzen, M., Veselinovic, T., Vernaleken, I., 2011. Therapeutic plasma concentrations of antidepressants and antipsychotics: lessons from PET imaging. Pharmacopsychiatry 44, 236–248.

Gryglewski, G., Lanzenberger, R., Kranz, G.S., Cumming, P., 2014. Meta-analysis of molecular imaging of serotonin transporters in major depression. J Cereb Blood Flow Metab 34, 1096–1103.

Javitt, D.C., Hashim, A., Sershen, H., 2005. Modulation of striatal dopamine release by glycine transport inhibitors. Neuropsychopharmacology 30, 649–656.

Kambeitz, J., Abi-Dargham, A., Kapur, S., Howes, O.D., 2014. Alterations in cortical and extrastriatal subcortical dopamine function in schizophrenia: systematic review and meta-analysis of imaging studies. Br J Psychiatry 204, 420–429.

Kapur, S., Zipursky, R.B., Remington, G., Jones, C., DaSilva, J., Wilson, A.A., et al., 1998. 5-HT2 and D2 receptor occupancy of olanzapine in schizophrenia: a PET investigation. Am J Psychiatry 155, 921–928.

Kapur, S., Zipursky, R., Jones, C., Remington, G., Houle, S., 2000. Relationship between dopamine D(2) occupancy, clinical response, and side effects: a double-blind PET study of first-episode schizophrenia. Am J Psychiatry 157, 514–520.

Keefe, R.S., Bilder, R.M., Davis, S.M., Harvey, P.D., Palmer, B.W., Gold, J.M., et al., 2007. Neurocognitive effects of antipsychotic medications in patients with chronic schizophrenia in the CATIE Trial. Arch Gen Psychiatry 64, 633–647.

Kegeles, L.S., Abi-Dargham, A., Zea-Ponce, Y., et al., 2000. Modulation of amphetamine-induced striatal dopamine release by ketamine in humans: implications for schizophrenia. Biol Psychiatry 48, 627–640.

Kegeles, L.S., Martinez, D., Kochan, L.D., Hwang, D.R., Huang, Y., Mawlawi, O., et al., 2002. NMDA antagonist effects on striatal dopamine release: positron emission tomography studies in humans. Synapse 43, 19–29.

Kegeles, L.S., Abi-Dargham, A., Frankle, W.G., Gil, R., Cooper, T.B., Slifstein, M., et al., 2010. Increased synaptic dopamine function in associative regions of the striatum in schizophrenia. Arch Gen Psychiatry 67, 231–239.

Kinon, B.J., Lieberman, J.A., 1996. Mechanisms of action of atypical antipsychotic drugs: a critical analysis. Psychopharmacology (Berl) 124, 2–34.

Knable, M.B., Hyde, T.M., Herman, M.M., Carter, J.M., Bigelow, L., Kleinman, J.E., 1994. Quantitative autoradiography of dopamine-D1 receptors, D2 receptors, and dopamine uptake sites in postmortem striatal specimens from schizophrenic patients. Biol Psychiatry 36, 827–835.

Kumakura, Y., Cumming, P., 2009. PET studies of cerebral levodopa metabolism: a review of clinical findings and modeling approaches. Neuroscientist 15, 635–650.

Kumakura, Y., Cumming, P., Vernaleken, I., Buchholz, H.G., Siessmeier, T., Heinz, A., et al., 2007. Elevated [18F] fluorodopamine turnover in brain of patients with schizophrenia: an [18F]fluorodopa/positron emission tomography study. J Neurosci 27, 8080–8087.

Laruelle, M., 2011. Measuring dopamine synaptic transmission with molecular imaging and pharmacological challenges: the state of the art. In: Gründer, G. (Ed.), Molecular Imaging in the Clinical Neurosciences Humana Press

Laruelle, M., Abi-Dargham, A., van Dyck, C.H., Rosenblatt, W., Zea-Ponce, Y., Zoghbi, S.S., et al., 1995. SPECT imaging of striatal dopamine release after amphetamine challenge. J Nucl Med 36, 1182–1190.

Laruelle, M., Abi-Dargham, A., van Dyck, C.H., Gil, R., D'Souza, C.D., Erdos, J., et al., 1996. Single photon emission computerized tomography imaging of amphetamine-induced dopamine release in drug-free schizophrenic subjects. Proc Natl Acad Sci USA 93, 9235–9240.

Laruelle, M., D'Souza, C.D., Baldwin, R.M., Abi-Dargham, A., Kanes, S.J., Fingado, C.L., et al., 1997. Imaging D2 receptor occupancy by endogenous dopamine in humans. Neuropsychopharmacology 17, 162–174.

Laruelle, M., Abi-Dargham, A., Gil, R., Kegeles, L., Innis, R., 1999. Increased dopamine transmission in schizophrenia: relationship to illness phases. Biol Psychiatry 46, 56–72.

Lindström, L.H., Gefvert, O., Hagberg, G., Lundberg, T., Bergström, M., Hartvig, P., et al., 1999. Increased dopamine synthesis rate in medial prefrontal cortex and striatum in schizophrenia indicated by L-(beta-11C) DOPA and PET. Biol Psychiatry 46, 681–688.

Maas, J.W., Bowden, C.L., Miller, A.L., Javors, M.A., Funderburg, L.G., Berman, N., et al., 1997. Schizophrenia, psychosis, and cerebral spinal fluid homovanillic acid concentrations. Schizophr Bull 23, 147–154.

Mamo, D., Graff, A., Mizrahi, R., Shammi, C.M., Romeyer, F., Kapur, S., 2007. Differential effects of aripiprazole on D(2), 5-HT(2), and 5-HT(1A) receptor occupancy in patients with schizophrenia: a triple tracer PET study. Am J Psychiatry 164, 1411–1417.

Meltzer, H.Y., Stahl, S.M., 1976. The dopamine hypothesis of schizophrenia: a review. Schizophr Bull 2, 19–76.

Meyer-Lindenberg, A., Miletich, R.S., Kohn, P.D., Esposito, G., Carson, R.E., Quarantelli, M., et al., 2002. Reduced prefrontal activity predicts exaggerated striatal dopaminergic function in schizophrenia. Nat Neurosci 5, 267–271.

Minzenberg, M.J., Laird, A.R., Thelen, S., Carter, C.S., Glahn, D.C., 2009. Meta-analysis of 41 functional neuroimaging studies of executive function in schizophrenia. Arch Gen Psychiatry 66, 811–822.

Mizrahi, R., Agid, O., Borlido, C., Suridjan, I., Rusjan, P., Houle, S., et al., 2011. Effects of antipsychotics on D3 receptors: a clinical PET study in first episode antipsychotic naive patients with schizophrenia using [11C]-(+)-PHNO. Schizophr Res 131, 63–68.

Mizrahi, R., Kenk, M., Suridjan, I., Boileau, I., George, T.P., McKenzie, K., et al., 2014. Stress-induced dopamine response in subjects at clinical high risk for schizophrenia with and without concurrent cannabis use. Neuropsychopharmacology 39, 1479–1489.

Narendran, R., Mason, N.S., Laymon, C.M., Lopresti, B.J., Velasquez, N.D., May, M.A., et al., 2010. A comparative evaluation of the dopamine D(2/3) agonist radiotracer [11C](-)-N-propyl-norapomorphine and antagonist [11C]raclopride to measure amphetamine-induced dopamine release in the human striatum. J Pharmacol Exp Ther 333, 533–539.

Nordström, A.L., Farde, L., Wiesel, F.A., Forslund, K., Pauli, S., Halldin, C., et al., 1993. Central D2-dopamine receptor occupancy in relation to antipsychotic drug effects: a double-blind PET study of schizophrenic patients. Biol Psychiatry 33, 227–235.

Nyberg, S., Farde, L., Halldin, C., 1997. A PET study of 5-HT2 and D2 dopamine receptor occupancy induced by olanzapine in healthy subjects. Neuropsychopharmacology 16, 1–7.

Nyberg, S., Eriksson, B., Oxenstierna, G., Halldin, C., Farde, L., 1999. Suggested minimal effective dose of risperidone based on PET-measured D2 and 5-HT2A receptor occupancy in schizophrenic patients. Am J Psychiatry 156, 869–875.

Okubo, Y., Suhara, T., Suzuki, K., Kobayashi, K., Inoue, O., Terasaki, O., et al., 1997. Decreased prefrontal dopamine D1 receptors in schizophrenia revealed by PET. Nature 385, 634–636.

Poels, E.M., Kegeles, L.S., Kantrowitz, J.T., Slifstein, M., Javitt, D.C., Lieberman, J.A., et al., 2014. Imaging glutamate in schizophrenia: review of findings and implications for drug discovery. Mol Psychiatry 19, 20–29.

Quednow, B.B., Treyer, V., Hasler, F., Dörig, N., Wyss, M.T., Burger, C., et al., 2012. Assessment of serotonin release capacity in the human brain using dexfenfluramine challenge and [18F]altanserin positron emission tomography. Neuroimage 59, 3922–3932.

Rausch, F., Mier, D., Eifler, S., Esslinger, C., Schilling, C., Schirmbeck, F., et al., 2014. Reduced activation in ventral striatum and ventral tegmental area during probabilistic decision-making in schizophrenia. Schizophr Res 156, 143–149.

Reith, J., Benkelfat, C., Sherwin, A., Yasuhara, Y., Kuwabara, H., Andermann, F., et al., 1994. Elevated dopa decarboxylase activity in living brain of patients with psychosis. Proc Natl Acad Sci USA 91, 11651–11654.

Richter, A., Petrovic, A., Diekhof, E.K., Trost, S., Wolter, S., Gruber, O., 2015. Hyperresponsivity and impaired prefrontal control of the mesolimbic reward system in schizophrenia. J Psychiatr Res 71, 8–15.

Saunders, R.C., Kolachana, B.S., Bachevalier, J., Weinberger, D.R., 1998. Neonatal lesions of the medial temporal lobe disrupt prefrontal cortical regulation of striatal dopamine. Nature 393, 169–171.

Seeman, P., Lee, T., Chau-Wong, M., Wong, K., 1976. Antipsychotic drug doses and neuroleptic/dopamine receptors. Nature 261, 717–719.

Seneca, N., Finnema, S.J., Farde, L., Gulyás, B., Wikström, H.V., Halldin, C., et al., 2006. Effect of amphetamine on dopamine D2 receptor binding in nonhuman primate brain: a comparison of the agonist radioligand [11C]MNPA and antagonist [11C]raclopride. Synapse 59, 260–269.

Slifstein, M., van de Giessen, E., Van Snellenberg, J., Thompson, J.L., Narendran, R., Gil, R., et al., 2015. Deficits in prefrontal cortical and extrastriatal dopamine release in schizophrenia: a positron emission tomographic functional magnetic resonance imaging study. JAMA Psychiatry 72, 316–324.

Smith, G.S., Schloesser, R., Brodie, J.D., Dewey, S.L., Logan, J., Vitkun, S.A., et al., 1998. Glutamate modulation of dopamine measured in vivo with positron emission tomography (PET) and 11C-raclopride in normal human subjects. Neuropsychopharmacology 18, 18–25.

Strzelecki, D., Podgórski, M., Kałużyńska, O., Gawlik-Kotelnicka, O., Stefańczyk, L., Kotlicka-Antczak, M., et al., 2015. Supplementation of antipsychotic treatment with sarcosine – GlyT1 inhibitor – causes changes of glutamatergic (1)NMR spectroscopy parameters in the left hippocampus in patients with stable schizophrenia. Neurosci Lett 606, 7–12.

Suridjan, I., Rusjan, P., Addington, J., Wilson, A.A., Houle, S., Mizrahi, R., 2013. Dopamine D2 and D3 binding in people at clinical high risk for schizophrenia, antipsychotic-naive patients and healthy controls while performing a cognitive task. J Psychiatry Neurosci 38, 98–106.

Talbot, P.S., Slifstein, M., Hwang, D.R., Huang, Y., Scher, E., Abi-Dargham, A., et al., 2012. Extended characterisation of the serotonin 2A (5-HT2A) receptor-selective PET radiotracer 11C-MDL100907 in humans: quantitative analysis, test-retest reproducibility, and vulnerability to endogenous 5-HT tone. Neuroimage 59, 271–285.

Taylor, S.F., Koeppe, R.A., Tandon, R., Zubieta, J.K., Frey, K.A., 2000. In vivo measurement of the vesicular monoamine transporter in schizophrenia. Neuropsychopharmacology 23, 667–675.

Van Rossum, J.M., 1966. The significance of dopamine-receptor blockade for the mechanism of action of neuroleptic drugs. Arch Int Pharmacodyn Ther 160, 492–494.

Verhoeff, N.P., Kapur, S., Hussey, D., Lee, M., Christensen, B., Psych, C., et al., 2001. A simple method to measure baseline occupancy of neostriatal dopamine D2 receptors by dopamine in vivo in healthy subjects. Neuropsychopharmacology 25, 213–223.

Vernaleken, I., Janouschek, H., Raptis, M., Hellmann, S., Veselinovic, T., Bröcheler, A., et al., 2010. Dopamine D2/3 receptor occupancy by quetiapine in striatal and extrastriatal areas. Int J Neuropsychopharmacol 13, 951–960.

Vernaleken, I., Kumakura, Y., Cumming, P., Buchholz, H.G., Siessmeier, T., Stoeter, P., et al., 2006. Modulation of [18F]fluorodopa (FDOPA) kinetics in the brain of healthy volunteers after acute haloperidol challenge. Neuroimage 30, 1332–1339.

Vernaleken, I., Klomp, M., Moeller, O., Raptis, M., Nagels, A., Rösch, F., et al., 2013. Vulnerability to psychotogenic effects of ketamine is associated with elevated D2/3-receptor availability. Int J Neuropsychopharmacol 16, 745–754.

Volkow, N.D., Wang, G.J., Fowler, J.S., Logan, J., Schlyer, D., Hitzemann, R., et al., 1994. Imaging endogenous dopamine competition with [11C]raclopride in the human brain. Synapse 16, 255–262.

Vollenweider, F.X., Vontobel, P., Oye, I., Hell, D., Leenders, K.L., 2000. Effects of (S)-ketamine on striatal dopamine: a [11C]raclopride PET study of a model psychosis in humans. J Psychiatr Res 34, 35–43.

Weinberger, D.R., 1987. Implications of normal brain development for the pathogenesis of schizophrenia. Arch Gen Psychiatry 44, 660–669.

Weinberger, D.R., Laruelle, M., 2002. Neurochemical and neuropharmacological imaging in schizophrenia. In: Davis, K.L., Charney, D.S., Coyle, J.T., Nemeroff, C. (Eds.), Neuropsychopharmacology: The Fifth Generation of Progress Lippincott, Williams, and Wilkins, Philadelphia, pp. 833–855.

Willeit, M., Ginovart, N., Graff, A., Rusjan, P., Vitcu, I., Houle, S., et al., 2008. First human evidence of d-amphetamine induced displacement of a D2/3 agonist radioligand: A [11C]-(+)-PHNO positron emission tomography study. Neuropsychopharmacology 33, 279–289.

Winblad, B., Bucht, G., Gottfries, C.G., Roos, B.E., 1979. Monoamines and monoamine metabolites in brains from demented schizophrenics. Acta Psychiatr Scand 60, 17–28.

Wolkin, A., Barouche, F., Wolf, A.P., Rotrosen, J., Fowler, J.S., Shiue, C.Y., et al., 1989. Dopamine blockade and clinical response: evidence for two biological subgroups of schizophrenia. Am J Psychiatry 146, 905–908.

Wong, D.F., Tauscher, J., Gründer, G., 2009. The role of imaging in proof of concept for CNS drug discovery and development. Neuropsychopharmacology 34, 187–203.

Wong, D.F., Wagner Jr, H.N., Tune, L.E., Dannals, R.F., Pearlson, G.D., Links, J.M., et al., 1986. Positron emission tomography reveals elevated D2 dopamine receptors in drug-naive schizophrenics. Science 234, 1558–1563.

CHAPTER 8

THE PSD: A MICRODOMAIN FOR CONVERGING MOLECULAR ABNORMALITIES IN SCHIZOPHRENIA

A. Banerjee[1], K.E. Borgmann-Winter[1,2], R. Ray[1] and C.-G. Hahn[1]

[1]*Department of Psychiatry, University of Pennsylvania, Philadelphia, PA, United States* [2]*The Children's Hospital of Philadelphia, University of Pennsylvania School of Medicine, Philadelphia, PA, United States*

CHAPTER OUTLINE

Introduction 125
The Postsynaptic Density, the Hub of Postsynaptic Signaling 126
    Dendritic Spines 126
    The Postsynaptic Density 127
    Signaling Pathways in the PSD 128
    Proteomic Landscape of the PSD 131
Evidence Supporting the Role of the PSD in Schizophrenia 133
    Genetic Evidence 133
    Postmortem Evidence for PSD Dysregulation in Schizophrenia 134
    Dysregulations in PSD Proteins and Their Signaling in Schizophrenia 134
PSD as a Microdomain for Converging Molecular Alterations in Schizophrenia 136
Summary and Future Directions 137
References 138

INTRODUCTION

An important feature of the pathophysiology of schizophrenia is that it is a complex trait disorder (Braff et al., 2007; Coyle, 2006; Ross et al., 2006) that involves multiple genetic and epigenetic factors impacting diverse domains of neurocognitive functions (Hall et al., 2015; Kennedy et al., 2005). In addition to this complexity, schizophrenia is a highly heterogeneous entity, because the illness may result from different combinations of etiologic components in any given individual (Modinos et al., 2013). Nonetheless, clinical manifestations can be classified into categories that are shared by many, suggesting that diverse etiologic underpinnings somehow find ways to be manifested by common phenotypes. An important question, then, is in regard to how such multifactorial and heterogeneous etiologic factors can lead to neuropsychiatric symptoms that are shared in common by many patients. This may be an important question fundamental to an etiologic understanding of virtually all complex trait disorders (Hall et al., 2015). One explanation could be that these various factors and their combinations somehow find the common final pathways, which are directly involved in the endophenotypic manifestation of the illness (Hall et al., 2015; Horvath and Mirnics, 2015; Modinos et al., 2013). Such convergence could occur at specific levels of biological processes, such as genes, epigenomic markings, proteins, pathways, and/or cell–cell interactions. At a molecular level, it could be via a single molecule, pathway, or subcellular locale (de Bartolomeis et al., 2014; Hashimoto et al., 2007) on which multiple dysregulations associated with schizophrenia converge.

This chapter focuses on the postsynaptic density (PSD) as a microdomain that serves as a point of convergence for various dysregulations, increasing the susceptibility for schizophrenia. This is prompted by recent evidence demonstrating that multiple proteins and pathways in the PSD are associated with schizophrenia via their common or rare variants (Fromer et al., 2014; Purcell et al., 2014; Schizophrenia Working Group of the Psychiatric Genomics Consortium, 2014) and are dysregulated in brains of patients (Banerjee et al., 2015; Hahn et al., 2006). The PSD is an electron-dense fibrous specialization (Baron et al., 2006; Blomberg et al., 1977; Kennedy, 1997) in which multiple receptor and nonreceptor signaling pathways interact and impact synaptic strength, plasticity, and connectivity (Kennedy, 1998; Kennedy et al., 2005). The PSD is housed in dendritic spines with endoplasmic reticulum, ribosomes, and mitochondria; therefore, the protein composition and function of the PSD could reflect changes in the intracellular microenvironment. Thus, the PSD may reflect intricate interplay between various signaling pathways therein and the intracellular microenvironment impacting synaptic plasticity. In this chapter, we summarize the evidence implicating the PSD in the pathophysiology of schizophrenia, describe the function and ultrastructure of the PSD, review the results of recent studies involving patient-derived tissues, and review future directions and potential for study of the PSD as a subcellular microdomain on which various etiologic factors converge.

THE POSTSYNAPTIC DENSITY, THE HUB OF POSTSYNAPTIC SIGNALING
DENDRITIC SPINES

Insights into the structure and function of dendritic spines have been derived from studies using electron microscopy (EM) of fixed tissues. Spines are membranous dynamic protrusions with various morphologies ranging from 0.2 to 2 μm, or up to 6 μm in some regions. Mature dendrites have numerous spines, 1–10 spines per 1 μm in length (Sorra and Harris, 2000). Typically, spines have a bulbous "head" connected to the dendrite by a thin neck, which is up to 1 μm long but is only 75–300 nm in diameter. The diameter of the neck is thought to modulate the diffusion of molecules between the spine and the parent dendrite (Adrian et al., 2014; Bourne and Harris, 2008; Hayashi and Majewska, 2005), although the precise mechanism by which this occurs is not well understood (MacGillavry and Hoogenraad, 2015). Spine shapes have classically been described as thin, stubby, or mushroom (Hering and Sheng, 2001); however, recent evidence suggests that these shapes are on a continuum and reflect the developmental stage and strength of the synapse (Kasai et al., 2003; Tada and Sheng, 2006). Newer imaging techniques have shown that the size of the head of the dendrite is correlated with the stability and strength of synaptic connections (Holtmaat et al., 2006; Matsuzaki et al., 2004).

The components of dendritic spines include the cytoskeletal elements and the PSD as well as organelles, including the smooth endoplasmic reticulum, endocytic vesicles, polyribosomes, and rare mitochondria (Sorra and Harris, 2000). Actin serves as the primary cytoskeletal component within dendritic spines, determining their size, shape, and stability (Kaech et al., 1997; Markham and Fifkova, 1986). The molecular construction of spines is modulated by a large variety of morphogenic factors such as actin binding proteins, small GTPases, cell surface receptors, adhesion molecules, and receptor tyrosine kinases (Sala and Segal, 2014).

The smooth endoplasmic reticulum extends from the dendritic shaft and stores intracellular calcium (Spacek and Harris, 1997). It also functions in the transport of lipids and proteins (Horton and Ehlers, 2004; Verkhratsky, 2005). Stacks of smooth endoplasmic reticulum, known as the spine apparatus, are found more commonly in the neck of large mushroom spines and are thought to function as a calcium store involved in synaptic function and plasticity (Spacek and Harris, 1997; Svoboda and Mainen, 1999). Polyribosomes and smooth vesicles are seen more frequently in spines with a spine apparatus (Spacek and Harris, 1997), as are endosomes, suggesting a mechanism of local protein degradation.

The PSD is located under the surface of the spine membrane on the top or side of the spine head (Kasai et al., 2003; Kennedy, 2000; Sheng and Kim, 2002; Siekevitz, 1985). The PSD, as the hub of postsynaptic signaling, is in close proximity to the intracellular organelles specialized in translation of mRNAs and posttranslational modification. This ultrastructural organization of dendritic spines offers a milieu in which synaptic activity translates into structural changes via interactions between the protein composition of the PSD and subcellular microorganelles.

THE POSTSYNAPTIC DENSITY

The PSD is an irregular, segmented, disc-like structure that is 200–800 nm wide and 30–50 nm thick and harbors various receptor signaling pathways critical for postsynaptic signaling (Carlin et al., 1980; Farley et al., 2015; Yun-Hong et al., 2011). It is associated with the postsynaptic membrane and has a direct physical linkage with the presynaptic membrane. The PSD comprises hundreds (if not more than a thousand) of proteins and harbors receptors, scaffolding proteins, and signaling proteins (Bayes et al., 2011). The molecular organization of the PSD has been illuminated by studies using EM as well as various protein quantification methods, including proteomic methods (Bayes et al., 2014; Blomberg et al., 1977; Carlin et al., 1980). Studies of the spatial organization of specific proteins within the PSD are important because they offer clues to the interactions between molecules in this microdomain. EM tomography studies have demonstrated that the PSD structure comprises vertical filaments containing membrane-associated guanylate kinase (MAGUK) family proteins that link surface receptors to other scaffolding molecules near the postsynaptic membrane (Chen et al., 2008). Consistent with previous studies identifying PSD-95 as the most abundant member of the MAGUK family of proteins in the PSD residing within 21 nm of the synaptic cleft, it has also been identified specifically in the vertical filaments forming a scaffold that interacts with transmembrane receptors and channels (Petralia et al., 2005; Valtschanoff and Weinberg, 2001). Horizontal filaments lying 10–20 nm from the postsynaptic membrane containing GKAP/SAPAP, Shank, and Homer (Hayashi et al., 2009) run between vertical filaments. Longer horizontal filaments link *N*-methyl-D-aspartate (NMDA) receptors with each other and the shorter filaments link both NMDAR and α-amino-3-hydroxy-5-methyl-4-isoxazolepropionic acid (AMPA) types of structures with each other (Chen et al., 2008).

Membrane receptors such as AMPA receptors and NMDARs protrude from the membrane and have a cytoplasmic tail. The AMPAR cytoplasmic tail is 50–100 AA in length compared to that of the NMDAR cytoplasmic tail, which is 600 AA in length, facilitating interactions with downstream signaling molecules, including those of cytoskeletal, scaffolding, and signaling proteins. AKAP-79/150 has been identified as an anchoring protein that binds protein kinase A (PKA), protein kinase C (PKC), and PP2B for phosphoregulation of the AMPAR (Tunquist et al., 2008) as well as NMDAR-mediated long-term depression (LTD) (Jurado et al., 2010). Additional postsynaptic signaling molecules are diverse, including kinase/phosphatases, GTPases, and adhesion molecules (Sheng and Hoogenraad, 2007).

The extracellular domains of NMDAR and AMPAR are associated with adhesion molecules (Chen et al., 2008). In addition, the distribution of these receptors within the PSD differs, with NMDA receptors concentrated in the middle and AMPA receptors located more to the periphery (Kharazia and Weinberg, 1997; Racca et al., 2000). Interestingly, the arrangement of the vertical filaments at the edges of the PSD allows for the possibility that AMPAR could transit in and out of the edges of the PSD (Chen et al., 2008). More recent studies have demonstrated that the distribution of NMDA and AMPA receptors is more heterogeneous than previously believed and varies between individual synapses (Dani et al., 2010).

The constellation and function of proteins within the PSD are dynamic, and changes therein are critically modulated by synaptic activity, occurring in timeframes ranging from minutes to hours. The quantity and function of these proteins within the PSD are modulated by processes such as phosphorylation (Collins et al., 2005; Sheng and Kim, 2002; Trinidad et al., 2006), ubiquitin–proteasome-mediated protein degradation (Colledge et al., 2003; Ehlers, 2003; Pak and Sheng, 2003), and local protein translation (Schuman et al., 2006). Another important mechanism regulating the function of receptor signaling pathways is receptor trafficking to and from the PSD. Known mechanisms for this include lateral movement between the synaptic and extrasynaptic domains as well as maintenance of and/or alterations in the balance between surface expression and internalization of receptors. As a whole, the abundance and function of proteins within the PSD are critical for formation of long-term potentiation (LTP) (Dosemeci et al., 2001; Hu et al., 1998; Inoue and Okabe, 2003; Malinow and Malenka, 2002). Importantly, there are both brain region (Cheng et al., 2006) and cell-type-specific differences in the relative abundance of proteins in the PSD. For example, synaptic GTPase-activating protein (SynGAP) is more prevalent in excitatory hippocampal neurons (Cheng et al., 2006), whereas CITRON, a Rho effector, is more prevalent in inhibitory neurons in the hippocampus (Zhang et al., 1999).

SIGNALING PATHWAYS IN THE PSD

The PSD constitutes the functional topography of the postsynaptic membrane in which receptor complexes and their pathways are brought together and thereby organized for effective signaling and interactions. Multiple receptor pathways that are critical for synaptic activity/plasticity reside in the PSD and include glutamatergic receptors, NMDAR, AMPA, and metabotropic glutamate receptors (mGluRs).

NMDA receptors

NMDA receptors reside in synaptic and extrasynaptic regions in postsynaptic neurons (Gill et al., 2015; Kennedy, 1997, 1998) and synaptic NMDA receptor complexes are highly concentrated in the PSD (O'Brien et al., 1998; Sheng, 2001). NMDA receptor complexes are hetero-multimers of GluN1 and GluN2 subunits. GuN1 is an obligatory subunit required for all NMDAR complexes and GluN2 subunits, whereas GluN2A, GluN2B, GluN2C, and GluN2D are associated with GluN1 in various combinations. These varying combinations partly result from brain region-specific expression of GluN2 subunits (Kew et al., 1998). GluN2A is expressed in almost all brain regions, whereas GluN2B is restricted to the forebrain, but GluN2C is expressed in cerebellum, thalamus, and olfactory bulb. A central component of scaffolding proteins in NMDA receptor complexes is PSD-95. PSD-95 is a member of the family of MAGUK proteins, which shares five protein-binding motifs, including three pentylenetetrazole (PDZ)-binding domains (Gomperts, 1996). These PDZ domains bind to GluN2 subunits, which serve as a scaffold for the subunit assembly. In addition, they associate with other signaling proteins, including nNOS and SynGap and thus mediate activation of downstream molecular events of NMDA receptor complexes.

Other important components of NMDA receptor complexes are kinases/phosphatases that regulate phosphorylation of GluN subunits and the activity of the ion channels (Kennedy, 1998; Salter and Kalia, 2004). Of these are CAMKII, Src/Fyn kinase, PKC, and PKA. CAMKII is a highly abundant protein in the PSD, which is a target for Ca^{2+} flowing through the NMDA receptor and is critical for synaptic plasticity (Chakravarthy et al., 1999). On activation, CAMKII binds to the PSD via its carboxyl-terminal domains and associates with the cytosolic tails of the GluN2 subunits, which serve as a docking site for this kinase in the PSD. As a result, the catalytic site of CAMKII is positioned such that it could be readily activated by Ca^{2+} coming through the channel and, in turn, could phosphorylate GluN2 subunits.

Members of the Src/Fyn family kinases (SFKs) are nonreceptor protein tyrosine kinases (PTKs) that serve various intracellular functions. Of these is their ability to phosphorylate tyrosine residues of NR2 subunits, which enhances the kinetics of the ion channels of the receptor complexes. SFKs are a crucial point of convergence for signaling pathways that enhance NMDA receptor activity and thus could serve as a hub for the control of NMDA receptors (Salter and Kalia, 2004). Recent studies have shown that Src kinase plays an important role in integrating various molecular alterations within NMDA receptor complexes in schizophrenia (Banerjee et al., 2015).

AMPA receptors

AMPA receptors (or GluRs), a class of ionotropic glutamate receptors, and their associated signaling partners are important regulators of synaptic plasticity in postsynaptic neurons (Henley and Wilkinson, 2013; Sheng and Kim, 2002). A functional AMPA receptor complex is a hetero-tetramer of GluRs, each of which contains four transmembrane domains and a long intracellular C-terminus. GluRs consist of four subunit types, 1 through 4, each of which is linked to specific intracellular signaling partners. AMPA receptor complexes containing GluR1 and GluR2 subunits are involved with synaptic activity–related transmission in adult brains. GluR2- and GluR3-containing receptors maintain a steady synaptic pool of AMPA receptors constitutively and are rapidly replaced by GluA1-containing types in an activity-dependent fashion. GluA4 subtypes are expressed in earlier stages of development (Henley and Wilkinson, 2013; Zhu et al., 2002).

The mechanisms of LTP/LTD induction involving AMPA receptors are highly regulated, partly via intricate interactions with NMDA receptor signaling. Activation of AMPA receptors induces sodium influx through the channels, which in turn overcomes the voltage-dependent Mg^{++} blockade of NMDA receptors. The calcium influx resulting from this triggers a series of signal transduction cascades involving kinases, phosphatases, and scaffolding proteins. The strength of depolarization and activation of specific groups of signaling molecules determine whether AMPA receptors are inserted into the membrane, which can be a basis for induction of LTP, or endocytosed from the membrane, causing LTD. CAMKII, MAPK, and PKA are involved with membrane insertion of AMPA receptor complexes in LTP via phosphorylation of GluR1, whereas PKC-dependent phosphorylation of GluR2 occurs during LTD removing AMPA receptor complexes from the postsynaptic membranes. In addition, calcineurine, PP2A, Sap97, MyosineVI, GRIP1, PICK1, stargazine, and PSD-95 are critical for AMPA receptor signaling because they regulate insertion or removal of AMPA receptors in the PSD, thereby maintaining synaptic plasticity (Borgdorff and Choquet, 2002; Henley and Wilkinson, 2013; Zhu et al., 2002).

mGluRs

Metabotropic glutamate receptors (mGluRs) are G-protein-coupled receptors (GPCRs) that are widely expressed in the central nervous system (Conn and Pin, 1997; Niswender and Conn, 2010; Rondard and Pin, 2015). Group I mGluRs (mGluRs 1 and 5) are mostly localized in postsynaptic neurons, whereas group II mGluRs (mGluRs 2 and 3) and group III mGluRs (mGluRs 4, 6, 7, and 8) (Niswender and Conn, 2010) are mostly localized at the presynaptic terminals. The group I mGluRs couple to G_q/G_{11} and activate phospholipase C_β, which results in the hydrolysis of phosphoinositides and generation of inositol 1,4,5-trisphosphate (IP_3) and diacyl-glycerol. This classical pathway leads to calcium mobilization and activation of PKC. Depending on the cell type or neuronal population, group I mGluRs can activate a range of downstream effectors, including the mitogen-activated protein kinase/extracellular receptor kinase (MAPK/ERK) pathway and the mammalian target of rapamycin (MTOR)/p70 S6 kinase pathway, which play important roles in the regulation of synaptic plasticity by group I mGluRs (Martinez-Lozada and Ortega, 2015; Niswender and Conn, 2010). Group II and group III mGluRs are coupled predominantly to $G_{i/o}$ proteins, which work through adenylyl cyclase inhibition and directly regulate ion channels and other downstream signaling partners via liberation of $G_{\beta\gamma}$ subunits. In addition, group II and group III mGluRs are also coupled to other signaling pathways, including activation of MAPK and phosphatidyl inositol 3-kinase PI3 kinase pathways (Iacovelli et al., 2002), providing further complexity regarding the mechanisms by which these receptors can regulate synaptic transmission.

mGluRs are associated with Homer–Shank complexes, a critical component of PSD scaffolds. Within the PSD scaffolds, Shanks (also called ProSAP, Synamon, CortBP, Spank, and SSTRIP) and Homers (also called Vesl, Cupidin, and PSD-Zip45) are located more toward the cytoplasmic face of the PSD compared to PSD-95 and other MAGUKs (Sheng, 2001). Shank has PDZ domains that facilitate associations with GKAP (guanylate kinase–associated protein) and SAM domains enabling associations between Shanks as the internal consensus PPXXF motif via which it binds to Homer (Tao-Cheng et al., 2015; Tu et al., 1999). The N-terminal region of Homer has a domain of at least 110 amino acids called the EVH1 domain, through which Homer selectively binds the C-terminal region of group I mGluRs (mGluR1 and mGluR5) (Tu et al., 1999). Homer proteins also contain a coiled–coil motif at their C-terminal region, which facilitates self-association (Sheng, 2001; Xiao et al., 1998) and forms multivalent complexes that can cross-talk with other proteins with PPXXF motif, including those of the IP3 receptor (IP3R). Thus, it is believed that Homer brings IP3R in close proximity to the mGluRs and intracellular calcium stores, thus the downstream signaling (Kennedy, 2000; Sheng, 2001; Xiao et al., 1998).

mGluRs are expressed widely in the prefrontal cortex (PFC) and striatum, which are regions often implicated in the pathophysiology of schizophrenia. It has been well documented that among the three mGluR subtypes, group I ($mGluR_5$) and group II ($mGluR_2$ and $mGluR_3$), are colocalized with NMDA receptors and possibly involved in the circuitry disruptions implicated in schizophrenia (Vinson and

Conn, 2012). Group I mGluRs have a positive regulatory function in excitatory NMDA receptor activity, whereas the other two classes function negatively by suppressing the intracellular cAMP level, which inhibits the export of neurotoxic glutamate from microglia and plays a neuroprotective role (Kritis et al., 2015).

PROTEOMIC LANDSCAPE OF THE PSD

Receptor signaling activity in the PSD is critically modulated by the protein composition in the microdomain (Kennedy, 2000; Sheng and Hoogenraad, 2007). Biochemical analysis of the PSD proteome has therefore been an important research avenue and has provided a great deal of insight into the PSD biology (Grant, 2006; Grant and Blackstock, 2001). Brain homogenates can be readily enriched for the PSD by the methods that were originally devised decades ago (Carlin et al., 1980; Phillips et al., 2001, 2005) and have been modified in various iterations (Goebel-Goody et al., 2009; Hahn et al., 2009). Typically, brain homogenates are fractionated by sucrose density gradient, yielding synaptosomes that can then be precipitated in the presence of 1% Triton-X 100 for enrichments for the PSD. Thus, subcellular fractions enriched for the PSD are obtained as Triton-X 100 insoluble synaptic membranes. More recently, Triton-X 100 precipitation has been fine-tuned by varying the pH, which can also permit separation of extrasynaptic membranes and synaptic vesicular fractions (Phillips et al., 2001, 2005).

Subcellular fractions enriched for the PSD have been analyzed by immunoblotting as well as 2D gel electrophoresis (Banerjee et al., 2015; Goebel-Goody et al., 2009; Hahn et al., 2009). Subsequently, mass spectrometry (MS) analyses became an important approach for its ability to assess multitudes of proteins in the PSD. These include matrix-assisted laser desorption/ionization time-of-flight (MALDI-TOF) MS and liquid chromatography coupled with tandem MS (Li et al., 2004; Walikonis et al., 2000). In addition, a highly quantitative method using selective reaction monitoring–based MS in conjunction with isotope-labeled standards now permits accurate and simultaneous assessment of more than hundreds of proteins in subcellular fractions (MacDonald et al., 2012). Macromolecular complexes constructed via protein–protein interactions account for the majority of the PSD proteome. Thus, the PSD proteome has been examined for major receptor complexes or scaffolding proteins such as GluN subunits or PSD-95 using immunoprecipitation capture of complexes (Banerjee et al., 2015; Hahn et al., 2009).

Husi and associates conducted MS and immunoblotting analyses of NMDA receptor complexes isolated from mouse brain (Husi and Grant, 2001; Husi et al., 2000). They found proteins organized into the groups of receptors, scaffolding proteins, signaling molecules, and cell adhesion and cytoskeletal proteins. Interestingly, NMDAR complexes were linked to cadherins and L1 cell adhesion molecules and were also associated with kinases, phosphatases, and GTPase-activating proteins (Fig. 8.1). Another protein complex that has been examined as a macromolecular complex in the PSD is captured by immunoprecipitation with antibodies for PSD-95. Dosemeci and colleagues conducted affinity purification of PSD-95 complexes in subcellular fractions derived from rat brains enriched for the PSD (Dosemeci et al., 2006, 2007). The PSD-95 immunoprecipitations included not only NMDAR complexes but also AMPA receptors in abundance. In addition, they found small G-protein regulators, cell adhesion molecules, and others, which is grossly similar to that found in the study by Husi and associates.

Peng and colleagues analyzed purified PSD fractions by liquid chromatography coupled with tandem MS (LC–MS/MS) (Peng et al., 2004). They identified 374 proteins in the PSD, which included receptors and ion channels, cell adhesion molecules, cytoskeletal molecules, translational molecules,

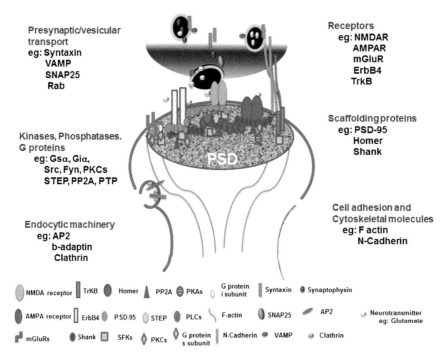

FIGURE 8.1

Molecular architecture of the postsynaptic density. The postsynaptic density is essentially a protein assembly that consists of multiple groups of proteins that are critical for postsynaptic function. These include: receptors such as glutamatergic and tyrosine kinase receptors; scaffolding proteins such as molecular infrastructure; kinases and phosphatases, which govern downstream signal transduction of the receptors; endocytic machinery, which modulate trafficking of molecules in and out of the PSD; adhesion molecules that maintain the apposition of presynaptic and postsynaptic membranes; and on the presynaptic side, vesicular transport proteins regulate the assembly, transport and release of neurotransmitters, which in turn trigger postsynaptic signaling via binding to the receptors on the postsynaptic side of the membrane.

kinases/phosphatases, as well as scaffold proteins. Utilizing a quantitative method with synthetic isotope-labeled peptides as internal standards, the same group further measured molar abundance of key PSD proteins, which led to a highly quantitative assessment of key PSD proteins and stoichiometric ratios between glutamate receptors, scaffolding proteins, and signaling molecules in the PSD (Cheng et al., 2006).

Because subcellular fractions often include contaminants, affinity purification of PSD-95 was further refined using targeted tandem affinity purification (TAP) tags. Fernandez and associates captured PSD-95 complexes by fusing the C-terminus of PSD-95 with TAP tags and conducting immunoaffinity purification (Fernandez et al., 2009). In this more stringent affinity purification of macromolecular complexes, they found 118 proteins comprising crucial functional groups, including glutamate receptors, scaffolding, and signaling proteins.

It was important to delineate the protein composition of the PSD derived from human brains compared to that of rodents as described. Hahn and associates made the first attempt to obtain subcellular fractions enriched for the PSD from human postmortem brain tissues by using a few variations of the previously developed methods as described (Hahn et al., 2009). They found that PSD enrichments can be obtained from human postmortem brains, with the purity verified by extensive immunoblotting analyses, EM, and MS analyses (MacDonald and Chafee, 2006). Notably, the results from the LC-MS/MS platform revealed groups of proteins (Hahn et al., 2009; MacDonald and Chafee, 2006) that were strikingly similar to those shown in rodent brains (Trinidad et al., 2006, 2008).

Adding to the outcomes of electrophysiological and signal transduction studies, proteomic analyses allow us to construct a comprehensive view of the architecture and function of this microdomain. The PSD proteome has been examined by multiple groups, and their findings highlight several groups of proteins as major constituents of the microdomain, including glutamate receptors as major neurotransmitter receptors and cytoskeletal proteins critical for the architecture of dendritic spines, such as adhesion molecules for synaptic adhesion, scaffolding proteins, kinases, and phosphatases (Fig. 8.1). These multitudes of proteins are organized into a hierarchy of protein structures—individual proteins, simple protein complexes, and macromolecular complexes (Grant, 2013). Some of these components are conserved during evolution and are connected to neurocognitive functions that are altered in various neuropsychiatric illnesses such as schizophrenia.

EVIDENCE SUPPORTING THE ROLE OF THE PSD IN SCHIZOPHRENIA
GENETIC EVIDENCE

One of the most remarkable advancements in schizophrenia research was identification of genetic variants that are associated with the illnesses (Hall et al., 2015; Harrison and Owen, 2003). Because of the recent advent of new technologies such as genome-wide microarrays and deep sequencing, it is now possible to conduct a genome-wide survey of common variants (Schizophrenia Working Group of the Psychiatric Genomics Consortium, 2014), copy number variants (CNVs), and mutations (Kirov et al., 2012; Fromer et al., 2014) and variants. Also important was the formation of the worldwide consortia, which led to surveying DNA derived from large populations of patients. Strikingly, the results from these studies involving common as well as rare variants consistently pointed to the genes that are highly enriched in and critical for the signaling of the PSD. Kirov and associates conducted CNV analysis on 66 schizophrenia proband–parent trios and found that rare de novo CNVs were significantly more frequent in cases than in controls and that de novo CNVs were significantly enriched within the PSD proteome. These striking changes were largely attributed to enrichment of the genes that are part of NMDAR or activity-regulated cytoskeleton-associated protein (ARC) postsynaptic signaling complexes. Based on exome sequencing results from 623 schizophrenia trios, another group (Fromer et al., 2014) also reported that small de novo mutations affecting up to a few nucleotides are over-represented in postsynaptic proteins also clustered around two pathways, NMDAR and ARC. Also, rare disruptive mutations were found across many genes and still showed further enrichment in proteins that interact with these complexes to modulate synaptic strength, such as those regulating actin filament dynamics (Fromer et al., 2014). In addition, another group (Purcell et al., 2014) conducted exome sequencing for more than 2500 patients with schizophrenia and controls and showed that polygenic burden arises from ion channels and ARC complexes in the PSD.

Studies of common genetic variants have consistently pointed to signaling pathways enriched in the PSD. In the pre-GWAS era, a number of candidate genes and their pathways were found to be associated with schizophrenia converged on NMDA receptor complexes (Harrison and Weinberger, 2005). These included nrg 1, erbB4, dystrobrevin 1, and DISC-1, all of which were known to impact signaling activity of the NMDA receptor pathway. More recently, the largest study of common variants in schizophrenia to date found 108 genomic loci of significance in genome-wide association studies (GWAS), which included the loci containing glutamate receptors, members of the calcium ion channel family of proteins, and many genes involved with synaptic plasticity (Schizophrenia Working Group of the Psychiatric Genomics Consortium, 2014). Together, these indicate that signaling pathways that comprise the major regulators are highly enriched in the PSD.

POSTMORTEM EVIDENCE FOR PSD DYSREGULATION IN SCHIZOPHRENIA

Dendritic spine alterations

Several groups have examined the density of dendritic spines in neocortical regions in human postmortem brains of individuals with schizophrenia. The earliest study using the Golgi method found that spine densities of pyramidal neurons in layer 3 of the PFC decreased by 63% and 59% in the frontal cortex and temporal cortex, respectively (Garey et al., 1998). This finding was replicated for pyramidal neurons in deep layer 3 of the dorsolateral PFC, but it was not observed in the visual cortex of individuals with schizophrenia (Glantz and Lewis, 2000). Spine densities in deep layer 3 of the primary and associated auditory cortices were also found to be decreased using a stereologic immunohistochemical measurement of a marker of dendritic spines (Sweet et al., 2009). However, spine densities examined in layers 5 and 6 in the dorsolateral PFC were unaltered in schizophrenia (Kolluri et al., 2005), leading to the conclusion that reductions in spine densities in schizophrenia are both layer- and region-specific.

The layer 3 pyramidal neurons function to integrate cortico-cortical and cortico-thalamic inputs (Kritzer and Goldman-Rakic, 1995; Melchitzky et al., 2001). It is therefore hypothesized that decreased spine densities in layer 3 may contribute to, or be a reflection of, altered excitatory inputs to these regions in schizophrenia (Glantz and Lewis, 2000; Kolluri et al., 2005; Sweet et al., 2009). Although antipsychotic exposure is a common potential confound in postmortem brain studies, neither rodent studies (Vincent et al., 1991; Wang and Deutch, 2008) nor a nonhuman primate study (Sweet et al., 2009) involving 6 months to 1 year of antipsychotic exposure found decreases in spine densities as reported in human postmortem studies of schizophrenia.

Studies of spine pathology in noncortical areas are more limited. Decreased spine size has been observed in the striatum (Roberts et al., 1996), and increased axo-spinous synaptic density has been observed in the caudate, but not the putamen, using EM (Roberts et al., 2005). Studies of mossy fiber axo-spinous synapse density in the hippocampus are mixed, with one negative (Kolomeets et al., 2005) and one reporting decreases (Kolomeets et al., 2007). Altogether, data for spine alterations in noncortical areas remain to be delineated.

DYSREGULATIONS IN PSD PROTEINS AND THEIR SIGNALING IN SCHIZOPHRENIA

PSD molecules can be categorized into a few functional groups, such as neurotransmitter receptors, scaffolding proteins, the kinases and phosphatases, and adhesion molecules. Multiple groups have examined the expression of PSD proteins and reported alterations in the postmortem brains of patients with schizophrenia compared to their matched controls. These investigations, based on in situ hybridization, quantitative RT-PCR, and Western blotting, have described interesting changes in PSD molecules in schizophrenia, although the results have not always been consistent between cohorts and methods. Within the NMDA receptor complex, the GluN1 subunit was found to be decreased in several areas of the cortex, namely the superior frontal gyrus (Sokolov, 1998), PFC (Beneyto and Meador-Woodruff, 2008), dorso-lateral prefrontal cortex (DLPFC) (Weickert et al., 2013; Errico et al., 2013), and temporal cortex (Humphries et al., 1996), whereas studies involving different postmortem brain cohorts found increases in the GluN1 subunit in the DLPFC, occipital (Dracheva et al., 2001), anterior cingulate (Kristiansen et al., 2010), and temporal cortices. Similarly, some investigators found GluN2A to be decreased in the anterior cingulate cortex (ACC) and DLPFC (Woo et al., 2004; Beneyto and Meador-Woodruff, 2008; Errico et al., 2013), whereas others observed increased GluN2A in the ACC (Woo et al., 2008) and occipital cortex (Dracheva et al., 2001). GluN2B protein has been found to be decreased in PFC (Errico et al., 2013) and DLPFC (Kristiansen et al., 2010) and increased in the dorsomedial thalamus of patients with schizophrenia (Clinton et al., 2006). GluN2C was reported to be decreased in the frontal pole (Akbarian et al., 1996) in PFC (Beneyto and Meador-Woodruff, 2008) and in DLPFC (Weickert et al., 2013), and GluN3A protein expression was found to be upregulated in DLPFC (Mueller and Meador-Woodruff, 2004) of patients with schizophrenia.

Among the scaffolding and cytoskeletal proteins, neurofilament-L (NF-L) in the DLPFC and both PSD-95 and PSD-93 mRNAs were found to be increased in the ACC, whereas all three molecules showed decreased protein levels (Kristiansen et al., 2006). In a separate study, no change was observed in PSD-95 in the DLPFC (Hahn et al., 2006) of patients with schizophrenia. SAP-97 protein level has been reported to be significantly decreased in the PFC of schizophrenia patients, along with a concomitant decrease in its binding partner GluA1 (Toyooka et al., 2002).

Among other susceptible schizophrenia risk genes located in or functionally associated with PSD and schizophrenia, NRG 1, a type I mRNA, was found to be increased in schizophrenia DLPFC (Hashimoto et al., 2004). This result was corroborated by another study that found increased NRG1 and its receptor ErbB4 proteins in DLPFC of schizophrenia patients (Chong et al., 2008).

Expression of G-protein-coupled receptor kinase, which uncouples GPCRs from G proteins and regulates downstream signaling, was found to be increased in the ACC in schizophrenia (Funk et al., 2014). Protein expression of serine racemase, which synthesizes D-serine, an NMDA receptor coagonist in the brain, was reported to be increased in the hippocampus in schizophrenia (Steffek et al., 2006).

Numerous pharmacologic, behavioral, and genetics studies support the NMDA receptor hypofunction hypothesis (Coyle et al., 2010; Javitt, 2007) as one of the leading postulates for the pathophysiology of schizophrenia. Yet, studies of protein expression levels of NMDA receptors and its subunits have been mixed, as noted. The challenge for the field has been to develop methods for examining NMDA receptor signaling in human brains. Our group first made the observation of decreases in the tyrosine phosphorylation NMDA receptor subunit 2 in the PSD of the DLPFC in patients with schizophrenia (Hahn et al., 2006). This was followed by the finding of decreased activity of PKC, Pyk2, and Src kinase in NMDAR complexes. Src kinase function was found to be decreased in schizophrenia independently of upstream changes in NMDARs, implicating postreceptor kinases in the pathophysiology of schizophrenia (Banerjee et al., 2015) (Fig. 8.2).

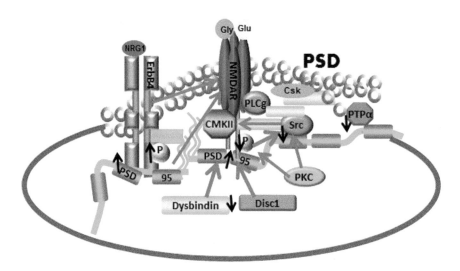

FIGURE 8.2

Various molecular abnormalities in the PSD converge on the NMDA receptor pathway in schizophrenia. A number of molecular pathways that have been implicated for the pathophysiology of schizophrenia reside in the PSD. Key molecules of these pathways are altered in the PSD, including erbB4, dysbindin-1, PSD-95, and rPTPa. ErbB4 and PSD-95 were found to be increased in NR complexes in the DLPFC of schizophrenia, which will decrease GluN2 tyrosine phosphorylation. rPTPa and dysbindin-1 were decreased, which are expected to decrease NR function. Importantly, these alterations were found in association with each protein with NMDA receptor complexes, pointing to protein–protein interactions as an important mechanism by which various molecular abnormalities converge on the NMDA receptor pathway.

PSD AS A MICRODOMAIN FOR CONVERGING MOLECULAR ALTERATIONS IN SCHIZOPHRENIA

The evidence summarized here point to the PSD as a subcellular locale in which various etiologic factors reside, interact, and converge on common final pathways. The next important question will be in regard to how such convergence occurs and what role the PSD serves in this context. Presently, we do not have a knowledge base from which to construct a model that can explain various mechanisms for convergence. Notably, however, recent studies on NMDA receptor hypofunction in schizophrenia have begun to offer insights into how abnormalities in various pathways could interact and impact a common final pathway in the PSD. By way of example, we discuss the impact of various molecular abnormalities on the NMDA receptor pathway in schizophrenia and consider possible underlying mechanisms that are orchestrated in the PSD (Fig. 8.2).

It has been proposed that NMDAR signaling could be a point of convergence for various candidate susceptibility genes for schizophrenia (Harrison and Owen, 2003; Harrison and Weinberger, 2005). Genetics studies have identified common and rare variants in GluN1 and GluN2, which are subunits of NMDA receptors and which may underpin altered NMDA receptor signaling in some individuals. Given a fairly low incidence of these genetic variants among patients, however, NMDA receptor hypofunction is likely to be precipitated by multiple other factors that are outside the pathway. There are multiple genes or pathways that have been implicated for schizophrenia and may impact NMDAR signaling via their interactions in the PSD. As shown in Fig. 8.2, these candidate pathways include neuregulin-1, erbB4, DISC-1, serine racemase, and dysbindin-1, each of which can affect NMDA receptor signaling at the postsynaptic side. Neuregulin–erbB4 signaling was found to be altered in the DLPFC of schizophrenia (Hahn et al., 2006), which can dysregulate NMDA receptor signaling. In addition, dysbindin-1 was decreased in the DLPFC and hippocampus, which can precipitate NMDA receptor hypofunction (Talbot et al., 2004).

How could these pathways converge on the NMDA receptor signaling in the PSD? The PSD is a protein assembly in which many signaling pathways are brought together and their interactions are thereby facilitated. These events can impact the recruitment and posttranslational modification or activation of key proteins in pathways and modulate their activities. An important modulator of these processes is the intermolecular proximity, for which protein–protein associations play a key role. Some of the susceptibility genes can more directly impact protein associations of NMDAR complexes (Fig. 8.2). DISC-1 associates with AKAP-9 and alpha-actinin-1 (Brandon, 2007), which are binding partners of NMDAR. Altered DISC-1 expression can therefore modulate the relationships of AKAP-9 and alpha-actinin-1 with other signaling proteins of NMDAR complexes (Camargo et al., 2007). Dysbindin-1, which was also found in the PSD (Talbot et al., 2006), can bind the exocyst protein sec8 (Camargo et al., 2007), which is critical to NMDAR trafficking. Thus, the decreased level of dysbindin observed in schizophrenia (Talbot et al., 2004; Weickert et al., 2004, 2008) can impact the protein association and trafficking of NMDAR.

Protein–protein interactions could also be a mechanism by which susceptibility pathways could interact and conspire to impact the NMDA receptor pathway. In recent studies, decreased tyrosine phosphorylation in NR2 subunits, as noted, was found to be mediated by hypoactivity of Src kinases (nonreceptor protein kinases) in the DLPFC of patients with schizophrenia (Banerjee et al., 2015). Src kinase activity is governed by rPTPa, CSK, erbB4, and PSD-95. These molecules are either directly or indirectly associated with Src in the PSD, which impacts their modulation of Src kinase. Decreased tyrosine phosphorylation of NR2 subunits was accompanied by decreased rPTPa and increased erbB4 and PSD-95 in NMDA receptor complexes.

SUMMARY AND FUTURE DIRECTIONS

Multiple lines of evidence point to the PSD as a microdomain in which various molecular alterations converge and conspire to form neurobiological substrates for schizophrenia. Genetics studies indicate that common and rare variants and mutations that are associated with schizophrenia are highly enriched among the genes comprising the PSD signaling compared to other pathways. Among the most consistent observations made in the postmortem brains of patients with schizophrenia are altered dendritic spines, which are unmistakable ultrastructural signatures of PSD dysregulations in the illness. Moreover, there is now evidence that PSD proteins and their transcripts as well as receptor signaling are dysregulated in postmortem brains of patients with schizophrenia. Understanding the mechanisms by which various etiologic factors interact and conspire to precipitate disease pathology is a challenging goal in the investigation of many complex trait disorders. As such, unpacking the mechanisms for convergence will be an important and necessary step in future studies.

Of particular interest are protein–protein associations leading to posttranslational modification of key signaling proteins. The PSD is a macromolecular protein assembly in which various receptor pathways are physically interconnected and thereby modulate each other. These complex interactions appear to occur in an orderly fashion, in part due to a hierarchical organization of protein complexes in the PSD; PSD proteins can interact with individual proteins, with simple protein complexes, or with macromolecular complexes (Grant, 2013). This hierarchy is built on protein associations that underpin the interactions among proteins and pathways. As noted in the example of Src alterations described, protein associations involving key signaling proteins can be molecular substrates for interactions and convergence between molecular alterations in the PSD.

The structure–function interplay between the PSD and dendritic spines is also important. Equipped with their own microorganelles, dendritic spines reflect overall activity of PSD signaling via ultrastructural changes, which in turn impact PSD signaling as a whole. Thus, ultrastructural changes in the dendritic spines have higher-order impacts on the PSD proteome and signaling. In this regard, further studies involving cytoskeletal and other structural proteins and their dysregulations in schizophrenia will be critically important.

Essential for synaptic transmission is the apposition of presynaptic and postsynaptic membranes, which are regulated by adhesion molecules in synapses. Alterations in adhesion molecules have been well established in postmortem studies of autism and other developmental disorders, and similar evidence for schizophrenia is beginning to emerge. How perturbations of synaptic adhesion can impact PSD proteome and signaling will be an important avenue.

Finally, brain region specificity will also be an important issue to be addressed. Increasing evidence indicates that different brain regions may have differing characteristics in protein composition and PSD signaling. Thus, alterations in the PSD signaling could differ between brain regions, which will offer insight into how similar sets of molecular alterations in the same subjects could lead to different outcomes via region-specific mechanisms for convergence.

REFERENCES

Adrian, M., Kusters, R., Wierenga, C.J., Storm, C., Hoogenraad, C.C., Kapitein, L.C., 2014. Barriers in the brain: resolving dendritic spine morphology and compartmentalization. Front. Neuroanat. 8, 142. http://dx.doi.org/10.3389/fnana.2014.00142.

Akbarian, S., Sucher, N.J., Bradley, D., Tafazzoli, A., Trinh, D., Hetrick, W.P., et al., 1996. Selective alterations in gene expression for NMDA receptor subunits in prefrontal cortex of schizophrenics. J. Neurosci. 16 (1), 19–30. Retrieved from: <http://www.ncbi.nlm.nih.gov/pubmed/8613785>.

Banerjee, A., Wang, H.Y., Borgmann-Winter, K.E., MacDonald, M.L., Kaprielian, H., Stucky, A., et al., 2015. Src kinase as a mediator of convergent molecular abnormalities leading to NMDAR hypoactivity in schizophrenia. Mol. Psychiatry 20 (9), 1091–1100. http://dx.doi.org/10.1038/mp.2014.115.

Baron, M.K., Boeckers, T.M., Vaida, B., Faham, S., Gingery, M., Sawaya, M.R., et al., 2006. An architectural framework that may lie at the core of the postsynaptic density. Science 311 (5760), 531–535. http://dx.doi.org/10.1126/science.1118995.

Bayes, A., van de Lagemaat, L.N., Collins, M.O., Croning, M.D., Whittle, I.R., Choudhary, J.S., et al., 2011. Characterization of the proteome, diseases and evolution of the human postsynaptic density. Nat. Neurosci. 14 (1), 19–21. http://dx.doi.org/10.1038/nn.2719.

Bayes, A., Collins, M.O., Galtrey, C.M., Simonnet, C., Roy, M., Croning, M.D., et al., 2014. Human post-mortem synapse proteome integrity screening for proteomic studies of postsynaptic complexes. Mol. Brain 7, 88. http://dx.doi.org/10.1186/s13041-014-0088-4.

Beneyto, M., Meador-Woodruff, J.H., 2008. Lamina-specific abnormalities of NMDA receptor-associated postsynaptic protein transcripts in the prefrontal cortex in schizophrenia and bipolar disorder. Neuropsychopharmacology 33 (9), 2175–2186. http://dx.doi.org/10.1038/sj.npp.1301604.

Blomberg, F., Cohen, R.S., Siekevitz, P., 1977. The structure of postsynaptic densities isolated from dog cerebral cortex. II. Characterization and arrangement of some of the major proteins within the structure. J. Cell Biol. 74 (1), 204–225. Retrieved from: <http://www.ncbi.nlm.nih.gov/pubmed/406264>.

Borgdorff, A.J., Choquet, D., 2002. Regulation of AMPA receptor lateral movements. Nature 417 (6889), 649–653. http://dx.doi.org/10.1038/nature00780.

Bourne, J.N., Harris, K.M., 2008. Balancing structure and function at hippocampal dendritic spines. Annu. Rev. Neurosci. 31, 47–67. http://dx.doi.org/10.1146/annurev.neuro.31.060407.125646.

Braff, D.L., Freedman, R., Schork, N.J., Gottesman, I.I., 2007. Deconstructing schizophrenia: an overview of the use of endophenotypes in order to understand a complex disorder. Schizophr. Bull. 33 (1), 21–32. http://dx.doi.org/10.1093/schbul/sbl049.

Brandon, N.J., 2007. Dissecting DISC1 function through protein-protein interactions. Biochem. Soc. Trans. 35 (Pt 5), 1283–1286. http://dx.doi.org/10.1042/BST0351283.

Camargo, L.M., Collura, V., Rain, J.C., Mizuguchi, K., Hermjakob, H., Kerrien, S., et al., 2007. Disrupted in Schizophrenia 1 Interactome: evidence for the close connectivity of risk genes and a potential synaptic basis for schizophrenia. Mol. Psychiatry 12 (1), 74–86. http://dx.doi.org/10.1038/sj.mp.4001880.

Carlin, R.K., Grab, D.J., Cohen, R.S., Siekevitz, P., 1980. Isolation and characterization of postsynaptic densities from various brain regions: enrichment of different types of postsynaptic densities. J. Cell Biol. 86 (3), 831–845. Retrieved from: <http://www.ncbi.nlm.nih.gov/pubmed/7410481>.

Chakravarthy, B., Morley, P., Whitfield, J., 1999. Ca2+-calmodulin and protein kinase Cs: a hypothetical synthesis of their conflicting convergences on shared substrate domains. Trends Neurosci. 22 (1), 12–16. Retrieved from: <http://www.ncbi.nlm.nih.gov/pubmed/10088994>.

Chen, X., Winters, C., Azzam, R., Li, X., Galbraith, J.A., Leapman, R.D., et al., 2008. Organization of the core structure of the postsynaptic density. Proc. Natl. Acad. Sci. U.S.A. 105 (11), 4453–4458. http://dx.doi.org/10.1073/pnas.0800897105.

Cheng, D., Hoogenraad, C.C., Rush, J., Ramm, E., Schlager, M.A., Duong, D.M., et al., 2006. Relative and absolute quantification of postsynaptic density proteome isolated from rat forebrain and cerebellum. Mol. Cell. Proteomics 5 (6), 1158–1170. http://dx.doi.org/10.1074/mcp.D500009-MCP200.

Chong, V.Z., Thompson, M., Beltaifa, S., Webster, M.J., Law, A.J., Weickert, C.S., 2008. Elevated neuregulin-1 and ErbB4 protein in the prefrontal cortex of schizophrenic patients. Schizophr. Res. 100 (1–3), 270–280. Retrieved from: <http://www.ncbi.nlm.nih.gov/entrez/query.fcgi?cmd=Retrieve&db=PubMed&dopt=Citation&list_uids=18243664>.

Clinton, S.M., Haroutunian, V., Meador-Woodruff, J.H., 2006. Up-regulation of NMDA receptor subunit and postsynaptic density protein expression in the thalamus of elderly patients with schizophrenia. J. Neurochem. 98 (4), 1114–1125. http://dx.doi.org/10.1111/j.1471-4159.2006.03954.x.

Colledge, M., Snyder, E.M., Crozier, R.A., Soderling, J.A., Jin, Y., Langeberg, L.K., et al., 2003. Ubiquitination regulates PSD-95 degradation and AMPA receptor surface expression. Neuron 40 (3), 595–607. Retrieved from: <http://www.ncbi.nlm.nih.gov/pubmed/14642282>.

Collins, M.O., Yu, L., Coba, M.P., Husi, H., Campuzano, I., Blackstock, W.P., et al., 2005. Proteomic analysis of in vivo phosphorylated synaptic proteins. J. Biol. Chem. 280 (7), 5972–5982. http://dx.doi.org/10.1074/jbc.M411220200.

Conn, P.J., Pin, J.P., 1997. Pharmacology and functions of metabotropic glutamate receptors. Annu. Rev. Pharmacol. Toxicol. 37, 205–237. http://dx.doi.org/10.1146/annurev.pharmtox.37.1.205.

Coyle, J.T., 2006. Glutamate and schizophrenia: beyond the dopamine hypothesis. Cell. Mol. Neurobiol. 26 (4–6), 365–384. http://dx.doi.org/10.1007/s10571-006-9062-8.

Coyle, J.T., Balu, D., Benneyworth, M., Basu, A., Roseman, A., 2010. Beyond the dopamine receptor: novel therapeutic targets for treating schizophrenia. Dialogues Clin. Neurosci. 12 (3), 359–382. Retrieved from: <http://www.ncbi.nlm.nih.gov/pubmed/20954431>.

Dani, A., Huang, B., Bergan, J., Dulac, C., Zhuang, X., 2010. Superresolution imaging of chemical synapses in the brain. Neuron 68 (5), 843–856. http://dx.doi.org/10.1016/j.neuron.2010.11.021.

de Bartolomeis, A., Latte, G., Tomasetti, C., Iasevoli, F., 2014. Glutamatergic postsynaptic density protein dysfunctions in synaptic plasticity and dendritic spines morphology: relevance to schizophrenia and other behavioral disorders pathophysiology, and implications for novel therapeutic approaches. Mol. Neurobiol. 49 (1), 484–511. http://dx.doi.org/10.1007/s12035-013-8534-3.

Dosemeci, A., Tao-Cheng, J.H., Vinade, L., Winters, C.A., Pozzo-Miller, L., Reese, T.S., 2001. Glutamate-induced transient modification of the postsynaptic density. Proc. Natl. Acad. Sci. U.S.A. 98 (18), 10428–10432. http://dx.doi.org/10.1073/pnas.181336998.

Dosemeci, A., Tao-Cheng, J.H., Vinade, L., Jaffe, H., 2006. Preparation of postsynaptic density fraction from hippocampal slices and proteomic analysis. Biochem. Biophys. Res. Commun. 339 (2), 687–694. Retrieved from: <http://www.ncbi.nlm.nih.gov/entrez/query.fcgi?cmd=Retrieve&db=PubMed&dopt=Citation&list_uids=16332460>.

Dosemeci, A., Makusky, A.J., Jankowska-Stephens, E., Yang, X., Slotta, D.J., Markey, S.P., 2007. Composition of the synaptic PSD-95 complex. Mol. Cell. Proteomics 6 (10), 1749–1760. Retrieved from: <http://www.ncbi.nlm.nih.gov/entrez/query.fcgi?cmd=Retrieve&db=PubMed&dopt=Citation&list_uids=17623647>, <http://www.mcponline.org/content/6/10/1749.full.pdf>.

Dracheva, S., Marras, S.A., Elhakem, S.L., Kramer, F.R., Davis, K.L., Haroutunian, V., 2001. N-methyl-D-aspartic acid receptor expression in the dorsolateral prefrontal cortex of elderly patients with schizophrenia. Am. J. Psychiatry 158 (9), 1400–1410. Retrieved from: <http://www.ncbi.nlm.nih.gov/entrez/query.fcgi?cmd=Retrieve&db=PubMed&dopt=Citation&list_uids=11532724>.

Ehlers, M.D., 2003. Activity level controls postsynaptic composition and signaling via the ubiquitin-proteasome system. Nat. Neurosci. 6 (3), 231–242. http://dx.doi.org/10.1038/nn1013.

Errico, F., Napolitano, F., Squillace, M., Vitucci, D., Blasi, G., de Bartolomeis, A., et al., 2013. Decreased levels of D-aspartate and NMDA in the prefrontal cortex and striatum of patients with schizophrenia. J. Psychiatr. Res. 47 (10), 1432–1437. http://dx.doi.org/10.1016/j.jpsychires.2013.06.013.

Farley, M.M., Swulius, M.T., Waxham, M.N., 2015. Electron tomographic structure and protein composition of isolated rat cerebellar, hippocampal and cortical postsynaptic densities. Neuroscience 304, 286–301. http://dx.doi.org/10.1016/j.neuroscience.2015.07.062.

Fernandez, E., Collins, M.O., Uren, R.T., Kopanitsa, M.V., Komiyama, N.H., Croning, M.D., et al., 2009. Targeted tandem affinity purification of PSD-95 recovers core postsynaptic complexes and schizophrenia susceptibility proteins. Mol. Syst. Biol. 5, 269. http://dx.doi.org/10.1038/msb.2009.27.

Fromer, M., Pocklington, A.J., Kavanagh, D.H., Williams, H.J., Dwyer, S., Gormley, P., et al., 2014. De novo mutations in schizophrenia implicate synaptic networks. Nature 506 (7487), 179–184. http://dx.doi.org/10.1038/nature12929.

Funk, A.J., Haroutunian, V., Meador-Woodruff, J.H., McCullumsmith, R.E., 2014. Increased G protein-coupled receptor kinase (GRK) expression in the anterior cingulate cortex in schizophrenia. Schizophr. Res. 159 (1), 130–135. http://dx.doi.org/10.1016/j.schres.2014.07.040.

Garey, L.J., Ong, W.Y., Patel, T.S., Kanani, M., Davis, A., Mortimer, A.M., et al., 1998. Reduced dendritic spine density on cerebral cortical pyramidal neurons in schizophrenia. J. Neurol. Neurosurg. Psychiatry 65 (4), 446–453. Retrieved from: <http://www.ncbi.nlm.nih.gov/pubmed/9771764>, <http://www.ncbi.nlm.nih.gov/pmc/articles/PMC2170311/pdf/v065p00446.pdf>.

Gill, I., Droubi, S., Giovedi, S., Fedder, K.N., Bury, L.A., Bosco, F., et al., 2015. Presynaptic NMDA receptors—dynamics and distribution in developing axons in vitro and in vivo. J. Cell. Sci. 128 (4), 768–780. http://dx.doi.org/10.1242/jcs.162362.

Glantz, L.A., Lewis, D.A., 2000. Decreased dendritic spine density on prefrontal cortical pyramidal neurons in schizophrenia. Arch. Gen. Psychiatry 57 (1), 65–73. Retrieved from: <http://www.ncbi.nlm.nih.gov/pubmed/10632234>, <http://archpsyc.jamanetwork.com/data/Journals/PSYCH/11803/yoa9030.pdf>.

Goebel-Goody, S.M., Davies, K.D., Alvestad Linger, R.M., Freund, R.K., Browning, M.D., 2009. Phospho-regulation of synaptic and extrasynaptic N-methyl-D-aspartate receptors in adult hippocampal slices. Neuroscience 158 (4), 1446–1459. http://dx.doi.org/10.1016/j.neuroscience.2008.11.006.

Gomperts, S.N., 1996. Clustering membrane proteins: it's all coming together with the PSD-95/SAP90 protein family. Cell 84 (5), 659–662. Retrieved from: <http://www.ncbi.nlm.nih.gov/pubmed/8625403>.

Grant, S.G., 2006. The synapse proteome and phosphoproteome: a new paradigm for synapse biology. Biochem. Soc. Trans. 34 (Pt 1), 59–63. http://dx.doi.org/10.1042/BST0340059.

Grant, S.G., 2013. SnapShot: organizational principles of the postsynaptic proteome. Neuron 80 (2), 534.e1. http://dx.doi.org/10.1016/j.neuron.2013.10.014.

Grant, S.G., Blackstock, W.P., 2001. Proteomics in neuroscience: from protein to network. J. Neurosci. 21 (21), 8315–8318. Retrieved from: <http://www.ncbi.nlm.nih.gov/pubmed/11606617>, <http://www.jneurosci.org/content/21/21/8315.full.pdf>.

Hahn, C.G., Wang, H.Y., Cho, D.S., Talbot, K., Gur, R.E., Berrettini, W.H., et al., 2006. Altered neuregulin 1-erbB4 signaling contributes to NMDA receptor hypofunction in schizophrenia. Nat. Med. 12 (7), 824–828. Retrieved from: <http://www.ncbi.nlm.nih.gov/entrez/query.fcgi?cmd=Retrieve&db=PubMed&dopt=Citation&list_uids=16767099>.

Hahn, C.G., Banerjee, A., Macdonald, M.L., Cho, D.S., Kamins, J., Nie, Z., et al., 2009. The post-synaptic density of human postmortem brain tissues: an experimental study paradigm for neuropsychiatric illnesses. PLoS. One. 4 (4), e5251. http://dx.doi.org/10.1371/journal.pone.0005251.

Hall, J., Trent, S., Thomas, K.L., O'Donovan, M.C., Owen, M.J., 2015. Genetic risk for schizophrenia: convergence on synaptic pathways involved in plasticity. Biol. Psychiatry 77 (1), 52–58. http://dx.doi.org/10.1016/j.biopsych.2014.07.011.

Harrison, P.J., Owen, M.J., 2003. Genes for schizophrenia? Recent findings and their pathophysiological implications. Lancet 361 (9355), 417–419. Retrieved from: <http://www.ncbi.nlm.nih.gov/entrez/query.fcgi?cmd=Retrieve&db=PubMed&dopt=Citation&list_uids=12573388>.

Harrison, P.J., Weinberger, D.R., 2005. Schizophrenia genes, gene expression, and neuropathology: on the matter of their convergence. Mol. Psychiatry 10 (1), 40–68. image 45. Retrieved from: <http://www.ncbi.nlm.nih.gov/entrez/query.fcgi?cmd=Retrieve&db=PubMed&dopt=Citation&list_uids=15263907>.

Hashimoto, R., Straub, R.E., Weickert, C.S., Hyde, T.M., Kleinman, J.E., Weinberger, D.R., 2004. Expression analysis of neuregulin-1 in the dorsolateral prefrontal cortex in schizophrenia. Mol. Psychiatry 9 (3), 299–307. http://dx.doi.org/10.1038/sj.mp.4001434.

Hashimoto, R., Tankou, S., Takeda, M., Sawa, A., 2007. Postsynaptic density: a key convergent site for schizophrenia susceptibility factors and possible target for drug development. Drugs Today (Barc) 43 (9), 645–654. http://dx.doi.org/10.1358/dot.2007.43.9.1088821.

Hayashi, M.K., Tang, C., Verpelli, C., Narayanan, R., Stearns, M.H., Xu, R.M., et al., 2009. The postsynaptic density proteins Homer and Shank form a polymeric network structure. Cell 137 (1), 159–171. http://dx.doi.org/10.1016/j.cell.2009.01.050.

Hayashi, Y., Majewska, A.K., 2005. Dendritic spine geometry: functional implication and regulation. Neuron 46 (4), 529–532. http://dx.doi.org/10.1016/j.neuron.2005.05.006.

Henley, J.M., Wilkinson, K.A., 2013. AMPA receptor trafficking and the mechanisms underlying synaptic plasticity and cognitive aging. Dialogues Clin. Neurosci. 15 (1), 11–27. Retrieved from: <http://www.ncbi.nlm.nih.gov/pubmed/23576886>.

Hering, H., Sheng, M., 2001. Dendritic spines: structure, dynamics and regulation. Nat. Rev. Neurosci. 2 (12), 880–888. http://dx.doi.org/10.1038/35104061.

Holtmaat, A., Wilbrecht, L., Knott, G.W., Welker, E., Svoboda, K., 2006. Experience-dependent and cell-type-specific spine growth in the neocortex. Nature 441 (7096), 979–983. http://dx.doi.org/10.1038/nature04783.

Horton, A.C., Ehlers, M.D., 2004. Secretory trafficking in neuronal dendrites. Nat. Cell. Biol. 6 (7), 585–591. http://dx.doi.org/10.1038/ncb0704-585.

Horvath, S., Mirnics, K., 2015. Schizophrenia as a disorder of molecular pathways. Biol. Psychiatry 77 (1), 22–28. http://dx.doi.org/10.1016/j.biopsych.2014.01.001.

Hu, B.R., Park, M., Martone, M.E., Fischer, W.H., Ellisman, M.H., Zivin, J.A., 1998. Assembly of proteins to postsynaptic densities after transient cerebral ischemia. J. Neurosci. 18 (2), 625–633. Retrieved from: <http://www.ncbi.nlm.nih.gov/pubmed/9425004>, <http://www.jneurosci.org/content/18/2/625.full.pdf>.

Humphries, C., Mortimer, A., Hirsch, S., de Belleroche, J., 1996. NMDA receptor mRNA correlation with antemortem cognitive impairment in schizophrenia. Neuroreport 7 (12), 2051–2055. Retrieved from: <http://www.ncbi.nlm.nih.gov/pubmed/8905723>.

Husi, H., Grant, S.G., 2001. Proteomics of the nervous system. Trends Neurosci. 24 (5), 259–266. Retrieved from: <http://www.ncbi.nlm.nih.gov/pubmed/11311377>, <http://www.sciencedirect.com/science/article/pii/S0166223600017926>.

Husi, H., Ward, M.A., Choudhary, J.S., Blackstock, W.P., Grant, S.G., 2000. Proteomic analysis of NMDA receptor-adhesion protein signaling complexes. Nat. Neurosci. 3 (7), 661–669. http://dx.doi.org/10.1038/76615.

Iacovelli, L., Bruno, V., Salvatore, L., Melchiorri, D., Gradini, R., Caricasole, A., et al., 2002. Native group-III metabotropic glutamate receptors are coupled to the mitogen-activated protein kinase/phosphatidylinositol-3-kinase pathways. J. Neurochem. 82 (2), 216–223. Retrieved from: <http://www.ncbi.nlm.nih.gov/pubmed/12124422>.

Inoue, A., Okabe, S., 2003. The dynamic organization of postsynaptic proteins: translocating molecules regulate synaptic function. Curr. Opin. Neurobiol. 13 (3), 332–340. Retrieved from: <http://www.ncbi.nlm.nih.gov/pubmed/12850218>, <http://www.sciencedirect.com/science/article/pii/S0959438803000771>.

Javitt, D.C., 2007. Glutamate and schizophrenia: phencyclidine, N-methyl-D-aspartate receptors, and dopamine-glutamate interactions. Int. Rev. Neurobiol. 78, 69–108. doi:10.1016/S0074-7742(06)78003-5.

Jurado, S., Biou, V., Malenka, R.C., 2010. A calcineurin/AKAP complex is required for NMDA receptor-dependent long-term depression. Nat. Neurosci. 13 (9), 1053–1055. http://dx.doi.org/10.1038/nn.2613.

Kaech, S., Fischer, M., Doll, T., Matus, A., 1997. Isoform specificity in the relationship of actin to dendritic spines. J. Neurosci. 17 (24), 9565–9572. Retrieved from: <http://www.ncbi.nlm.nih.gov/pubmed/9391011>.

Kasai, H., Matsuzaki, M., Noguchi, J., Yasumatsu, N., Nakahara, H., 2003. Structure-stability-function relationships of dendritic spines. Trends Neurosci. 26 (7), 360–368. doi:10.1016/S0166-2236(03)00162-0.

Kennedy, M.B., 1997. The postsynaptic density at glutamatergic synapses. Trends Neurosci. 20 (6), 264–268. Retrieved from: <http://www.ncbi.nlm.nih.gov/pubmed/9185308>.

Kennedy, M.B., 1998. Signal transduction molecules at the glutamatergic postsynaptic membrane. Brain Res. Brain Res. Rev. 26 (2-3), 243–257. Retrieved from: <http://www.ncbi.nlm.nih.gov/pubmed/9651538>.

Kennedy, M.B., 2000. Signal-processing machines at the postsynaptic density. Science 290 (5492), 750–754. Retrieved from: <http://www.ncbi.nlm.nih.gov/pubmed/11052931>, <http://www.sciencemag.org/content/290/5492/750.full.pdf>.

Kennedy, M.B., Beale, H.C., Carlisle, H.J., Washburn, L.R., 2005. Integration of biochemical signalling in spines. Nat. Rev. Neurosci. 6 (6), 423–434. http://dx.doi.org/10.1038/nrn1685.

Kew, J.N., Richards, J.G., Mutel, V., Kemp, J.A., 1998. Developmental changes in NMDA receptor glycine affinity and ifenprodil sensitivity reveal three distinct populations of NMDA receptors in individual rat cortical neurons. J. Neurosci. 18 (6), 1935–1943. Retrieved from: <http://www.ncbi.nlm.nih.gov/pubmed/9482779>.

Kharazia, V.N., Weinberg, R.J., 1997. Tangential synaptic distribution of NMDA and AMPA receptors in rat neocortex. Neurosci. Lett. 238 (1-2), 41–44. Retrieved from: <http://www.ncbi.nlm.nih.gov/pubmed/9464650>, <http://www.sciencedirect.com/science/article/pii/S030439409700846X>.

Kirov, G., Pocklington, A.J., Holmans, P., Ivanov, D., Ikeda, M., Ruderfer, D., et al., 2012. De novo CNV analysis implicates specific abnormalities of postsynaptic signalling complexes in the pathogenesis of schizophrenia. Mol. Psychiatry 17 (2), 142–153. http://dx.doi.org/10.1038/mp.2011.154.

Kolluri, N., Sun, Z., Sampson, A.R., Lewis, D.A., 2005. Lamina-specific reductions in dendritic spine density in the prefrontal cortex of subjects with schizophrenia. Am. J. Psychiatry 162 (6), 1200–1202. http://dx.doi.org/10.1176/appi.ajp.162.6.1200.

Kolomeets, N.S., Orlovskaya, D.D., Rachmanova, V.I., Uranova, N.A., 2005. Ultrastructural alterations in hippocampal mossy fiber synapses in schizophrenia: a postmortem morphometric study. Synapse 57 (1), 47–55. http://dx.doi.org/10.1002/syn.20153.

Kolomeets, N.S., Orlovskaya, D.D., Uranova, N.A., 2007. Decreased numerical density of CA3 hippocampal mossy fiber synapses in schizophrenia. Synapse 61 (8), 615–621. http://dx.doi.org/10.1002/syn.20405.

Kristiansen, L.V., Beneyto, M., Haroutunian, V., Meador-Woodruff, J.H., 2006. Changes in NMDA receptor subunits and interacting PSD proteins in dorsolateral prefrontal and anterior cingulate cortex indicate abnormal regional expression in schizophrenia. Mol. Psychiatry 11 (8), 737–747. 705. doi:10.1038/sj.mp.4001844.

Kristiansen, L.V., Patel, S.A., Haroutunian, V., Meador-Woodruff, J.H., 2010. Expression of the NR2B-NMDA receptor subunit and its Tbr-1/CINAP regulatory proteins in postmortem brain suggest altered receptor processing in schizophrenia. Synapse 64 (7), 495–502. http://dx.doi.org/10.1002/syn.20754.

Kritis, A.A., Stamoula, E.G., Paniskaki, K.A., Vavilis, T.D., 2015. Researching glutamate—induced cytotoxicity in different cell lines: a comparative/collective analysis/study. Fron. Cell. Neurosci. 9. http://dx.doi.org/10.3339/fricel.2015.00091.

Kritzer, M.F., Goldman-Rakic, P.S., 1995. Intrinsic circuit organization of the major layers and sublayers of the dorsolateral prefrontal cortex in the rhesus monkey. J. Comp. Neurol. 359 (1), 131–143. http://dx.doi.org/10.1002/cne.903590109.

Li, K.W., Hornshaw, M.P., Van Der Schors, R.C., Watson, R., Tate, S., Casetta, B., et al., 2004. Proteomics analysis of rat brain postsynaptic density. Implications of the diverse protein functional groups for the integration of synaptic physiology. J. Biol. Chem. 279 (2), 987–1002. http://dx.doi.org/10.1074/jbc.M303116200.

MacDonald 3rd, A.W., Chafee, M.V., 2006. Translational and developmental perspective on N-methyl-D-aspartate synaptic deficits in schizophrenia. Dev. Psychopathol. 18 (3), 853–876. Retrieved from: <http://www.ncbi.nlm.nih.gov/entrez/query.fcgi?cmd=Retrieve&db=PubMed&dopt=Citation&list_uids=17152404>.

MacDonald, M.L., Ciccimaro, E., Prakash, A., Banerjee, A., Seeholzer, S.H., Blair, I.A., et al., 2012. Biochemical fractionation and stable isotope dilution liquid chromatography-mass spectrometry for targeted and microdomain-specific protein quantification in human postmortem brain tissue. Mol. Cell. Proteomics 11 (12), 1670–1681. http://dx.doi.org/10.1074/mcp.M112.021766.

MacGillavry, H.D., Hoogenraad, C.C., 2015. The internal architecture of dendritic spines revealed by super-resolution imaging: what did we learn so far? Exp. Cell. Res. 335 (2), 180–186. http://dx.doi.org/10.1016/j.yexcr.2015.02.024.

Malinow, R., Malenka, R.C., 2002. AMPA receptor trafficking and synaptic plasticity. Annu. Rev. Neurosci. 25, 103–126. http://dx.doi.org/10.1146/annurev.neuro.25.112701.142758.

Markham, J.A., Fifkova, E., 1986. Actin filament organization within dendrites and dendritic spines during development. Brain Res. 392 (1-2), 263–269. Retrieved from: <http://www.ncbi.nlm.nih.gov/pubmed/3708380>.

Martinez-Lozada, Z., Ortega, A., 2015. Glutamatergic transmission: a matter of three. Neural. Plast. 2015, 787396. http://dx.doi.org/10.1155/2015/787396.

Matsuzaki, M., Honkura, N., Ellis-Davies, G.C., Kasai, H., 2004. Structural basis of long-term potentiation in single dendritic spines. Nature 429 (6993), 761–766. http://dx.doi.org/10.1038/nature02617.

Melchitzky, D.S., Gonzalez-Burgos, G., Barrionuevo, G., Lewis, D.A., 2001. Synaptic targets of the intrinsic axon collaterals of supragranular pyramidal neurons in monkey prefrontal cortex. J. Comp. Neurol. 430 (2), 209–221. Retrieved from: <http://www.ncbi.nlm.nih.gov/pubmed/11135257>, <http://onlinelibrary.wiley.com/doi/10.1002/1096-9861(20010205)430:2%3C209::AID-CNE1026%3E3.0.CO;2-%23/abstract>.

Modinos, G., Iyegbe, C., Prata, D., Rivera, M., Kempton, M.J., Valmaggia, L.R., et al., 2013. Molecular genetic gene-environment studies using candidate genes in schizophrenia: a systematic review. Schizophr. Res. 150 (2-3), 356–365. http://dx.doi.org/10.1016/j.schres.2013.09.010.

Mueller, H.T., Meador-Woodruff, J.H., 2004. NR3A NMDA receptor subunit mRNA expression in schizophrenia, depression and bipolar disorder. Schizophr. Res. 71 (2-3), 361–370. http://dx.doi.org/10.1016/j.schres.2004.02.016.

Niswender, C.M., Conn, P.J., 2010. Metabotropic glutamate receptors: physiology, pharmacology, and disease. Annu. Rev. Pharmacol. Toxicol. 50, 295–322. http://dx.doi.org/10.1146/annurev.pharmtox.011008.145533.

O'Brien, R.J., Lau, L.F., Huganir, R.L., 1998. Molecular mechanisms of glutamate receptor clustering at excitatory synapses. Curr. Opin. Neurobiol. 8 (3), 364–369. Retrieved from: <http://www.ncbi.nlm.nih.gov/pubmed/9687358>.

Pak, D.T., Sheng, M., 2003. Targeted protein degradation and synapse remodeling by an inducible protein kinase. Science 302 (5649), 1368–1373. http://dx.doi.org/10.1126/science.1082475.

Peng, J., Kim, M.J., Cheng, D., Duong, D.M., Gygi, S.P., Sheng, M., 2004. Semiquantitative proteomic analysis of rat forebrain postsynaptic density fractions by mass spectrometry. J. Biol. Chem. 279 (20), 21003–21011. http://dx.doi.org/10.1074/jbc.M400103200.

Petralia, R.S., Sans, N., Wang, Y.X., Wenthold, R.J., 2005. Ontogeny of postsynaptic density proteins at glutamatergic synapses. Mol. Cell. Neurosci. 29 (3), 436–452. http://dx.doi.org/10.1016/j.mcn.2005.03.013.

Phillips, G.R., Huang, J.K., Wang, Y., Tanaka, H., Shapiro, L., Zhang, W., et al., 2001. The presynaptic particle web: ultrastructure, composition, dissolution, and reconstitution. Neuron 32 (1), 63–77. Retrieved from: <http://www.ncbi.nlm.nih.gov/entrez/query.fcgi?cmd=Retrieve&db=PubMed&dopt=Citation&list_uids=11604139>.

Phillips, G.R., Florens, L., Tanaka, H., Khaing, Z.Z., Fidler, L., Yates, J.R., et al., 2005. Proteomic comparison of two fractions derived from the transsynaptic scaffold. J. Neurosci. Res. 81 (6), 762–775. Retrieved from: <http://www.ncbi.nlm.nih.gov/entrez/query.fcgi?cmd=Retrieve&db=PubMed&dopt=Citation&list_uids=16047384>.

Purcell, S.M., Moran, J.L., Fromer, M., Ruderfer, D., Solovieff, N., Roussos, P., et al., 2014. A polygenic burden of rare disruptive mutations in schizophrenia. Nature 506 (7487), 185–190. http://dx.doi.org/10.1038/nature12975.

Racca, C., Stephenson, F.A., Streit, P., Roberts, J.D., Somogyi, P., 2000. NMDA receptor content of synapses in stratum radiatum of the hippocampal CA1 area. J. Neurosci. 20 (7), 2512–2522. Retrieved from: <http://www.ncbi.nlm.nih.gov/pubmed/10729331>, <http://www.jneurosci.org/content/20/7/2512.full.pdf>.

Roberts, R.C., Conley, R., Kung, L., Peretti, F.J., Chute, D.J., 1996. Reduced striatal spine size in schizophrenia: a postmortem ultrastructural study. Neuroreport 7 (6), 1214–1218. Retrieved from: <http://www.ncbi.nlm.nih.gov/pubmed/8817535>.

Roberts, R.C., Roche, J.K., Conley, R.R., 2005. Synaptic differences in the postmortem striatum of subjects with schizophrenia: a stereological ultrastructural analysis. Synapse 56 (4), 185–197. http://dx.doi.org/10.1002/syn.20144.

Rondard, P., Pin, J.P., 2015. Dynamics and modulation of metabotropic glutamate receptors. Curr. Opin. Pharmacol. 20, 95–101. http://dx.doi.org/10.1016/j.coph.2014.12.001.

Ross, C.A., Margolis, R.L., Reading, S.A., Pletnikov, M., Coyle, J.T., 2006. Neurobiology of schizophrenia. Neuron 52 (1), 139–153. http://dx.doi.org/10.1016/j.neuron.2006.09.015.

Sala, C., Segal, M., 2014. Dendritic spines: the locus of structural and functional plasticity. Physiol. Rev. 94 (1), 141–188. http://dx.doi.org/10.1152/physrev.00012.2013.

Salter, M.W., Kalia, L.V., 2004. Src kinases: a hub for NMDA receptor regulation. Nat. Rev. Neurosci. 5 (4), 317–328. http://dx.doi.org/10.1038/nrn1368.

Schizophrenia Working Group of the Psychiatric Genomics Consortium, 2014. Biological insights from 108 schizophrenia-associated genetic loci. Nature 511 (7510), 421–427. http://dx.doi.org/10.1038/nature13595.

Schuman, E.M., Dynes, J.L., Steward, O., 2006. Synaptic regulation of translation of dendritic mRNAs. J. Neurosci. 26 (27), 7143–7146. http://dx.doi.org/10.1523/JNEUROSCI.1796-06.2006.

Sheng, M., 2001. Molecular organization of the postsynaptic specialization. Proc. Natl. Acad. Sci. U.S.A. 98 (13), 7058–7061. http://dx.doi.org/10.1073/pnas.111146298.

Sheng, M., Kim, M.J., 2002. Postsynaptic signaling and plasticity mechanisms. Science 298 (5594), 776–780. http://dx.doi.org/10.1126/science.1075333.

Sheng, M., Hoogenraad, C.C., 2007. The postsynaptic architecture of excitatory synapses: a more quantitative view. Annu. Rev. Biochem. 76, 823–847. http://dx.doi.org/10.1146/annurev.biochem.76.060805.160029.

Siekevitz, P., 1985. The postsynaptic density: a possible role in long-lasting effects in the central nervous system. Proc. Natl. Acad. Sci. U.S.A. 82 (10), 3494–3498. Retrieved from: <http://www.ncbi.nlm.nih.gov/pubmed/2987929>.

Sokolov, B.P., 1998. Expression of NMDAR1, GluR1, GluR7, and KA1 glutamate receptor mRNAs is decreased in frontal cortex of "neuroleptic-free" schizophrenics: evidence on reversible up-regulation by typical neuroleptics. J. Neurochem. 71 (6), 2454–2464. Retrieved from: <http://www.ncbi.nlm.nih.gov/pubmed/9832144>.

Sorra, K.E., Harris, K.M., 2000. Overview on the structure, composition, function, development, and plasticity of hippocampal dendritic spines. Hippocampus 10 (5), 501–511. doi:10.1002/1098-1063(2000)10:5< 501::AID-HIPO1> 3.0.CO;2-T.

Spacek, J., Harris, K.M., 1997. Three-dimensional organization of smooth endoplasmic reticulum in hippocampal CA1 dendrites and dendritic spines of the immature and mature rat. J. Neurosci. 17 (1), 190–203. Retrieved from: <http://www.ncbi.nlm.nih.gov/pubmed/8987748>, <http://www.jneurosci.org/content/17/1/190.full.pdf>.

Steffek, A.E., Haroutunian, V., Meador-Woodruff, J.H., 2006. Serine racemase protein expression in cortex and hippocampus in schizophrenia. Neuroreport 17 (11), 1181–1185. http://dx.doi.org/10.1097/01.wnr.0000230512.01339.72.

Svoboda, K., Mainen, Z.F., 1999. Synaptic [Ca2+]: intracellular stores spill their guts. Neuron 22 (3), 427–430. Retrieved from: <http://www.ncbi.nlm.nih.gov/pubmed/10197523>, <http://ac.els-cdn.com/S0896627300806984/1-s2.0-S0896627300806984-main.pdf?_tid=2f12498e-8c1e-11e5-b97c-00000aacb361&acdnat=1447649877_a61cb27405e229adda6f219c35eed5a7>.

Sweet, R.A., Henteleff, R.A., Zhang, W., Sampson, A.R., Lewis, D.A., 2009. Reduced dendritic spine density in auditory cortex of subjects with schizophrenia. Neuropsychopharmacology 34 (2), 374–389. http://dx.doi.org/10.1038/npp.2008.67.

Tada, T., Sheng, M., 2006. Molecular mechanisms of dendritic spine morphogenesis. Curr. Opin. Neurobiol. 16 (1), 95–101. http://dx.doi.org/10.1016/j.conb.2005.12.001.

Talbot, K., Eidem, W.L., Tinsley, C.L., Benson, M.A., Thompson, E.W., Smith, R.J., et al., 2004. Dysbindin-1 is reduced in intrinsic, glutamatergic terminals of the hippocampal formation in schizophrenia. J. Clin. Invest. 113 (9), 1353–1363. http://dx.doi.org/10.1172/JCI20425.

Talbot, K., Cho, D.S., Ong, W.Y., Benson, M.A., Han, L.Y., Kazi, H.A., et al., 2006. Dysbindin-1 is a synaptic and microtubular protein that binds brain snapin. Hum. Mol. Genet. 15 (20), 3041–3054. Retrieved from: <http://www.ncbi.nlm.nih.gov/entrez/query.fcgi?cmd=Retrieve&db=PubMed&dopt=Citation&list_uids=16980328>.

Tao-Cheng, J.H., Yang, Y.J., Reese, T.S., Dosemeci, A., 2015. Differential Distribution of Shank and GKAP at the Postsynaptic Density. PLoS One. 10 (3).http://dx.doi.org/10.1371/journal.pone.0118750.

Toyooka, K., Iritani, S., Makifuchi, T., Shirakawa, O., Kitamura, N., Maeda, K., et al., 2002. Selective reduction of a PDZ protein, SAP-97, in the prefrontal cortex of patients with chronic schizophrenia. J. Neurochem. 83 (4),

797–806. Retrieved from: <http://www.ncbi.nlm.nih.gov/entrez/query.fcgi?cmd=Retrieve&db=PubMed&dopt=Citation&list_uids=12421351>.

Trinidad, J.C., Specht, C.G., Thalhammer, A., Schoepfer, R., Burlingame, A.L., 2006. Comprehensive identification of phosphorylation sites in postsynaptic density preparations. Mol. Cell. Proteomics 5 (5), 914–922. http://dx.doi.org/10.1074/mcp.T500041-MCP200.

Trinidad, J.C., Thalhammer, A., Specht, C.G., Lynn, A.J., Baker, P.R., Schoepfer, R., et al., 2008. Quantitative analysis of synaptic phosphorylation and protein expression. Mol. Cell. Proteomics 7 (4), 684–696. Retrieved from: <http://www.ncbi.nlm.nih.gov/entrez/query.fcgi?cmd=Retrieve&db=PubMed&dopt=Citation&list_uids=18056256>, <http://www.mcponline.org/content/7/4/684.full.pdf>.

Tu, J.C., Xiao, B., Naisbitt, S., Yuan, J.P., Petralia, R.S., Brakeman, P., et al., 1999. Coupling of mGluR/Homer and PSD-95 complexes by the shank family of postsynaptic density proteins. Neuron 23 (3), 583–592. doi:10.1016/S0896-6273(00)80810-7.

Tunquist, B.J., Hoshi, N., Guire, E.S., Zhang, F., Mullendorff, K., Langeberg, L.K., et al., 2008. Loss of AKAP150 perturbs distinct neuronal processes in mice. Proc. Natl. Acad. Sci. U.S.A. 105 (34), 12557–12562. http://dx.doi.org/10.1073/pnas.0805922105.

Valtschanoff, J.G., Weinberg, R.J., 2001. Laminar organization of the NMDA receptor complex within the postsynaptic density. J. Neurosci. 21 (4), 1211–1217. Retrieved from: <http://www.ncbi.nlm.nih.gov/pubmed/11160391>.

Verkhratsky, A., 2005. Physiology and pathophysiology of the calcium store in the endoplasmic reticulum of neurons. Physiol. Rev. 85 (1), 201–279. http://dx.doi.org/10.1152/physrev.00004.2004.

Vincent, S.L., McSparren, J., Wang, R.Y., Benes, F.M., 1991. Evidence for ultrastructural changes in cortical axodendritic synapses following long-term treatment with haloperidol or clozapine. Neuropsychopharmacology 5 (3), 147–155. Retrieved from: <http://www.ncbi.nlm.nih.gov/pubmed/1755930>.

Vinson, P.N., Conn, P.J., 2012. Metabotropic glutamate receptors as therapeutic targets for schizophrenia. Neuropharmacology 62 (3), 1461–1472. http://dx.doi.org/10.1016/j.neuropharm.2011.05.005.

Walikonis, R.S., Jensen, O.N., Mann, M., Provance Jr., D.W., Mercer, J.A., Kennedy, M.B., 2000. Identification of proteins in the postsynaptic density fraction by mass spectrometry. J. Neurosci. 20 (11), 4069–4080. Retrieved from: <http://www.ncbi.nlm.nih.gov/pubmed/10818142>.

Wang, H.D., Deutch, A.Y., 2008. Dopamine depletion of the prefrontal cortex induces dendritic spine loss: reversal by atypical antipsychotic drug treatment. Neuropsychopharmacology 33 (6), 1276–1286. http://dx.doi.org/10.1038/sj.npp.1301521.

Weickert, C.S., Straub, R.E., McClintock, B.W., Matsumoto, M., Hashimoto, R., Hyde, T.M., et al., 2004. Human dysbindin (DTNBP1) gene expression in normal brain and in schizophrenic prefrontal cortex and midbrain. Arch. Gen. Psychiatry 61 (6), 544–555. http://dx.doi.org/10.1001/archpsyc.61.6.54461/6/544.

Weickert, C.S., Rothmond, D.A., Hyde, T.M., Kleinman, J.E., Straub, R.E., 2008. Reduced DTNBP1 (dysbindin-1) mRNA in the hippocampal formation of schizophrenia patients. Schizophr. Res. 98 (1–3), 105–110. http://dx.doi.org/10.1016/j.schres.2007.05.041.

Weickert, C.S., Fung, S.J., Catts, V.S., Schofield, P.R., Allen, K.M., Moore, L.T., et al., 2013. Molecular evidence of N-methyl-D-aspartate receptor hypofunction in schizophrenia. Mol. Psychiatry 18 (11), 1185–1192. http://dx.doi.org/10.1038/mp.2012.137.

Woo, T.U., Walsh, J.P., Benes, F.M., 2004. Density of glutamic acid decarboxylase 67 messenger RNA-containing neurons that express the N-methyl-D-aspartate receptor subunit NR2A in the anterior cingulate cortex in schizophrenia and bipolar disorder. Arch. Gen. Psychiatry 61 (7), 649–657. http://dx.doi.org/10.1001/archpsyc.61.7.649.

Woo, T.U., Shrestha, K., Lamb, D., Minns, M.M., Benes, F.M., 2008. N-methyl-D-aspartate receptor and calbindin-containing neurons in the anterior cingulate cortex in schizophrenia and bipolar disorder. Biol. Psychiatry 64 (9), 803–809. http://dx.doi.org/10.1016/j.biopsych.2008.04.034.

Xiao, B., Tu, J.C., Petralia, R.S., Yuan, J.P., Doan, A., Breder, C.D., et al., 1998. Homer regulates the association of group 1 metabotropic glutamate receptors with multivalent complexes of Homer-related, synaptic proteins. Neuron 21 (4), 707–716. doi:10.1016/S0896-6273(00)80588-7.

Yun-Hong, Y., Chih-Fan, C., Chia-Wei, C., Yen-Chung, C., 2011. A study of the spatial protein organization of the postsynaptic density isolated from porcine cerebral cortex and cerebellum. Mol. Cell. Proteomics 10 (10) M110.007138. doi:10.1074/mcp.M110.007138.

Zhang, W., Vazquez, L., Apperson, M., Kennedy, M.B., 1999. Citron binds to PSD-95 at glutamatergic synapses on inhibitory neurons in the hippocampus. J. Neurosci. 19 (1), 96–108. Retrieved from: <http://www.ncbi.nlm.nih.gov/pubmed/9870942>, <http://www.jneurosci.org/content/19/1/96.full.pdf>.

Zhu, J.J., Qin, Y., Zhao, M., Van Aelst, L., Malinow, R., 2002. Ras and Rap control AMPA receptor trafficking during synaptic plasticity. Cell 110 (4), 443–455. Retrieved from: <http://www.ncbi.nlm.nih.gov/pubmed/12202034>.

CHAPTER 9

TARGETING COGNITIVE DEFICITS IN SCHIZOPHRENIA VIA GABA$_A$ RECEPTORS: FOCUS ON α2, α3, AND α5 RECEPTOR SUBTYPES

U. Rudolph[1,2] and H. Möhler[3,4]

[1]Laboratory of Genetic Neuropharmacology, McLean Hospital, Belmont, MA, United States
[2]Department of Psychiatry, Harvard Medical School, Boston, MA, United States
[3]Institute of Pharmacology, University of Zurich, Zurich, Switzerland
[4]Department of Chemistry and Applied Biosciences, Swiss Federal Institute of Technology (ETH), Zurich, Switzerland

CHAPTER OUTLINE

Excitatory and Inhibitory Tuning 150
Multiplicity of GABA$_A$ Receptors 150
Maturation of GABA$_A$ Receptors 151
GABA$_A$ Receptor Subtypes in Cortical Period Plasticity 152
GABA$_A$ Receptor Genetics and Schizophrenia 152
Deficits of GABAergic Cortical Interneurons and Impaired Cortical Oscillations in Schizophrenia 152
Physiological and Pharmacological Functions of GABA$_A$ Receptor Subtypes 154
Attempts to Enhance Cognition via α2/α3 GABA$_A$ Receptor Modulation 155
Inhibition of Dopaminergic Neurons by α3-Containing GABA$_A$ Receptors 156
Indirect Inhibition of Dopaminergic Neurons by α5-Containing GABA$_A$ Receptors 157
Multispecific Modulation of α2-, α3-, and α5-Containing GABA$_A$ Receptors 158
Proof-of-Concept Clinical Study With the Partial Positive Allosteric Modulator Bretazenil 159
α5-Negative Allosteric Modulators and Cognition Enhancement 159
References 160

The Neurobiology of Schizophrenia. DOI: http://dx.doi.org/10.1016/B978-0-12-801829-3.00017-3
© 2016 Elsevier Inc. All rights reserved.

EXCITATORY AND INHIBITORY TUNING

In schizophrenia, genetic variations and developmental abnormalities are thought to bias molecular processes of synaptic assembly and plasticity toward abnormal connectivity of cortical circuitry with repercussions throughout life. In this neurodevelopmental hypothesis of schizophrenia, the expression of psychosis, negative symptoms, and abnormal cognitive functions are considered to be manifestations of alterations in the function of canonical cortical systems (Harrison and Weinberger, 2005; Weinberger and Levitt, 2011). Current thinking regarding how these systems malfunction has centered on an unbalance between excitatory and inhibitory tuning of cortical microcircuitry (Krystal and Moghaddam, 2011; Lewis and Gonzalez-Burgos, 2008; Weinberger and Levitt, 2011; Winterer and Weinberger, 2004).

As a polygenetic brain disorder, more than 100 schizophrenia-associated genetic loci and rare disruptive mutations with low penetrance are known that are thought to contribute in various individual combinations to the disease. Many of these genes are involved in postsynaptic specialization of excitatory neurons and dopamine neurons (Schizophrenia Working Group of the Psychiatric Genomics Consortium, 2014; Fromer et al., 2014; McCarroll and Hyman, 2013; Purcell et al., 2014) but also participate at various stages of GABA circuitry maturation, including the genesis of GABA neurons (eg, DISC1, NRG1), their migration and settling (DISC1, NRG1, ERBB4), synapse formation (eg, NRXN1, DISC1, NRG1, TGF4), GABA synthesis (GAD1), and GABA becoming an inhibitory neurotransmitter (CHRNA7, NRG1) (Fazzari et al., 2010; Jaaro-Peled et al., 2009; Liu et al., 2006; Yang et al., 2013).

The ability of the NMDA receptor antagonist ketamine to induce thought disorders in humans reminiscent of schizophrenia led to the hypothesis of an NMDA receptor–linked hypofunction contributing to the disease pathology. In the past, therapeutic approaches based on the glutamate hypothesis have focused on mGluR5 agonists, mGluR2 antagonists, and inhibitors of glycine transporter 1 as molecular targets (Krystal and Moghaddam, 2011). A deficit of the GABA system in schizophrenia was most apparent in immunohistochemical postmortem studies, indicating that distinct cortical GABA interneurons were dysfunctional. This deficit is thought to contribute to the disruption of high-frequency neural oscillations and an impairment of cognitive functions. Concomitantly, specific $GABA_A$ receptor subtypes show changes in expression. Furthermore, because the monoaminergic neurons are under GABAergic control, the GABA system provides a link to the classical hyperdopaminergic hypothesis of schizophrenia (Yee et al., 2005). Therapeutic attempts initially focused on a ligand acting on α2 and α3 GABA receptor subtypes. However, in the future, negative allosteric modulators acting selectively on α5 subunit–containing $GABA_A$ receptors (here termed α5 $GABA_A$ receptors) may provide a new mechanistic approach to overcome cognitive deficits (Mohler, 2009, 2011, 2012, 2015; Rudolph and Knoflach, 2011; Rudolph and Mohler, 2014). The present article focuses on the role of $GABA_A$ receptors in neural development and plasticity and on their therapeutic potential in ameliorating cognitive deficits in schizophrenia.

MULTIPLICITY OF $GABA_A$ RECEPTORS

In cortical circuits, a rich diversity of GABA interneurons control the activity of principal cells by modulating signal integration, controlling spike timing, sculpting neuronal rhythms, selecting cell assemblies, and implementing brain states (Klausberger and Somogyi, 2008). Specific interneurons innervate distinct subcellular domains of pyramidal cells and act in discrete time windows. To accommodate the spatio-temporal patterns of neural signaling, the diversity of GABAergic interneurons is paralleled by a diversity of $GABA_A$ receptors, which is based on differences in their heteropentameric subunit assembly, cell type–specific expression, channel kinetics, affinity for GABA, rate of desensitization, and the ability for transient chemical modification such as phosphorylation. Based on the presence of seven subunit families comprising at least 19 subunits in the CNS (α1–6, β1–3, γ1–3, δ, ϵ, θ, μ, ρ1–3), the receptors display an extraordinary structural heterogeneity. Receptors containing the α1, α2, α3, or α5 subunits in combination with a β2 or β3 subunit and the γ2 subunit are most prevalent in the brain, forming heteropentamers of 2α, 2β, and 1γ subunit. These receptors are benzodiazepine-sensitive with the drug binding site located at the interface between the α and γ2 subunits. The major receptor subtype (2xα1, 2xβ2, 1xγ2) is present in nearly all brain regions. Receptors containing the α2 or α3 subunit, frequently co-expressed with the β3 and γ2 subunits, are considerably less abundant and are highly expressed in areas where the α1 subunit is absent or present at low levels, such as the axon initial segment of hippocampal pyramidal cells (α2, β3, γ2) or cholinergic forebrain neurons or dopaminergic midbrain neurons (α3, β2, γ2). Receptors containing the α_5 subunit (most likely α5, β3, γ2) are of minor abundance in the brain but are significantly expressed mainly in hippocampus. The subunits γ1 and γ3 characterize very small receptor populations.

Receptors containing the α4 or α6 subunit are benzodiazepine-insensitive and are expressed at low abundance, except that α4 receptors are prominent in thalamus and dentate gyrus and α6 receptors in cerebellar granule cells (for review, see Table 1 in Rudolph and Mohler, 2014; see also Fritschy and Mohler, 1995; Fritschy and Panzanelli, 2014; Olsen and Sieghart, 2008; Rudolph and Knoflach, 2011; Mohler, 2015).

MATURATION OF $GABA_A$ RECEPTORS

GABA neurons that seed cortical structures arise from the ganglionic eminence and integrate into the cerebral cortex where they settle, develop processes, and form synaptic contacts. $GABA_A$ receptors are expressed at early stages by neural precursor cells and during neuronal differentiation and have been proposed to regulate cell proliferation, migration, and differentiation, possibly though GABAergic depolarization and calcium influx as a second messenger (Lu et al., 2014). In situ hybridization data in embryonic brain indicate the presence of α2, α3, β2, β3, and γ2 $GABA_A$ receptor subunits. Because this stage precedes synaptogenesis, these receptors are thought to mediate tonic, presumably depolarizing, GABA currents. With increasing expression of KCC2 and the ensuing reduction of the intracellular chloride concentration gradient in mature neurons, GABA induces hyperpolarizing currents (Dehorter et al., 2012). The α5 subunit is strongly expressed perinatally, and its expression decreases during synaptogenesis, except in regions where it remains abundant in adult brain (hippocampus, olfactory bulb, brainstem) (Fritschy and Panzanelli, 2014). α1 $GABA_A$ receptors are upregulated in a region-specific manner during synaptogenesis and become the predominant subtype in adult CNS (Fritschy and Panzanelli, 2014). By their fast response and decay kinetics, the α1 receptors accelerate mIPSC kinetics during brain development and endow postsynaptic neurons with fast response kinetics required for the input of fast-spiking interneurons such as parvalbumin (PV)-positive basket cells, which are involved in the generation of high-frequency oscillations. Subunits that define the main extrasynaptic $GABA_A$

receptors of adult brain (α4, α6, δ) are absent or weakly expressed in fetal brain but upregulated at late neuronal maturation. In the human neocortex, the GABA system develops during the second half of pregnancy and in infancy (Fillman et al., 2010; Xu et al., 2011).

GABA$_A$ RECEPTOR SUBTYPES IN CORTICAL PERIOD PLASTICITY

A striking example of GABA$_A$ receptor–mediated regulation of cortical development is the observation of critical period plasticity. The time of onset of the critical period of the visual system, during which sensory deprivation causes lasting structural and functional alterations, can be delayed or advanced by reducing or enhancing GABA transmission (Fagiolini and Hensch, 2000). This effect requires a highly specific cortical circuit (Katagiri et al., 2007), by which PV-positive basket cells control the output of principal cells through activating α1 GABA$_A$ receptors. Evidence for the key role of these receptors was provided by diazepam being able to advance the onset of the cortical period window in wild-type mice but not in mice with diazepam-insensitive α1 GABA$_A$ receptors [α1 (H101R)] (Fagiolini et al., 2004). Conversely, enhancing nicotinic acetylcholine receptor function allowed re-opening of the critical period window in adulthood, an effect that can be blocked by diazepam (Morishita et al., 2010), demonstrating that a balance between excitation and inhibition, rather than the action of a single transmitter, determines the opening and closing of critical period windows.

GABA$_A$ RECEPTOR GENETICS AND SCHIZOPHRENIA

Molecular genetic studies linked SNPs in the gene that codes for the β2 subunit of the GABA$_A$ receptor (GABRAB2) with schizophrenia (Liu et al., 2005; Zhao et al., 2007). Other subunit genes were only weakly implicated in the heritable risk of schizophrenia with polymorphisms in the α1, α6, and β2 subunit genes (Petryshen et al., 2005, p. 1057). In addition, a chromosomal duplication in a locus encoding four GABA$_A$ receptor subunits (4q12; α2, α4, β1, γ1) was associated with neurodevelopmental disorders (Polan et al., 2014). Also, a preliminary association between the GABA$_B$ receptor 1 gene (GABRB1) and schizophrenia was reported (Zai et al., 2005; Zhao et al., 2007).

DEFICITS OF GABAERGIC CORTICAL INTERNEURONS AND IMPAIRED CORTICAL OSCILLATIONS IN SCHIZOPHRENIA

Gamma oscillations (30–100 Hz) are thought to underlie complex cognitive processing in human brain (Klausberger and Somogyi, 2008; Roux and Buzsaki, 2015). Fast-spiking PV-expressing GABA neurons, which are critical for generating gamma frequency oscillations in the cortex, emerge late and, in mice, are not prominent before early adulthood (Okaty et al., 2009). In schizophrenia, oscillatory neural activity is abnormal, as illustrated by EEG time-frequency maps (Ferrarelli et al., 2008; Javitt et al., 2008). For instance, in a test of visual Gestalt perception (Fig. 9.1; Spencer et al., 2004), controls and schizophrenia patients were asked to respond with a button press according to whether an illusionary square was present on a screen. In controls, the visual Gestalt stimulus elicited a γ-band oscillation

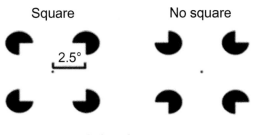

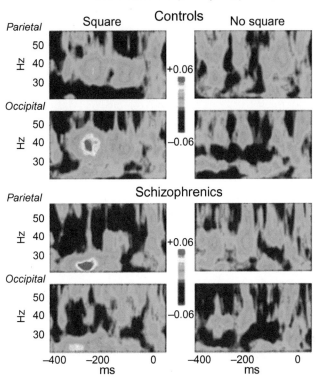

FIGURE 9.1

Perception and cognition indexed by neuronal synchrony. Stimulus-locked time-frequency maps of phase-locking values for controls and schizophrenia patients. Subjects fixated on a central cross on a computer screen and responded with a button press according to whether an illusory square (left) was present or absent (right). The oscillation is thought to reflect the neuronal mechanism involved in linking the elements of the illusory square into a coherent percept. Color scales indicate phase-locking values.

The figure is reproduced from Spencer, K.M., et al., 2004. Neural synchrony indexes disordered perception and cognition in schizophrenia, Proc Natl Acad Sci USA 101, 17288–17293. Copyright (2004) National Academy of Sciences, USA.

(~40 Hz) in the occipital cortex. The fact that this pattern was elicited only by a Gestalt pattern and was phase-locked to the reaction time suggested that it could reflect the neural mechanism involved in linking the four elements of the illusory square into a coherent percept. Although both controls and schizophrenics display γ-band oscillations in Gestalt recognition, the frequency of the oscillation was lower in schizophrenics (23 Hz), and an abnormal parietal oscillation suggests difficulties in efficiently integrating the information (Fig. 9.1). The neural networks in schizophrenics are apparently not able to support high-frequency synchronization. The deficit of fast neural oscillations may reflect a basic pathophysiological mechanism in schizophrenia, potentially attributed to dysfunctional GABAergic interneurons (Javitt et al., 2008).

In postmortem studies of schizophrenia, gene products that regulate GABA transmission showed strikingly and consistently reduced levels of expression affecting both the synthesis (GAD_{67}) and reuptake (GAT1) of GABA (Harrison et al., 2011; Lewis and Gonzalez-Burgos, 2008). On the cellular level, GABA neuronal deficits have been reported in mainly four subsets of GABA interneurons: basket cells, Chandelier neurons, and interneurons containing somatostatin/NPY or CCK. GABA interneurons containing calretinin are not affected. In PV-positive neurons, the level of PV mRNA was reduced and approximately half of the PV neurons lacked GAD_{67} mRNA (Hashimoto et al., 2003). Chandelier neurons show atrophy of axon terminals, a reduction of PV protein, and reduced expression of GAD and GAT1. Concomitantly, their postsynaptic target, the axon initial segment of pyramidal neurons, showed a marked increase of the $GABA_A$ α2 subunit expression (Volk et al., 2002; Hashimoto et al., 2003; Lewis et al., 2005; Lewis and Gonzalez-Burgos, 2008). These changes are not found in other psychiatric disorders and do not result from antipsychotic medication, as shown in monkeys (Hashimoto et al., 2003). Basket cells containing CCK, which target the periosomatic region, are abnormal, as shown by the reduced CCK mRNA level (Hashimoto et al., 2008; Lewis and Gonzalez-Burgos, 2008). The GABA neurons expressing somatostatin/NPY target the distal dendrites of pyramidal cells. The expression of both neuropeptides was reduced, as were dendritic $GABA_A$ receptor subunits (α1, α4, δ) (Lewis and Gonzalez-Burgos, 2008).

PHYSIOLOGICAL AND PHARMACOLOGICAL FUNCTIONS OF $GABA_A$ RECEPTOR SUBTYPES

The differential distribution of $GABA_A$ receptor subunits in the CNS suggests that receptor subtypes may have differential functions. Clinically used benzodiazepines like diazepam and chlordiazepoxide bind nonselectively to $GABA_A$ receptors containing the α1, α2, α3, and α5 subunits (Fig. 9.2). They are used for their hypnotic, anxiolytic, anticonvulsant, and muscle relaxant actions. Problems include development of tolerance as well as dependence and abuse liability. It would thus be desirable to identify which effects of benzodiazepines are mediated by which $GABA_A$ receptor subtype to design novel drugs with more selective actions, such as nonsedating anxiolytics. One major problem is that truly subtype-specific modulators of $GABA_A$ receptor subtypes are not available. Using mice with knock-in point-mutated diazepam-insensitive $GABA_A$ receptors, it was found that the sedative action is mediated by α1-containing $GABA_A$ receptors (McKernan et al., 2000; Rudolph et al., 1999), whereas the anxiolytic-like action is mediated by α2-containing $GABA_A$ receptors (Low et al., 2000). Thus, an α2-selective PAM not modulating α1-containing $GABA_A$ receptors might be a nonsedating anxiolytic, and this assumption was confirmed with TPA023 (MK-0777) in humans (Atack, 2010). The muscle

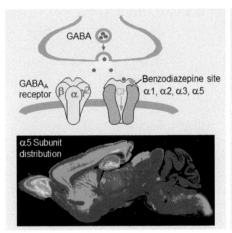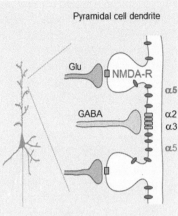

FIGURE 9.2

Scheme of benzodiazepine-sensitive $GABA_A$ receptors. Top left: Benzodiazepine-sensitive receptor subtypes are distinguished by the type of α subunit (α1, α2, α3, α5). Bottom left: Regional immunohistochemical distribution of α5 $GABA_A$ receptors. Right: Scheme of the distribution of $GABA_A$ receptor subtypes in hippocampal pyramidal cell dendrites. In balancing excitatory input, phasic GABAergic inhibition is mediated via α2 and α3 $GABA_A$ receptors, whereas extrasynaptic α5 $GABA_A$ receptors, located at the base of the spines and the adjacent dendritic shaft, mediate tonic inhibition. Reducing α5 $GABA_A$ receptor function genetically or pharmacologically facilitates excitatory transmission (Glu, glutamate) and enhances cognitive performance.

The figure is reproduced from Mohler, H., 2015. The legacy of the benzodiazepine receptor: from flumazenil to enhancing cognition in Down syndrome and social interaction in autism. Adv Pharmacol 72, 1–36, with permission.

relaxant action of diazepam is mediated by α2- and α3-containing $GABA_A$ receptors (Crestani et al., 2001), and the antihyperalgesic action is mediated by α2-, α3-, and α5-containing $GABA_A$ receptors (Knabl et al., 2008). Compounds essentially based on this "blueprint" are currently in development in the pharmaceutical industry. In the following, we review the role that different $GABA_A$ receptor subtypes might play in the pathophysiology of schizophrenia and/or as potential drug targets.

ATTEMPTS TO ENHANCE COGNITION VIA α2/α3 $GABA_A$ RECEPTOR MODULATION

According to the glutamate hypothesis of schizophrenia (Coyle, 1996), hypofunction of NMDA receptors on PV-positive interneurons is an essential step in the pathophysiology of schizophrenia, resulting in decreased activity in PV-positive GABAergic interneurons. This view is in line with findings in mice in which the NR1 subunit has been ablated, specifically in PV-positive GABAergic neurons that developed morphological and behavioral changes reminiscent of schizophrenia (Belforte et al., 2010). In this scenario, the increase in the expression of the α2 subunit on the axon initial segment (AIS) of pyramidal neurons found in patients may be interpreted as an insufficient compensatory upregulation to counteract the presumed decreased activity of the PV-positive GABAergic Chandelier interneurons that synapse onto the axon initial segment. If this interpretation is correct, then positive allosteric modulation of α2-containing GABA$_A$ receptors might be a promising therapeutic option. Studies involving animals provide support for this view. A knockdown of the α2 subunit in the frontal cortex of mice resulted in reduced γ power, prepulse inhibition deficits, and working memory deficits (Hines et al., 2013). Furthermore, in rhesus monkeys, ketamine-induced, and thus potentially schizophrenia-like, memory deficits were reduced by TPA023 (also known as MK-0777) (Castner et al., 2010), which is an α2/α3-selective positive allosteric GABA$_A$ receptor modulator (Atack et al., 2006).

In a first clinical proof of concept, randomized, placebo-controlled, single-dose MK-0777, double-blind, single-site study involving patients with chronic schizophrenia ($N=15$), the compound was associated with increased frontal γ band power during the Preparing to Overcome Prepotency task and with improved performance on the N-back, AX Continuous Performance Test, and Preparing to Overcome Prepotency tasks; however, it had no effect on the brief psychiatric rating scale (BPRS) or on the Repeatable Battery for the Assessment of Neuropsychological Status scores, except for an improvement in the Repeatable Battery for the Assessment of Neuropsychological Status delayed memory index (Lewis et al., 2008). Overall, this study indicated an improvement of electrophysiological and at least some behavioral measures, thus providing initial support for the concept that positive allosteric modulation of α2-containing GABA$_A$ receptors may improve prefrontal function in patients with schizophrenia. A second study that was a placebo-controlled, double-dose MK-0777, double-blind, multicenter study ($N = 60$) did not find significant group differences in the primary outcome measure, the MATRICS Consensus Cognitive Battery composite score (Buchanan et al., 2011). A secondary analysis revealed that patients randomized to placebo performed significantly better on visual memory and reasoning/problem-solving tests. There were no significant group differences in the AX-Continuous Performance Test, the N-Back d prime scores, or the UCSD Performance-Based Skills Assessment and Schizophrenia Cognition Rating total scores. Overall, this study did not demonstrate an MK0777-induced improvement of cognitive impairments in patients with schizophrenia. When interpreting the results from the cited clinical studies, it should be taken into consideration that the MK-0777 is only a weak and partial positive allosteric modulator, with a relative efficacy (determined on recombinant receptors) of 0.11 ± 0.03 at α2-containing GABA$_A$ receptors compared to the classical benzodiazepine chlordiazepoxide (CDP). Although it is completely inactive at α1-containing GABA$_A$ receptors (relative efficacy compared to CDP: 0.00 ± 0.03), it also has some efficacy at α3-containing (relative efficacy compared to CDP: 0.21 ± 0.03) and α5-containing GABA$_A$ receptors (relative efficacy compared to CDP: 0.05 ± 0.02) (Atack et al., 2006). The possibility that α3-containing and/or perhaps even α5-containing GABA$_A$ receptors may contribute to effects seen with MK-0777 cannot be formally excluded. More importantly, in view of the low relative efficacy of MK-0777 at α2-containing GABA$_A$ receptors, it is conceivable that an α2-selective compound with higher relative efficacy at this receptor subtype than MK-0777 might have clearly demonstrable effects on electrophysiological and cognitive parameters.

INHIBITION OF DOPAMINERGIC NEURONS BY α3-CONTAINING GABA$_A$ RECEPTORS

Overactivity of the dopaminergic system in the brain is considered to be a contributing factor to the development and symptomatology of schizophrenia. In dopaminergic neurons in the ventral tegmental area, α3 is the preferentially expressed GABA$_A$ receptor subunit (Tan et al., 2010). By disrupting the gene encoding the α3 subunit (α3KO mice), the GABA-induced whole-cell current in mid-brain dopaminergic neurons was reduced (Yee et al., 2005). In global α3 knockout mice, the whole-cell current from these neurons was reduced, spontaneous locomotor activity was slightly increased, and a marked deficit of prepulse inhibition of the acoustic startle reflex was apparent, indicating a sensorimotor information processing deficit (Yee et al., 2005). Moreover, this prepulse inhibition deficit was completely normalized by the antipsychotic D2-receptor antagonist haloperidol (Yee et al., 2005). In contrast, the amphetamine-induced hyperlocomotion was not altered in the α3KO mice (Yee et al., 2005). These results suggest that drug-induced enhancement of α3GABA$_A$ receptors may attenuate a hyperdopaminergic phenotype, a common feature among psychiatric conditions.

INDIRECT INHIBITION OF DOPAMINERGIC NEURONS BY α5-CONTAINING GABA$_A$ RECEPTORS

Although the α5 subunit of the GABA$_A$ receptor is expressed at relatively low levels in several brain regions, its expression is particularly strong in the hippocampus (Fritschy and Mohler, 1995). When we generated α5(H105R) mice, we noticed that, for reasons that are not understood, the expression of α5 in the hippocampus was reduced by approximately 20–30% (Crestani et al., 2002). These mice display increased trace fear conditioning (shock following 1 s or 20 s after the end of the tone), which is hippocampus-dependent, but unaltered delay fear conditioning (shock administered during the last 500 ms of the tone or immediately after the tone), which is hippocampus-independent (Crestani et al., 2002; Yee et al., 2004). It is of relevance here that these findings demonstrate an alteration in hippocampal function of these mice. Likewise, α5 global knockout mice display improved performance in the Morris water maze (ie, a reduced latency to find the platform, which is also known to be a hippocampus-dependent task) (Collinson et al., 2002). Furthermore, the α5 (H105R) mice display slightly increased locomotor activity and reduced prepulse inhibition of acoustic startle (Hauser et al., 2005). Moreover, these mice also display a latent inhibition deficit. Latent inhibition is a form of selective learning; it is restricted conditioning to a stimulus that is repeatedly presented without any reinforcement contingencies. A single pre-exposure to the taste-conditioned stimulus (CS), here sucrose, prior to pairing it with LiCl-induced nausea (US) reduces the conditioned aversion against the CS in wild-type mice, but not in α5 (H105R) mice (Gerdjikov et al., 2008). Although a prepulse inhibition deficit and a latent inhibition deficit are not specific for any neuropsychiatric disorder, both can be present in patients with schizophrenia. In any case, the findings listed demonstrate that α5-containing GABA$_A$ receptors are involved in cognitive functions, specifically in hippocampus-dependent tasks. Furthermore, α5-containing GABA$_A$ receptors modulate γ oscillations (Glykys et al., 2008; Towers et al., 2004). It is thus conceivable, but yet unproven, that positive modulation of α5-containing GABA$_A$ receptors would at least partially normalize dysfunctional changes in oscillations and, as a consequence, in cognition.

We already mentioned the predominantly hippocampal location of α5-containing GABA$_A$ receptors (Fig. 9.2, lower left panel). In a rat model of schizophrenia, methylazoxymethanol acetate (MAM) administered on gestational day 17 leads to behavioral abnormalities consistent with schizophrenia, including impairments in prepulse inhibition, latent inhibition, spatial working memory with extended delay, extradimensional set-shifting, and increased behavioral responses to amphetamine (Gill and Grace, 2014). In the ventral hippocampus of MAM rats, there is a loss of PV interneurons and a corresponding disruption of sensory-evoked gamma activity and shifts in baseline local field potential power in specific frequency bands (Lodge et al., 2009; Lodge and Grace, 2007). Hyperactivity in the ventral hippocampus has been suggested to underlie the hyper-responsivity of the dopamine system in the MAM rat model because inactivation of the ventral hippocampus with tetrodotoxin reversed the elevated dopamine neuron population activity and normalized the increased amphetamine-induced locomotion (Lodge and Grace, 2007). SH-053-2′F-R-CH$_3$ is an α5-selective positive allosteric modulator, which has at least approximately 7.5-times higher affinity at recombinant α5-containing GABA$_A$ receptors compared to α1-, α2-, or α3-containing GABA$_A$ receptors; efficacy relative to diazepam was 0.08 at α1-containing, 0.07 at α2-containing, 0.29 at α3-containing, and 0.48 at α5-containing GABA$_A$ receptors at 100 nM, and 0.25 at α1-containing, 0.19 at α2-containing, 0.18 at α3-containing, and 1.18 at α5-containing GABA$_A$ receptors at 1 μM (Fischer et al., 2010). Although at the concentration of 100 nM the selectivity for α5-containing GABA$_A$ receptors is quite remarkable, at 1 μM it is more limited.

SH-053-2′F-R-CH$_3$, administered systemically or locally into the ventral hippocampus, reduces the hyperactivity of the dopamine system and of the locomotor response to amphetamine in MAM rats and reduces the number of spontaneously active dopaminergic neurons in the VTA and excitatory responses of the ventral hippocampal neurons to entorhinal cortex stimulation (Gill et al., 2011). Thus, SH-053-2′F-R-CH$_3$ could represent a novel therapeutic approach to schizophrenia, targeting abnormal hippocampal output, and thus normalizing the activity of the dopaminergic neurons in the VTA. It has not formally been shown whether these effects are really or exclusively mediated by α5-containing GABA$_A$ receptors. A potential drawback comes from the observation that when MAM rats are withdrawn from haloperidol, resulting in reduced spontaneous dopamine activity and enhanced locomotor response to amphetamine, they are unresponsive to SH-053-2′F-R-CH$_3$ or ventral hippocampal inactivation by tetrodotoxin, and it has been suggested that testing novel compounds with this mechanism of action on chronically treated schizophrenia patients may be ineffective (Gill et al., 2014).

Previous evidence for a potential role of α5-containing GABA$_A$ receptors as a drug target for schizophrenia came from the study of the *reeler* mouse model, with a 50% downregulation of reelin expression. The sensorimotor deficits in these mice can be reversed by imidazenil, which is a GABA$_A$ receptor–positive allosteric modulator with some (limited) selectivity for α5-containing GABA$_A$ receptors (Auta et al., 2008; Costa et al., 2002). Another interesting observation linking α5-containing GABA$_A$ receptors to schizophrenia is that in a positron emission tomography study with six drug-naïve and five drug-free patients with schizophrenia and 12 healthy controls, [^{11}C]Ro15-4513 (a nonselective compound with high affinity for α5-containing GABA$_A$ receptors) binding in the prefrontal cortex and hippocampus was negatively correlated with negative symptom scores in patients; however, it should be noted that there was no significant difference in binding potential (Asai et al., 2008).

MULTISPECIFIC MODULATION OF α2-, α3-, AND α5-CONTAINING GABA$_A$ RECEPTORS

Although development of drugs for highly selective targets has been a very successful strategy overall, the question remains whether multispecific "dirty" drugs may be more useful for specific applications. We have previously outlined that positive allosteric modulation of α2-, α3-, and α5-containing GABA$_A$ receptors are predicted to have desired therapeutic effects in schizophrenia, with α2- and perhaps α5-containing receptors improving cognition and α3- and α5-containing receptors reducing the activity of dopaminergic neurons in the VTA. Because α1-containing GABA$_A$ receptors mediate sedation

(Rudolph et al., 1999; McKernan et al., 2000) and the disinhibition of dopaminergic neurons in the VTA (via α1-containing GABA$_A$ receptors on GABAergic interneurons in the VTA) and reinforcing properties (Engin et al., 2014; Tan et al., 2010), it is best for this receptor subtype to be avoided. Thus, α2/α3/α5-multiselective partial positive allosteric modulators may provide an interesting approach. One example is L-838,417; it was developed by Merck and shown to have anxiolytic-like, but not sedative, actions in rodents and primates (McKernan et al., 2000; Rowlett et al., 2005). To our knowledge, only absolute (McKernan et al., 2000), but not relative, efficacies at individual GABA$_A$ receptor subtypes compared to chlordiazepoxide or diazepam have been reported. This compound was not developed further due to poor pharmacokinetic properties. However, CTP-354 (formerly C-21191), a new chemical entity based on L-838,417 with selective deuterium integration and improved pharmacokinetics, was recently tested in a Phase 1 study, where it did not cause sedation or ataxia (http://ir.concertpharma.com/releasedetail.cfm?releaseid=875736). It remains to be determined whether this or other compounds may provide any clinical benefits for patients with schizophrenia.

PROOF-OF-CONCEPT CLINICAL STUDY WITH THE PARTIAL POSITIVE ALLOSTERIC MODULATOR BRETAZENIL

An early proof-of-concept study for antipsychotic action of partial GABA$_A$ receptor agonists comes from a small, 6-week, open label study with bretazenil. This compound is a partial positive allosteric modulator at α1-, α2-, α3-, and α5-containing receptors (Costa and Guidotti, 1996) and was developed as an anxiolytic. The idea was that because a lower receptor occupancy is required for anxiolysis compared to sedation, a nonsubtype-selective partial positive allosteric modulator might be anxiolytic but not or only mildly sedative. Notably, bretazenil was efficacious as a monotherapy in 44% of neuroleptic-free patients in an intention-to-treat analysis in a 6-week open label trial (Delini-Stula and Berdah-Tordjman, 1996). The excellent tolerability of bretazenil and the lack of extrapyramidal side effects of clinically used antipsychotics like haloperidol indicate that partial allosteric GABA$_A$ receptor modulators may be useful as antipsychotics.

α5-NEGATIVE ALLOSTERIC MODULATORS AND COGNITION ENHANCEMENT

In the hippocampus, α5-GABA$_A$ receptors are located extrasynaptically at the base of the dendritic spines that receive glutamatergic input (Fig. 9.2), suggesting that α5-GABA$_A$ receptors may exert control over this input. The observation that mice with a partial or complete deletion of α5-containing GABA$_A$ receptors display better performance in some cognitive tasks (Crestani et al., 2002; Collinson et al., 2002) led to the development of α5-selective negative allosteric modulators as potential cognitive enhancers, and, indeed, in rodents such compounds provided cognitive improvements. For example, Ro4938581 reverses scopolamine-induced impairment during a delayed match-to-position task (Ballard et al., 2009). It has been discovered that in a mouse model of Down syndrome, the Ts65Dn mice display deficits in hippocampal synaptic plasticity (reduced long-term potentiation (LTP) and increased long-term depression (LTD)), and plasticity appears to be obstructed by excessive inhibition in the dentate gyrus. A low dose of the GABA$_A$ receptor antagonist, pentylenetetrazole (PTZ), which is a convulsant at higher doses, normalized hippocampal LTP, novel object recognition, T maze alternation, and spatial cognition (Fernandez et al., 2007). The α5-seletive partial negative allosteric modulator α5IA reversed deficits in spatial reference learning in the Morris water maze and in novel object recognition (Braudeau et al., 2011). Likewise, the α5-selective partial negative allosteric modulator Ro4938581, which is more selective than α5IA, improved deficits in hippocampal LTP and neurogenesis and rescued performance in the Morris water maze without affecting sensorimotor abilities or motor coordination (rotarod performance test), spontaneous motor activity, seizure induction, or induction of anxiety (open field test); it even showed anxiolytic-like actions in both Ts65Dn and control mice (Martinez-Cue et al., 2013). Thus, there is a wealth of evidence that α5-selective NAMs may be able to reverse at least some cognitive deficits in patients with Down syndrome. F. Hoffmann-La Roche entered RG1662 into clinical trials for patients with Down syndrome (see RG1662 http://clinicaltrials.gov).

The observation that reduced α5 function may lead to cognitive deficits in a rat model of schizophrenia while increased α5-mediated inhibition may lead to a cognitive deficit in a mouse model of Down syndrome suggests that deviations of α5 activity in either direction may lead to cognitive deficits and that normalization of these deviations may have therapeutic benefits. This notion is also supported by a recent study which has found that α5-containing $GABA_A$ receptor-mediated tonic inhibition in dentate gyrus granule cells plays an important role in ensuring normal cognitive functioning under high- but not under low-interference conditions (Engin et al., 2015).

REFERENCES

Asai, Y., Takano, A., Ito, H., Okubo, Y., Matsuura, M., Otsuka, A., et al., 2008. $GABA_A$/Benzodiazepine receptor binding in patients with schizophrenia using [11C]Ro15-4513, a radioligand with relatively high affinity for α5 subunit. Schizophr Res 99 (1–3), 333–340.

Atack, J.R., 2010. $GABA_A$ receptor α2/α3 subtype-selective modulators as potential nonsedating anxiolytics. Curr Top Behav Neurosci 2, 331–360.

Atack, J.R., Wafford, K.A., Tye, S.J., Cook, S.M., Sohal, B., Pike, A., et al., 2006. TPA023 [7-(1,1-dimethylethyl)-6-(2-ethyl-2H-1,2,4-triazol-3-ylmethoxy)-3-(2-fluorophenyl)-1,2,4-triazolo[4,3-b]pyridazine], an agonist selective for α2- and α3-containing GABAA receptors, is a nonsedating anxiolytic in rodents and primates. J Pharm Exp Ther 316 (1), 410–422.

Auta, J., Impagnatiello, F., Kadriu, B., Guidotti, A., Costa, E., 2008. Imidazenil: a low efficacy agonist at α1- but high efficacy at α5-$GABA_A$ receptors fail to show anticonvulsant cross tolerance to diazepam or zolpidem. Neuropharmacology 55 (2), 148–153.

Ballard, T.M., Knoflach, F., Prinssen, E., Borroni, E., Vivian, J.A., Basile, J., et al., 2009. RO4938581, a novel cognitive enhancer acting at $GABA_A$ α5 subunit-containing receptors. Psychopharmacology 202 (1–3), 207–223.

Belforte, J.E., Zsiros, V., Sklar, E.R., Jiang, Z., Yu, G., Li, Y., et al., 2010. Postnatal NMDA receptor ablation in corticolimbic interneurons confers schizophrenia-like phenotypes. Nat Neurosci 13 (1), 76–83.

Braudeau, J., Delatour, B., Duchon, A., Pereira, P.L., Dauphinot, L., de Chaumont, F., et al., 2011. Specific targeting of the $GABA_A$ receptor α5 subtype by a selective inverse agonist restores cognitive deficits in Down syndrome mice. J Psychopharmacol 25 (8), 1030–1042.

Buchanan, R.W., Keefe, R.S., Lieberman, J.A., Barch, D.M., Csernansky, J.G., Goff, D.C., et al., 2011. A randomized clinical trial of MK-0777 for the treatment of cognitive impairments in people with schizophrenia. Biol Psychiatry 69 (5), 442–449.

Castner, S.A., Arriza, J.L., Roberts, J.C., Mrzljak, L., Christian, E.P., Williams, G.V., 2010. Reversal of ketamine-induced working memory impairments by the $GABA_A$α2/3 agonist TPA023. Biol Psychiatry 67 (10), 998–1001.

Collinson, N., Kuenzi, F.M., Jarolimek, W., Maubach, K.A., Cothliff, R., Sur, C., et al., 2002. Enhanced learning and memory and altered GABAergic synaptic transmission in mice lacking the α5 subunit of the GABA$_A$ receptor. J Neurosci 22 (13), 5572–5580.

Costa, E., Guidotti, A., 1996. Benzodiazepines on trial: a research strategy for their rehabilitation. Trends Pharmacol Sci 17 (5), 192–200.

Costa, E., Davis, J., Pesold, C., Tueting, P., Guidotti, A., 2002. The heterozygote reeler mouse as a model for the development of a new generation of antipsychotics. Curr Opin Pharmacol 2 (1), 56–62.

Coyle, J.T., 1996. The glutamatergic dysfunction hypothesis for schizophrenia. Harv Rev Psychiatry 3 (5), 241–253.

Crestani, F., Low, K., Keist, R., Mandelli, M., Mohler, H., Rudolph, U., 2001. Molecular targets for the myorelaxant action of diazepam. Mol Pharmacol 59 (3), 442–445.

Crestani, F., Keist, R., Fritschy, J.M., Benke, D., Vogt, K., Prut, L., et al., 2002. Trace fear conditioning involves hippocampal α5 GABA(A) receptors. Proc Natl Acad Sci USA 99 (13), 8980–8985.

Dehorter, N., Vinay, L., Hammond, C., Ben-Ari, Y., 2012. Timing of developmental sequences in different brain structures: physiological and pathological implications. Eur J Neurosci 35 (12), 1846–1856.

Delini-Stula, A., Berdah-Tordjman, D., 1996. Antipsychotic effects of bretazenil, a partial benzodiazepine agonist in acute schizophrenia – a study group report. J Psychiatr Res 30 (4), 239–250.

Engin, E., Bakhurin, K.I., Smith, K.S., Hines, R.M., Reynolds, L.M., Tang, W., et al., 2014. Neural basis of benzodiazepine reward: requirement for α2 containing GABA$_A$ receptors in the nucleus accumbens. Neuropsychopharmacology 39 (8), 1805–1815.

Engin, E., Zarnowska, E.D., Benke, D., Tsvetkov, E., Sigal, M., Keist, R., et al., 2015. Tonic inhibitory control of dentate gyrus granule cells by alpha5-containing GABAA receptors reduces memory interference. J. Neurosci. 35 (40), 13698–13712.

Fagiolini, M., Hensch, T.K., 2000. Inhibitory threshold for critical-period activation in primary visual cortex. Nature 404 (6774), 183–186.

Fagiolini, M., Fritschy, J.M., Low, K., Mohler, H., Rudolph, U., Hensch, T.K., 2004. Specific GABA$_A$ circuits for visual cortical plasticity. Science 303 (5664), 1681–1683.

Fazzari, P., Paternain, A.V., Valiente, M., Pla, R., Lujan, R., Lloyd, K., et al., 2010. Control of cortical GABA circuitry development by Nrg1 and ErbB4 signalling. Nature 464 (7293), 1376–1380.

Fernandez, F., Morishita, W., Zuniga, E., Nguyen, J., Blank, M., Malenka, R.C., et al., 2007. Pharmacotherapy for cognitive impairment in a mouse model of Down syndrome. Nat Neurosci 10 (4), 411–413.

Ferrarelli, F., Massimini, M., Peterson, M.J., Riedner, B.A., Lazar, M., Murphy, M.J., et al., 2008. Reduced evoked gamma oscillations in the frontal cortex in schizophrenia patients: a TMS/EEG study. Am J Psychiatry 165 (8), 996–1005.

Fillman, S.G., Duncan, C.E., Webster, M.J., Elashoff, M., Weickert, C.S., 2010. Developmental co-regulation of the beta and gamma GABA$_A$ receptor subunits with distinct alpha subunits in the human dorsolateral prefrontal cortex. Int J Dev Neurosci 28 (6), 513–519.

Fischer, B.D., Licata, S.C., Edwankar, R.V., Wang, Z.J., Huang, S., He, X., et al., 2010. Anxiolytic-like effects of 8-acetylene imidazobenzodiazepines in a rhesus monkey conflict procedure. Neuropharmacology 59 (7–8), 612–618.

Fritschy, J.M., Mohler, H., 1995. GABA$_A$-receptor heterogeneity in the adult rat brain: differential regional and cellular distribution of seven major subunits. J Comp Neurol 359 (1), 154–194.

Fritschy, J.M., Panzanelli, P., 2014. GABA$_A$ receptors and plasticity of inhibitory neurotransmission in the central nervous system. Eur J Neurosci 39 (11), 1845–1865.

Fromer, M., Pocklington, A.J., Kavanagh, D.H., Williams, H.J., Dwyer, S., Gormley, P., et al., 2014. De novo mutations in schizophrenia implicate synaptic networks. Nature 506 (7487), 179–184.

Gerdjikov, T.V., Rudolph, U., Keist, R., Mohler, H., Feldon, J., Yee, B.K., 2008. Hippocampal α5 subunit-containing GABA A receptors are involved in the development of the latent inhibition effect. Neurobiol Learn Mem 89 (2), 87–94.

Gill, K.M., Grace, A.A., 2014. The role of α5 GABA$_A$ receptor agonists in the treatment of cognitive deficits in schizophrenia. Curr Pharm Des 20, 5069–5076.

Gill, K.M., Lodge, D.J., Cook, J.M., Aras, S., Grace, A.A., 2011. A novel α5GABA(A)R-positive allosteric modulator reverses hyperactivation of the dopamine system in the MAM model of schizophrenia. Neuropsychopharmacology 36 (9), 1903–1911.

Gill, K.M., Cook, J.M., Poe, M.M., Grace, A.A., 2014. Prior antipsychotic drug treatment prevents response to novel antipsychotic agent in the methylazoxymethanol acetate model of schizophrenia. Schizophr Bull 40, 341–350.

Glykys, J., Mann, E.O., Mody, I., 2008. Which GABA(A) receptor subunits are necessary for tonic inhibition in the hippocampus? J Neurosci 28 (6), 1421–1426.

Harrison, P.J., Weinberger, D.R., 2005. Schizophrenia genes, gene expression, and neuropathology: on the matter of their convergence. Mol Psychiatry 10, 40–68.

Harrison, P.J., Lewis, D.A., Kleinman, J.L., 2011. Neuropathology of schizophrenia. In: Weinberger, D.R., Harrison, P.J. (Eds.), Schizophrenia Wiley Blackwell, pp. 372–392.

Hashimoto, T., Volk, D.W., Eggan, S.M., Mirnics, K., Pierri, J.N., Sun, Z., et al., 2003. Gene expression deficits in a subclass of GABA neurons in the prefrontal cortex of subjects with schizophrenia. J Neurosci 23 (15), 6315–6326.

Hashimoto, T., Arion, D., Unger, T., Maldonado-Aviles, J.G., Morris, H.M., Volk, D.W., et al., 2008. Alterations in GABA-related transcriptome in the dorsolateral prefrontal cortex of subjects with schizophrenia. Mol Psychiatry 13 (2), 147–161.

Hauser, J., Rudolph, U., Keist, R., Mohler, H., Feldon, J., Yee, B.K., 2005. Hippocampal α5 subunit-containing GABA$_A$ receptors modulate the expression of prepulse inhibition. Mol Psychiatry 10 (2), 201–207.

Hines, R.M., Hines, D.J., Houston, C.M., Mukherjee, J., Haydon, P.G., Tretter, V., et al., 2013. Disrupting the clustering of GABA$_A$ receptor α2 subunits in the frontal cortex leads to reduced-power and cognitive deficits. Proc Natl Acad Sci USA 110 (41), 16628–16633.

Jaaro-Peled, H., Hayashi-Takagi, A., Seshadri, S., Kamiya, A., Brandon, N.J., Sawa, A., 2009. Neurodevelopmental mechanisms of schizophrenia: understanding disturbed postnatal brain maturation through neuregulin-1-ErbB4 and DISC1. Trends Neurosci 32 (9), 485–495.

Javitt, D.C., Spencer, K.M., Thaker, G.K., Winterer, G., Hajos, M., 2008. Neurophysiological biomarkers for drug development in schizophrenia. Nat Rev Drug Discov 7 (1), 68–83.

Katagiri, H., Fagiolini, M., Hensch, T.K., 2007. Optimization of somatic inhibition at critical period onset in mouse visual cortex. Neuron 53 (6), 805–812.

Klausberger, T., Somogyi, P., 2008. Neuronal diversity and temporal dynamics: the unity of hippocampal circuit operations. Science 321 (5885), 53–57.

Knabl, J., Witschi, R., Hosl, K., Reinold, H., Zeilhofer, U.B., Ahmadi, S., et al., 2008. Reversal of pathological pain through specific spinal GABA$_A$ receptor subtypes. Nature 451 (7176), 330–334.

Krystal, J.H., Moghaddam, B., 2011. Contributions of Glutamate and GABA Systems to the Neurobiology and Treatment of Schizophrenia, third ed. Wiley Blackwell, 433–461.

Lewis, D.A., Gonzalez-Burgos, G., 2008. Neuroplasticity of neocortical circuits in schizophrenia. Neuropsychopharmacology 33 (1), 141–165.

Lewis, D.A., Hashimoto, T., Volk, D.W., 2005. Cortical inhibitory neurons and schizophrenia. Nat.Rev Neurosci 6 (4), 312–324.

Lewis, D.A., Cho, R.Y., Carter, C.S., Eklund, K., Forster, S., Kelly, M.A., et al., 2008. Subunit-selective modulation of GABA type A receptor neurotransmission and cognition in schizophrenia. Am J Psychiatry 165, 1585–1593.

Liu, J., Shi, Y., Tang, W., Guo, T., Li, D., Yang, Y., et al., 2005. Positive association of the human GABA$_A$-receptor beta 2 subunit gene haplotype with schizophrenia in the Chinese Han population. Biochem Biophys Res Commun 334 (3), 817–823.

REFERENCES

Liu, Z., Neff, R.A., Berg, D.K., 2006. Sequential interplay of nicotinic and GABAergic signaling guides neuronal development. Science 314 (5805), 1610–1613.

Lodge, D.J., Grace, A.A., 2007. Aberrant hippocampal activity underlies the dopamine dysregulation in an animal model of schizophrenia. J Neurosci 27 (42), 11424–11430.

Lodge, D.J., Behrens, M.M., Grace, A.A., 2009. A loss of parvalbumin-containing interneurons is associated with diminished oscillatory activity in an animal model of schizophrenia. J Neurosci 29 (8), 2344–2354.

Low, K., Crestani, F., Keist, R., Benke, D., Brunig, I., Benson, J.A., et al., 2000. Molecular and neuronal substrate for the selective attenuation of anxiety. Science 290 (5489), 131–134.

Lu, J.C., Hsiao, Y.T., Chiang, C.W., Wang, C.T., 2014. $GABA_A$ receptor-mediated tonic depolarization in developing neural circuits. Mol Neurobiol 49 (2), 702–723.

Martinez-Cue, C., Martinez, P., Rueda, N., Vidal, R., Garcia, S., Vidal, V., et al., 2013. Reducing $GABA_A$ $\alpha 5$ receptor-mediated inhibition rescues functional and neuromorphological deficits in a mouse model of down syndrome. J Neurosci 33 (9), 3953–3966.

McCarroll, S.A., Hyman, S.E., 2013. Progress in the genetics of polygenic brain disorders: significant new challenges for neurobiology. Neuron 80 (3), 578–587.

McKernan, R.M., Rosahl, T.W., Reynolds, D.S., Sur, C., Wafford, K.A., Atack, J.R., et al., 2000. Sedative but not anxiolytic properties of benzodiazepines are mediated by the $GABA_A$ receptor $\alpha 1$ subtype. Nat Neurosci 3 (6), 587–592.

Mohler, H., 2009. Role of $GABA_A$ receptors in cognition. Biochem Soc Trans 37 (Pt 6), 1328–1333.

Mohler, H., 2011. The rise of a new GABA pharmacology. Neuropharmacology 60 (7–8), 1042–1049.

Mohler, H., 2012. Cognitive enhancement by pharmacological and behavioral interventions: the murine Down syndrome model. Biochem Pharmacol 84 (8), 994–999.

Mohler, H., 2015. The legacy of the benzodiazepine receptor: from flumazenil to enhancing cognition in Down syndrome and social interaction in autism. Adv Pharmacol 72, 1–36.

Morishita, H., Miwa, J.M., Heintz, N., Hensch, T.K., 2010. Lynx1, a cholinergic brake, limits plasticity in adult visual cortex. Science 330 (6008), 1238–1240.

Okaty, B.W., Miller, M.N., Sugino, K., Hempel, C.M., Nelson, S.B., 2009. Transcriptional and electrophysiological maturation of neocortical fast-spiking GABAergic interneurons. J Neurosci 29 (21), 7040–7052.

Olsen, R.W., Sieghart, W., 2008. International Union of Pharmacology. LXX. Subtypes of gamma-aminobutyric acid(A) receptors: classification on the basis of subunit composition, pharmacology, and function. Update. Pharmacol Rev 60 (3), 243–260.

Petryshen, T.L., Middleton, F.A., Tahl, A.R., Rockwell, G.N., Purcell, S., Aldinger, K.A., et al., 2005. Genetic investigation of chromosome 5q GABAA receptor subunit genes in schizophrenia. Mol Psychiatry 10 (12), 1074–1088, 1057.

Polan, M.B., Pastore, M.T., Steingass, K., Hashimoto, S., Thrush, D.L., Pyatt, R., et al., 2014. Neurodevelopmental disorders among individuals with duplication of 4p13 to 4p12 containing a $GABA_A$ receptor subunit gene cluster. Eur J Human Genet 22 (1), 105–109.

Purcell, S.M., Moran, J.L., Fromer, M., Ruderfer, D., Solovieff, N., Roussos, P., et al., 2014. A polygenic burden of rare disruptive mutations in schizophrenia. Nature 506 (7487), 185–190.

Roux, L., Buzsaki, G., 2015. Tasks for inhibitory interneurons in intact brain circuits. Neuropharmacology 88, 10–23.

Rowlett, J.K., Platt, D.M., Lelas, S., Atack, J.R., Dawson, G.R., 2005. Different $GABA_A$ receptor subtypes mediate the anxiolytic, abuse-related, and motor effects of benzodiazepine-like drugs in primates. Proc Natl Acad Sci USA 102 (3), 915–920.

Rudolph, U., Knoflach, F., 2011. Beyond classical benzodiazepines: novel therapeutic potential of $GABA_A$ receptor subtypes. Nat Rev Drug Discov 10 (9), 685–697.

Rudolph, U., Mohler, H., 2014. $GABA_A$ receptor subtypes: therapeutic potential in Down syndrome, affective disorders, schizophrenia, and autism. Annu Rev Pharmacol Toxicol 54, 483–507.

Rudolph, U., Crestani, F., Benke, D., Brunig, I., Benson, J.A., Fritschy, J.M., et al., 1999. Benzodiazepine actions mediated by specific γ-aminobutyric acid$_A$ receptor subtypes. Nature 401 (6755), 796–800.

Schizophrenia Working Group of the Psychiatric Genomics Consortium, 2014. Biological insights from 108 schizophrenia-associated genetic loci. Nature 511, 421–427.

Spencer, K.M., Nestor, P.G., Perlmutter, R., Niznikiewicz, M.A., Klump, M.C., Frumin, M., et al., 2004. Neural synchrony indexes disordered perception and cognition in schizophrenia. Proc Natl Acad Sci USA 101 (49), 17288–17293.

Tan, K.R., Brown, M., Labouebe, G., Yvon, C., Creton, C., Fritschy, J.M., et al., 2010. Neural bases for addictive properties of benzodiazepines. Nature 463 (7282), 769–774.

Towers, S.K., Gloveli, T., Traub, R.D., Driver, J.E., Engel, D., Fradley, R., et al., 2004. α5 subunit-containing GABA$_A$ receptors affect the dynamic range of mouse hippocampal kainate-induced gamma frequency oscillations in vitro. J Physiol 559 (Pt 3), 721–728.

Volk, D.W., Pierri, J.N., Fritschy, J.M., Auh, S., Sampson, A.R., Lewis, D.A., 2002. Reciprocal alterations in pre- and postsynaptic inhibitory markers at chandelier cell inputs to pyramidal neurons in schizophrenia. Cereb Cortex 12 (10), 1063–1070.

Weinberger, D.R., Levitt, P., 2011. Neurodevelopmental origin of schizophrenia. In: Weinberger, D.R., Harrison, P.J. (Eds.), Schizophrenia, third ed. Wiley Blackwell, pp. 393–412.

Winterer, G., Weinberger, D.R., 2004. Genes, dopamine and cortical signal-to-noise ratio in schizophrenia. Trends Neurosci 27 (11), 683–690.

Xu, G., Broadbelt, K.G., Haynes, R.L., Folkerth, R.D., Borenstein, N.S., Belliveau, R.A., et al., 2011. Late development of the GABAergic system in the human cerebral cortex and white matter. J Neuropathol Exp Neurol 70 (10), 841–858.

Yang, J.M., Zhang, J., Chen, X.J., Geng, H.Y., Ye, M., Spitzer, N.C., et al., 2013. Development of GABA circuitry of fast-spiking basket interneurons in the medial prefrontal cortex of erbb4-mutant mice. J Neurosci 33 (50), 19724–19733.

Yee, B.K., Hauser, J., Dolgov, V.V., Keist, R., Mohler, H., Rudolph, U., et al., 2004. GABA receptors containing the α5 subunit mediate the trace effect in aversive and appetitive conditioning and extinction of conditioned fear. Eur J Neurosci 20 (7), 1928–1936.

Yee, B.K., Keist, R., von Boehmer, L., Studer, R., Benke, D., Hagenbuch, N., et al., 2005. A schizophrenia-related sensorimotor deficit links α3-containing GABA$_A$ receptors to a dopamine hyperfunction. Proc Natl Acad Sci USA 102 (47), 17154–17159.

Zai, G., King, N., Wong, G.W., Barr, C.L., Kennedy, J.L., 2005. Possible association between the gamma-aminobutyric acid type B receptor 1 (GABBR1) gene and schizophrenia. Eur Neuropsychopharmacol 15 (3), 347–352.

Zhao, X., Qin, S., Shi, Y., Zhang, A., Zhang, J., Bian, L., et al., 2007. Systematic study of association of four GABAergic genes: glutamic acid decarboxylase 1 gene, glutamic acid decarboxylase 2 gene, GABA(B) receptor 1 gene and GABA(A) receptor subunit beta2 gene, with schizophrenia using a universal DNA microarray. Schizophr Res 93 (1–3), 374–384.

PART IV

THE BIOCHEMICAL BASIS OF SCHIZOPHRENIA

CHAPTER 10

METABOLOMICS OF SCHIZOPHRENIA

D. Rujescu and I. Giegling

Department of Psychiatry, Psychotherapy, and Psychosomatics,
Martin Luther University Halle-Wittenberg, Halle, Germany

CHAPTER OUTLINE

Introduction 167
Lipidomics in Schizophrenia 169
Energy Metabolism in Schizophrenia 172
Outlook 174
References 174

INTRODUCTION

Schizophrenia is a complex mental disorder affecting 0.5–1% of the population. It is characterized by disintegration of thought processes and of emotional responsiveness, most commonly manifesting as hallucinations, paranoid delusions, or disorganized speech and thinking, and is accompanied by social or occupational dysfunction. It mostly presents with several episodes, has an early age of onset (mostly during early adulthood), and tends to become chronic (Nasrallah et al., 2011; Keshavan et al., 2011). Approximately 30% of patients with schizophrenia require support throughout their lives and have a high burden of disease. Using a summary measure of population health, called the disability-adjusted life-year (DALY; a time-based measure that combines in a single indicator years of life lost from premature death [YLLs] and years of life lived with a disability [YLDs]), the estimates from the Global Burden of Disease study indicate that neuropsychiatric disorders contribute to more than 10.4% of all DALYs and more than 28.2% of all YLDs, summing up to more than one-quarter of all diseases. Within 20 years (1990–2010), the DALYs for schizophrenia increased by 43.6%, ranking schizophrenia 13th among all diseases in central Europe for YLD (eg, before ischemic heart disease, which ranks 15th) (Vos et al., 2012; Murray et al., 2012).

Major causes for long and expensive hospitalization are nonresponse to a specific drug as well as severe side effects. Predictive characterization of nonresponders may minimize prolonged exposure to suboptimal or ineffective treatment strategies (Ascher-Svanum et al., 2008). The way to meet the challenge for a better and more efficient treatment is to select the most efficient medication as early as possible, to avoid severe adverse drug reactions (ADRs), and, as a consequence, to reduce hospitalization to a minimum. Unfortunately, schizophrenia is still only treated based on clinical impression and clinical guidelines without taking into account predictive biomarkers or pharmacogenetic profiles. Until now, there have been no valid predictive algorithms for schizophrenia treatment based on genetic or biomarker signatures.

Historically, antipsychotic medication is split into first-generation treatments, the "typical" antipsychotics (eg, haloperidol), which act principally on the dopamine system, and second-generation treatments, the "atypical" antipsychotics (eg, clozapine, olanzapine), which act on multiple neurotransmitter receptors. The first-generation treatments can induce motor side effects like tardive dyskinesia, akathisia, or parkinsonism (Giegling et al., 2012, 2013). To avoid these side effects, much hope was given to the second-generation antipsychotics, which were thought to lower the ADR rate. However, unfortunately, these often have side effects of severe weight gain, diabetes, and metabolic syndrome (Flanagan, 2008).

Metabolomics could be a new powerful approach to find not only pathophysiological causes of the disease but also causes for side effects like weight gain. Metabolomics is the most recent of the "omics" disciplines and offers potent tools for detection of metabolic pathways and networks characterizing conditions such as schizophrenia (He et al., 2012).

Metabolites are small molecules that are chemically altered during metabolism; as a result, they offer a functional pattern of the cellular state. In contrast to genes and proteins, metabolites act as direct signatures of biochemical activity and are easier to correlate with phenotype. Therefore, metabolomics has become an influential strategy that has been accepted for clinical diagnostics (Patti et al., 2012).

In contrast to classical biochemical methodologies strictly focused on single metabolites, metabolomics accumulates quantitative data regarding a larger series of metabolites in an attempt to delineate a complete depiction of metabolism and/or metabolic fluctuations linked to the relevant pathological condition (Dharuri et al., 2014). Metabolomics provides analytical tools that can simultaneously measure thousands of elements contained in a biological sample. This analytical feature must then be combined with bioinformatic tools that can detect a molecular signal for a disease among millions of data. Ideally, metabolomics will finally produce a comprehensive map of the modulation of metabolic pathways, and, therefore, of the interaction of proteins with environmental factors, including drug exposure.

The major application of this is represented by biomarker discovery. Such biomarkers refer to changes of endogenous metabolite variation associated with specific diseases when compared to a control group. These biomarkers might be used for diagnostic purposes, choice of therapy, assessment of the treatment effect, and monitoring the course of the disease.

Metabolic markers in diagnosis are thought to be one of the most fascinating categories of biomarkers. Because a biomarker should be detected and measured in a sample obtained using noninvasive procedures, body fluids, including plasma/serum, urine, saliva, and, in some measure, cerebrospinal fluid (CSF), are thought to be ideal sources for biomarker monitoring (Nordström and Lewensohn, 2010). Metabolomics makes available novel potent strategies to investigate, simultaneously and quantitatively, hundreds of these crucial molecules (Kristal et al., 2007) and allows exploring metabolites within significant pathways to clarify their involvement in central nervous system disorders and represent more clearly the mechanisms of action of drugs targeting such pathways (Fig. 10.1).

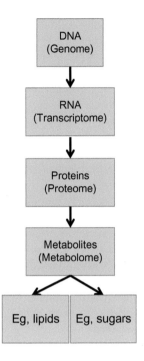

FIGURE 10.1

Different "omics" approaches in schizophrenia.

LIPIDOMICS IN SCHIZOPHRENIA

Psychiatric diseases have been discovered to be associated with metabolic pathway disturbances that could be reproduced in metabolomic profiles. In particular, in schizophrenia, various abnormalities in membrane composition, neurotransmitter, signal transduction, antioxidant system, and immune activities have been reported among other alterations (Mahadik and Yao, 2006; Kaddurah-Daouk and Krishnan, 2009). Evidence exists that phospholipids, which contribute to the structure and function of membranes, are compromised in schizophrenia (Horrobin, 1998; Berger et al., 2002; Mahadik and Yao, 2006). Lipids and their constituent fatty acids that provide scaffolding for several key functional systems—such as neurotransmitter receptor binding, signal transduction, transmembrane ion channels, prostanoid synthesis, and mitochondrial electron transport systems—could induce the pathogenesis of schizophrenia (Lieberman and Koreen, 1993).

Lipids are organic compounds with long-chain hydrocarbon molecules that are soluble in organic solvents but are not soluble in water. Some examples of lipids include long-chain hydrocarbons, alcohols, aldehydes, fatty acids, their derivatives, fat-soluble vitamins (A, D, E, and K), carotenoids, and sterols. Approximately half of the dry weight of the brain is made of lipids. Lipids are important in many brain functions, including membrane composition, signal transduction, and biological messenger. Thus, changes in the concentrations of brain lipids may reflect physiopathologic processes (Fonteh et al., 2006). Prior studies on lipid metabolics show changes linked to schizophrenia (Mahadik and Evans, 2003). Data according to variations in lipid metabolism during schizophrenia were in line with Horrobin's hypothesis (Horrobin and Huang, 1983) and former analyses of red blood cells (Peet et al., 1995), erythrocytes (Keshavan et al., 1993), postmortem brains (Yao et al., 2000), and the brain (Potwarka et al., 1999).

Therefore, lipidomics is one major focus. Several techniques using mass spectrometry (MS) are accessible for qualitative and quantitative examination of major lipids in complex samples (Wenk, 2010). The first reports of MS-based studies of complex lipid mixtures via ionization-based methods date back to the 1990s (Han and Gross, 1994; Kim et al., 1994). MS supported the further development of lipidomics.

In a study by Kaddurah-Daouk et al. (2007), a dedicated lipidomics platform (Lipomics Technologies, Inc., West Sacramento, CA, USA) (Watkins et al., 2002) was used to quantify approximately 300 lipid metabolites, mainly structural and energetic lipids, across seven lipid sets to assess lipid alterations in schizophrenia before and after therapy with three atypical antipsychotics (olanzapine, risperidone, and aripiprazole). The depiction of lipid signatures was obtained from plasma of 50 patients with schizophrenia before and after 2–3 weeks. At baseline and before drug administration, the most important alterations were reported in two phospholipids groups, phosphotidylcholine (PC) and phosphotidylethanolamine (PE). This finding indicates that phospholipids implicated in membrane structure/function appear to be compromised in schizophrenia. In particular, biochemical defects were observed within the ω-3 and ω-6 subgroups in PC and PE. Furthermore, shifts between saturated fatty acids and polyunsaturated fatty acids (PUFAs) were described (Kaddurah-Daouk et al., 2007). The effects of the antipsychotic were then assessed by comparing metabolic profiles at baseline to that during post-treatment. Notably, each drug exhibited a distinctive signature. Amounts of PE that were lowered at baseline in schizophrenia patients had increased after administration of all three drugs. However, olanzapine and risperidone had an impact on a broader range of lipids than did aripiprazole, with nearly 50 lipids elevated after treatment, but not after aripiprazole administration. The alterations generated in the lipidome due to aripiprazole were negligible; these data are in line with its restricted metabolic side effects. In addition, elevated levels of triacylglycerols and reduced amounts of free fatty acids were revealed following olanzapine and risperidone treatment, but not after aripiprazole treatment. These data indicate the existence of peripheral effects potentially linked to the metabolic side effects reported for these drugs and emphasize hepatic lipases as probable pharmacological targets. Finally, baseline lipid changes were found to be associated with acute treatment response (Kaddurah-Daouk et al., 2007).

Another metabolomics-based strategy based on lipidomics in psychotic disorders (schizophrenia, affective psychoses, and other nonaffective psychoses) was created by Orešič and colleagues (2011). Two analytical systems were utilized: (1) a global lipidomics platform covering phospholipids, sphingolipids, and neutral lipids and (2) a platform for small polar metabolite-based MS exploring amino acids, free fatty acids, various other organic acids, sterols, and sugars. After grouping both lipidomic and metabolomic data into subsets, schizophrenia patients were found to have significantly higher metabolite concentrations in six lipid clusters—primarily including saturated and longer-chain triglycerides—and in two small-molecule clusters, mostly represented by the following: (1) branched chain amino acids, phenylalanine, and tyrosine and (2) proline, glutamic, lactic, and pyruvic acids (Orešič et al., 2011).

Interestingly, the same authors detected lipidomic profiles in twin pairs discordant for schizophrenia. They discovered higher serum levels of triglycerides and greater insulin resistance in schizophrenic twins versus age-matched and gender-matched healthy twins. The integration of lipidomic results with magnetic resonance imaging data unveiled a major correlation between reduced gray matter density and augmented triglycerides (Orešič et al., 2012).

Plasmalogens as a subclass of glycerophospholipids are ubiquitous constituents of cellular membranes and serum lipoproteins. Kaddurah-Daouk and colleagues (2012) reported a comparison of plasma plasmalogen levels across 20 drug-naïve patients experiencing first psychotic episodes, 20 recently unmedicated patients experiencing psychotic relapses after failing to comply with prescribed medications, and 17 matched healthy control subjects. Multiple plasma phosphatidylcholine and phosphatidylethanolamine plasmalogen levels were significantly lower in first-episode patients and patients with recurrent disease compared to healthy controls. Therefore, reduced plasmalogen levels appear to be a trait evident at the onset of psychotic illness and after multiple psychotic relapses. It is implied that reductions in plasmalogen levels are not related to antipsychotic treatment but rather are due to the illness itself. Reduced plasmalogen levels suggest impairments in membrane structure and function in patients with schizophrenia that might happen early in development (Kaddurah-Daouk et al., 2012).

McEvoy et al. (2013) performed a study by using a lipidomics platform with 300 lipid metabolites (Watkins et al., 2002). The sample consisted of 20 medication-naïve patients showing a first episode of schizophrenia, 20 showing chronic schizophrenia (recurrent episode), and 29 race-matched healthy controls. Lipid metabolic signatures were assessed and compared between study groups and within groups before and after atypical antipsychotics therapy (ie, risperidone and aripiprazole). Compared to healthy controls, first-episode patients displayed substantial downregulation of many n3 PUFAs within the main phospholipid classes, PC and PE. However, levels of n6 phospholipids were comparable between both groups. Thus, these results suggested an early occurrence of changes in n3 lipid metabolism in the pathology. On the contrary, chronic schizophrenia patients did not have any major variation in n3 or n6 phospholipids compared to controls. This suggested that either disease advancement or previous drug therapy enhanced the early modifications in lipid metabolic pathways connected to schizophrenia. Another significant discovery was that, in first-episode patients, antipsychotics caused important alterations (pretreatments vs post-treatment) in both the n3 and n6 lipid levels. However, in the chronic schizophrenia group, drug administration caused marginal fluctuations in phospholipids. As a result, the effects of risperidone and aripiprazole on lipid metabolism were changed, either by disease progression or by previous pharmacological therapy (McEvoy et al., 2013).

Wood et al. (2014) performed a shotgun lipidomic analysis of more than 700 lipids across 26 lipid subclasses in the frontal cortex of schizophrenia subjects and hippocampus of G72/G30 transgenic mice. They demonstrated that glycosphingolipids and choline plasmalogens, structural lipid pools in myelin, are significantly elevated in the frontal cortex obtained from patients with schizophrenia and the hippocampus of G72/G30 transgenic mice. These results suggest that structural lipid alterations in oligodendrocyte glycosynapses are responsible for dysconnectivity in schizophrenia (Wood et al., 2014). The same group further investigated a lipidomics analysis of choline and ethanolamine plasmalogens in the plasma and platelets of 23 patients with schizophrenia and 27 age-matched controls. Plasma levels of both choline and ethanolamine plasmalogens were decreased by 23% to 45% in patients with schizophrenia. In platelets from patients with schizophrenia, ethanolamine plasmalogens also were decreased but choline plasmalogen levels were increased. Levels of docosahexaenoic acid (DHA) were decreased by approximately 30% in plasma and in platelets. These results suggest that alterations in lipid transport or lipid remodeling/metabolism of plasmalogens are present in schizophrenia and that changes in the steady-state levels of these complex lipid pools may be involved in altered neuronal function (Wood et al., 2015).

Two metaanalyses further support the importance of lipidomics in schizophrenia. Hoen et al. (2013) performed a systematic review and metaanalysis for docosapentaenoic acid (DPA), docosahexaenoic acid (DHA), linoleic acid (LA), and arachidonic acid (AA). They identified 18 studies that compared PUFA in the erythrocyte cell membrane between patients with schizophrenia and controls. A total of 642 patients (169 were antipsychotic-naïve) and 574 controls participated in these studies. Suggestive evidence was shown for the levels of DPA (C22:5n3) and DHA (C22:6n3), which were decreased in patients currently being treated with antipsychotic medication and in antipsychotic-naïve patients. Furthermore, the findings suggest that the levels of LA (C18:2n6) are decreased in the medicated subgroup, but not in the antipsychotic-naive group. Finally, they found decreased levels of AA (C20:4n6) most convincingly in antipsychotic-naïve patients (Hoen et al., 2013; Van der Kemp et al., 2012).

Lipids detected in liquid chromatography are common in liver-produced very-low-density lipoproteins (VLDLs) and are linked to insulin resistance (Kotronen et al., 2009). Consistent with these findings, schizophrenia patients are more likely insulin-resistant and exhibit increased fasting serum insulin amounts (Orešič et al., 2011). Therefore, schizophrenia is expected to present insulin resistance, improved hepatic VLDL biosynthesis (Kotronen and Yki-Jarvinen, 2008), and augmented serum levels of specific triglycerides. These results are strengthened by data reporting anomalous insulin release and response (Fernandez-Egea et al., 2009; Guest et al., 2010, 2011), irregular glucose tolerance, and risk of diabetes (Kirkpatrick et al., 2012, 2013) in first-episode, drug-naïve patients with schizophrenia. A metaanalysis including 25 studies with 145,718 individuals with schizophrenia and 4,343,407 controls showed a 9.5% prevalence of type 2 diabetes in people with schizophrenia (95% CI, 7.0–12.8). The pooled relative risk was 1.82 (95% CI, 1.56–2.13). People with schizophrenia have at least double the risk of developing type 2 diabetes according to recognized type 2 diabetes criteria. Proactive lifestyle and screening programs should be given clinical priority (Stubbs et al., 2015). Finally, genes associated with the risk of both schizophrenia and diabetes have been discovered (Hansen et al., 2011; Liu et al., 2013).

ENERGY METABOLISM IN SCHIZOPHRENIA

The link between dysregulated pathways associated with energy metabolism and schizophrenia has been demonstrated based on indications from proteins, transcripts, and metabolites (Khaitovich et al., 2008). Increased amounts of pyruvate, a significant intermediate of glucose metabolism, suggest increased energy demands in schizophrenia (Buchsbaum et al., 2007). In schizophrenia, the brain energy supply is limited as a result of mitochondrial dysfunction and, therefore, of inhibited glucose metabolism. For this reason, the brain is supposed to modify, in some measure, its energy demand from glucose toward ketone bodies as an unconventional energy source. Hence, fatty acids in the liver are mobilized and catabolized to release the required ketone bodies.

A metabolomic study by Xuan and colleagues (2011) examined global changes in metabolic signatures of serum samples from untreated Chinese Han patients with schizophrenia before and after 8-week risperidone monotherapy. The aim was the identification of possible biomarkers linked to schizophrenia and risperidone therapy (Xuan et al., 2011). Twenty-two metabolic markers contributing to the differentiation between schizophrenic patients and healthy controls were detected. Citrate, palmitic acid, myo-inositol, and allantoin showed the highest discriminatory power. Twenty markers accounting for the discrimination between pretreatment and post-treatment patients were found, with myo-inositol, uric acid, and tryptophan exhibiting the highest discriminatory power. Overall, the detected markers indicated disruption at the level of energy metabolism, antioxidant system, neurotransmitter metabolism, fatty acid biosynthesis, and phospholipid metabolism in schizophrenic patients, which could be, in some measure, regularized after risperidone treatment (Xuan et al., 2011).

Serum glutamate levels were higher in all individuals with psychoses compared to healthy controls in a study by Orešič et al. (2011), thus leading to speculation that glutamate-related metabolic aberrations are part of a common pathway across psychoses (Cherlyn et al., 2010). Serum proline upregulation was considered typical of schizophrenia; in this connection, polymorphisms in the Proline Dehydrogenase (Oxidase) 1 (PRODH) gene encoding a mitochondrial proline dehydrogenase involved in proline catabolism correlate with schizophrenia risk (Liu et al., 2002; Kempf et al., 2008). PRODH gene functional variants responsible for the decrease of proline oxidase activity and hyperprolinemia are related to amplified risk of schizophrenia and alterations in fronto-striatal structure/function (Kempf et al., 2008; Bender et al., 2005).

In a study by He and colleagues (2012), 103 metabolites (acylcarnitines, amino acids, glycerophospholipids, sphingolipids, and hexose) were tested in plasma samples from 265 schizophrenic patients and 216 healthy controls (He et al., 2012). Compared with healthy controls, levels of five metabolites were found to be significantly altered in schizophrenic patients (p ranged from 2.9×10^{-8} to 2.5×10^{-4}) and in neuroleptics-free probands, respectively. These metabolites include four amino acids (arginine, glutamine, histidine, and ornithine) and one lipid (PC ae C38:6) and are suggested to be candidate biomarkers for schizophrenia. To explore the genetic susceptibility on the associated metabolic pathways, the authors constructed a molecular network connecting these five aberrant metabolites with 13 schizophrenia risk genes. The results implicated aberrations in biosynthetic pathways linked to glutamine and arginine metabolism and associated signaling pathways as genetic risk factors, which may contribute to patho-mechanisms and memory deficits associated with schizophrenia (He et al., 2012).

Interestingly, a metabolomics study on MK-801-treated mice—a validated model for single symptoms of psychosis (Rujescu et al., 2006)—showed further evidence for the involvement of glutamine. Clear distinctions between the group treated with MK-801 (phencyclidine PCP) and the control group in both the cortex and hippocampus were found. The change of a series of metabolites accounted for the separation, such as glutamate, glutamine, citrate, and succinate. Most of these metabolites were in a pathway characterized by downregulated glutamate synthesis and disturbed Krebs cycle. Analysis further confirmed the involvement of energy metabolism abnormality induced by MK-801 treatment (Sun et al., 2013); the results of the study by Wesseling et al. (2013) demonstrated support for this. They reported findings from the first comprehensive label-free liquid chromatography – mass spectrometry proteomic-based and proton nuclear magnetic resonance–based metabonomic profiling of the rat frontal cortex after chronic phencyclidine (PCP) intervention, which induces schizophrenia-like symptoms. The findings were compared with results from proteomic profiling of postmortem prefrontal cortex from schizophrenia patients and with relevant findings in the literature. Metabonomic profiling revealed changes in the levels of glutamate, glutamine, glycine, pyruvate, and the Ca_{2+} regulator taurine. Effects on similar pathways were also identified in the prefrontal cortex tissue from human schizophrenia subjects (Wesseling et al., 2013).

Yang et al. (2013) reported another metabolic profiling study involving 112 schizophrenic patients and 110 healthy subjects. A panel of serum markers consisting of glycerate, eicosenoic acid, β-hydroxybutyrate, pyruvate, and cystine was identified as an effective diagnostic tool that achieved an area under the receiver-operating characteristic curve (AUC) of 0.945 in the training samples (62 patients and 62 controls) and 0.895 in the test samples (50 patients and 48 controls). Furthermore, a composite panel by the addition of urine β-hydroxybutyrate to the serum panel achieved more satisfactory accuracy, reaching an AUC of 1 in both the training set and the test set. Multiple fatty acids and ketone bodies were found to be significantly increased in the serum and urine of patients, suggesting an upregulated fatty acid catabolism, presumably resulting from insufficiency of the glucose supply in the brain of schizophrenia patients (Yang et al., 2013). The antioxidant glutathione (GSH) is critical for cellular detoxification of reactive oxygen species in brain cells (Dringen and Hirrlinger, 2003). Cystine is the preferred form of cysteine for the synthesis of GSH in cells participating in the immune activities. The decrease in serum levels of cystine might suggest a disrupted GSH system in the brain under elevated oxidative stress in neurological disorders (Yang et al., 2013).

OUTLOOK

The screening of small molecules in biological fluids as well as in cells, tissues, and organs is considered an important and rapidly growing element of the "new biology" (Quinones and Kaddurah-Daouk, 2009; Kaddurah-Daouk, 2006; Kaddurah-Daouk et al., 2008). The development of cutting-edge technologies, all capable of precisely measuring hundreds of thousands of small molecules in biological samples, is expected to significantly improve our understanding of disease pathophysiology and to make possible the discovery of biomarkers for various disorders. Studies of neuropsychiatric disorders, such as schizophrenia, have used CSF, plasma/serum, urine, erythrocytes, or postmortem brain tissues to define panels of metabolic markers discriminating patients from healthy controls. Although postmortem brain and CSF samples are certainly preferred, more accessible tissues such as plasma and serum are typically utilized in clinical practice.

Nevertheless, some possible drawbacks should be considered. First, age, sex, nutritional status, and time of sampling might affect the metabolite composition of the selected fluid. Fluctuations in these physiologically normal metabolic patterns could theoretically mask molecular alterations due to the pathology. In addition, potential markers could be diluted in the moderately large volumes of body fluids. In summary, aspects such as nutritional/physiological status, sample nature, cohort size and heterogeneity, and analytical sensitivity should be prudently evaluated before starting a metabolomics study intended to investigate clinically relevant biomarkers (Nordström and Lewensohn, 2010).

The application of metabolomics-based systems will permit simultaneous measurement of several metabolites in crucial interacting pathways in schizophrenia. As a result, the novel emerging biomarkers might provide relevant clinical data. The understanding of metabolic disruptions linking schizophrenia, biochemical pathways, and treatment effect–response should provide new insights into schizophrenia pathophysiology and novel strategies for therapeutic monitoring and outcome. The major goal should be to combine metabolomics, genomics, epigenetics, proteomics, and imaging, as well as clinical, epidemiological, and environmental factors of the largest samples within systems medicine. The hope is that such approaches will generate new and urgently needed prediction algorithms. These algorithms could make it possible to detect the patient's personal risk for a disease or side effect prior to treatment initiation.

REFERENCES

Ascher-Svanum, H., Nyhuis, A.W., Faries, D.E., Kinon, B.J., Baker, R.W., Shekhar, A., 2008. Clinical, functional, and economic ramifications of early nonresponse to antipsychotics in the naturalistic treatment of schizophrenia. Schizophr. Bull. 34 (6), 1163–1171.

REFERENCES

Bender, H.U., Almashanu, S., Steel, G., Hu, C.A., Lin, W.W., Willis, A., et al., 2005. Functional consequences of PRODH missense mutations. Am. J. Hum. Genet. 6, 409–420.

Berger, G.E., Wood, S.J., Pantelis, C., Velakoulis, D., Wellard, R.M., McGorry, P.D., 2002. Implications of lipid biology for the pathogenesis of schizophrenia. Aust. N. Z. J. Psychiatry 36, 355–366.

Buchsbaum, M.S., Buchsbaum, B.R., Hazlett, E.A., Haznedar, M.M., Newmark, R., Tang, C.Y., et al., 2007. Relative glucose metabolic rate higher in white matter in patients with schizophrenia. Am. J. Psychiatry 164, 1072–1081.

Cherlyn, S.Y., Woon, P.S., Liu, J.J., Ong, W.Y., Tsai, G.C., Sim, K., 2010. Genetic association studies of glutamate, GABA and related genes in schizophrenia and bipolar disorder: a decade of advance. Neurosci. Biobehav. Rev. 34, 958–977.

Dharuri, H., Demirkan, A., van Klinken, J.B., Mook-Kanamori, D.O., van Duijn, C.M., 't Hoen, P.A., et al., 2014. Genetics of the human metabolome, what is next? Biochim. Biophys. Acta 1842 (10), 1923–1931.

Dringen, R., Hirrlinger, J., 2003. Glutathione pathways in the brain. Biol. Chem. 384 (4), 505–516.

Fernandez-Egea, E., Bernardo, M., Donner, T., Conget, I., Parellada, E., Justicia, A., et al., 2009. Metabolic profile of antipsychotic-naive individuals with non-affective psychosis. Br. J. Psychiatry 194, 434–438.

Flanagan, R.J., 2008. Side effects of clozapine and some other psychoactive drugs. Curr. Drug Saf. 3 (2), 115–122.

Fonteh, A.N., Harrington, R.J., Huhmer, A.F., Biringer, R.G., Riggins, J.N., Harrington, M.G., 2006. Identification of disease markers in human cerebrospinal fluid using lipidomic and proteomic methods. Dis. Markers 22 (1–2), 39–64. Review.

Giegling, I., Porcelli, S., Balzarro, B., Andrisano, C., Schäfer, M., Möller, H.J., et al., 2012. Antipsychotic response in the first week predicts later efficacy. Neuropsychobiology 66 (2), 100–105.

Giegling, I., Balzarro, B., Porcelli, S., Schäfer, M., Hartmann, A.M., Friedl, M., et al., 2013. Influence of ANKK1 and DRD2 polymorphisms in response to haloperidol. Eur. Arch. Psychiatry Clin. Neurosci. 263 (1), 65–74.

Guest, P.C., Wang, L., Harris, L.W., Burling, K., Levin, Y., Ernst, A., et al., 2010. Increased levels of circulating insulin-related peptides in first-onset, antipsychotic naïve schizophrenia patients. Mol. Psychiatry 15, 118–119.

Guest, P.C., Schwarz, E., Krishnamurthy, D., Harris, L.W., Leweke, F.M., Rothermundt, M., et al., 2011. Altered levels of circulating insulin and other neuroendocrine hormones associated with the onset of schizophrenia. Psychoneuroendocrinology 36, 1092–1096.

Han, X., Gross, R.W., 1994. Electrospray ionization mass spectroscopic analysis of human erythrocyte plasma membrane phospholipids. Proc. Natl. Acad. Sci. U.S.A. 91, 10635–10639.

Hansen, T., Ingason, A., Djurovic, S., Melle, I., Fenger, M., Gustafsson, O., et al., 2011. At-risk variant in TCF7L2 for type II diabetes increases risk of schizophrenia. Biol. Psychiatry 70, 59–63.

He, Y., Yu, Z., Giegling, I., Xie, L., Hartmann, A.M., Prehn, C., et al., 2012. Schizophrenia shows a unique metabolomics signature in plasma. Transl. Psychiatry 2, e149.

Hoen, W.P., Lijmer, J.G., Duran, M., Wanders, R.J., van Beveren, N.J., de Haan, L., 2013. Red blood cell polyunsaturated fatty acids measured in red blood cells and schizophrenia: a meta-analysis. Psychiatry Res. (207), 1–12.

Horrobin, D.F., 1998. The membrane phospholipid hypothesis as a biochemical basis for the neurodevelopmental concept of schizophrenia. Schizophr. Res. 30, 193–208.

Horrobin, D.F., Huang, Y.S., 1983. Schizophrenia: the role of abnormal essential fatty acid and prostaglandin metabolism. Med. Hypotheses 10, 329–336.

Kaddurah-Daouk, R., 2006. Metabolic profiling of patients with schizophrenia. PLoS Med. 3, e363.

Kaddurah-Daouk, R., Krishnan, K.R., 2009. Metabolomics: a global biochemical approach to the study of central nervous system diseases. Neuropsychopharmacology 34, 173–186.

Kaddurah-Daouk, R., McEvoy, J., Baillie, R.A., Lee, D., Yao, J.K., Doraiswamy, P.M., et al., 2007. Metabolomic mapping of atypical antipsychotic effects in schizophrenia. Mol. Psychiatry 12, 934–945.

Kaddurah-Daouk, R., Kristal, B.S., Weinshilboum, R.M., 2008. Metabolomics: a global biochemical approach to drug response and disease. Annu. Rev. Pharmacol. Toxicol. 48, 653–683.

Kaddurah-Daouk, R., McEvoy, J., Baillie, R., Zhu, H., Yao, J.K., Nimgaonkar, V.L., et al., 2012. Impaired plasmalogens in patients with schizophrenia. Psychiatry Res. 198 (3), 347–352.

Kempf, L., Nicodemus, K.K., Kolachana, B., Vakkalanka, R., Verchinski, B.A., Egan, M.F., et al., 2008. Functional polymorphisms in PRODH are associated with risk and protection for schizophrenia and fronto-striatal structure and function. PLoS Genet. 4, e1000252.

Keshavan, M.S., Mallinger, A.G., Pettegrew, J.W., Dippold, C., 1993. Erythrocyte membrane phospholipids in psychotic patients. Psychiatry Res. 49, 89–95.

Keshavan, M.S., Nasrallah, H.A., Tandon, R., 2011. Schizophrenia, "Just the Facts" 6. Moving ahead with the schizophrenia concept: from the elephant to the mouse. Schizophr. Res. 127 (1–3), 3–13.

Khaitovich, P., Lockstone, H.E., Wayland, M.T., Tsang, T.M., Jayatilaka, S.D., Guo, A.J., et al., 2008. Metabolic changes in schizophrenia and human brain evolution. Genome. Biol. 9, R124.

Kim, H.Y., Wang, T.C., Ma, Y.C., 1994. Liquid chromatography/mass spectrometry of phospholipids using electrospray ionization. Anal. Chem. 66, 3977–3982.

Kirkpatrick, B., 2013. Understanding the physiology of schizophrenia. J. Clin. Psychiatry 74 (3), e05.

Kirkpatrick, B., Miller, B.J., Garcia-Rizo, C., Fernandez-Egea, E., Bernardo, M., 2012. Is abnormal glucose tolerance in antipsychotic-naive patients with nonaffective psychosis confounded by poor health habits? Schizophr. Bull. 38, 280–284.

Kotronen, A., Yki-Jarvinen, H., 2008. Fatty liver: a novel component of the metabolic syndrome. Arterioscler. Thromb. Vasc. Biol. 28, 27–38.

Kotronen, A., Velagapudi, V.R., Yetukuri, L., Westerbacka, J., Bergholm, R., Ekroos, K., et al., 2009. Saturated fatty acids ontaining triacylglycerols are better markers of insulin resistance than total serum triacylglycerol concentrations. Diabetologia 52, 684–690.

Kristal, B.S., Kaddurah-Daouk, R., Beal, M.F., Matson, W.R., 2007. Metabolomics: concept and potential neuroscience application. In: Lajtha, A., Gibson, G., Dienel, G. (Eds.), Handbook of Neurochemistry and Molecular Neurobiology. Brain Energetics. Integration of Molecular and Cellular Processes, third ed. Springer-Verlag New York Inc., New York, NY, pp. 889–912.

Lieberman, J.A., Koreen, A.R., 1993. Neurochemistry and neuroendocrinology of schizophrenia, a selective review. Schizophr. Bull. 19, 371–429.

Liu, H., Heath, S.C., Sobin, C., Roos, J.L., Galke, B.L., Blundell, M.L., et al., 2002. Genetic variation at the 22q11 PRODH2/DGCR6 locus presents an unusual pattern and increases susceptibility to schizophrenia. Proc. Natl. Acad. Sci. U.S.A. 99, 3717–3722.

Liu, Y., Li, Z., Zhang, M., Deng, Y., Yi, Z., Shi, T., 2013. Exploring the pathogenetic association between schizophrenia and type 2 diabetes mellitus diseases based on pathway analysis. BMC Med. Genomics 6 (Suppl. 1), S17.

Mahadik, S.P., Evans, D.R., 2003. Is schizophrenia a metabolic brain disorder? Membrane phospholipid dysregulation and its therapeutic implications. Psychiatr. Clin. North Am. 26, 85–102.

Mahadik, S.P., Yao, J.K., 2006. Phospholipids in schizophrenia. In: Lieberman, J.A., Stroup, T.S., Perkins, D.O. (Eds.), Textbook of Schizophrenia. American Psychiatric Publishing Inc, Washington DC, pp. 117–135.

McEvoy, J., Baillie, R.A., Zhu, H., Buckley, P., Keshavan, M.S., Nasrallah, H.A., et al., 2013. Lipidomics reveals early metabolic changes in subjects with schizophrenia: effects of atypical antipsychotics. PLoS One 8, e68717.

Murray, C.J., Vos, T., Lozano, R., Naghavi, M., Flaxman, A.D., Michaud, C., et al., 2012. Disability-adjusted life years (DALYs) for 291 diseases and injuries in 21 regions, 1990–2010: a systematic analysis for the Global Burden of Disease study 2010. Lancet 380 (9859), 2197–2223.

Nasrallah, H., Tandon, R., Keshavan, M., 2011. Beyond the facts in schizophrenia: closing the gaps in diagnosis, pathophysiology, and treatment. Epidemiol. Psychiatr. Sci. 20 (4), 317–327. Review.

Nordström, A., Lewensohn, R., 2010. Metabolomics: moving to the clinic. J. Neuroimmune. Pharmacol. 5, 4–17.

Orešič, M., Tang, J., Seppänen-Laakso, T., Mattila, I., Saarni, S.E., Saarni, S.I., et al., 2011. Metabolome in schizophrenia and other psychotic disorders: a general population-based study. Genome Med. 3, 19.

Orešič, M., Seppänen-Laakso, T., Sun, D., Tang, J., Therman, S., Viehman, R., et al., 2012. Phospholipids and insulin resistance in psychosis: a lipidomics study of twin pairs discordant for schizophrenia. Genome Med. 4, 1.

Patti, G.J., Yanes, O., Siuzdak, G., 2012. Innovation: metabolomics: the apogee of the omics trilogy. Nat. Rev. Mol. Cell. Biol. 13, 263–269.

Peet, M., Laugharne, J., Rangarajan, N., Horrobin, D., Reynolds, G., 1995. Depleted red cell membrane essential fatty acids in drug-treated schizophrenic patients. J. Psychiatr. Res. 29, 227–232.

Potwarka, J.J., Drost, D.J., Williamson, P.C., Carr, T., Canaran, G., Rylett, W.J., et al., 1999. A 1H-decoupled 31P chemical shift imaging study of medicated schizophrenic patients and healthy controls. Biol. Psychiatry 45, 687–693.

Quinones, M.P., Kaddurah-Daouk, R., 2009. Metabolomics tools for identifying biomarkers for neuropsychiatric diseases. Neurobiol. Dis. 35, 165–176.

Rujescu, D., Bender, A., Keck, M., Hartmann, A.M., Ohl, F., Raeder, H., et al., 2006. A pharmacological model for psychosis based on N-methyl-D-aspartate receptor hypofunction: molecular, cellular, functional and behavioral abnormalities. Biol. Psychiatry 59 (8), 721–729.

Stubbs, B., Vancampfort, D., De Hert, M., Mitchell, A.J., 2015. The prevalence and predictors of type two diabetes mellitus in people with schizophrenia: a systematic review and comparative meta-analysis. Acta Psychiatr. Scand. 132 (2), 144–157.

Sun, L., Li, J., Zhou, K., Zhang, M., Yang, J., Li, Y., et al., 2013. Metabolomic analysis reveals metabolic disturbance in the cortex and hippocampus of subchronic MK-801 treated rats. PLoS One 8 (4), e60598.

van der Kemp, W.J., Klomp, D.W., Kahn, R.S., Luijten, P.R., Hulshoff Pol, H.E., 2012. A meta-analysis of the polyunsaturated fatty acid composition of erythrocyte membranes in schizophrenia. Schizophr. Res. 141, 153–161.

Vos, T., Flaxman, A.D., Naghavi, M., Lozano, R., Michaud, C., Ezzati, M., et al., 2012. Years lived with disability (YLDs) for 1160 sequelae of 289 diseases and injuries 1990–2010: a systematic analysis for the Global Burden of Disease study 2010. Lancet 380 (9859), 2163–2196.

Watkins, S.M., Reifsnyder, P.R., Pan, J.H., German, J.B., Leiter, E.H., 2002. Lipid metabolome-wide effects of the PPARgamma agonist rosiglitazone. J. Lipid Res. 43, 1809–1817.

Wenk, M.R., 2010. Lipidomics: new tools and applications. Cell 143, 888–895.

Wesseling, H., Chan, M.K., Tsang, T.M., Ernst, A., Peters, F., Guest, P.C., et al., 2013. A combined metabonomic and proteomic approach identifies frontal cortex changes in a chronic phencyclidine rat model in relation to human schizophrenia brain pathology. Neuropsychopharmacology 38 (12), 2532–2544.

Wood, P.L., Filiou, M.D., Otte, D.M., Zimmer, A., Turck, C.W., 2014. Lipidomics reveals dysfunctional glycosynapses in schizophrenia and the G72/G30 transgenic mouse. Schizophr. Res. 159 (2–3), 365–369.

Wood, P.L., Unfried, G., Whitehead, W., Phillipps, A., Wood, J.A., 2015. Dysfunctional plasmalogen dynamics in the plasma and platelets of patients with schizophrenia. Schizophr. Res. 161 (2–3), 506–510.

Xuan, J., Pan, G., Qiu, Y., Yang, L., Su, M., Liu, Y., et al., 2011. Metabolomic profiling to identify potential serum biomarkers for schizophrenia and risperidone action. J. Proteome. Res. 10, 5433–5443.

Yang, J., Chen, T., Sun, L., Zhao, Z., Qi, X., Zhou, K., et al., 2013. Potential metabolite markers of schizophrenia. Mol. Psychiatry 18, 67–78.

Yao, J.K., Leonard, S., Reddy, R.D., 2000. Membrane phospholipid abnormalities in postmortem brains from schizophrenic patients. Schizophr. Res. 42, 7–17.

CHAPTER 11

THE ROLE OF INFLAMMATION AND THE IMMUNE SYSTEM IN SCHIZOPHRENIA

N. Müller[1], E. Weidinger[1], B. Leitner[1] and M.J. Schwarz[2]

[1]*Department of Psychiatry and Psychotherapy, Ludwig Maximilian University, Munich, Germany*
[2]*Institute of Laboratory Medicine, Medical Center of Ludwig Maximilian University (LMU), Munich, Germany*

CHAPTER OUTLINE

Introduction	179
Inflammatory Mechanisms in the CNS	180
Microglia as an Important Cellular Basis of Inflammation in the CNS	180
Kindling and Sensitization of the Immune Response: The Basis for the Stress-Induced Inflammatory Response in Psychiatric Disorders	181
The Vulnerability-Stress-Inflammation Model of Schizophrenia	181
The Immune Dysbalance in Schizophrenia Is Associated With Chronic Inflammation	182
The Impact of Inflammation on Neurotransmitters in Schizophrenia	182
The Possible Role of Infection in Schizophrenia	183
CNS Volume Loss in Imaging Studies: A Consequence of an Inflammatory Process?	184
Cyclooxygenase-2 Inhibition as an Anti-Inflammatory Therapeutic Approach in Schizophrenia	184
Further Immune-Related Substances in the Therapy of Schizophrenia	185
Methodological Aspects of the Response to Immune-Based Therapy in Schizophrenia	185
Conclusions	186
Acknowledgment	188
Statement of Conflict of Interest	188
References	188

INTRODUCTION

We humans are constantly being assaulted by infectious agents, noxious chemicals, and physical traumata. Fortunately, we have evolved a complex process, the inflammatory response, to help fight and clear infection, remove damaging chemicals, and repair damaged tissue (O'Neill, 2008). The harmful effects of inflammation can be observed in many infectious and autoimmune diseases. The interactions between environmental factors and genetically encoded components of the inflammatory response determine whether the outcome will be health or disease.

As in other sites of the body, inflammation in the central nervous system (CNS) has a dual role (ie, it may be neuroprotective or neurotoxic) (Hohlfeld et al., 2007). Although acute inflammation in the CNS (eg, acute encephalitis) leads to life-threatening states within hours or days, chronic inflammation might be associated with impairment over months, years, or a lifetime.

As an example, multiple sclerosis (MS) is an inflammatory disease of the CNS that shows a relapsing–remitting course and, in a certain percentage of patients, also a chronic, progressive course. Parallels between MS and schizophrenia, which also often shows a chronic course, have repeatedly been highlighted as arguments for similar pathogenetic mechanisms in these disorders (Hanson and Gottesman, 2005).

The concept of smoldering inflammation implies that CNS inflammation drives the disease process in both acute and chronic stages (Kutzelnigg et al., 2005). During acute inflammation, the peripheral immune system interacts closely with the CNS, which is invaded by macrophages and B and T cells; in chronic processes, the immune response in the CNS is thought to be increasingly secluded from the peripheral immune system (compartmentalization of the inflammatory process) (Meinl et al., 2008; Kerschensteiner et al., 2009). Chronic MS, for example, is primarily characterized by disseminated activation of microglial cells.

There are numerous descriptions of an association between infection and chronic inflammation of the CNS and schizophrenia (Anderson et al., 2013). For example, symptoms of schizophrenia have been described in the encephalitic form of MS (Felgenhauer, 1990), in viral CNS infection with herpes simplex virus type 1 (HSV-1) (Chiveri et al., 2003), HSV-2 (Oommen et al., 1982), and measles (Hiroshi et al., 2003), and also in autoimmune processes such as poststreptococcal disorders (Mercadante et al., 2000; Teixeira et al., 2007; Kerbeshian et al., 2007; Bechter et al., 2007), lupus erythematodes, and scleroderma (Müller et al., 1992; Müller et al., 1993; van Dam, 1991; Nikolich-Zugich, 2008).

INFLAMMATORY MECHANISMS IN THE CNS

Inflammation in the CNS is mediated by pro-inflammatory cytokines, microglial cells (resident macrophages in the brain), astrocytes, and invading immune cells such as monocytes, macrophages, and T or B lymphocytes. Although a well-regulated inflammatory process is essential for tissue homeostasis and proper function, an excessive inflammatory response can be the source of additional injury to host cells. Uncontrolled inflammation may be either the result of infectious agents (eg, bacteria, viruses) or a reaction to neuronal lesions from traumata, a genetic defect or environmental toxins.

MICROGLIA AS AN IMPORTANT CELLULAR BASIS OF INFLAMMATION IN THE CNS

Microglia comprise approximately 15% of the total CNS cells and are the primary component of the intrinsic immune system in the CNS, where they provide the first line of defense after injury or disease and are the principal component of neuroinflammation.

Microglia can be activated in different ways:

1. A systemic inflammatory challenge triggers microglia activation, resulting in the release of pro-inflammatory cytokines in the CNS, which can mediate sickness behavior (Dantzer, 2001) and

other mental states. Microglia play a role in the synthesis of these central cytokines (van Dam, 1991). Cytokine release can be sustained in the absence of the triggering signal for a period of approximately 10 months, thus contributing to a chronic inflammatory state.

2. Microglia are sensitized or primed (Perry, 2007) by different stimuli, including neurodegeneration (Cunningham et al., 2005), aging (Godbout and Johnson, 2009), and stress (Sparkman and Johnson, 2008). This process of sensitization or priming results in the elicitation of an exaggerated immune response to a weak stimulus. After priming, a second stimulus (eg, low systemic inflammation or stress) may lead to microglia proliferation and increased production of pro-inflammatory cytokines by microglia (Frank et al., 2007). This exaggerated cytokine response may result in acute changes in behavior by exacerbating or re-exacerbating an inflammatory pathology in the CNS.

KINDLING AND SENSITIZATION OF THE IMMUNE RESPONSE: THE BASIS FOR THE STRESS-INDUCED INFLAMMATORY RESPONSE IN PSYCHIATRIC DISORDERS

The immune response and the release of cytokines can become more sensitized to activating stimuli by a kindling process: the initial immune response (ie, the release of cytokines and other mediators of immune activation) is initiated as a result of exposure to a certain stimulus; thereafter, either re-exposure to the same stimulus (eg, stress or infection) is associated with an increased release of cytokines or a weaker stimulus is necessary for the same activation process. This sensitization or kindling may be due to the memory function of the acquired immune system (Sparkman and Johnson, 2008; Furukawa et al., 1998). Stress-associated release of IL-6 was shown to reactivate (prenatally) conditioned processes (Zhou et al., 1993). In healthy persons, a second stimulus (eg, systemic inflammation, stress) led to immune activation associated with cellular proliferation and increased production and release of pro-inflammatory cytokines (Frank et al., 2007). This mechanism is a key mechanism for triggering immune activation and inflammation (eg, the stress-induced immune activation leading to psychopathological symptoms). A sensitization process in the immune system is in accordance with the view that after an infection during early childhood, re-infection, or another stimulation of the immune system in later stages of life might be associated with a boosted release of sensitized cytokines, resulting in neurotransmitter disturbances.

Sensitization phenomena play a role in stress-related, cytokine-induced, neurotransmitter-mediated behavioral abnormalities (ie, the cytokine response to a stimulus increases while the intensity of the stimulus decreases) (Sparkman and Johnson, 2008). In animal experiments, however, cytokines promote greater neurotransmitter responses when the animals are re-exposed to the cytokine (Anisman and Merali, 2003), for example, TNF-α (Hayley et al., 2002). In the CNS, the stress-induced activation and proliferation of microglia may mediate these cytokine effects (Nair and Bonneau, 2006).

THE VULNERABILITY-STRESS-INFLAMMATION MODEL OF SCHIZOPHRENIA

The vulnerability–stress model of psychiatric disorders, first postulated for schizophrenia more than 30 years ago (Zubin and Spring, 1977), focuses on the role of physical and mental stress in triggering a psychotic episode. In schizophrenia, an increased vulnerability of the offspring was shown, in addition to genetic vulnerability, if an inflammatory response was induced in the mother during the second trimester of pregnancy or in the offspring during later stages of CNS development.

The underlying mechanisms of the co-occurrence of stress and inflammation were studied in animal experiments, and stress was repeatedly shown to be associated with an increase in pro-inflammatory cytokines (Sparkman and Johnson, 2008). The genetic risk contribution in the context of pathogen–host defense is evident (Raison and Miller, 2013).

The specific influence on neurotransmitter systems of the inflammatory mechanisms described in schizophrenia is discussed here. Moreover, the modulation of glutamatergic neurotransmission is highlighted. Glutamate is the most abundant neurotransmitter in the CNS and is differentially involved via cytokine-directed tryptophan/kynurenine metabolism in schizophrenia, presumably but not exclusively mediated by NMDA receptors. Genetic factors of the kynurenine metabolites (and other factors) play a role (Claes et al., 2011).

THE IMMUNE DYSBALANCE IN SCHIZOPHRENIA IS ASSOCIATED WITH CHRONIC INFLAMMATION

Degradation products of inflammatory substances have been described in schizophrenic brain tissue (Körschenhausen et al., 1996) and in the CSF of approximately 50% of schizophrenia patients (Wildenauer et al., 1991).

Regarding the cytokine pattern in schizophrenia, blunted type 1 and (compensatory) increased type 2 cytokine patterns have been repeatedly observed in unmedicated schizophrenia patients (Müller and Schwarz, 2006). These findings point to an imbalance of the type 1 and type 2 immune responses in schizophrenia. Overviews on the imbalance in schizophrenia of the type 1, type 2, pro-inflammatory, and anti-inflammatory immune systems as well as innate immunity, including the monocyte/macrophage system, have recently been published and indicate that an inflammatory process plays an important role in the pathophysiology of at least a subgroup of schizophrenia patients (Müller and Schwarz, 2010; Potvin et al., 2008). Accordingly, the first pilot experiences with the type 1-stimulating substance interferon-gamma (IFN-γ) as a therapeutic approach in schizophrenia are encouraging (Grüber et al., 2014).

THE IMPACT OF INFLAMMATION ON NEUROTRANSMITTERS IN SCHIZOPHRENIA

Over the past five decades, research on the neurobiology of schizophrenia has focused overwhelmingly on disturbances of dopaminergic neurotransmission (Carlsson, 1988). There is no doubt that a dysfunction of the dopamine system is involved in the pathogenesis of schizophrenia, although the mechanism is not clear and antidopaminergic antipsychotic drugs still show unsatisfactory therapeutic effects.

IL-1ß, which can induce the conversion of rat mesencephalic progenitor cells into a dopaminergic phenotype (Ling et al., 1998; Kabiersch et al., 1998; Potter et al., 1999), and IL-6, which is highly effective in decreasing the survival of fetal brain serotonergic neurons (Jarskog et al., 1997), seem to have an important influence on the development of the neurotransmitter systems involved in schizophrenia, although the specificity of these cytokines is a matter of discussion. Maternal immune stimulation during pregnancy was shown to increase the number of mesencephalic dopaminergic neurons in the fetal brain; the increase was probably associated with a dopaminergic excess in the midbrain (Winter et al., 2009). Persistent pathogens might be key factors that drive imbalances of the immune reaction (Nikolich-Zugich, 2008). Nevertheless, many questions about how immunity and immune pathology are involved in virus infections remain unanswered (Rouse and Sehrawat, 2010).

Much evidence seems to indicate that a lack of glutamatergic neurotransmission, mediated via NMDA antagonism, is a key mechanism in the pathophysiology of schizophrenia (Müller and Schwarz, 2007). The only NMDA receptor antagonist known to occur naturally in the human CNS is kynurenic acid (Stone, 1993), one of at least three neuroactive intermediate products of the kynurenine pathway. A predominant type 2 immune response inhibits the enzyme indoleamine 2,3-dioxygenase (IDO), resulting in increased production of kynurenic acid in schizophrenia and in NMDA receptor antagonism (Müller and Schwarz, 2007; Müller et al., 2011). The recent findings of NMDA receptor antibodies in approximately 10% of acute (unmedicated) schizophrenia patients are especially interesting in this regard (Steiner et al., 2013; Vincent and Bien, 2008).

Discrepancies in the findings regarding kynurenic acid in schizophrenia, however, have to be discussed. Elevated kynurenic acid has mainly been described in the CSF (Erhardt et al., 2001; Linderholm et al., 2010), in the brains of schizophrenia patients (Schwarcz et al., 2001; Sathyasaikumar et al., 2011), and in animal models of schizophrenia (Olsson et al., 2009). However, no increased kynurenic acid levels were observed in the peripheral blood of first-episode schizophrenia patients (Condray et al., 2011) and other groups of schizophrenia patients (Myint et al., 2011). Antipsychotic medication, however, influences kynurenine metabolites and has to be regarded as an interfering variable (Condray et al., 2011; Myint et al., 2011; Ceresoli-Borroni et al., 2006).

THE POSSIBLE ROLE OF INFECTION IN SCHIZOPHRENIA

Animal models of schizophrenia show that stimulation of the maternal immune system by viral agents leads to typical symptoms in the offspring (Meyer and Feldon, 2009; Meyer et al., 2011).

Evidence for prenatal or perinatal exposure to infections as a risk factor for schizophrenia has been obtained from animal models (Buka et al., 2001; Westergaard et al., 1999), and studies involving humans have been performed regarding several viruses (Pearce, 2001; Buka et al., 2008; Brown et al., 2004a). Increased risk for schizophrenia in the offspring was also observed after respiratory infections (Brown et al., 2000; Sorensen et al., 2009) and genital and reproductive tract infections (Sorensen et al., 2009; Babulas et al., 2006). Infection of mothers with *Toxoplasma gondii* was also described to be a risk factor (Brown et al., 2005).

Infections before birth increase the risk for later schizophrenia (Boksa, 2008; Gattaz et al., 2004; Brown, 2008; Dalman et al., 2008), as do infections—particularly CNS infections—during later stages of brain development. Antibody titers against viruses have been examined in the sera of schizophrenia patients for many years (Yolken and Torrey, 1995). The results, however, have been inconsistent because interfering factors were not controlled for. Antibody levels are associated with the medication state, a finding that partly explains the previous controversial results (Leweke et al., 2004). In one of our own studies, higher titers of different pathogens were found in schizophrenia patients than in controls, a phenomenon that we called the infectious index (Krause et al., 2010).

Prenatal immune activation, infection-triggered or not, is an important risk factor for schizophrenia (Meyer et al., 2011). In humans, increased maternal levels of the pro-inflammatory cytokine interleukin-8 (IL-8) during pregnancy were shown to be associated with an increased risk for schizophrenia in the offspring, whatever the reason for the increase in IL-8 (Brown et al., 2004b). Moreover, increased maternal IL-8 levels during pregnancy were also significantly related to decreased brain volume (ie, lower volumes of the right posterior cingulum and left entorhinal cortex and higher volumes of the ventricles in the schizophrenic offspring) (Ellman et al., 2010).

However, a recent study, the first large-scale epidemiological study in psychiatry, showed that severe infections and autoimmune disorders additively increase the risk of schizophrenia and schizophrenia spectrum disorders (Benros et al., 2011). Infections of the parents, including intrauterine infections, were not confirmed as definite risk factors (Benros et al., 2011; Benros et al., 2012). Because the sensitivity was not high, despite the large scale of the study, it may have clearly identified only the "tip of the iceberg" of risk factors (Benros et al., 2012).

CNS VOLUME LOSS IN IMAGING STUDIES: A CONSEQUENCE OF AN INFLAMMATORY PROCESS?

Gross inflammatory changes have not been found in neuroimaging or neuropathological studies of schizophrenia. However, there is no doubt that decreased CNS volume can be observed as early as the first episode, and a progressive loss in CNS volume occurs during the further course of the disease (Gogtay et al., 2008; Steen et al., 2006; Job et al., 2006; Chakos et al., 2005). Moreover, a relationship was described between volume loss and an increased genetic risk for higher production of the immune marker IL-1β (Meisenzahl et al., 2001); the relationship between maternal IL-8 levels and CNS volume was mentioned previously (Ellman et al., 2010).

The ligand PK 11195 is used in positron emission tomography (PET) to estimate microglial activation (Versijpt et al., 2003). In schizophrenia, an increased expression of PK 11195 was shown to be a marker of an inflammatory process in the CNS (van Berckel et al., 2008; Doorduin et al., 2009). Moreover, positive correlations were also observed between expression of the microglial activation marker DAA1106 and both schizophrenia positive symptoms and duration of the disease (Takano et al., 2010).

CYCLOOXYGENASE-2 INHIBITION AS AN ANTI-INFLAMMATORY THERAPEUTIC APPROACH IN SCHIZOPHRENIA

Modern anti-inflammatory agents have been explored in schizophrenia. The cyclooxygenase-2 (COX-2) inhibitor celecoxib was studied in a prospective, randomized, double-blind study of acute exacerbations of schizophrenia. The patients receiving celecoxib add-on to risperidone showed a statistically significantly better outcome than the patients receiving risperidone alone; the clinical effects of COX-2 inhibition in schizophrenia were especially pronounced in cognition (Müller et al., 2005). The efficacy of therapy with a COX-2 inhibitor seems most pronounced in the first years of the schizophrenic disease process (Müller, 2010; Müller et al., 2010). A recent study also demonstrated a beneficial effect of acetylsalicylic acid in schizophrenic spectrum disorders (Laan et al., 2010). A meta-analysis of the clinical effects of nonsteroidal anti-inflammatory drugs in schizophrenia revealed significant effects on schizophrenic total, positive, and negative symptoms (Sommer et al., 2012), whereas another meta-analysis found a significant benefit only in schizophrenia patients with a short duration of disease or during the first manifestation of schizophrenia (Nitta et al., 2013).

FURTHER IMMUNE-RELATED SUBSTANCES IN THE THERAPY OF SCHIZOPHRENIA

Minocycline, an antibiotic and inhibitor of microglia activation, is an interesting substance for the treatment of schizophrenia. The improvement of cognition by minocycline has been described in animal models of schizophrenia (Mizoguchi et al., 2008) and in two double-blind, placebo-controlled, add-on therapy trials involving schizophrenia patients (Levkovitz et al., 2010; Chaudhry et al., 2012). In clinical studies, positive effects on schizophrenic negative symptoms were noted as well (Chaudhry et al., 2012). Case reports documented positive effects of minocycline on the whole symptom spectrum in schizophrenia (Ahuja and Carroll, 2007).

Acetylcysteine (ACC) and other substances, including omega-3 fatty acids, that have anti-inflammatory and other effects also provide some benefit to schizophrenia patients (overview: Sommer et al., 2014).

The first pilot experiences with cytokine IFN-γ, which stimulates the monocytic type 1 immune response, as a therapeutic approach in schizophrenia are encouraging (Grüber et al., 2014), although side effects, including unwanted immune effects, have to be carefully monitored and the results are only preliminary. However, such a hypothesis-driven therapeutic approach allows interesting perspectives for the development of therapeutic substances based on etiopathology.

METHODOLOGICAL ASPECTS OF THE RESPONSE TO IMMUNE-BASED THERAPY IN SCHIZOPHRENIA

That schizophrenia is a syndrome and that different pathological mechanisms may play a role in the disorder seem to indicate that immune pathology is restricted to a subgroup of patients. Although several biological markers, including immune markers, have been thought to reflect subgroups of schizophrenia, so far no marker is established for an immune-related schizophrenia. Accordingly, an immune-based therapeutic approach might be effective only in a subgroup of patients or, put another way, immune-based therapy can be expected to show only a small therapeutic effect in an unselected group of schizophrenia patients. Another relevant point is that all clinical studies using immune-based treatment are studies of add-ons to an established standard therapy with antipsychotics. For ethical reasons, this design has to be used until the add-on substance has a proven effect in schizophrenia. To show superiority over an effective antipsychotic in monotherapy, however, the add-on substance has to have a huge additional effect before it reaches statistical significance over placebo and the antipsychotic. Regarding high and increasing placebo response rates in schizophrenia studies (Rutherford et al., 2014), anti-inflammatory substances need high effect sizes to show statistical superiority in double-blind, randomized, placebo-controlled studies. Moreover, schizophrenia patients participating in clinical studies, especially in studies with a new, unproven therapeutic approach, often show an unfavorable and sometimes therapy-resistant course of the disorder (ie, several of the studies may include a negative selection of severely ill patients).

The aforementioned methodological aspects may explain the difficulties in showing a convincing effect of anti-inflammatory drugs in schizophrenia.

CONCLUSIONS

The possible influence on the pathogenesis of schizophrenia of an immunological process resulting in inflammation has long been neglected. Increasing evidence for a role of pro-inflammatory cytokines in schizophrenia, the strong influence of pro-inflammatory and anti-inflammatory cytokines on tryptophan/kynurenine metabolism (Fig. 11.1), the related influence of cytokines on glutamatergic neurotransmission, the results of imaging studies, genetic findings, and, last but not least, the therapeutic effect of anti-inflammatory drugs all support the view that the recent increased focus of schizophrenia research on psychoneuroimmunology and inflammation is justified. However, one has to consider that immunological research is susceptible to artifacts (interfering variables such as medication, smoking, stress, sleep, and others play an important role and cannot always be controlled). This is exemplified by stress, which, according to the vulnerability–stress model (Fig. 11.2), is not only a condition sine qua non in schizophrenia but also a confounding factor in research of the immune system and inflammatory processes. The situation is similar for neuroimaging studies: volume loss might be the result of

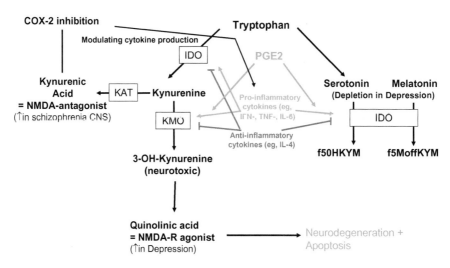

FIGURE 11.1

Vulnerability-stress-inflammation hypothesis of schizophrenia.

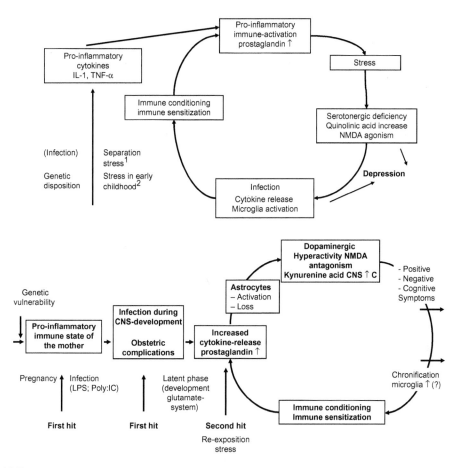

FIGURE 11.2

The Tryptophan/kynurenine metabolism: possible implications for psychiatric disorders.

different pathological processes other than inflammatory ones. Nevertheless, the results of these studies are encouraging, and further studies should focus on the relationship between inflammatory markers in the blood and CSF and volume loss in the CNS. Moreover, the influence of different disease stages in schizophrenia might also have been neglected. The syndrome of schizophrenia is thought to have different underlying pathological processes. Inflammation, however, also includes different stages and processes ranging from acute to chronic inflammation, including an autoimmune process.

These considerations show that even though much further research is necessary to clarify the role of the immune system in schizophrenia, recent findings encourage continued emphasis on this fascinating field.

ACKNOWLEDGMENT

Parts of this manuscript have been published before (Müller and Bechter, 2013; Müller, 2014). The authors thank Jacquie Klesing, Board-certified Editor in the Life Sciences (ELS), for assistance with editing the manuscript.

STATEMENT OF CONFLICT OF INTEREST

None to declare.

REFERENCES

Ahuja, N., Carroll, B.T., 2007. Possible anti-catatonic effects of minocycline in patients with schizophrenia. Prog. Neuropsychopharmacol. Biol. Psychiatry 31, 968–969.

Anderson, G., Berk, M., Dodd, S., Bechter, K., Altamura, A.C., Dell'osso, B., et al., 2013. Immuno-inflammatory, oxidative and nitrosative stress, and neuroprogressive pathways in the etiology, course and treatment of schizophrenia. Prog. Neuropsychopharmacol. Biol. Psychiatry 42, 1–4. http://dx.doi.org/10.1016/j.pnpbp.2012.10.008. Epub;%2012 October 18., 1–4.

Anisman, H., Merali, Z., 2003. Cytokines, stress and depressive illness: brain-immune interactions. Ann. Med. 35, 2–11.

Babulas, V., Factor-Litvak, P., Goetz, R., Schaefer, C.A., Brown, A.S., 2006. Prenatal exposure to maternal genital and reproductive infections and adult schizophrenia. Am. J. Psychiatry 163, 927–929.

Bechter, K., Bindl, A., Horn, M., Schreiner, V., 2007. Therapy-resistant depression with fatigue. A case of presumed streptococcal-associated autoimmune disorder. Nervenarzt 78 (3), 338. 340–338, 341.

Benros, M.E., Mortensen, P.B., Eaton, W.W., 2012. Autoimmune diseases and infections as risk factors for schizophrenia. Ann. N. Y. Acad. Sci. 1262, 56–66. http://dx.doi.org/10.1111/j.1749-6632.2012.06638.x.

Benros, M.E., Nielsen, P.R., Nordentoft, M., Eaton, W.W., Dalton, S.O., Mortensen, P.B., 2011. Autoimmune diseases and severe infections as risk factors for schizophrenia: a 30-year population-based register study. Am. J. Psychiatry 168, 1303–1310.

Boksa, P., 2008. Maternal infection during pregnancy and schizophrenia. J. Psychiatry Neurosci. 33, 183–185.

Brown, A.S., 2008. The risk for schizophrenia from childhood and adult infections. Am. J. Psychiatry 165, 7–10.

Brown, A.S., Cohen, P., Susser, E.S., Greenwald, M.A., 2000. Nonaffective psychosis after prenatal exposure to rubella. Am. J. Psychiatry 157, 438–443.

Brown, A.S., Begg, M.D., Gravenstein, S., Schaefer, C.A., Wyatt, R.J., Bresnahan, M., et al., 2004a. Serologic evidence of prenatal influenza in the etiology of schizophrenia. Arch. Gen. Psychiatry 61, 774–780.

Brown, A.S., Hooton, J., Schaefer, C.A., Zhang, H., Petkova, E., Babulas, V., et al., 2004b. Elevated maternal interleukin-8 levels and risk of schizophrenia in adult offspring. Am. J. Psychiatry 161, 889–895.

Brown, A.S., Schaefer, C.A., Quesenberry Jr., C.P., Liu, L., Babulas, V.P., Susser, E.S., 2005. Maternal exposure to toxoplasmosis and risk of schizophrenia in adult offspring. Am. J. Psychiatry 162, 767–773.

Buka, S.L., Tsuang, M.T., Torrey, E.F., Klebanoff, M.A., Bernstein, D., Yolken, R.H., 2001. Maternal infections and subsequent psychosis among offspring. Arch. Gen. Psychiatry 58, 1032–1037.

Buka, S.L., Cannon, T.D., Torrey, E.F., Yolken, R.H., 2008. Maternal exposure to herpes simplex virus and risk of psychosis among adult offspring. Biol. Psychiatry 63, 809–815.

Carlsson, A., 1988. The current status of the dopamine hypothesis of schizophrenia. Neuropsychopharmacology 1, 179–186.

Ceresoli-Borroni, G., Rassoulpour, A., Wu, H.Q., Guidetti, P., Schwarcz, R., 2006. Chronic neuroleptic treatment reduces endogenous kynurenic acid levels in rat brain. J. Neural. Transm. 113 (10), 1355–1365.

Chakos, M.H., Schobel, S.A., Gu, H., Gerig, G., Bradford, D., Charles, C., et al., 2005. Duration of illness and treatment effects on hippocampal volume in male patients with schizophrenia. Br. J. Psychiatry 186, 26–31.

Chaudhry, I.B., Hallak, J., Husain, N., Minhas, F., Stirling, J., Richardson, P., et al., 2012. Minocycline benefits negative symptoms in early schizophrenia: a randomised double-blind placebo-controlled clinical trial in patients on standard treatment. J. Psychopharmacol. 26, 1185–1193.

Chiveri, L., Sciacco, M., Prelle, A., 2003. Schizophreniform disorder with cerebrospinal fluid PCR positivity for herpes simplex virus type 1. Eur. Neurol. 50, 182–183.

Claes, S., Myint, A.M., Domschke, K., Del-Favero, J., Entrich, K., Engelborghs, S., et al., 2011. The kynurenine pathway in major depression: haplotype analysis of three related functional candidate genes. Psychiatry Res. 188, 355–360.

Condray, R., Dougherty, G.G., Keshavan, M.S., Reddy, R.D., Haas, G.L., Montrose, D.M., et al., 2011. 3-Hydroxykynurenine and clinical symptoms in first-episode neuroleptic-naive patients with schizophrenia. Int. J. Neuropsychopharmacol. 14, 756–767.

Cunningham, C., Wilcockson, D.C., Campion, S., Lunnon, K., Perry, V.H., 2005. Central and systemic endotoxin challenges exacerbate the local inflammatory response and increase neuronal death during chronic neurodegeneration. J. Neurosci. 25, 9275–9284.

Dalman, C., Allebeck, P., Gunnell, D., Harrison, G., Kristensson, K., Lewis, G., et al., 2008. Infections in the CNS during childhood and the risk of subsequent psychotic illness: a cohort study of more than one million Swedish subjects. Am. J. Psychiatry 165, 59–65.

Dantzer, R., 2001. Cytokine-induced sickness behavior: where do we stand? Brain Behav. Immun. 15, 7–24.

Doorduin, J., de Vries, E.F., Willemsen, A.T., de Groot, J.C., Dierckx, R.A., Klein, H.C., 2009. Neuroinflammation in schizophrenia-related psychosis: a PET study. J. Nucl. Med. 50, 1801–1807.

Ellman, L.M., Deicken, R.F., Vinogradov, S., Kremen, W.S., Poole, J.H., Kern, D.M., et al., 2010. Structural brain alterations in schizophrenia following fetal exposure to the inflammatory cytokine interleukin-8. Schizophr. Res. 121, 46–54.

Erhardt, S., Blennow, K., Nordin, C., Skogh, E., Lindstrom, L.H., Engberg, G., 2001. Kynurenic acid levels are elevated in the cerebrospinal fluid of patients with schizophrenia. Neurosci. Lett. 313, 96–98.

Felgenhauer, K., 1990. Psychiatric disorders in the encephalitic form of multiple sclerosis. J. Neurol. 237, 11–18.

Frank, M.G., Baratta, M.V., Sprunger, D.B., Watkins, L.R., Maier, S.F., 2007. Microglia serve as a neuroimmune substrate for stress-induced potentiation of CNS pro-inflammatory cytokine responses. Brain Behav. Immun. 21, 47–59.

Furukawa, H., del Rey, A., Monge-Arditi, G., Besedovsky, H.O., 1998. Interleukin-1, but not stress, stimulates glucocorticoid output during early postnatal life in mice. Ann. N. Y. Acad. Sci. 840, 117–122.

Gattaz, W.F., Abrahao, A.L., Foccacia, R., 2004. Childhood meningitis, brain maturation and the risk of psychosis. Eur. Arch. Psychiatry Clin. Neurosci. 254, 23–26.

Godbout, J.P., Johnson, R.W., 2009. Age and neuroinflammation: a lifetime of psychoneuroimmune consequences. Immunol. Allergy Clin. North Am. 29, 321–337.

Gogtay, N., Lu, A., Leow, A.D., Klunder, A.D., Lee, A.D., Chavez, A., et al., 2008. Three-dimensional brain growth abnormalities in childhood-onset schizophrenia visualized by using tensor-based morphometry. Proc. Natl. Acad. Sci. U.S.A. 105, 15979–15984.

Grüber, L., Bunse, T., Weidinger, E., Reichard, H., Müller, N., 2014. Adjunctive recombinant human interferon gamma-1b for treatment-resistant schizophrenia in 2 patients. J. Clin. Psychiatry 75 (11), 1266–1267.

Hanson, D.R., Gottesman, I.I., 2005. Theories of schizophrenia: a genetic-inflammatory-vascular synthesis. BMC Med. Genet. 6, 7.

Hayley, S., Wall, P., Anisman, H., 2002. Sensitization to the neuroendocrine, central monoamine and behavioural effects of murine tumor necrosis factor-alpha: peripheral and central mechanisms. Eur. J. Neurosci. 15, 1061–1076.

Hiroshi, H., Seiji, K., Toshihiro, K., Nobuo, K., 2003. [An adult case suspected of recurrent measles encephalitis with psychiatric symptoms]. Seishin Shinkeigaku Zasshi 105, 1239–1246.

Hohlfeld, R., Kerschensteiner, M., Meinl, E., 2007. Dual role of inflammation in CNS disease. Neurology 68, S58–S63.

Jarskog, L.F., Xiao, H., Wilkie, M.B., Lauder, J.M., Gilmore, J.H., 1997. Cytokine regulation of embryonic rat dopamine and serotonin neuronal survival in vitro. Int. J. Dev. Neurosci. 15, 711–716.

Job, D.E., Whalley, H.C., McIntosh, A.M., Owens, D.G., Johnstone, E.C., Lawrie, S.M., 2006. Grey matter changes can improve the prediction of schizophrenia in subjects at high risk. BMC Med. 4, 29.

Kabiersch, A., Furukawa, H., del, R.A., Besedovsky, H.O., 1998. Administration of interleukin-1 at birth affects dopaminergic neurons in adult mice. Ann. N.Y. Acad. Sci. 840, 123–127.

Kerbeshian, J., Burd, L., Tait, A., 2007. Chain reaction or time bomb: a neuropsychiatric-developmental/neurodevelopmental formulation of tourettisms, pervasive developmental disorder, and schizophreniform symptomatology associated with PANDAS. World J. Biol. Psychiatry 8, 201–207.

Kerschensteiner, M., Meinl, E., Hohlfeld, R., 2009. Neuro-immune crosstalk in CNS diseases. Neuroscience 158, 1122–1132.

Körschenhausen, D.A., Hampel, H.J., Ackenheil, M., Penning, R., Müller, N., 1996. Fibrin degradation products in post mortem brain tissue of schizophrenics: a possible marker for underlying inflammatory processes. Schizophr. Res. 19, 103–109.

Krause, D., Matz, J., Weidinger, E., Wagner, J., Wildenauer, A., Obermeier, M., et al., 2010. The association of infectious agents and schizophrenia. World J. Biol. Psychiatry 11, 739–743.

Kutzelnigg, A., Lucchinetti, C.F., Stadelmann, C., Bruck, W., Rauschka, H., Bergmann, M., et al., 2005. Cortical demyelination and diffuse white matter injury in multiple sclerosis. Brain 128, 2705–2712.

Laan, W., Grobbee, D.E., Selten, J.P., Heijnen, C.J., Kahn, R.S., Burger, H., 2010. Adjuvant aspirin therapy reduces symptoms of schizophrenia spectrum disorders: results from a randomized, double-blind, placebo-controlled trial. J. Clin. Psychiatry 71, 520–527.

Levkovitz, Y., Mendlovich, S., Riwkes, S., Braw, Y., Levkovitch-Verbin, H., Gal, G., et al., 2010. A double-blind, randomized study of minocycline for the treatment of negative and cognitive symptoms in early-phase schizophrenia. J. Clin. Psychiatry 71, 138–149.

Leweke, F.M., Gerth, C.W., Koethe, D., Klosterkotter, J., Ruslanova, I., Krivogorsky, B., et al., 2004. Antibodies to infectious agents in individuals with recent onset schizophrenia. Eur. Arch. Psychiatry Clin. Neurosci. 254, 4–8.

Linderholm, K.R., Skogh, E., Olsson, S.K., Dahl, M.L., Holtze, M., Engberg, G., et al., 2010. Increased levels of kynurenine and kynurenic acid in the CSF of patients with schizophrenia. Schizophr. Bull. 38 (3), 426–432.

Ling, Z.D., Potter, E.D., Lipton, J.W., Carvey, P.M., 1998. Differentiation of mesencephalic progenitor cells into dopaminergic neurons by cytokines. Exp. Neurol. 149, 411–423.

Meinl, E., Krumbholz, M., Derfuss, T., Junker, A., Hohlfeld, R., 2008. Compartmentalization of inflammation in the CNS: a major mechanism driving progressive multiple sclerosis. J. Neurol. Sci. 274, 42–44.

Meisenzahl, E.M., Rujescu, D., Kirner, A., Giegling, I., Kathmann, N., Leinsinger, G., et al., 2001. Association of an interleukin-1beta genetic polymorphism with altered brain structure in patients with schizophrenia. Am. J. Psychiatry 158, 1316–1319.

Mercadante, M.T., Busatto, G.F., Lombroso, P.J., Prado, L., Rosario-Campos, M.C., do Valle, R., et al., 2000. The psychiatric symptoms of rheumatic fever. Am. J. Psychiatry 157, 2036–2038.

Meyer, U., Feldon, J., 2009. Prenatal exposure to infection: a primary mechanism for abnormal dopaminergic development in schizophrenia. Psychopharmacology (Berl) 206, 587–602.

Meyer, U., Schwarz, M.J., Müller, N., 2011. Inflammatory processes in schizophrenia: a promising neuroimmunological target for the treatment of negative/cognitive symptoms and beyond. Pharmacol. Ther. 132, 96–110.

Mizoguchi, H., Takuma, K., Fukakusa, A., Ito, Y., Nakatani, A., Ibi, D., et al., 2008. Improvement by minocycline of methamphetamine-induced impairment of recognition memory in mice. Psychopharmacology (Berl) 196, 233–241.

Müller, N., 2010. COX-2 inhibitors as antidepressants and antipsychotics: clinical evidence. Curr. Opin. Investig. Drugs 11, 31–42.

Müller, N., 2014. Immunology of schizophrenia. Neuroimmunomodulation 21, 109–116.

Müller, N., Bechter, K., 2013. The mild encephalitis concept for psychiatric disorders revisited in the light of current psychoneuroimmunological findings. Neurol. Psychiatry Brain Res. 19, 87–101.

Müller, N., Gizycki-Nienhaus, B., Günther, W., Meurer, M., 1992. Depression as a cerebral manifestation of scleroderma: immunological findings in serum and cerebrospinal fluid. Biol. Psychiatry 31, 1151–1156.

Müller, N., Gizycki-Nienhaus, B., Botschev, C., Meurer, M., 1993. Cerebral involvement of scleroderma presenting as schizophrenia-like psychosis. Schizophr. Res. 10, 179–181.

Müller, N., Riedel, M., Schwarz, M.J., Engel, R.R., 2005. Clinical effects of COX-2 inhibitors on cognition in schizophrenia. Eur. Arch. Psychiatry Clin. Neurosci. 255, 149–151.

Müller, N., Krause, D., Dehning, S., Musil, R., Schennach-Wolff, R., Obermeier, M., et al., 2010. Celecoxib treatment in an early stage of schizophrenia: results of a randomized, double-blind, placebo-controlled trial of celecoxib augmentation of amisulpride treatment. Schizophr. Res. 121, 119–124.

Müller, N., Myint, A.M., Schwarz, M.J., 2011. Kynurenine pathway in schizophrenia: pathophysiological and therapeutic aspects. Curr. Pharm. Des. 17, 130–136.

Müller, N., Schwarz, M.J., 2006. Neuroimmune-endocrine crosstalk in schizophrenia and mood disorders. Expert Rev. Neurother. 6, 1017–1038.

Müller, N., Schwarz, M.J., 2007. The immunological basis of glutamatergic disturbance in schizophrenia: towards an integrated view. J. Neurotransmission, 269–280.

Müller, N., Schwarz, M.J., 2010. Immune system and schizophrenia. Curr. Immunol. Rev. 6, 213–220.

Myint, A.M., Schwarz, M.J., Verkerk, R., Mueller, H.H., Zach, J., Scharpe, S., et al., 2011. Reversal of imbalance between kynurenic acid and 3-hydroxykynurenine by antipsychotics in medication-naive and medication-free schizophrenic patients. Brain Behav. Immun. 25, 1576–1581.

Nair, A., Bonneau, R.H., 2006. Stress-induced elevation of glucocorticoids increases microglia proliferation through NMDA receptor activation. J. Neuroimmunol. 171, 72–85.

Nikolich-Zugich, J., 2008. Ageing and life-long maintenance of T-cell subsets in the face of latent persistent infections. Nat. Rev. Immunol. 8, 512–522.

Nitta, M., Kishimoto, T., Müller, N., Weiser, M., Davidson, M., Kane, J.M., et al., 2013. Adjunctive use of nonsteroidal anti-inflammatory drugs for schizophrenia: a meta-analytic investigation of randomized controlled trials. Schizophr Bull. 39 (6), 1230–1241.

Olsson, S.K., Andersson, A.S., Linderholm, K.R., Holtze, M., Nilsson-Todd, L.K., Schwieler, L., et al., 2009. Elevated levels of kynurenic acid change the dopaminergic response to amphetamine: implications for schizophrenia. Int. J. Neuropsychopharmacol. 12, 501–512.

O'Neill, L.A., 2008. How frustration leads to inflammation. Science 320, 619–620.

Oommen, K.J., Johnson, P.C., Ray, C.G., 1982. Herpes simplex type 2 virus encephalitis presenting as psychosis. Am. J. Med. 73, 445–448.

Pearce, B.D., 2001. Schizophrenia and viral infection during neurodevelopment: a focus on mechanisms. Mol. Psychiatry 6, 634–646.

Perry, V.H., 2007. Stress primes microglia to the presence of systemic inflammation: implications for environmental influences on the brain. Brain Behav. Immun. 21, 45–46.

Potter, E.D., Ling, Z.D., Carvey, P.M., 1999. Cytokine-induced conversion of mesencephalic-derived progenitor cells into dopamine neurons. Cell Tissue Res. 296, 235–246.

Potvin, S., Stip, E., Sepehry, A.A., Gendron, A., Bah, R., Kouassi, E., 2008. Inflammatory cytokine alterations in schizophrenia: a systematic quantitative review. Biol. Psychiatry 63, 801–808.

Raison, C.L., Miller, A.H., 2013. The evolutionary significance of depression in Pathogen Host Defense (PATHOS-D). Mol. Psychiatry 18, 15–37.

Rouse, B.T., Sehrawat, S., 2010. Immunity and immunopathology to viruses: what decides the outcome? Nat. Rev. Immunol. 10, 514–526.

Rutherford, B.R., Pott, E., Tandler, J.M., Wall, M.M., Roose, S.P., Lieberman, J.A., 2014. Placebo response in antipsychotic clinical trials: a meta-analysis. JAMA Psychiatry 71, 1409–1421.

Sathyasaikumar, K.V., Stachowski, E.K., Wonodi, I., Roberts, R.C., Rassoulpour, A., McMahon, R.P., et al., 2011. Impaired kynurenine pathway metabolism in the prefrontal cortex of individuals with schizophrenia. Schizophr. Bull. 37, 1147–1156.

Schwarcz, R., Rassoulpour, A., Wu, H.Q., Medoff, D., Tamminga, C.A., Roberts, R.C., 2001. Increased cortical kynurenate content in schizophrenia. Biol. Psychiatry 50, 521–530.

Sommer, I.E., de Westrhenen, L., Begemann, M., Kahn, R.S., 2012. Nonsteroidal anti-inflammatory drugs in schizophrenia: ready for practice or a good start? A meta-analysis. J. Clin. Psychiatry 73, 414–419.

Sommer, I.E., van Westrhenen, R., Begemann, M.J., de Witte, L.D., Leucht, S., Kahn, R.S., 2014. Efficacy of anti-inflammatory agents to improve symptoms in patients with schizophrenia: an update. Schizophr. Bull. 40, 181–191.

Sorensen, H.J., Mortensen, E.L., Reinisch, J.M., Mednick, S.A., 2009. Association between prenatal exposure to bacterial infection and risk of schizophrenia. Schizophr. Bull. 35, 631–637.

Sparkman, N.L., Johnson, R.W., 2008. Neuroinflammation associated with aging sensitizes the brain to the effects of infection or stress. Neuroimmunomodulation 15, 323–330.

Steen, R.G., Mull, C., McClure, R., Hamer, R.M., Lieberman, J.A., 2006. Brain volume in first-episode schizophrenia: systematic review and meta-analysis of magnetic resonance imaging studies. Br. J. Psychiatry 188, 510–518.

Steiner, J., Walter, M., Glanz, W., Sarnyai, Z., Bernstein, H.G., Vielhaber, S., et al., 2013. Increased prevalence of diverse N-methyl-D-aspartate glutamate receptor antibodies in patients with an initial diagnosis of schizophrenia: specific relevance of IgG NR1a antibodies for distinction from N-methyl-D-aspartate glutamate receptor encephalitis. JAMA Psychiatry 70, 271–278.

Stone, T.W., 1993. Neuropharmacology of quinolinic and kynurenic acids. Pharmacol. Rev. 45, 309–379.

Takano, A., Arakawa, R., Ito, H., Tateno, A., Takahashi, H., Matsumoto, R., et al., 2010. Peripheral benzodiazepine receptors in patients with chronic schizophrenia: a PET study with [11C]DAA1106. Int. J. Neuropsychopharmacol. 13, 943–950.

Teixeira Jr., A.L., Maia, D.P., Cardoso, F., 2007. Psychosis following acute Sydenham's chorea. Eur. Child Adolesc. Psychiatry 16, 67–69.

van Berckel, B.N., Bossong, M.G., Boellaard, R., Kloet, R., Schuitemaker, A., Caspers, E., et al., 2008. Microglia activation in recent-onset schizophrenia: a quantitative (R)-[11C]PK11195 positron emission tomography study. Biol. Psychiatry 64, 820–822.

van Dam, A.P., 1991. Diagnosis and pathogenesis of CNS lupus. Rheumatol. Int. 11, 1–11.

Versijpt, J.J., Dumont, F., Van Laere, K.J., Decoo, D., Santens, P., Audenaert, K., et al., 2003. Assessment of neuroinflammation and microglial activation in Alzheimer's disease with radiolabelled PK11195 and single photon emission computed tomography. A pilot study. Eur. Neurol. 50, 39–47.

Vincent, A., Bien, C.G., 2008. Anti-NMDA-receptor encephalitis: a cause of psychiatric, seizure, and movement disorders in young adults. Lancet Neurol. 7, 1074–1075.

Westergaard, T., Mortensen, P.B., Pedersen, C.B., Wohlfahrt, J., Melbye, M., 1999. Exposure to prenatal and childhood infections and the risk of schizophrenia: suggestions from a study of sibship characteristics and influenza prevalence. Arch. Gen. Psychiatry 56, 993–998.

Wildenauer, D.B., Körschenhausen, D., Hoechtlen, W., Ackenheil, M., Kehl, M., Lottspeich, F., 1991. Analysis of cerebrospinal fluid from patients with psychiatric and neurological disorders by two-dimensional electrophoresis: identification of disease-associated polypeptides as fibrin fragments. Electrophoresis 12, 487–492.

Winter, C., Djodari-Irani, A., Sohr, R., Morgenstern, R., Feldon, J., Juckel, G., et al., 2009. Prenatal immune activation leads to multiple changes in basal neurotransmitter levels in the adult brain: implications for brain disorders of neurodevelopmental origin such as schizophrenia. Int. J. Neuropsychopharmacol. 12, 513–524.

Yolken, R.H., Torrey, E.F., 1995. Viruses, schizophrenia, and bipolar disorder. Clin. Microbiol. Rev. 8, 131–145.

Zhou, D., Kusnecov, A.W., Shurin, M.R., DePaoli, M., Rabin, B.S., 1993. Exposure to physical and psychological stressors elevates plasma interleukin 6: relationship to the activation of hypothalamic-pituitary-adrenal axis. Endocrinology 133, 2523–2530.

Zubin, J., Spring, B., 1977. Vulnerability—a new view of schizophrenia. J. Abnorm. Psychol. 86, 103–126.

CHAPTER 12

PROTEOMICS OF SCHIZOPHRENIA

M.P. Coba

*Department of Psychiatry and Behavioral Sciences, Zilkha Neurogenetic Institute,
Keck School of Medicine, University of Southern California, Los Angeles, CA, United States*

CHAPTER OUTLINE

Biomarkers 196
Postmortem Studies 196
Protein Interaction Pathways and Signaling Networks 197
Embryonic Brain 202
Conclusions 204
References 205

Recent advances in psychiatric genetics have started to unveil the genetic architecture of schizophrenia (SCZ). In this polygenic disorder, we were able to collect a large amount of candidate risk factors reported in genome-wide association studies (GWAS), single nucleotide variants (SNV), and copy number variants (CNV) (Costain et al., 2013; Fromer et al., 2014; Kirov et al., 2009, 2011; Purcell et al., 2014; Ruderfer et al., 2011; Tam et al., 2009; Xu et al., 2012). However, there is still a gap in understanding how these gene products might work in connected signaling pathways. In this chapter, we aim to explore how current advances in proteomic assays can help us understand signaling mechanisms altered in schizophrenia.

As a first step, we must define what we understand regarding proteomic studies, the range of applications, and the kind of information that we can expect from them.

The term "proteome" was first described in the mid-1990s by Marc Wilkins and can be defined as "the total set" of expressed proteins by an organism, cell, or tissue under defined conditions (Wilkins et al., 1996). This concept can be further extended to proteomes describing subcellular organelles or post-translational modifications such us phosphorylation (eg, phosphoproteome). There are now several emerging proteomic platforms that can measure hundreds or thousands of molecules in a single biological sample; these methods generally represent "snapshots" of cellular states without regard for regulatory or dynamic mechanisms. These methods include solid support-based platforms such as peptide, protein, and antibody arrays, fluorescence-activated cell sorting (FACS), fluorescent bead-based methods, and a variety of mass spectrometry (MS)-based technologies. Although each one has specific applications and different degrees of throughput, MS has proved to be the most versatile and comprehensive platform for a wide range of high-throughput proteomic assays. The implementation of MS strategies evolved from early assays separating and isolating proteins by 2D gel electrophoresis followed by protein identification by MS to high-resolution chromatographic techniques, including multidimensional separation and ultra-performance liquid chromatography (Yates et al., 2009). Together with developments in MS instrumentation and techniques (eg, sample preparation, front-end separation, ionization, data acquisition, and data analysis), this made possible the establishment of shotgun (top-down) proteomic methods. These methods use enzymatic digestion of the proteome of interest and simultaneous identification and quantitation of the resulting peptides by sensitive and accurate MS analyses (Wolters et al., 2001).

Proteomics studies of SCZ can be broadly grouped into three main categories: (1) biomarkers studies; (2) analysis of postmortem samples; and (3) animal/cellular models studies integrating risk factor signaling mechanisms.

Here, we focus on proteomic methods directed to the study of signaling networks that might be represented in human genetic studies of SCZ, and we briefly outline recent results of the use of proteomics assays in the search of biomarkers and analysis of postmortem tissue.

BIOMARKERS

A biomarker can be described as an indicator of molecular changes in body tissues and fluids that can be used as a clinical indicator of health and disease. However, it has been disputed what a biomarker for SCZ might indicate, their use, and if they might really exist (Bahn et al., 2011; Gottschalk et al., 2013; Guest et al., 2013; Schwarz and Bahn, 2008; Weickert et al., 2013). Although a variety of proteomics approaches have been used in the search of biomarkers for SCZ, these disease reporters have proved to be elusive. However, studies suggested that some schizophrenia patients might have alterations in sets of biomolecules with the capacity to become biomarkers, such as serum concentrations of cytokines and molecules involved in inflammatory response (Muller and Schwarz, 2006), metabolic and hormonal pathways such us hypothalamic–pituitary–adrenal (HPA), and insulin signaling (Focking et al., 2011; Guest et al., 2010; Hasnain et al., 2010; Pennington et al., 2008; Ryan et al., 2003). A large number of proteins that seem to be altered in blood samples from schizophrenia patients are related to inflammatory, hormonal, and metabolic pathways. Therefore, it has been proposed that proteomics studies involving blood samples from SCZ patients can be used for stratification of patients based on these biosignatures and might be of use for prediction of patient-specific responses to drugs (Guest et al., 2014; Guest et al., 2013; Martins-de-Souza et al., 2011).

POSTMORTEM STUDIES

Postmortem brain tissues from subjects with schizophrenia have been extensively studied. However, there are a number of well-known problems that have been considered, such us age, sex, tissue pH, postmortem interval, brain regions, and patients history including previous exposure to medication, drugs of abuse, diagnostic criteria, and others (McCullumsmith et al., 2014). As a consequence, there are no large-scale proteomics studies of postmortem tissues; no more than a few dozen samples at a time are being studied. Moreover, because of the heterogeneity of this polygenic disorder, we need to consider how many samples might be needed to obtain representative outcomes. Furthermore, any biochemical study with only a single time point will convincingly indicate only a snapshot at a determined time, without the possibility of analysis of the dynamic changes in the proteome, not only during development but also by synaptic activity. Therefore, proteome studies in postmortem tissues might provide information regarding general protein changes in a subset of samples. They can also be used together with patient sequencing data as a way to stratify samples by human genetics. However, these assays are difficult to translate into molecular mechanisms altered in SCZ.

Proteomic studies of postmortem samples resulted in identification of changes in proteins involved in a variety of functions, including metabolic pathways, neuronal structure, transport, oxidative stress, trafficking, and signal transduction (Nascimento and Martins-de-Souza 2015; Guest et al., 2014; Martins-de-Souza et al., 2010). These changes were described in a variety of brain regions, with samples covering different ages, sex, treatments, and medication; therefore, with the consequent problems of interpretation within these reports, the most replicated changes have been observed in structural proteins, including cytoskeletal components of the intermediate filament family of proteins and myelination-pathway-related proteins. Significant changes have also been replicated in oxidative stress and apoptosis-related proteins and glial proteins.

Analysis of PSD extracts by label-free liquid chromatography–mass spectrometry (LC-MS/MS) in a small sample of 20 SCZ cases and controls reported changes in total levels of 143 proteins (Focking et al., 2015). These changes were reported to be not associated with abnormal PSD size or alterations in synaptic connections (Focking et al., 2015). Although this result indicates certain disregulation in the protein composition in PSD-enriched fractions in samples of SCZ postmortem brain tissues, it is not clear if these changes are involved in mechanisms associated with the pathophysiology of the disease or and if they are related to common signaling networks. However, human genetics studies have convincingly argued for a role of PSD in the pathophysiology of disease, and a role for PSD signaling in SCZ is discussed.

PROTEIN INTERACTION PATHWAYS AND SIGNALING NETWORKS

Human genetic studies had amassed a collection of common and rare variants in the form of candidate risk factors. Within these catalogues, a number of studies described a role or enrichment of rare variants (SNVs and CNVs) at the PSD (Fromer et al., 2014; Kirov et al., 2011). These results are based on proteomics studies that have identified the protein composition and PSD protein complexes such as NMDAR, PSD95, and Arc (Collins et al., 2006; Fernandez et al., 2009; Kirov et al., 2011). However, the protein–protein interaction organization of the PSD is far from clear. The PSD is a functional specialization of the postsynaptic membrane localized at the tip of the dendritic spines of excitatory synapses. The mammalian postsynaptic proteome (PSP) comprises ~1500 proteins organized into multiprotein complexes and molecular networks, including scaffolding and neurotransmitter receptor protein complexes (Bayes et al., 2011; Coba et al., 2009; Collins et al., 2005, 2006; Fernandez et al., 2009). Within this signaling machinery, a major problem is determining how these large sets of molecules organize to modulate synaptic function. A first attempt using immune purification and mass spectrometry approaches was able to identify a number of PSD protein complexes. These datasets were then used to map SCZ candidate genes to NMDAR, Arc, and Psd95 protein complexes (Fromer et al., 2014; Kirov et al., 2011). However, the analysis of protein–protein interactions from MS-derived assays also generates some problems. A common misconception in the analysis of protein complexes obtained by immunopurification and MS methods is assigning the role of "interactors" to all the proteins identified in a protein complex. It is worth highlighting that this technology cannot determine binary protein interactions. Attempts to overcome this problem have been made through the use of cross-linking agents. Several protocols showed feasibility of determining not only the direct protein–protein interactions but also the region involved in binding (Hoopmann et al., 2010; Schmidt and Robinson, 2014; Weisbrod et al., 2013). These methods have been proven to be useful in the determination of overexpressed protein or "small" protein complexes. However it is still a major challenge to translate them to the analysis of large and heterogeneous protein complexes with different degrees of stoichiometry, such as found in the PSD.

Another common incorrect interpretation of protein interaction datasets is assigning the role of "scaffold protein" to the target protein that has been immunopurified based on the identification of a number of "protein interactors" by MS methods. Although proteins isolated by immunopurification techniques can recover hundreds of protein interactions, this alone does not convincingly imply a role as a scaffold. As a brief summary, scaffold molecules present often structural or protein-clustering functions, often show a multimodular composition of protein domain, can homo-heterodimerize, and might form binary protein interactions with different proteins at the same time. This scenario cannot be established solely by the isolation of protein complexes, in which we can determine dozens of individual binary pairs of protein interactions without a protein being able to cluster a high number of proteins at a given time. For example, we can describe the case of a protein kinase with hundreds of substrates. Although these are all protein–protein interactions, we do not describe it as a scaffold protein. They do not represent what is considered a protein complex and can be better described as a collection of binary interactions. However, it is common to see these terms assigned for members of the PSD, where targets isolated by MS immunopurification methods are usually considered scaffold proteins, making the interpretation of the role of these molecules in signaling mechanisms related to SCZ more difficult.

A schematic view of the most studied and abundant scaffold proteins at the synapse can be described within a core protein interaction **structure** (Fig. 12.1). This can be viewed in three main groups of scaffold proteins. First, there is an "upper layer" of scaffold proteins associated with glutamate receptor ion channels; this group is enriched in members of the membrane-associated guanylate kinases (MAGUK), including the disk large-associated homologues 1–4 (DLGs). Second, there is a "bottom layer" of scaffold components, including the SH3 and multiple ankyrin repeat domain proteins (**SHANKs 1–3**). Third, there is an intermediate layer of scaffold molecules functioning as a link between the "upper and lower layers," including the disk large-associated protein 1 (**Dlgap1**) and the connector enhancer of kinase suppressor of ras 2 (**Cnksr2**), among others. Within these schematic structures, it is believed that a subset of protein interactions from PSD scaffolds might be enriched in common and rare SCZ risk variants (Fig. 12.2). Therefore, this suggests that a subset of PSD protein interactions might be enough to represent the core of "SCZ risk" at the postsynaptic site, instead of a more general PSD proteome.

Protein–protein interaction studies can give a sense of the organization of signaling pathways. However, they generally include large datasets with difficult interpretation. One way forward is to consider the protein domain composition and the architecture of protein domains within protein complexes. Organization of PSD protein complexes has been reported to be highly dependent of the protein domain composition of protein complexes and regulated by post-translational modifications (PTMs) such as phosphorylation. Protein domains are considered to be basic units of biological function

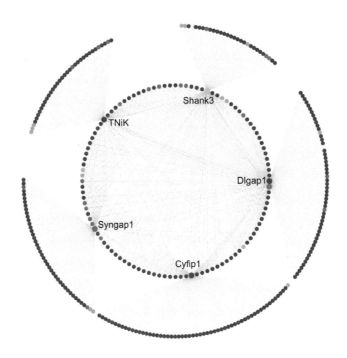

FIGURE 12.1

Clustering of TNiK, Shank3, Cnksr2, Syngap1, Cyfip1, and Dlgap1 protein interactions. Colors represent SCZ risk factors identified in SCZ human genetic studies for: GWAS (pink, single candidate within risk locus; orange, with multiple candidates within risk locus); de novo single nucleotide variants (red); de novo copy number variants (purple); genes identified by OMIM as a schizophrenia susceptibility locus (blue); and GWAS plus CNV (yellow/green). The representation of core components of protein complexes is shown by circular clustering using Cytoscape 3.2.1 (Shannon et al., 2003).

(Jin et al., 2009) responsible for enzymatic functions and cellular localization with a variety of molecular interactions that modulate signaling processes (Scott and Pawson, 2009). These multidomain architectures can be used as a measure of the cellular functions and protein interactions for a determined target protein (Apic et al., 2001; Fig. 12.3). Because protein–protein interactions are highly dependent on the composition and the architecture of protein domains (Jin and Pawson, 2012; Pawson and Scott, 1997), they can be informative of the broad spectrum of protein functions encoded by these protein complexes. Bioinformatics analysis shows that PSD complexes enriched in SCZ risk factors group protein families containing proteins with multidomain architectures, including SH3, PDZ, SAM, PH, ANK, GuKc, SPEC, and GKAP, among other domains engaged in protein–protein interaction functions.

The association of proteins in protein complexes is often used as a common regulatory mechanism to ensure that signaling components encounter their intracellular partners at the right place and time (Pawson and Scott, 1997; Scott and Pawson, 2009). Although protein scaffolds can use these specialized protein interaction modules (protein domains) as a key mechanism to achieve specificity, the combination of domain architectures within PSD complexes might allow a degree of protein

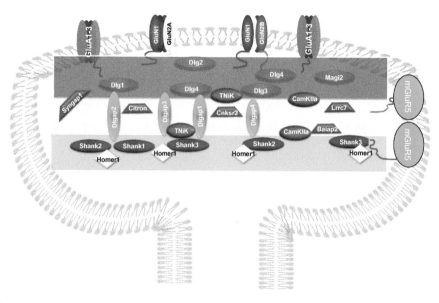

FIGURE 12.2

Cartoon shows a schematic view of the main scaffold proteins at the PSD organized in three main layers: pink (top), scaffold molecules binding ionotropic glutamate receptors; yellow (middle), scaffold proteins connecting top and bottom layers; and green (bottom), lower layer of scaffold proteins. Enzymes and different abundant scaffold molecules are distributed within these core components and are represented in purple and blue.

interactions and cross-talk between a number of PSD signaling pathways. Thus, mutations in SCZ risk factors might impair PSD protein interactions through multiple layers or from different nodes in the network. Highly connected domains might represent functional centers (Wuchty, 2001) and allow the communication of different components of the network (Basu et al., 2008). Therefore, SCZ risk factors might modulate protein complexes enriched in domain hubs and disrupt the domain architecture of the synaptic signaling network.

It is believed that the modulation of these clusters of protein interactions is regulated by synaptic activity through post-translational modification (Chung et al., 2004; Morabito et al., 2004; Scannevin and Huganir, 2000). Reversible post-translational modification (PTM) is a versatile way to modulate protein function. Historically, the most extensively studied PTM is phosphorylation, which is known to regulate every aspect of the signaling machinery (Cohen, 2002; Pawson and Scott, 2005). In the synapse, protein phosphorylation has been described to regulate a variety of events from synaptic plasticity to learning/memory and cognition (Abel et al., 1997; Kandel, 2001; O'Dell et al., 1992; Silva et al., 1992) The widespread phosphorylation-dependent regulation of protein–protein interactions might modulate different subsets of interactions and reshape the PSD (Coba et al., 2008, 2009).

Protein phosphorylation can modulate a number of protein domains and protein ligand domains, shaping the architecture of synaptic complexes (Chung et al., 2004; Coba et al., 2009). Protein scaffolds such as DLGs, DLGAPs, and SHANKs use their multimodular protein domain architecture to

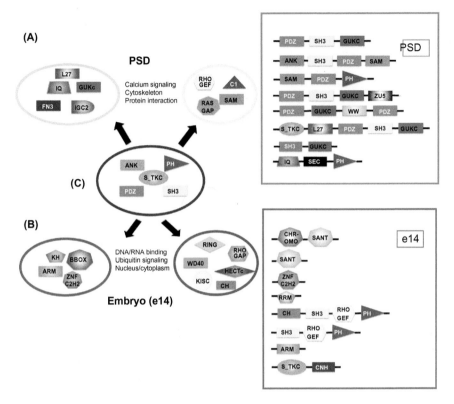

FIGURE 12.3

Cartoon represents the most abundant protein domains present in PSD protein (A) and e14 (B) protein complexes. Both sets of protein domains use a combination of common (C) protein domains to generate a variety of protein domain architectures (right) in PSD (top) and e14 (bottom) molecular complexes. Protein domains are annotated according to the SMART database (Letunic et al., 2015).

cluster and localize a set of PSD protein domains, defining groups of protein complexes. These protein complexes are likely to be modulated by synaptic activity through a variety of post-translational modifications (Coba et al., 2009).

Current studies have proposed the existence of more than 1500 phosphorylation sites at the synapse (Coba et al., 2009; Collins et al., 2005; Munton et al., 2007; Trinidad et al., 2006); however, how many of them might be relevant for synaptic function and their role within the PSD signaling machinery are not known. Therefore, their role in signaling pathways related to SCZ has not been pursued. This task seems to present a number of difficulties. Different patterns of PSD protein phosphorylation have been observed in synaptosomal preparations and NMDAR stimulation of mouse hippocampal slices (Coba et al., 2009; Munton et al., 2007; Trinidad et al., 2008). A number of studies have proposed that a combinatorial output of phosphorylation sites might be used to interpret the temporal profiling of activation of different neurotransmitter receptor activation states. Furthermore, the discovery of a large amount of phosphorylation sites does not correspond to the identification of the protein kinase/kinases that are able to phosphorylate these sites. Moreover, as for any MS proteomic assay, the need to achieve a level of protein enrichment that allows phospho-site quantitation limits the application of these methods. As a consequence, there is no information on how these processes might operate in wild-type or disease (animal/cellular) models. Therefore, there is still much work to accomplish to try to understand phosphorylation networks at the synapse and how they might impact synaptic function in SCZ models.

EMBRYONIC BRAIN

Although proteomics assays have been used to group a number of SCZ risk factors at the PSD and in a discrete number of protein complexes, these studies generally refer to risk factor enrichment in these fractions. This leaves a large group of common and rare SCZ candidate genes that do not fit within this "synaptic" scenario. Because many of these gene products are not present at postsynaptic or presynaptic sites, it is a difficult task to search for where and when their signaling networks might function. As a neurodevelopmental disorder, it is expected that risk factors might be distributed within developmental signaling networks. Genomics and transcriptomic analyses proposed a role of SCZ risk factors in fetal prefrontal cortex development (Gulsuner et al., 2013), with functions in transcriptional regulation, neuronal migration, transport, and cortical neurogenesis. These groups of functions represent a more heterogeneous group than synaptic signaling networks, and they can be distributed along a variety of spatio-temporal signaling cascades. In accordance with a role of SCZ risk factors in fetal cortical processes, proteomic assays have been able to cluster a number of SCZ risk factors in a number protein interactions in embryonic (e14) mouse cortex. These molecules have been related not only to SCZ but also to other psychiatric disorders like ASD, OCD, and BP (Buxbaum et al., 2012; Georgieva et al., 2014; Greenwood et al., 2012; Krumm et al., 2015; Mattheisen et al., 2015; O'Roak et al., 2012; Pinto et al., 2014). Analysis of the protein domain composition of these molecular complexes shows an enrichment of ubiquitin, DNA (Fig. 12.3), and binding domains, including AAA, ZNF_C2H2, BBOX, BBC, KH, RING, ARM, and KISC, suggesting an early role in chromatin remodeling, cellular proliferation, and centrosomal organization processes. Therefore, these protein–protein interactions are associated with wide cellular localization, including nucleus, centrosome, and cytoplasm. The functional network of these protein interactions resembles one of the most studied candidate risk factors for SCZ: disrupted in schizophrenia 1 (DISC1).

The DISC1 interactome was first described in yeast two-hybrid (DISC1) screens (Camargo et al., 2007) and included a variety of protein interactions, including TNiK, AKAP9, and Pde4dip. Through its interacting partners, DISC1 was reported to have a role in the modulation of centrosomal dynamics, cytoskeletal function, intracellular transport, and cell–cycle/division.

Recently, hiPSC-derived neurons from patients with SCZ and other psychiatric diseases have been used to start to address molecular mechanisms associated to these disorders. Phenotypic differences have been observed in hNPCs derived from patients with SCZ (Brennand et al., 2015; Hartley et al.,

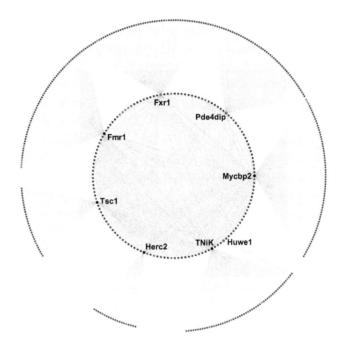

FIGURE 12.4

Circular clustering of protein complexes immunoisolated from e14 mice prefrontal cortex and identified by HPLC-MS/MS showing TNiK, Fmr1, Mycbp2, Tsc1, Herc2, Huwe1, Pde4dip, and Fxr1 protein interactions. Color code indicates the distribution of SCZ risk factors in GWAS, SNVs, and CNVs (see Fig. 12.2). Clustering of protein complexes was performed with Cytoscape 3.2.1 (Shannon et al., 2003).

2015). Proteomic characterization of SCZ hiPSC-derived neuronal progenitor cells (NPC) using stable isotope labeling by amino acids in cell culture (SILAC) was able to quantitate perturbations in cellular adhesion and oxidative stress pathways (Brennand et al., 2015). As observed for e14 protein interactions the core component of this network included a number of members of the centrosome machinery. This suggests that the SCZ common and rare variants in mouse e14 cortex can be mapped to a group of molecular complexes present in nucleus, cytoplasm, Golgi, and centrosome localization in a set of molecular processes modulating cell proliferation and neuronal development (Fig. 12.4). The centrosome is the major center for microtubular array organization and nucleation, regulating cell division, differentiation, and migration (Higginbotham and Gleeson, 2007). Mutations in structural components of the centrosome or in the pericentriolar material (PCM) have been associated with a variety of neurodevelopmental disorders, including SCZ (Kuijpers and Hoogenraad, 2011). Analysis of online Mendelian inheritance in man (OMIM) categories within components of these networks shows that they are largely associated with developmental disorders linked to centrosome function, including lissencephaly, cortical dysplasia, macrocephaly, microcephaly, and abnormal cellular proliferation. This set of functions have already been proposed to be modulated by components of the DISC1 interactome

(Bradshaw et al., 2008; Brandon and Sawa, 2011; Ishizuka et al., 2011; Morris et al., 2003; Wang et al., 2010). These molecules involve not only structural components but also the necessary signaling machinery, including protein synthesis, degradation, and modulation by post-translational modification. Similarly, as observed in PSD protein complexes, SCZ risk factors might impact core components of molecular functions that are modulated by hundreds of proteins grouped in a variety of protein interactions. Therefore, future studies of NPC derived from SCZ patients might focus on addressing the role of SCZ risk factors in the modulation and composition of the molecular machinery involved in sets of functions. Moreover, protein interactions described in molecular complexes at e14 and NPC, which modulate cellular proliferation and differentiation processes, might still be relevant in mature neurons modulating adult neurogenesis. A number of molecules associated with protein complexes in e14 and hNPC have been found to modulate adult neurogenesis, including TNiK (Coba et al., 2012), Cntnb1 (Clelland et al., 2009), Fmr1 (Guo et al., 2011), and Huwe1 (Zhao et al., 2009; Aimone et al., 2010). Although a role of adult neurogenesis (Ouchi et al., 2013; Reif et al., 2006; Toro and Deakin, 2007) in SCZ still remains to be established, analysis of protein interactions within these sets of signaling complexes might be used as molecular markers for a set of SCZ endophenotypes.

CONCLUSIONS

The combination of human genetics and proteomics datasets indicates a number of signaling processes that are more frequently targeted in SCZ. These molecular processes do not function in isolation, but rather in combination within hundreds of molecules in protein interacting networks. Therefore, we need tools that allow us to study the function of sets of genes. Systems biology approaches are well suited to study the relationship between sets of molecules. Here, we evaluated the use of MS-based proteomics in three areas, biomarkers, postmortem, and signaling mechanisms and emphasized the role of proteomic assays in addressing signaling networks that might be altered in SCZ. These studies must be expanded to map out the physical components and interactions of a system, but we also need to map how information propagates through this system in response to perturbations.

Protein complexes are dynamic and SCZ risk factors might be associated with different spatiotemporal dynamics; therefore, analysis of SCZ protein interactions will need to consider not only who the interactor is but also where and when this interaction occurs. This poses the problems of determining the role of SCZ risk factors in these networks and defining their role in the pathophysiology of SCZ within their appropriate molecular context.

Human genetics studies will provide more candidate risk factors that might integrate within these networks; therefore, beyond addressing molecular mechanisms within risk pathways (eg, cellular proliferation, immune system, PSD, centrosome dynamics), future studies will also need to address if and how these risk pathways contribute to susceptibility for the disease. It will be important to determine if it is enough to accumulate risk factors within one risk pathway, or if they need to be combined within different developmental networks. Although single-cell interactomics is not yet an option, cell-type specificity and brain regions will also need to be considered to analyze the properties of protein–protein interaction networks.

Determining the cellular networks where SCZ risk factors are functionally active might contribute to defining polygenic risk scores and developing strategies to stratify individuals who are vulnerable to schizophrenia.

REFERENCES

Abel, T., Nguyen, P.V., Barad, M., Deuel, T.A., Kandel, E.R., Bourtchouladze, R., 1997. Genetic demonstration of a role for PKA in the late phase of LTP and in hippocampus-based long-term memory. Cell 88 (5), 615–626. Retrieved from http://www.ncbi.nlm.nih.gov/pubmed/9054501.

Aimone, J.B., Deng, W., Gage, F.H., 2010. Adult neurogenesis: integrating theories and separating functions. Trends Cogn Sci 14 (7), 325–337. doi:S1364-6613(10)00088-4 [pii] 10.1016/j.tics.2010.04.003.

Apic, G., Gough, J., Teichmann, S.A., 2001. An insight into domain combinations. Bioinformatics 17 (Suppl. 1), S83–S89. Retrieved from http://www.ncbi.nlm.nih.gov/pubmed/11472996.

Bahn, S., Noll, R., Barnes, A., Schwarz, E., Guest, P.C., 2011. Challenges of introducing new biomarker products for neuropsychiatric disorders into the market. Int Rev Neurobiol 101, 299–327. http://dx.doi.org/10.1016/B978-0-12-387718-5.00012-2.

Basu, M.K., Carmel, L., Rogozin, I.B., Koonin, E.V., 2008. Evolution of protein domain promiscuity in eukaryotes. Genome Res 18 (3), 449–461. http://dx.doi.org/10.1101/gr.6943508.

Bayes, A., van de Lagemaat, L.N., Collins, M.O., Croning, M.D., Whittle, I.R., Choudhary, J.S., et al., 2011. Characterization of the proteome, diseases and evolution of the human postsynaptic density. Nat Neurosci 14 (1), 19–21. Retrieved from http://www.ncbi.nlm.nih.gov/entrez/query.fcgi?cmd=Retrieve&db=PubMed&dopt=Citation&list_uids=21170055.

Bradshaw, N.J., Ogawa, F., Antolin-Fontes, B., Chubb, J.E., Carlyle, B.C., Christie, S., et al., 2008. DISC1, PDE4B, and NDE1 at the centrosome and synapse. Biochem Biophys Res Commun 377 (4), 1091–1096. Retrieved from http://www.ncbi.nlm.nih.gov/entrez/query.fcgi?cmd=Retrieve&db=PubMed&dopt=Citation&list_uids=18983980.

Brandon, N.J., Sawa, A., 2011. Linking neurodevelopmental and synaptic theories of mental illness through DISC1. Nat Rev Neurosci 12 (12), 707–722. http://dx.doi.org/10.1038/nrn3120.

Brennand, K., Savas, J.N., Kim, Y., Tran, N., Simone, A., Hashimoto-Torii, K., et al., 2015. Phenotypic differences in hiPSC NPCs derived from patients with schizophrenia. Mol Psychiatry 20 (3), 361–368. http://dx.doi.org/10.1038/mp.2014.22.

Buxbaum, J.D., Daly, M.J., Devlin, B., Lehner, T., Roeder, K., State, M.W., et al., 2012. The autism sequencing consortium: large-scale, high-throughput sequencing in autism spectrum disorders. Neuron 76 (6), 1052–1056. http://dx.doi.org/10.1016/j.neuron.2012.12.008.

Camargo, L.M., Collura, V., Rain, J.C., Mizuguchi, K., Hermjakob, H., Kerrien, S., et al., 2007. Disrupted in schizophrenia 1 interactome: evidence for the close connectivity of risk genes and a potential synaptic basis for schizophrenia. Mol Psychiatry 12 (1), 74–86. Retrieved from http://www.ncbi.nlm.nih.gov/entrez/query.fcgi?cmd=Retrieve&db=PubMed&dopt=Citation&list_uids=17043677.

Chung, H.J., Huang, Y.H., Lau, L.F., Huganir, R.L., 2004. Regulation of the NMDA receptor complex and trafficking by activity-dependent phosphorylation of the NR2B subunit PDZ ligand. J Neurosci 24 (45), 10248–10259. http://dx.doi.org/10.1523/jneurosci.0546-04.2004.

Clelland, C.D., Choi, M., Romberg, C., Clemenson Jr., G.D., Fragniere, A., Tyers, P., et al., 2009. A functional role for adult hippocampal neurogenesis in spatial pattern separation. Science 325 (5937), 210–213. Retrieved from http://www.ncbi.nlm.nih.gov/entrez/query.fcgi?cmd=Retrieve&db=PubMed&dopt=Citation&list_uids=19590004.

Coba, M.P., Valor, L.M., Kopanitsa, M.V., Afinowi, N.O., Grant, S.G., 2008. Kinase networks integrate profiles of N-methyl-D-aspartate receptor-mediated gene expression in hippocampus. J Biol Chem 283 (49), 34101–34107. Retrieved from http://www.ncbi.nlm.nih.gov/entrez/query.fcgi?cmd=Retrieve&db=PubMed&dopt=Citation&list_uids=18815127.

Coba, M.P., Pocklington, A.J., Collins, M.O., Kopanitsa, M.V., Uren, R.T., Swamy, S., et al., 2009. Neurotransmitters drive combinatorial multistate postsynaptic density networks. Sci Signal 2 (68), ra19. http://dx.doi.org/10.1126/scisignal.2000102.

Coba, M.P., et al., 2012. TNiK is required for postsynaptic and nuclear signaling pathways and cognitive function. The Journal of neuroscience: the official journal of the Society for Neuroscience 32, 13987–13999. http://dx.doi.org/10.1523/JNEUROSCI.2433-12.2012.

Cohen, P., 2002. The origins of protein phosphorylation. Nat Cell Biol 4 (5), E127–E130. http://dx.doi.org/10.1038/ncb0502-e127.

Collins, M.O., Yu, L., Coba, M.P., Husi, H., Campuzano, I., Blackstock, W.P., et al., 2005. Proteomic analysis of in vivo phosphorylated synaptic proteins. J Biol Chem 280 (7), 5972–5982. http://dx.doi.org/10.1074/jbc.M411220200.

Collins, M.O., Husi, H., Yu, L., Brandon, J.M., Anderson, C.N., Blackstock, W.P., et al., 2006. Molecular characterization and comparison of the components and multiprotein complexes in the postsynaptic proteome. J Neurochem 97 (Suppl. 1), 16–23. Retrieved from http://www.ncbi.nlm.nih.gov/entrez/query.fcgi?cmd=Retrieve&db=PubMed&dopt=Citation&list_uids=16635246.

Costain, G., Lionel, A.C., Merico, D., Forsythe, P., Russell, K., Lowther, C., et al., 2013. Pathogenic rare copy number variants in community-based schizophrenia suggest a potential role for clinical microarrays. Hum Mol Genet 22 (22), 4485–4501. http://dx.doi.org/10.1093/hmg/ddt297.

Fernandez, E., Collins, M.O., Uren, R.T., Kopanitsa, M.V., Komiyama, N.H., Croning, M.D., et al., 2009. Targeted tandem affinity purification of PSD-95 recovers core postsynaptic complexes and schizophrenia susceptibility proteins. Mol Syst Biol 5, 269. Retrieved from http://www.ncbi.nlm.nih.gov/entrez/query.fcgi?cmd=Retrieve&db=PubMed&dopt=Citation&list_uids=19455133.

Focking, M., Dicker, P., English, J.A., Schubert, K.O., Dunn, M.J., Cotter, D.R., 2011. Common proteomic changes in the hippocampus in schizophrenia and bipolar disorder and particular evidence for involvement of cornu ammonis regions 2 and 3. Arch Gen Psychiatry 68 (5), 477–488. http://dx.doi.org/10.1001/archgenpsychiatry.2011.43.

Focking, M., Lopez, L.M., English, J.A., Dicker, P., Wolff, A., Brindley, E., et al., 2015. Proteomic and genomic evidence implicates the postsynaptic density in schizophrenia. Mol Psychiatry 20 (4), 424–432. http://dx.doi.org/10.1038/mp.2014.63.

Fromer, M., Pocklington, A.J., Kavanagh, D.H., Williams, H.J., Dwyer, S., Gormley, P., et al., 2014. De novo mutations in schizophrenia implicate synaptic networks. Nature. http://dx.doi.org/10.1038/nature12929.

Georgieva, L., Rees, E., Moran, J.L., Chambert, K.D., Milanova, V., Craddock, N., et al., 2014. De novo CNVs in bipolar affective disorder and schizophrenia. Hum Mol Genet 23 (24), 6677–6683. http://dx.doi.org/10.1093/hmg/ddu379.

Gottschalk, M.G., Schwarz, E., Bahn, S., 2013. Biomarker research in neuropsychiatry: challenges and potential. Fortschr Neurol Psychiatr 81 (5), 243–249. http://dx.doi.org/10.1055/s-0033-1335235.

Greenwood, T.A., Akiskal, H.S., Akiskal, K.K., Bipolar Genome, S., Kelsoe, J.R., 2012. Genome-wide association study of temperament in bipolar disorder reveals significant associations with three novel Loci. Biol Psychiatry 72 (4), 303–310. http://dx.doi.org/10.1016/j.biopsych.2012.01.018.

Guest, P.C., Wang, L., Harris, L.W., Burling, K., Levin, Y., Ernst, A., et al., 2010. Increased levels of circulating insulin-related peptides in first-onset, antipsychotic naive schizophrenia patients. Mol Psychiatry 15 (2), 118–119. http://dx.doi.org/10.1038/mp.2009.81.

Guest, P.C., Gottschalk, M.G., Bahn, S., 2013. Proteomics: improving biomarker translation to modern medicine? Genome Med 5 (2), 17. http://dx.doi.org/10.1186/gm421.

Guest, P.C., Chan, M.K., Gottschalk, M.G., Bahn, S., 2014. The use of proteomic biomarkers for improved diagnosis and stratification of schizophrenia patients. Biomark Med 8 (1), 15–27. http://dx.doi.org/10.2217/bmm.13.83.

Gulsuner, S., Walsh, T., Watts, A.C., Lee, M.K., Thornton, A.M., Casadei, S., et al., 2013. Spatial and temporal mapping of de novo mutations in schizophrenia to a fetal prefrontal cortical network. Cell 154 (3), 518–529. http://dx.doi.org/10.1016/j.cell.2013.06.049.

Guo, W., Allan, A.M., Zong, R., Zhang, L., Johnson, E.B., Schaller, E.G., et al., 2011. Ablation of Fmrp in adult neural stem cells disrupts hippocampus-dependent learning. Nat Med 17 (5), 559–565. http://dx.doi.org/10.1038/nm.2336.

Hartley, B.J., Tran, N., Ladran, I., Reggio, K., Brennand, K.J., 2015. Dopaminergic differentiation of schizophrenia hiPSCs. Mol Psychiatry 20 (5), 549–550. http://dx.doi.org/10.1038/mp.2014.194.

Hasnain, M., Fredrickson, S.K., Vieweg, W.V., Pandurangi, A.K., 2010. Metabolic syndrome associated with schizophrenia and atypical antipsychotics. Curr Diab Rep 10 (3), 209–216. http://dx.doi.org/10.1007/s11892-010-0112-8.

Higginbotham, H.R., Gleeson, J.G., 2007. The centrosome in neuronal development. Trends Neurosci 30 (6), 276–283. http://dx.doi.org/10.1016/j.tins.2007.04.001.

Hoopmann, M.R., Weisbrod, C.R., Bruce, J.E., 2010. Improved strategies for rapid identification of chemically cross-linked peptides using protein interaction reporter technology. J Proteome Res 9 (12), 6323–6333. http://dx.doi.org/10.1021/pr100572u.

Ishizuka, K., Kamiya, A., Oh, E.C., Kanki, H., Seshadri, S., Robinson, J.F., et al., 2011. DISC1-dependent switch from progenitor proliferation to migration in the developing cortex. Nature 473 (7345), 92–96. doi:nature09859 [pii] 10.1038/nature09859.

Jin, J., Pawson, T., 2012. Modular evolution of phosphorylation-based signalling systems. Philos Trans R Soc Lond B Biol Sci 367 (1602), 2540–2555. http://dx.doi.org/10.1098/rstb.2012.0106.

Jin, J., Xie, X., Chen, C., Park, J.G., Stark, C., James, D.A., et al., 2009. Eukaryotic protein domains as functional units of cellular evolution. Sci Signal 2 (98), ra76. http://dx.doi.org/10.1126/scisignal.2000546.

Kandel, E.R., 2001. The molecular biology of memory storage: a dialogue between genes and synapses. Science 294 (5544), 1030–1038. Retrieved from http://www.ncbi.nlm.nih.gov/entrez/query.fcgi?cmd=Retrieve&db=PubMed&dopt=Citation&list_uids=11691980.

Kirov, G., Grozeva, D., Norton, N., Ivanov, D., Mantripragada, K.K., Holmans, P., et al., 2009. Support for the involvement of large copy number variants in the pathogenesis of schizophrenia. Hum Mol Genet 18 (8), 1497–1503. http://dx.doi.org/10.1093/hmg/ddp043.

Kirov, G., Pocklington, A.J., Holmans, P., Ivanov, D., Ikeda, M., Ruderfer, D., et al., 2011. De novo CNV analysis implicates specific abnormalities of postsynaptic signalling complexes in the pathogenesis of schizophrenia. Mol Psychiatry. http://dx.doi.org/10.1038/mp.2011.154.

Krumm, N., Turner, T.N., Baker, C., Vives, L., Mohajeri, K., Witherspoon, K., et al., 2015. Excess of rare, inherited truncating mutations in autism. Nat Genet 47 (6), 582–588. http://dx.doi.org/10.1038/ng.3303.

Kuijpers, M., Hoogenraad, C.C., 2011. Centrosomes, microtubules and neuronal development. Mol Cell Neurosci 48 (4), 349–358. http://dx.doi.org/10.1016/j.mcn.2011.05.004.

Letunic, I., Doerks, T., Bork, P., 2015. SMART: recent updates, new developments and status in 2015. Nucleic Acids Res 43 (Database issue), D257–D260. http://dx.doi.org/10.1093/nar/gku949.

Martins-de-Souza, D., Schmitt, A., Roder, R., Lebar, M., Schneider-Axmann, T., Falkai, P., et al., 2010. Sex-specific proteome differences in the anterior cingulate cortex of schizophrenia. J Psychiatr Res 44 (14), 989–991. http://dx.doi.org/10.1016/j.jpsychires.2010.03.003.

Martins-de-Souza, D., Guest, P.C., Vanattou-Saifoudine, N., Harris, L.W., Bahn, S., 2011. Proteomic technologies for biomarker studies in psychiatry: advances and needs. Int Rev Neurobiol 101, 65–94. http://dx.doi.org/10.1016/B978-0-12-387718-5.00004-3.

Mattheisen, M., Samuels, J.F., Wang, Y., Greenberg, B.D., Fyer, A.J., McCracken, J.T., et al., 2015. Genome-wide association study in obsessive-compulsive disorder: results from the OCGAS. Mol Psychiatry 20 (3), 337–344. http://dx.doi.org/10.1038/mp.2014.43.

McCullumsmith, R.E., Hammond, J.H., Shan, D., Meador-Woodruff, J.H., 2014. Postmortem brain: an underutilized substrate for studying severe mental illness. Neuropsychopharmacology 39 (1), 65–87. http://dx.doi.org/10.1038/npp.2013.239.

Morabito, M.A., Sheng, M., Tsai, L.H., 2004. Cyclin-dependent kinase 5 phosphorylates the N-terminal domain of the postsynaptic density protein PSD-95 in neurons. J Neurosci 24 (4), 865–876. http://dx.doi.org/10.1523/JNEUROSCI.4582-03.2004.

Morris, J.A., Kandpal, G., Ma, L., Austin, C.P., 2003. DISC1 (disrupted-in-schizophrenia 1) is a centrosome-associated protein that interacts with MAP1A, MIPT3, ATF4/5 and NUDEL: regulation and loss of interaction with mutation. Hum Mol Genet 12 (13), 1591–1608. Retrieved from http://www.ncbi.nlm.nih.gov/entrez/query.fcgi?cmd=Retrieve&db=PubMed&dopt=Citation&list_uids=12812986.

Muller, N., Schwarz, M., 2006. Schizophrenia as an inflammation-mediated dysbalance of glutamatergic neurotransmission. Neurotox Res 10 (2), 131–148. Retrieved from http://www.ncbi.nlm.nih.gov/pubmed/17062375.

Munton, R.P., Tweedie-Cullen, R., Livingstone-Zatchej, M., Weinandy, F., Waidelich, M., Longo, D., et al., 2007. Qualitative and quantitative analyses of protein phosphorylation in naive and stimulated mouse synaptosomal preparations. Mol Cell Proteomics 6 (2), 283–293. Retrieved from http://www.ncbi.nlm.nih.gov/entrez/query.fcgi?cmd=Retrieve&db=PubMed&dopt=Citation&list_uids=17114649.

Nascimento, J.M., Martins-de-Souza, D., 2015. The proteome of schizophrenia. npj Schizophrenia.

O'Dell, T.J., Grant, S.G., Karl, K., Soriano, P.M., Kandel, E.R., 1992. Pharmacological and genetic approaches to the analysis of tyrosine kinase function in long-term potentiation. Cold Spring Harb Symp Quant Biol 57, 517–526. Retrieved from http://www.ncbi.nlm.nih.gov/entrez/query.fcgi?cmd=Retrieve&db=PubMed&dopt=Citation&list_uids=1339688.

O'Roak, B.J., Vives, L., Girirajan, S., Karakoc, E., Krumm, N., Coe, B.P., et al., 2012. Sporadic autism exomes reveal a highly interconnected protein network of de novo mutations. Nature 485 (7397), 246–250. http://dx.doi.org/10.1038/nature10989.

Ouchi, Y., Banno, Y., Shimizu, Y., Ando, S., Hasegawa, H., Adachi, K., et al., 2013. Reduced adult hippocampal neurogenesis and working memory deficits in the Dgcr8-deficient mouse model of 22q11.2 deletion-associated schizophrenia can be rescued by IGF2. J Neurosci 33 (22), 9408–9419. http://dx.doi.org/10.1523/jneurosci.2700-12.2013.

Pawson, T., Scott, J.D., 1997. Signaling through scaffold, anchoring, and adaptor proteins. Science 278 (5346), 2075–2080. Retrieved from http://www.ncbi.nlm.nih.gov/pubmed/9405336.

Pawson, T., Scott, J.D., 2005. Protein phosphorylation in signaling – 50 years and counting. Trends Biochem Sci 30 (6), 286–290. http://dx.doi.org/10.1016/j.tibs.2005.04.013.

Pennington, K., Beasley, C.L., Dicker, P., Fagan, A., English, J., Pariante, C.M., et al., 2008. Prominent synaptic and metabolic abnormalities revealed by proteomic analysis of the dorsolateral prefrontal cortex in schizophrenia and bipolar disorder. Mol Psychiatry 13 (12), 1102–1117. http://dx.doi.org/10.1038/sj.mp.4002098.

Pinto, D., Delaby, E., Merico, D., Barbosa, M., Merikangas, A., Klei, L., et al., 2014. Convergence of genes and cellular pathways dysregulated in autism spectrum disorders. Am J Hum Genet 94 (5), 677–694. http://dx.doi.org/10.1016/j.ajhg.2014.03.018.

Purcell, S.M., Moran, J.L., Fromer, M., Ruderfer, D., Solovieff, N., Roussos, P., et al., 2014. A polygenic burden of rare disruptive mutations in schizophrenia. Nature. http://dx.doi.org/10.1038/nature12975.

Reif, A., Fritzen, S., Finger, M., Strobel, A., Lauer, M., Schmitt, A., et al., 2006. Neural stem cell proliferation is decreased in schizophrenia, but not in depression. Mol Psychiatry 11 (5), 514–522. Retrieved from http://www.ncbi.nlm.nih.gov/entrez/query.fcgi?cmd=Retrieve&db=PubMed&dopt=Citation&list_uids=16415915.

Ruderfer, D.M., Kirov, G., Chambert, K., Moran, J.L., Owen, M.J., O'Donovan, M.C., et al., 2011. A family-based study of common polygenic variation and risk of schizophrenia. Mol Psychiatry 16 (9), 887–888. http://dx.doi.org/10.1038/mp.2011.34.

Ryan, M.C., Collins, P., Thakore, J.H., 2003. Impaired fasting glucose tolerance in first-episode, drug-naive patients with schizophrenia. Am J Psychiatry 160 (2), 284–289. Retrieved from http://www.ncbi.nlm.nih.gov/pubmed/12562574.

Scannevin, R.H., Huganir, R.L., 2000. Postsynaptic organization and regulation of excitatory synapses. Nat Rev Neurosci 1 (2), 133–141. http://dx.doi.org/10.1038/35039075.

Schmidt, C., Robinson, C.V., 2014. A comparative cross-linking strategy to probe conformational changes in protein complexes. Nat Protoc 9 (9), 2224–2236. http://dx.doi.org/10.1038/nprot.2014.144.

Schwarz, E., Bahn, S., 2008. The utility of biomarker discovery approaches for the detection of disease mechanisms in psychiatric disorders. Br J Pharmacol 153 (Suppl. 1), S133–S136. http://dx.doi.org/10.1038/sj.bjp.0707658.

Scott, J.D., Pawson, T., 2009. Cell signaling in space and time: where proteins come together and when they're apart. Science 326 (5957), 1220–1224. http://dx.doi.org/10.1126/science.1175668.

Shannon, P., Markiel, A., Ozier, O., Baliga, N.S., Wang, J.T., Ramage, D., et al., 2003. Cytoscape: a software environment for integrated models of biomolecular interaction networks. Genome Res 13 (11), 2498–2504. http://dx.doi.org/10.1101/gr.1239303.

Silva, A.J., Stevens, C.F., Tonegawa, S., Wang, Y., 1992. Deficient hippocampal long-term potentiation in alpha-calcium-calmodulin kinase II mutant mice. Science 257 (5067), 201–206. Retrieved from http://www.ncbi.nlm.nih.gov/pubmed/1378648.

Tam, G.W., Redon, R., Carter, N.P., Grant, S.G., 2009. The role of DNA copy number variation in schizophrenia. Biol Psychiatry Retrieved from http://www.ncbi.nlm.nih.gov/entrez/query.fcgi?cmd=Retrieve&db=PubMed&dopt=Citation&list_uids=19748074.

Toro, C.T., Deakin, J.F., 2007. Adult neurogenesis and schizophrenia: a window on abnormal early brain development? Schizophr Res 90 (1-3), 1–14. Retrieved from http://www.ncbi.nlm.nih.gov/entrez/query.fcgi?cmd=Retrieve&db=PubMed&dopt=Citation&list_uids=17123784.

Trinidad, J.C., Specht, C.G., Thalhammer, A., Schoepfer, R., Burlingame, A.L., 2006. Comprehensive identification of phosphorylation sites in postsynaptic density preparations. Mol Cell Proteomics, 5 (5), 914–922. Retrieved from http://www.ncbi.nlm.nih.gov/entrez/query.fcgi?cmd=Retrieve&db=PubMed&dopt=Citation&list_uids=16452087.

Trinidad, J.C., Thalhammer, A., Specht, C.G., Lynn, A.J., Baker, P.R., Schoepfer, R., et al., 2008. Quantitative analysis of synaptic phosphorylation and protein expression. Mol Cell Proteomics 7 (4), 684–696. Retrieved from http://www.ncbi.nlm.nih.gov/entrez/query.fcgi?cmd=Retrieve&db=PubMed&dopt=Citation&list_uids=18056256.

Wang, Q., Charych, E.I., Pulito, V.L., Lee, J.B., Graziane, N.M., Crozier, R.A., et al., 2010. The psychiatric disease risk factors DISC1 and TNIK interact to regulate synapse composition and function. Mol Psychiatry, 1–18. Retrieved from http://www.ncbi.nlm.nih.gov/entrez/query.fcgi?cmd=Retrieve&db=PubMed&dopt=Citation&list_uids=20838393.

Weickert, C.S., Weickert, T.W., Pillai, A., Buckley, P.F., 2013. Biomarkers in schizophrenia: a brief conceptual consideration. Dis Markers 35 (1), 3–9. http://dx.doi.org/10.1155/2013/510402.

Weisbrod, C.R., Chavez, J.D., Eng, J.K., Yang, L., Zheng, C., Bruce, J.E., 2013. In vivo protein interaction network identified with a novel real-time cross-linked peptide identification strategy. J Proteome Res 12 (4), 1569–1579. http://dx.doi.org/10.1021/pr3011638.

Wilkins, M.R., Sanchez, J.C., Gooley, A.A., Appel, R.D., Humphery-Smith, I., Hochstrasser, D.F., et al., 1996. Progress with proteome projects: why all proteins expressed by a genome should be identified and how to do it. Biotechnol Genet Eng Rev 13, 19–50. Retrieved from http://www.ncbi.nlm.nih.gov/pubmed/8948108.

Wolters, D.A., Washburn, M.P., Yates 3rd, J.R., 2001. An automated multidimensional protein identification technology for shotgun proteomics. Anal Chem 73 (23), 5683–5690. Retrieved from http://www.ncbi.nlm.nih.gov/pubmed/11774908.

Wuchty, S., 2001. Scale-free behavior in protein domain networks. Mol Biol Evol 18 (9), 1694–1702. Retrieved from http://www.ncbi.nlm.nih.gov/pubmed/11504849.

Xu, B., Ionita-Laza, I., Roos, J.L., Boone, B., Woodrick, S., Sun, Y., et al., 2012. De novo gene mutations highlight patterns of genetic and neural complexity in schizophrenia. Nat Genet 44 (12), 1365–1369. http://dx.doi.org/10.1038/ng.2446.

Yates, J.R., Ruse, C.I., Nakorchevsky, A., 2009. Proteomics by mass spectrometry: approaches, advances, and applications. Annu Rev Biomed Eng 11, 49–79. http://dx.doi.org/10.1146/annurev-bioeng-061008-124934.

Zhao, X., D'Arca, D., Lim, W.K., Brahmachary, M., Carro, M.S., Ludwig, T., et al., 2009. The N-Myc-DLL3 cascade is suppressed by the ubiquitin ligase Huwe1 to inhibit proliferation and promote neurogenesis in the developing brain. Dev Cell 17 (2), 210–221. http://dx.doi.org/10.1016/j.devcel.2009.07.009.

PART V

THE ELECTRO-PHYSIOLOGICAL BASIS OF SCHIZOPHRENIA

CHAPTER 13

EEG AND MEG PROBES OF SCHIZOPHRENIA PATHOPHYSIOLOGY

E. Neustadter[1], K. Mathiak[2,3] and B.I. Turetsky[4]

[1]Yale University School of Medicine, New Haven, CT, United States [2]Department of Psychiatry, Psychotherapy, and Psychosomatics, Medical School, RWTH Aachen University, Aachen, Germany [3]Jülich Aachen Research Alliance-Translational Brain Medicine, RWTH Aachen University, Aachen, Germany [4]Neuropsychiatry Program, Department of Psychiatry, Perelman School of Medicine, University of Pennsylvania, Philadelphia, PA, United States

CHAPTER OUTLINE

Introduction	213
Measures of Inhibitory Failure	**215**
P50 Auditory Sensory Gating	215
Prepulse Inhibition of Startle	218
Measures of Aberrant Salience Detection	**220**
Mismatch Negativity	220
P300	223
Measures of Abnormal Gamma Band Oscillations	**225**
Measuring Neural Oscillations	226
Neurobiology of Gamma Oscillations	226
Gamma Oscillations and Schizophrenia	228
Conclusion	**229**
References	**230**

INTRODUCTION

Schizophrenia is a complex neuropsychiatric disorder characterized by positive symptoms (eg, hallucinations and delusions), negative symptoms (eg, flattened affect), and disorganized thoughts and behavior. It is also associated with a variety of neurocognitive deficits, ranging from higher-order working memory and executive function to early preattentive information processing. Electroencephalography (EEG) and magnetoencephalography (MEG) have been widely used to probe the neurophysiological mechanisms of cognition and information processing in both healthy and clinical populations. The aim of this chapter is to review current EEG and MEG findings in schizophrenia to demonstrate the utility of these measures as quantitative pathophysiological disease biomarkers and to highlight their potential clinical applications.

EEG is noninvasive methodology that records brain electric potentials on the scalp surface that arise from underlying cortical pyramidal neurons. The EEG changes regularly over time, in an oscillatory manner, and reflects both intrinsic neural activity and responses to specific internal and external events. EEG is a productive methodology to study neurocognitive function because it represents the coordinated neural activity that underlies perceptual and cognitive processes. Unlike other neuroimaging modalities, it provides a real-time measure of neural activity with millisecond-level temporal resolution, enabling it to capture complex neural dynamics at very early stages of sensory perception and information processing. Nevertheless, electroencephalography is not without methodological shortcomings. Because EEG only reflects the summed activity of similarly oriented cortical neurons, it may only capture a portion of the relevant neural activity associated with a particular process. EEG also suffers from poor spatial resolution due to differences in electrical conductivity between scalp, skull, and intracranial structures, which introduce distortions. Consequently, the sources of changing scalp-recorded electrical activity cannot be precisely localized within particular brain regions.

Magnetoencephalography (MEG), in contrast, measures the magnetic fields that are induced by neural currents. These exit the body without further distortion. Therefore, this technique offers better spatial resolution. Despite reflecting the same underlying physical principles, practice shows that EEG and MEG also exhibit different sensitivities to different neural processes. MEG is particularly well suited for studying neural currents running tangential to the skull, such as in auditory or somatosensory cortices. Even then, however, the observed fields are on the order of femtotesla (10^{-15}T), requiring sensitive supra-conducting quantum interference device (SQUID) detectors and expensive shielding chambers for their measurement, which renders MEG much less accessible for routine clinical applications. Thus, MEG often focuses on relatively narrow research questions, such as the neural mechanisms underlying the etiology of abnormal sensory gating in schizophrenia, at the level of the auditory cortex.

Traditionally, electrophysiological research in schizophrenia has focused on the assessment of event-related potentials (ERPs), which are understood as time-locked voltage changes in response to a stimulus event, reflecting a specific neuronal or psychological process (Kappenman and Luck, 2012). An alternative and complimentary approach considers EEG activity as the summation of superimposed oscillations that reflect synchronous neural activity within and across functional brain networks (Winterer and McCarley, 2011). Whereas the ERP approach assesses how stimuli cause voltage deflections, the event-related oscillation approach examines how stimuli cause event-related spectral perturbations that alter the dynamics of endogenous neural oscillations.

In this chapter, we review those ERP and neural oscillation measures that have been most widely applied to the study of neurocognitive dysfunction in schizophrenia. Although schizophrenia is associated with a range of neuropsychological deficits, a breakdown in the processes that regulate the inflow of information from the environment appears critical. Successful processing of sensory inputs requires the ability to inhibit responses to redundant or noninformative stimuli and, reciprocally, to allocate attentional resources to novel or salient stimuli (Turetsky et al., 2007). Both of these processes are impaired in schizophrenia, and electrophysiological measures of both inhibitory failure (ie, P50 auditory sensory gating and prepulse inhibition of the startle response (PPI)) and aberrant salience detection (ie, mismatch negativity (MMN) and the auditory P300 event-related potential) are discussed. More recently, abnormal oscillations, particularly in the gamma range frequency (>30Hz), which are associated with a range of higher-order cognitive processes, have been implicated in the pathophysiology of schizophrenia (Uhlhaas and Singer, 2010). Therefore, research on abnormal gamma oscillations is also reviewed. A more detailed examination of gamma oscillation deficits within the context of animal models of schizophrenia can be found in an accompanying chapter by White and Siegel in this text. For each electrophysiological measure, the following are discussed: theoretical framework, methodological considerations, current understanding of the underlying neurobiology, and the nature of the abnormality in schizophrenia and its putative role as a pathophysiological biomarker for the disorder.

MEASURES OF INHIBITORY FAILURE
P50 AUDITORY SENSORY GATING

P50 is an evoked potential component that occurs approximately 50ms after the presentation of an auditory stimulus. P50 sensory gating is assessed through a paired-click paradigm, wherein the amplitude of the P50 response to the second of two auditory clicks is reduced (Adler et al., 1982). This amplitude reduction is understood as a brain mechanism that filters out, or suppresses, neural responses to repetitive or irrelevant stimuli. The standard paired-click paradigm uses a conditioning stimulus (S1) and a test stimulus (S2) presented 500ms apart, with a 10-s interval between click pairs to ensure that P50 suppression in one trial does not affect the next. The amplitudes of the P50 responses to S1 and S2 are identified, and P50 gating is then assessed by calculating the S2:S1 amplitude ratio. A low S2:S1 ratio thus signifies a robust gating response, whereas a higher ratio reflects deficient auditory sensory gating. A somewhat arbitrary cutoff of <0.5 has been proposed as a "normal" gating response (Potter et al., 2006).

One important consideration in interpreting differences in gating ratios between controls and patients is the relative contributions of S1 and S2 amplitudes. Higher S2:S1 ratios resulting from larger S2 responses are indicative of impaired gating, whereas similar ratios due to small S1 responses may reflect impaired initial processing of the unconditioned auditory stimulus rather than a gating deficiency. Given that patients with schizophrenia often exhibit smaller S1 (with or without larger S2) responses, some have raised concerns about the use of this ratio parameter as a reliable index of gating deficits associated with the disorder (Olincy et al., 2010; Fig. 13.1).

Neurobiology of P50 suppression

P50 suppression appears to be mediated by a widely distributed neural network, including the hippocampus, the temporo-parietal region (Brodmann's areas 22 and 2), the prefrontal cortex (Brodmann's areas 6 and 24) (Grunwald et al., 2003), and the primary auditory cortex (Reite et al., 1987). A significant role for the hippocampus and its cholinergic inputs from the medial septal nucleus in gating the P50 response is supported by evidence from animal models and pharmacological and genetic linkage studies. Hippocampal CA3-CA4 interneurons receive cholinergic inputs from the medial septal nucleus via low-affinity alpha7-nicotinic receptors (Javitt and Freedman, 2015). Nicotinic antagonists targeting these receptors disrupt P20-N40 suppression, the rodent analog of human P50 gating (Luntz-Leybman et al., 1992), and, conversely, a selective alpha7-nicotinic partial agonist increases sensory gating in DBA mice, a strain that normally exhibits no gating response (Stevens et al., 1998). Acute administration of this same nicotinic agonist also normalized S2:S1 gating ratios in schizophrenia patients (Olincy et al., 2006). Similar transient normalization of gating was previously observed after cigarette smoking (Adler et al., 1993). Family studies have also demonstrated a genetic linkage in humans between P50 suppression and the *CHRNA7* alpha7-nicotinic receptor gene locus (Freedman et al., 1997).

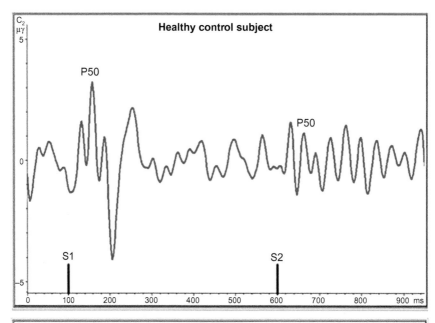
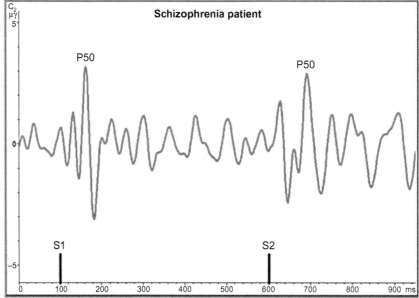

FIGURE 13.1

P50 responses to paired clicks (S1 and S2) in a healthy control subject (top) and a schizophrenia patient (bottom).

These findings suggest the following mechanistic model of P50 suppression. When paired clicks are presented, S1 is processed normally through the hippocampus. However, S1 also causes concomitant activation of CA3 GABAergic interneurons through alpha7-nicotinic receptor stimulation from the septal nucleus. The release of GABA from these interneurons inhibits postsynaptic pyramidal neurons in CA3. Thus, when S2 is presented and processed through the auditory pathway, it is suppressed in the CA3-CA4 region of the hippocampus due to the persistent inhibitory activity initiated by S1.

P50 suppression and schizophrenia

Many studies have shown that patients with schizophrenia have impaired sensory gating, as evidenced by increased S2:S1 ratios compared to healthy controls. Given our current understanding of the mechanisms underlying P50 gating, and given the ability of nicotinic agonists to transiently normalize it, this abnormality presumably reflects aberrant cholinergic transmission in the hippocampus. A meta-analysis of 20 studies found the effect size for this deficit to be high (pooled standardized effect size = -1.56), such that the magnitude of the P50 gating deficit is comparable to the most robust findings in other neuroimaging and electrophysiological studies in schizophrenia (Bramon et al., 2004).

Although impaired P50 gating has been related to cognitive deficits in patients, its association with clinical symptomology is less clear. P50 suppression deficits have been associated with reduced performance on neuropsychological measures of attention (Erwin et al., 1998). Smith et al. (2010) found associations between P50 gating ratios and measures of attention, working memory, and verbal and visual long-delay memory. However, this association was fully accounted for by S1 amplitude, suggesting that these deficits were manifestations of poor information encoding as opposed to impaired sensory gating. More recently, Smucny et al. (2013) found that increased P50 ratios correlated with the magnitude of noise-induced increases in reaction time in an attention task. This correlation was fully accounted for by magnitude of the P50 amplitude to S2, and not S1, suggesting that distractibility in patients is directly associated with impaired P50 suppression. However, in a review of clinical correlates of P50 sensory gating abnormalities, Potter et al. (2006) failed to find a consistent relationship between P50 suppression and symptomology. Furthermore, Boutros et al. (2009) and Olincy et al. (2010) failed to find any relationship between P50 suppression and clinical subtype or clinical status in patients with schizophrenia.

Deficits of the neuromagnetic analogue of P50 suppression can be replicated with MEG in schizophrenia populations. Importantly, MEG can effectively separate responses from the left and right auditory cortex, but EEG cannot. Doing so revealed a lateralization of the characteristic schizophrenia deficit. The disorder produced more pronounced deficits in the left hemisphere that were associated with positive symptoms, whereas negative symptoms were associated with right hemispheric dysfunction (Thoma et al., 2003). Analyses of neuromagnetic oscillations associated with the P50 paradigm indicated that reduced alpha band activity (~10 to 13 Hz) before onset of the second stimulus predicted the subsequent neuromagnetic P50 response and associated energy consumption (Popov et al., 2011b; Mathiak et al., 2011). A specific cognitive training program was also able to normalize the neuromagnetic suppression index (Popov et al., 2011a). However, given the quantitative limitations of MEG, these intriguing research findings from relatively small samples need to be confirmed in larger populations.

In summary, current evidence suggests that P50 suppression is a surrogate marker of a discrete neurocognitive process that functions to limit sensory distractibility and is underpinned by specific genetic and neural processes (ie, those mediating cholinergic pathways in the hippocampus). Given that patients' P50 gating deficits are not clearly related to acute symptomology and given that similar deficits are observed in clinically unaffected first-degree relatives (Siegel et al., 1984), the deficit may constitute a vulnerability marker, or endophenotype, of schizophrenia, as opposed to an index of current disease status (Potter et al., 2006).

PREPULSE INHIBITION OF STARTLE

Prepulse inhibition of the startle response (PPI) is a normal inhibitory phenomenon whereby a weak prestimulus, or prepulse, reduces the magnitude of the musculoskeletal response to a subsequent intense startling stimulus. PPI is an automatic preconscious process that is evident across species and is thought to help animals navigate their environment by minimizing the allocation of attentional resources to nonsalient stimuli. As such, PPI is a highly heritable phenotypic measure—one that is deficient across multiple neuropsychiatric disorders (Braff et al., 2001) such that impaired individuals demonstrate a persistent, strong response to startling stimuli even when they are preceded by prepulses. PPI is one of the most robust and reproducible markers of inhibitory dysfunction in schizophrenia (Swerdlow et al., 2001), and it has been widely used as a translational model to probe the genetic and neurobiological underpinnings of this information-processing deficit.

PPI methodology

The PPI paradigm is a variant of several paired-stimulus paradigms that assess automatic, preattentive sensory gating in which the motor and/or perceptual response to a target stimulus is inhibited when preceded by a conditioning stimulus. Typically, in human studies, acoustic prepulses are delivered between 30 and 120ms prior to a startling white noise burst, and responses are quantified by the magnitude of orbicularis oculi contraction (eye blink), as measured on the electromyogram. Prepulse inhibition is quantified by calculating the %PPI, which is the relative decrease in eye blink magnitude when the startling stimulus is preceded by a prepulse compared to the startle stimulus alone. A lower %PPI reflects reduced or impaired sensorimotor gating. Importantly, despite both conceptual and methodological similarities between PPI and P50, the two measures appear to be uncorrelated across schizophrenia subjects, suggesting that they denote entirely independent neural processes (Braff et al., 2007).

PPI is a relative stable neurobiological marker and exhibits robust test–retest reliability (interclass correlations >0.90) in healthy subjects over the course of several months (Cadenhead et al., 1999). At the same time, PPI is very sensitive to changes in stimulus parameters and testing conditions such as stimulus type, intensities of prepulse and pulse, and timing of prepulse and pulse delivery. PPI is also sensitive to a variety of state and trait variables, including medications (Swerdlow et al., 2006) and cigarette smoking (Braff et al., 2001).

Neurobiology of PPI

Evidence from human and rodent studies suggests that PPI originates in the reticular brainstem and engages a large subcortical and cortical network, implicating cortico-striato-pallido-pontine (CSPP) circuitry (Swerdlow et al., 2001). Although PPI is generated in the pons without any forebrain requirement, it can be reduced or eliminated by pharmacological manipulations (eg, D1 receptor blockage) in the prefrontal cortex (Ellenbroek et al., 1996), suggesting that distal cortical regions can substantially regulate or modulate PPI.

Javitt and Freedman (2015) recently provided the following mechanistic pathway for PPI: a pulse stimulates pontine reticular formation neurons to activate motor neurons to produce a startle response. A prepulse activates circuitry from the hippocampus that inhibits subsequent startle responses via the nucleus accumbens through inhibition of pedunculopontine (PPTg) neurons that regulate the excitability of reticular formation neurons during startle. In schizophrenia, PPI deficits are thought to result from disruptions in this regulatory circuitry, such that decreased hippocampal inhibition and increased dopaminergic activation converging on the nucleus accumbens result in maintained startle responses to pulses following weak prepulse stimuli.

Evidence from rodent and human studies indicates that PPI is a substantially heritable trait (Braff, 2010), and evidence from a human twin study suggests that more than 50% of the variance could be accounted for by genetic factors. Mapping of quantitative trait loci in rodents has identified multiple genes spanning several chromosomes (eg, Joober et al., 2002). Although there are certainly multiple genes that similarly regulate PPI in humans, only a subset of these are likely to be relevant to the abnormality as it presents in schizophrenia. It is notable, in this regard, that an association analysis of 94 candidate genes and schizophrenia endophenotypes observed a relationship between the NRGL-ERBB4 risk genes and PPI deficits in schizophrenia (Greenwood et al., 2012).

PPI and schizophrenia

Many studies have confirmed the presence of PPI deficits across the schizophrenia spectrum (Braff, 2010), including patients with schizophrenia, individuals with schizotypal personality disorder, unaffected first-degree unaffected relatives, and clinical ultra-high-risk (UHR) individuals. The presence of these deficits in clinically unaffected samples without overt psychosis suggests that PPI is a neurobiological vulnerability or risk marker, as opposed to a biomarker, for frank schizophrenia. Considering PPI as a heritable risk marker is consistent with findings indicating that PPI deficits have weak or no correlations with schizophrenia symptomology as measured by SANS or SAPS ratings (Swerdlow et al., 2006). Swerdlow et al. (2006) also reported no correlations between PPI deficits and specific neurocognitive functions, but they did observe an association with global measures of functional impairment. The relationship between impaired PPI and global functioning, in the absence of any associations with specific symptoms, may be explained in part by the effect of PPI deficits on the subjective severity of symptoms and social cognition. For example, Kumari et al. (2008) found that marked PPI deficits in patients were associated not with the presence of auditory hallucinations per se, but with the subjective experience of having no control over their occurrence. Thus, PPI may mediate the relationship between clinical symptoms and functional impairment.

As noted, PPI deficits are not unique to schizophrenia. However, although abnormal sensorimotor gating is reflective of disrupted CSPP circuitry across multiple neuropsychiatric disorders, this is unlikely due to a single common pathophysiological mechanism. With regard to schizophrenia, many regions regulating PPI map onto structures previously implicated in the disease, such as the prefrontal cortex and the hippocampus (Javitt and Freedman, 2015). Genetic and pharmacological studies of PPI are also consistent with several neurotransmitter systems implicated in schizophrenic pathology. The NRGL–ERBB4 complex, which controls glutamatergic synapse maturation and plasticity, has been genetically linked to PPI deficits in schizophrenia (Greenwood et al., 2012). Also, dopamine agonists have been shown to produce "schizophrenia-like" PPI deficits in humans and rodents (Braff, 2010), and PPI is increased in patients using atypical antipsychotics (Swerdlow et al., 2006). Still, although it is known that atypical antipsychotics act on D2 receptors, it is unclear if these are enhancing PPI by optimizing functionally intact gating mechanisms, or if they are actually normalizing or "treating" the primary dysfunctional mechanism. Although such questions remain, PPI continues to be a promising endophenotypic marker for studies designed to further elucidate the genetic and pathophysiological nature of forebrain circuitry dysfunction in schizophrenia.

MEASURES OF ABERRANT SALIENCE DETECTION
MISMATCH NEGATIVITY

In the auditory domain, mismatch negativity (MMN) is a negative deflection in the evoked potential that occurs when a series of identical repetitive auditory stimuli is interrupted by an oddball, or deviant, stimulus that is different in some physical characteristic such as pitch or latency. The MMN occurs 50–150 ms after stimulus presentation and represents a preattentive form of sensory discrimination that relies on an automatic echoic memory comparison to detect a "mismatch" between standard and deviant stimuli within an auditory stream. Functionally, MMN is thought to reflect an early-phase sensory discrimination mechanism that enables organisms to reorient themselves to, and attend to, salient or novel stimuli in the auditory environment (Belger et al., 2012).

MMN is a widely used measure of cortical auditory function and is a sensitive marker of N-methyl D-aspartate (NMDA) receptor functioning (Javitt and Freedman, 2015). Clinically, MMN has been utilized across a range of neurological and neuropsychiatric disorders including schizophrenia, dyslexia, stroke, autism, and epilepsy to probe abnormalities in frontotemporal brain systems (for review, see Näätänen et al., 2012). In schizophrenia research, MMN has also shown exciting promise as a potential neurophysiological endophenotye (Light et al., 2015), as a quantitative biomarker to improve prediction of the onset of psychosis in high-risk individuals (Belger et al., 2012) and to track and predict responses to novel treatment interventions (Light and Swerdlow, 2014).

MMN measurement

In a typical MMN paradigm, an unchanging "standard" tone is presented in relativity rapid succession (eg, 500 ms interstimulus interval) to ensure that an echoic memory trace is still active when the subsequent tone is presented (Turetsky et al., 2007). Attention is specifically directed *away* from the tones by having the subject watch a silent movie or perform a visual task. Approximately 10% of the tones are, at random, "deviant," such that they differ in one or more physical characteristics (eg, pitch, duration, loudness) from the standard tone. The MMN deflection is observed approximately 150–200 ms after the deviant tone, with a maximal response at frontocentral recording sites. MMN is usually assessed from a difference wave generated by subtracting the averaged standard evoked potential response from the averaged deviant response (Fig. 13.2).

Methodologically, auditory MMN possesses several characteristics that make it a promising biomarker for schizophrenia research. MMN has been recorded in several mammal species, including monkeys (eg, Javitt et al., 1992) and mice (eg, Umbricht et al., 2005), and it has recently been utilized as a neurophysiological index of sensory and cognitive function in a nonhuman primate model of schizophrenia (Gil-da-Costa et al., 2013). MMN can be easily assessed in humans and, as a repeated measure, demonstrates high test–retest reliability over 1 year (interclass correlation coefficient ~0.90)

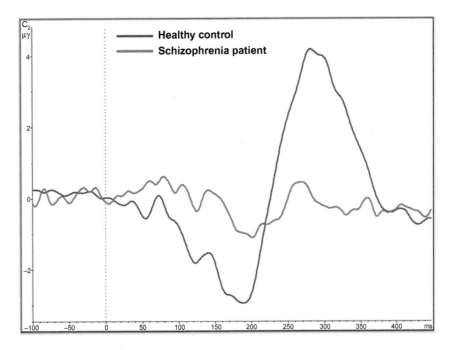

FIGURE 13.2

Difference waveforms showing MMN response in a schizophrenia patient and a healthy control subject.

in both healthy controls and patients with schizophrenia (Light et al., 2012). Also, because MMN is generated preconsciously, it requires minimal task demands and can be reliably assessed in low-functioning patients.

Neurobiology of MMN

Converging evidence from magnetoencephalography, ERP source modeling, fMRI, and PET imaging suggests that MMN is generated bilaterally in the primary and secondary auditory cortices, with perhaps a later contribution from the inferior frontal cortex. It is suggested that auditory and frontal processes, respectively, constitute two subcomponents of the MMN response; the earlier superior temporal gyrus response reflects a more basic automatic detection of physical change in the auditory stimulus, whereas the frontal response reflects subsequent reorientation or shifting of attention (Näätänen et al., 2012, Belger et al., 2012).

Pharmacological studies demonstrate the critical importance of intact NMDA receptor functioning in generating MMN. In monkeys, both competitive and noncompetitive NMDA antagonists specifically inhibit MMN generation without affecting prior obligatory activity in the primary auditory cortex (Javitt et al., 1996). Similarly, the NMDA antagonist ketamine reduces MMN amplitude in healthy human subjects (Umbricht et al., 2000). Notably, baseline MMN amplitude has also been shown to predict which otherwise healthy subjects will develop transient psychotic symptoms after ketamine administration (Umbricht et al., 2002), thus suggesting a link between the NMDA-mediated MMN response and the NMDA model of psychosis.

Mechanistically, NMDA antagonists are thought to act, in this regard, by blocking NMDA receptors located specifically on fast-spiking GABAergic interneurons. The effect of this receptor blockade is the loss of recurrent GABAergic inhibition of pyramidal cells. This leads to a state of neuronal disinhibition within the cortical microcircuitry. Such disruption in the excitation–inhibition balance results in neuronal hyperexcitation that, on a phenomenological level, results in the degradation of stimulus-selective responses and the imprecision of stored information (Murray et al., 2014). One might understand this as reduced signal-to-noise during stimulus processing, such that salient changes in stimulus characteristics are more difficult to detect.

Mismatch negativity and schizophrenia

MMN deficits in schizophrenia were first reported by Shelley et al. (1991) and have since been demonstrated to be a robust reproducible deficit associated with the disorder. In a meta-analysis including 32 MMN studies, Umbricht and Krljes (2005) found the effect of the deficit to be large (ES = 0.99). The same study found MMN to changes in stimulus duration to be more impaired than MMN to changes in pitch. Additionally, MMN to pitch deviance significantly correlated with duration of illness, suggesting that it tracks ongoing neuropathological changes in the auditory cortex associated with disease progression.

MMN has also been proposed as a putative schizophrenia endophenotype (Light et al., 2015) because MMN deficits are heritable (Hall et al., 2006a,b), evident in both clinically unaffected family members (Jessen et al., 2001) and young people at risk of psychosis (Schreiber et al., 1992), and not consistently related to clinical symptomology (Umbricht and Krljes, 2005; Light et al., 2012). However, in patients, MMN deficits are associated with global impairments in functioning (Light and Braff, 2005). In light of the fact that MMN impairments are associated with behavioral deficits in auditory discrimination and tone matching, Javitt and Freedman (2015) suggest that MMN deficits likely impact global functioning by impairing patients' abilities to orient to important environmental and social stimuli (eg, auditory modulations conveying pertinent emotional information). Neurobiologically, MMN deficits in schizophrenia were found to be associated with reduced activation in medial frontal brain regions, including the cingulate cortex and medial frontal gyri, as well as the auditory cortex. This is consistent with growing evidence that neurocognitive deficits are associated with frontal brain abnormalities in the disorder.

As for P50 suppression, MEG is able to measure a neuromagnetic analogue of MMN. As an extension of the EEG method, bilateral sources can be calculated. In particular, co-localization of MEG with fMRI sources is possible. This approach has revealed disturbed hemispheric lateralization in patients with schizophrenia (Kircher et al., 2004). In a direct comparison with EEG, MEG was found to be more sensitive to MMN deficits in schizophrenia, yielding effect sizes of up to 1.5 for impaired temporal processing (Thönnessen et al., 2008). MEG findings support the view that impaired MMN in schizophrenia can be linked to dysfunction at the level of the auditory cortex.

More recently, MMN has also demonstrated exciting clinical utility for patients and those at clinical high risk. Bodatsch et al. (2011) found that at baseline, individuals at clinical high risk who converted to psychosis have lower MMN amplitudes than those who did not. Furthermore, Perez et al. (2014b) found that MMN predicted the time to psychosis onset in clinical high-risk converters. In patients, phonetic MMN predicted degree of improvement after social skills acquisition (Kawakubo et al., 2007), and Perez et al. (2014a) found that patients with higher pretraining MMN demonstrated greater improvements after targeted cognitive training (TCT). Thus, in addition to its utility as a research tool to probe the genetic and neurobiological underpinnings of information-processing deficits in schizophrenia, MMN also shows promise as a quantitative clinical biomarker to predict and track individuals at clinical high risk and to index treatment responses to novel psychosocial and neurocognitive interventions.

P300

The P300 event-related potential indexes higher-order cognitive processes associated with updating mental representations of the external environment and is elicited by infrequent (task-relevant or novel) stimuli. Like MMN, P300 is usually assessed through variations of the oddball paradigm; however, whereas the former assesses preattentive cognitive information processing, P300 generation relies on a subject's active attention to task demands (Polich, 2007). Moreover, although MMN reflects a sensory-based mismatch detection process, theoretical accounts of P300 posit that P300 is generated through an attention-driven comparison process in working memory (Polich, 2012). P300 does not represent a unitary neurocognitive process; instead, it is a composite ERP waveform that represents temporally distributed activity in frontal and temporal-parietal brain regions related to attention and memory. The P300 wave is usually deconstructed into two functional components, the P3a and P3b (Squire et al., 1975). The P3a is an earlier frontal response reflecting the engagement of attentional processes, whereas the P3b is a later parietal scalp maximum that is thought to reflect stimulus evaluation associated with environmental representation and memory maintenance. P300 has been studied extensively in clinical contexts, and abnormalities in P300 generation reflect underlying deficits in salience detection (Turetsky et al., 2015), attentional resource allocation, and higher-order information processing (McCarley et al., 2002).

P300 methodology

P300 is primarily assessed using an oddball paradigm in either the auditory or the visual sensory modality. A common experimental procedure entails subjects physically or mentally responding to an infrequent stimulus designated the "target," which is randomly embedded in a background of "standard" stimuli. Target stimuli elicit a large positive-going potential with a maximum latency of 300–400 ms and whose scalp distribution increases in magnitude from medial frontal to parietal electrode sites. P3a and P3b subcomponents can be distinguished by manipulating the experimental conditions; P3a can be elicited by novel or deviant stimuli even in the absence of a detection task, whereas the P3b is only produced when subjects are instructed to actively detect target stimuli (Javitt et al., 2008). One common task used to assess these subcomponents is the three-stimulus oddball, in which subjects respond solely to infrequent target tones embedded in a stimulus train that also includes standard tones and infrequent novel "distractor" stimuli. In this task, ERP responses to targets reflect P3b, whereas ERP responses to distractors reflect P3a.

P300 amplitudes demonstrate high test–retest reliability in both control subjects and schizophrenia patients (Mathalon et al., 2000). One methodological concern with severely ill patients is their level of cooperation and capacity to attend to target stimuli for the duration of the task. Another is that P300 amplitude is sensitive to several state and trait characteristics. P300 is smaller in men and has reduced amplitude and prolonged latency in the elderly. P300 amplitude is also sensitive to level of fatigue, tonic exercise, and acute nicotine use (for review of biological and cognitive determinants of P300, see Polich and Kok, 1995).

Neurobiology of P300

Generic P300 is produced by several simultaneous bilateral generators that include both cortical and subcortical structures (Winterer and McCarley, 2011). Findings from intracranial recordings, fMRI, and EEG source localization implicate the hippocampus (Halgren et al., 1980), thalamus (Yingling and Hosobuchi, 1984), prefrontal cortex (Winterer et al., 2001), and parietal and temporal lobes (Linden, 2005). Current mechanistic models suggest that the composite P300 waveform reflects electrical activity generated by interactions between frontal, hippocampal, and temporal-parietal regions. Specifically, perception of deviance initiates frontal lobe activity related to attentional reorientation. P3a reflects those processes, likely generated in the anterior cingulate, by which a novel stimulus sufficiently engages attentional resources, such that stimulus representation in working memory is updated. P3b, elicited only by a salient target stimulus, reflects further processing in temporal-parietal regions, wherein the updated stimulus representation undergoes further memory storage operations initiated in the hippocampus (Polich, 2007).

Although considerable progress has been made, the precise neural basis of P300 generation is still largely unknown. The dual-transmitter hypothesis (Polich, 2007) posits that the frontal activity responsible for P3a generation is mediated by dopaminergic activity, whereas temporal-parietal processes indexed by P3b are associated with norepinephrine activity. Support for the role of dopamine comes from findings that patients with Parkinson disease with reduced dopamine levels produce smaller p3a responses (Stanzione et al., 1991). Evidence for the role of norepinephrine derives primarily from pharmacological studies of animals implicating the neuromodulatory locus coeruleus–norepinephrine system (Nieuwenhuis et al., 2005) and is in agreement with the functional anatomy of dense NE inputs to the temporal-parietal cortex. A more recent pharmacological study of healthy humans subjects (Watson et al., 2009) found that ketamine, a glutamate receptor antagonist, and thiopental, a GABA-A receptor agonist, both reduced P3a and P3b amplitudes, suggesting that NMDA and GABA receptor systems also modulate P300 generation.

P300 and schizophrenia

Abnormalities associated with P300 are some of the most robust, consistent, and reproducible electrophysiological deficits in schizophrenia. Meta-analyses have found both reduced amplitude and prolonged latency of the P300 waveform in schizophrenia patients compared to healthy controls (eg, Bramon et al., 2004). Although these P300 deficits have been shown to vary with symptomology and to demonstrate some state-dependent modulation (Mathalon et al., 2000), they are relatively stable and primarily exhibit trait-like characteristics of a disease biomarker (Turetsky et al., 2015). Meta-analyses have found that the large effect size for auditory P300 amplitude deficits (~0.85) was not significantly influenced by either level of symptomology or medication status (Fig. 13.3).

P300 amplitude exhibits relatively strong genetic heritability of approximately ~0.60 (O'Connor et al., 1994). Consistent with this, deficits are also found in healthy first-degree relatives of schizophrenia patients (Turetsky et al., 2000), in people with schizotypal personality disorder (Gassab et al., 2006), and in young people at clinical high risk for psychosis (Bramon et al., 2008). This suggests that P300 amplitude decrements are a viable genetic endophenotype for schizophrenia. However, recent evidence also suggests that P300 amplitude may have clinical utility as a sensitive predictor of transition to psychosis (Nieman et al., 2014; van Tricht et al., 2010). Although P300 is already reduced during the prepsychotic prodromal state, the magnitude of this decrement indicates both the likelihood of conversion to overt psychosis and the relative proximity of this occurrence in an individual at high risk.

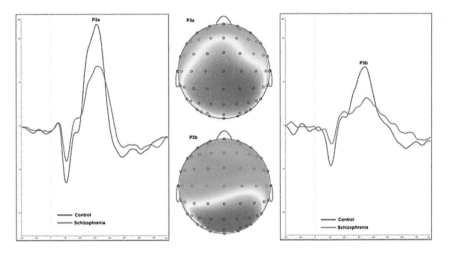

FIGURE 13.3

P3a and P3b waveforms and associated scalp topographies in schizophrenia patients and healthy control subjects.

It is worth noting that P300 abnormalities are not unique to the schizophrenia spectrum, and P300 deficits have also been reported in Alzheimer disease, ADHD, alcoholism, unipolar depression, and bipolar disorder. However, given the multifactorial nature of the waveform, it is likely that different pathological mechanisms underlie P300 abnormalities in different disorders. In chronic schizophrenia patients, it has been shown that the magnitude of P300 deficits is correlated with reduced gray matter in the left posterior superior temporal gyrus (STG), and the same association has been observed in first-episode patients with schizophrenia, but not in those with affective psychosis (McCarley et al., 2002). These findings are in agreement with EEG source localization studies associating P300 abnormalities deficits in schizophrenia with broad reduced activity in left hemispheric regions, including the prefrontal cortex, posterior cingulum, and temporal lobe (Winterer et al., 2001). Taken together, these findings suggest that lateralized left hemisphere P300 deficits may be relatively specific to schizophrenia and that the magnitude of P300 abnormalities may be a relatively specific index of structural abnormalities in the temporal lobe associated with disease progression.

MEASURES OF ABNORMAL GAMMA BAND OSCILLATIONS

Neural oscillations are a fundamental mechanism that enables the synchronization of neural activity within and across brain regions and promotes the precise temporal coordination of neural processes underlying cognition, memory, perception, and behavior. Neural oscillations are an emergent property of neural networks generated by coordinated synaptic transmission across neuronal populations (Ford et al., 2007). Historically, oscillations are measured by EEG and divided into five frequency bands: delta (0–4 Hz), theta (4–8 Hz), alpha (8–12 Hz), beta (12–30 Hz), and gamma (>30 Hz). These frequency ranges reflect the activity of different oscillation generators and differentially contribute to various cognitive functions (Başar et al., 2001). Oscillations in high frequency ranges (eg, beta and gamma) tend to promote short distance synchronization within local cortical networks, whereas lower frequency oscillations (eg, theta and alpha) support long distance synchronization.

Gamma oscillations are of particular relevance in schizophrenia research because they underlie a host of sensory and cognitive processes, such as perceptual feature binding, selective attention, stimulus salience, working memory, and executive function, that are dysfunctional in schizophrenia and refractory to treatment (Gandal et al., 2012). We first review paradigms by which neural oscillations are measured, followed by an overview of the neurobiology of gamma oscillations and the nature of gamma synchrony deficits in schizophrenia.

MEASURING NEURAL OSCILLATIONS

Electroencephalography and magnetoencephalography record variations, over time, of electrical voltage and magnetic field strength, respectively. Neural oscillations are derived by converting these time series measurements into equivalent information about amplitude and phase at various frequencies using spectral power and decomposition techniques such as fast-Fourier transform (FFT) and wavelet analysis. Oscillations can be recorded under different experimental conditions, resulting in several different categories of oscillations: resting state, entrained, evoked, and induced. Resting state oscillations refer to those generated while the subject is not engaged in any experimental task, and these are typically recorded separately under "eyes open" and "eyes closed" conditions. Entrained oscillations, or steady-state evoked potentials (SSEPs), are elicited by repetitive auditory or visual stimuli presented at a fixed frequency. SSEPs probe the brain's ability to generate endogenous oscillations that are entrained or synchronized with the frequency of the external stimulus. Gamma oscillations, typically, are elicited by auditory stimuli presented at a frequency of 40 Hz. Evoked oscillations are phase-locked responses (ie, synchronized in time) to an individual stimulus, whereas induced oscillations are responses that are reliably elicited by a stimulus but are not time-locked to it. Evoked oscillations usually occur earlier than induced oscillations and represent bottom-up sensory processing, whereas induced oscillations reflect higher-order cognitive processing (Uhlhaas and Singer, 2010). By selecting appropriate behavioral paradigms, it is possible to elicit, in turn, each of these subsets of neural oscillations (Figs. 13.4 and 13.5).

NEUROBIOLOGY OF GAMMA OSCILLATIONS

Gamma synchronization is a fundamental activity of neural populations in a variety of species from insects to mammals and occurs in all cortical and subcortical structures (Phillips and Uhlhaas, 2015). As such, gamma oscillations offer an opportunity to study concurrent processes in both human clinical samples and associated animal models. The chapter by White and Siegel, in this text, illustrates the power of applying such convergent methods across species. Gamma oscillations measured at the scalp primarily reflect oscillations generated in the thalamus, hippocampus, and cortex (Başar and Bullock, 1992). Gamma oscillations can arise from the intrinsic properties of specialized "pacemaker cells" or as an emergent property of neuronal networks. For example, Cunningham et al. (2004) found that fast rhythmic bursting (FRB) neurons, which intrinsically generate gamma frequency output, can produce neocortical gamma oscillations by providing large-scale pacemaker input to gap-junctionally connected pyramidal cells. Alternatively, and of importance to schizophrenia, gamma oscillations are also

MEASURES OF ABNORMAL GAMMA BAND OSCILLATIONS 227

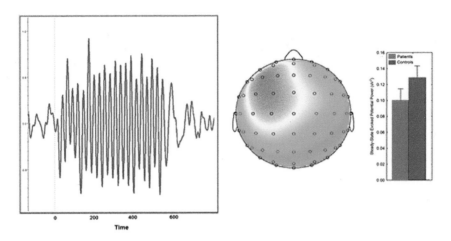

FIGURE 13.4

SSEP response elicited by auditory clicks presented at 40Hz for 500ms. Left: EEG entrainment to clicks. Middle: Associated brain topography. Right: Attenuated response magnitude in schizophrenia.

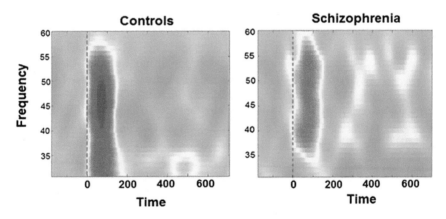

FIGURE 13.5

Early evoked gamma band response to an individual auditory stimulus in schizophrenia patients and controls.

generated in both the hippocampus and neocortex by a network of fast-spiking parvalbumin-positive GABAergic interneurons that produce rhythmic inhibitory postsynaptic potentials (IPSPs) in pyramidal neurons. Recent optogenetic studies have found that selective excitation of fast-spiking inhibitory interneurons generates gamma oscillations, whereas the loss of inhibitory activity of parvalbumin-positive interneurons disrupts gamma oscillations in vivo (Sohal et al., 2009).

Gamma oscillations are modulated by several neurotransmitters, including GABA, glutamate, and acetylcholine. The synchronous inhibitory output of gamma-generating fasting-spiking interneurons is mediated by GABA receptors, and both direct and indirect pharmacological manipulation of GABAergic signaling modulate gamma activity (Whittington et al., 1998). Parvalbumin-positive fast-spiking GABAergic interneurons receive excitatory glutamatergic inputs, and both NMDA and AMPA ionotropic glutamate receptors have been shown to influence gamma generation (Whittington et al., 1998; Pinault, 2008), likely through altering excitatory input to these interneurons. Pharmacological and genetic evidence also suggests a modulatory role for cholinergic transmission (Phillips and Uhlhaas, 2015). Nicotine increases 40-Hz auditory SSEPs (Sivarao et al., 2013), and a microdeletion at 15q13.3, a region that includes the gene for the α7 nicotinic receptor, reduces the 40-Hz auditory SSEP response (Fejgin et al., 2014). It is the loss of GABAergic inhibition in localized neural networks, possibly mediated through dysregulation of excitatory postsynaptic NMDA receptors on the cell membrane of fast-spiking interneurons, which underlies current thinking regarding the so-called glutamate hypothesis of schizophrenia. In this context, gamma oscillations, as the intrinsic emergent output of these inhibitory neural networks, can be thought of as a fundamental component of schizophrenia pathophysiology (Gonzalez-Burgos et al., 2015).

GAMMA OSCILLATIONS AND SCHIZOPHRENIA

Consistent with this hypothesis, gamma oscillation abnormalities have been reported in schizophrenia in both resting and behavioral paradigms (for review, see Uhlhaas and Singer, 2010). Kwon et al. (1999) first reported reduced entrained SSEP oscillations to 40-Hz auditory stimulation, with normal SSEP responses at lower stimulation frequencies. Deficits in 40-Hz steady-state evoked potentials have since been extensively replicated (eg, Hamm et al., 2011). Schizophrenia patients also demonstrate deficits in the amplitude and phase synchrony of early (<150 ms) evoked gamma-band responses to simple auditory stimuli, which are thought to reflect basic perceptual processing (eg, Roach and Mathalon, 2008). Although several studies have failed to replicate this finding (see Gandal et al., 2012 for discussion of conflicting results), the largest study to date from the multisite Consortium on the Genetics of Schizophrenia Endophenotypes (COGS) demonstrates robust deficits in both schizophrenia patients and unaffected first-degree family members, with an estimated heritability of ~0.64 (Turetsky et al., 2011). Abnormalities in gamma activity associated with working memory (Barr et al., 2010), cognitive control (Cho et al., 2006), and perceptual feature-binding (Spencer et al., 2003) have also been reported.

Studies have also reported associations between gamma activity and symptoms, such that different patterns of gamma disturbances differentially correlate with different aspects of schizophrenia symptomology. Reduced evoked gamma band and 40-Hz auditory SSEP responses have been linked to measures of psychomotor poverty, disorganization, and negative symptoms (Hamm et al., 2011). Interestingly, positive symptoms, such as reality distortion and hallucinations, have been associated with enhanced (ie, more normalized) gamma activity in circumscribed brain regions (Spencer et al., 2009). Local increases in gamma activity in the left auditory cortex in association with hallucinations are consistent with theories of cortical hyperexcitability (Spencer et al., 2009) and may be associated with long-range desynchronization across distal brain regions (Uhlhaas and Singer, 2010).

Abnormal gamma activity in schizophrenia may result from gross anatomical deficits and/or dysregulations of several neurotransmitter systems. Schizophrenia is associated with reductions in gray matter volume; therefore, reduced synaptic connectivity may contribute to reduced gamma synchrony. This hypothesis has recently been directly studied by Edgar et al. (2014). Patients exhibited reduced

40-Hz auditory SSEPs in the superior temporal gyrus compared to healthy controls. Controls, but not patients, demonstrated correlations between 40-Hz SSEPs and age-related changes in STG cortical thickness. The authors suggest that steady-state gamma abnormalities in schizophrenia are linked to damage in STG gray matter associated with disease pathology. Altered gamma activity may also result from deficient GABAergic transmission and is consistent with a large body of evidence demonstrating reduced GABA production in parvalbumin-positive inhibitory neurons (Lewis et al., 2005) responsible for generating cortical gamma oscillations. Several authors (eg, Gandal et al., 2012; Uhlhaas and Singer, 2010) also suggest that GABAergic deficits associated with impaired gamma synchrony may result from upstream reduction in glutamate signaling and NMDA hypofunction. For example, NMDA antagonists modulate baseline gamma power (Hong et al., 2010). Plus, a recent double-blind, placebo-controlled study found that an MGluR2-positive allosteric modulator, which enhances glutamate transmission, effectively normalized 40-Hz SSEP deficits in schizophrenia patients (Turetsky et al., 2013). Overall, abnormal gamma oscillations are a pervasive deficit in schizophrenia. Further research elucidating the precise pathophysiology of this deficit will help forge the link between cortical disconnectivity, clinical symptoms, and perceptual and cognitive deficits associated with the disorder.

CONCLUSION

This chapter demonstrates that electrophysiological measures provide a valuable resource to probe neurocognitive deficits in schizophrenia. Compared to other neuroimaging methodologies, electrophysiology provides greater temporal resolution, enabling it to elucidate the earliest stages of information processing in health and disease. Current findings suggest that cognitive dysfunction in schizophrenia includes pervasive abnormalities in sensory processing. These include deficits in automatic inhibitory failure, as evidenced through P50 and PPI, and impaired salience detection, as evidenced through MMN and P300. Evidence for altered neural oscillations and synchrony in the gamma range also implicates a basic deficit in the temporal coordination of neural activity relevant for cognition. Furthermore, these electrophysiological measures provide a window into the specific neuronal dysfunctions underlying these cognitive deficits, implicating several brain regions (including tempo-parietal cortex, frontal cortex, and the hippocampus) and neurotransmitters (including dopamine, acetylcholine, GABA, and glutamate). As such, these physiological markers of aberrant information processing provide a critical bridge between clinical/behavioral phenomenology and cellular/molecular mechanisms.

The neurophysiological deficits discussed are present in patients and in clinically high-risk individuals and clinically unaffected family members. Thus, neurophysiological measures can be utilized as biomarkers of genetic risk in an effort to deconstruct the complex genetic etiology of the schizophrenia spectrum. It is worth noting that although neurocognitive deficits are pervasive in schizophrenia, different electrophysiological markers likely index genetically and neurobiological distinct substrates. In a twin study, Hall et al. (2006a,b) found little evidence for genetic or environmental associations between P300, P50 suppression, and MMN, suggesting they are mechanistically distinct and influenced by different sets of genes. Even within schizophrenia, measures within a single conceptual framework, such as inhibition, can diverge, as evidenced by the lack of correlation between P50 and PPI in patients (Braff et al., 2007). This suggests that patients who are impaired in several of these measures may have a higher genetic load for the disorder, whereas those who are only impaired in a single measure may reflect a distinct genetic subtype not reflected in current diagnostic practice (Turetsky et al., 2007).

Further research is needed to more closely chart the longitudinal progression of these electrophysiological deficits from childhood through first episode of psychosis and subsequent progression of the illness. Consistent with the endophenotype concept, these neurophysiological measures are heritable and relatively independent of the clinical state. However, recent research suggests that some measures such as P300 and MMN also appear to predict conversion in clinically high-risk individuals and to track illness progression (Perez et al., 2014a,b; Nieman et al., 2014). Future studies involving humans can be supplemented by animal models, wherein the ease of genetic and pharmacological can further elucidate the mechanistic origins of these deficits and chart the relationship between electrophysiological measures and neurodevelopment.

The cognitive and perceptual deficits of schizophrenia are refractory to treatment with current first- and second-generation antipsychotics, which do not explicitly target the neural mechanisms that underlie these deficits (Javitt and Freedman, 2015; Gandal et al., 2012). These neurophysiological measures have the potential to serve as new biomarkers, denoting specific pathophysiological mechanisms that can be targeted for future drug development (eg, Javitt et al., 2008). Although still unproven, it is reasonable to expect that treatments that systematically target and ameliorate these neural-based information processing deficits will have beneficial downstream effects on higher-order cognition, behavior, and social functioning. In contrast with this predominant "deficit approach," Light and Swerdlow (2014) propose an alternative strategy in which neurophysiological biomarkers can be used to identify "spared" neurocognitive functions that could be exploited for novel therapeutic interventions. Such an approach also appears promising. For example, patients demonstrating greater PPI at baseline benefited more from cognitive-behavioral therapy (CBT) than those with lower (more impaired) PPI at baseline (Kumari et al., 2012). Similarly, patients with less impaired MMN response at baseline demonstrated greater improvements after cognitive training (Perez et al., 2014a). These findings suggest that, in addition to providing insight into genetic and neurobiological underpinnings of cognitive dysfunction in schizophrenia, electrophysiological measures can also be used as "biomarkers of health" (Light and Swerdlow, 2014) to identify patients who will likely be responsive to specific treatment interventions and to develop novel therapies that build upon intact cognitive processes.

REFERENCES

Adler, L.E., Pachtman, E., Franks, R.D., Pecevich, M., Waldo, M.C., Freedman, R., 1982. Neurophysiological evidence for a defect in neuronal mechanisms involved in sensory gating in schizophrenia. Biol. Psychiatry 17 (6), 639–654.

Adler, L.E., Hoffer, L.D., Wiser, A., Freedman, R., 1993. Normalization of auditory physiology by cigarette smoking in schizophrenic patients. Am. J. Psychiatry 150 (12), 1856–1861.

Barr, M.S., Farzan, F., Tran, L.C., Chen, R., Fitzgerald, P.B., Daskalakis, Z.J., 2010. Evidence for excessive frontal evoked gamma oscillatory activity in schizophrenia during working memory. Schizophr. Res. 121 (1–3), 146–152. http://dx.doi.org/10.1016/j.schres.2010.05.023.

Başar, E., Bullock, T.H., 1992. Induced Rhythms in the Brain. Birkhäuser, Boston, MA.

Basar, E., Basar-Eroglu, C., Karakas, S., Schurmann, M., 2001. Gamma, alpha, delta, and theta oscillations govern cognitive processes. Int. J. Psychophysiol. 39 (2-3), 241–248.

Belger, A., Yucel, G.H., Donkers, F.C., 2012. In search of psychosis biomarkers in high-risk populations: is the mismatch negativity the one we've been waiting for? Biol. Psychiatry 71 (2), 94–95. http://dx.doi.org/10.1016/j.biopsych.2011.11.009.

Bodatsch, M., Ruhrmann, S., Wagner, M., Muller, R., Schultze-Lutter, F., Frommann, I., et al., 2011. Prediction of psychosis by mismatch negativity. Biol. Psychiatry 69 (10), 959–966. http://dx.doi.org/10.1016/j.biopsych.2010.09.057.

Boutros, N.N., Brockhaus-Dumke, A., Gjini, K., Vedeniapin, A., Elfakhani, M., Burroughs, S., et al., 2009. Sensory-gating deficit of the N100 mid-latency auditory evoked potential in medicated schizophrenia patients. Schizophr. Res. 113 (2-3), 339–346. http://dx.doi.org/10.1016/j.schres.2009.05.019.

Braff, D.L., 2010. Prepulse inhibition of the startle reflex: a window on the brain in schizophrenia. Curr. Top. Behav. Neurosci. 4, 349–371.

Braff, D.L., Geyer, M.A., Swerdlow, N.R., 2001. Human studies of prepulse inhibition of startle: normal subjects, patient groups, and pharmacological studies. Psychopharmacology (Berl) 156 (2–3), 234–258.

Braff, D.L., Light, G.A., Swerdlow, N.R., 2007. Prepulse inhibition and P50 suppression are both deficient but not correlated in schizophrenia patients. Biol. Psychiatry 61 (10), 1204–1207. http://dx.doi.org/10.1016/j.biopsych.2006.08.015.

Bramon, E., Rabe-Hesketh, S., Sham, P., Murray, R.M., Frangou, S., 2004. Meta-analysis of the P300 and P50 waveforms in schizophrenia. Schizophr. Res. 70 (2-3), 315–329. http://dx.doi.org/10.1016/j.schres.2004.01.004.

Bramon, E., Shaikh, M., Broome, M., Lappin, J., Berge, D., Day, F., et al., 2008. Abnormal P300 in people with high risk of developing psychosis. Neuroimage 41 (2), 553–560. http://dx.doi.org/10.1016/j.neuroimage.2007.12.038.

Cadenhead, K.S., Carasso, B.S., Swerdlow, N.R., Geyer, M.A., Braff, D.L., 1999. Prepulse inhibition and habituation of the startle response are stable neurobiological measures in a normal male population. Biol. Psychiatry 45 (3), 360–364.

Cho, R.Y., Konecky, R.O., Carter, C.S., 2006. Impairments in frontal cortical gamma synchrony and cognitive control in schizophrenia. Proc. Natl. Acad. Sci. USA 103 (52), 19878–19883. http://dx.doi.org/10.1073/pnas.0609440103.

Cunningham, M.O., Whittington, M.A., Bibbig, A., Roopun, A., LeBeau, F.E., Vogt, A., et al., 2004. A role for fast rhythmic bursting neurons in cortical gamma oscillations in vitro. Proc. Natl. Acad. Sci. USA 101 (18), 7152–7157. http://dx.doi.org/10.1073/pnas.0402060101.

Edgar, J.C., Chen, Y.H., Lanza, M., Howell, B., Chow, V.Y., Heiken, K., et al., 2014. Cortical thickness as a contributor to abnormal oscillations in schizophrenia? Neuroimage Clin. 4, 122–129. http://dx.doi.org/10.1016/j.nicl.2013.11.004.

Ellenbroek, B.A., Budde, S., Cools, A.R., 1996. Prepulse inhibition and latent inhibition: the role of dopamine in the medial prefrontal cortex. Neuroscience 75 (2), 535–542.

Erwin, R.J., Turetsky, B.I., Moberg, P., Gur, R.C., Gur, R.E., 1998. P50 abnormalities in schizophrenia: relationship to clinical and neuropsychological indices of attention. Schizophr. Res. 33 (3), 157–167.

Fejgin, K., Nielsen, J., Birknow, M.R., Bastlund, J.F., Nielsen, V., Lauridsen, J.B., et al., 2014. A mouse model that recapitulates cardinal features of the 15q13.3 microdeletion syndrome including schizophrenia- and epilepsy-related alterations. Biol. Psychiatry 76 (2), 128–137. http://dx.doi.org/10.1016/j.biopsych.2013.08.014.

Ford, J.M., Krystal, J.H., Mathalon, D.H., 2007. Neural synchrony in schizophrenia: from networks to new treatments. Schizophr. Bull. 33 (4), 848–852. http://dx.doi.org/10.1093/schbul/sbm062.

Freedman, R., Coon, H., Myles-Worsley, M., Orr-Urtreger, A., Olincy, A., Davis, A., et al., 1997. Linkage of a neurophysiological deficit in schizophrenia to a chromosome 15 locus. Proc. Natl. Acad. Sci. USA 94 (2), 587–592.

Gandal, M.J., Edgar, J.C., Klook, K., Siegel, S.J., 2012. Gamma synchrony: towards a translational biomarker for the treatment-resistant symptoms of schizophrenia. Neuropharmacology 62 (3), 1504–1518. http://dx.doi.org/10.1016/j.neuropharm.2011.02.007.

Gassab, L., Mechri, A., Dogui, M., Gaha, L., d'Amato, T., Dalery, J., et al., 2006. Abnormalities of auditory event-related potentials in students with high scores on the Schizotypal Personality Questionnaire. Psychiatry Res. 144 (2–3), 117–122. http://dx.doi.org/10.1016/j.psychres.2004.09.010.

Gil-da-Costa, R., Stoner, G.R., Fung, R., Albright, T.D., 2013. Nonhuman primate model of schizophrenia using a noninvasive EEG method. Proc. Natl. Acad. Sci. USA 110 (38), 15425–15430. http://dx.doi.org/10.1073/pnas.1312264110.

Gonzalez-Burgos, G., Cho, R.Y., Lewis, D.A., 2015. Alterations in cortical network oscillations and parvalbumin neurons in schizophrenia. Biol. Psychiatry. http://dx.doi.org/10.1016/j.biopsych.2015.03.010.

Greenwood, T.A., Light, G.A., Swerdlow, N.R., Radant, A.D., Braff, D.L., 2012. Association analysis of 94 candidate genes and schizophrenia-related endophenotypes. PLoS One 7 (1), e29630. http://dx.doi.org/10.1371/journal.pone.0029630.

Grunwald, T., Boutros, N.N., Pezer, N., von Oertzen, J., Fernandez, G., Schaller, C., et al., 2003. Neuronal substrates of sensory gating within the human brain. Biol. Psychiatry 53 (6), 511–519.

Halgren, E., Squires, N.K., Wilson, C.L., Rohrbaugh, J.W., Babb, T.L., Crandall, P.H., 1980. Endogenous potentials generated in the human hippocampal formation and amygdala by infrequent events. Science 210 (4471), 803–805.

Hall, M.H., Schulze, K., Bramon, E., Murray, R.M., Sham, P., Rijsdijk, F., 2006a. Genetic overlap between P300, P50, and duration mismatch negativity. Am. J. Med. Genet. B Neuropsychiatr. Genet. 141B (4), 336–343. http://dx.doi.org/10.1002/ajmg.b.30318.

Hall, M.H., Schulze, K., Rijsdijk, F., Picchioni, M., Ettinger, U., Bramon, E., et al., 2006b. Heritability and reliability of P300, P50 and duration mismatch negativity. Behav. Genet. 36 (6), 845–857. http://dx.doi.org/10.1007/s10519-006-9091-6.

Hamm, J.P., Gilmore, C.S., Picchetti, N.A., Sponheim, S.R., Clementz, B.A., 2011. Abnormalities of neuronal oscillations and temporal integration to low- and high-frequency auditory stimulation in schizophrenia. Biol. Psychiatry 69 (10), 989–996. http://dx.doi.org/10.1016/j.biopsych.2010.11.021.

Hong, L.E., Summerfelt, A., Buchanan, R.W., O'Donnell, P., Thaker, G.K., Weiler, M.A., et al., 2010. Gamma and delta neural oscillations and association with clinical symptoms under subanesthetic ketamine. Neuropsychopharmacology 35 (3), 632–640. http://dx.doi.org/10.1038/npp.2009.168.

Javitt, D.C., Freedman, R., 2015. Sensory processing dysfunction in the personal experience and neuronal machinery of schizophrenia. Am. J. Psychiatry 172 (1), 17–31. http://dx.doi.org/10.1176/appi.ajp.2014.13121691.

Javitt, D.C., Schroeder, C.E., Steinschneider, M., Arezzo, J.C., Vaughan Jr., H.G., 1992. Demonstration of mismatch negativity in the monkey. Electroencephalogr. Clin. Neurophysiol. 83 (1), 87–90.

Javitt, D.C., Steinschneider, M., Schroeder, C.E., Arezzo, J.C., 1996. Role of cortical N-methyl-D-aspartate receptors in auditory sensory memory and mismatch negativity generation: implications for schizophrenia. Proc. Natl. Acad. Sci. USA 93 (21), 11962–11967.

Javitt, D.C., Spencer, K.M., Thaker, G.K., Winterer, G., Hajos, M., 2008. Neurophysiological biomarkers for drug development in schizophrenia. Nat. Rev. Drug Discov. 7 (1), 68–83. http://dx.doi.org/10.1038/nrd2463.

Jessen, F., Fries, T., Kucharski, C., Nishimura, T., Hoenig, K., Maier, W., et al., 2001. Amplitude reduction of the mismatch negativity in first-degree relatives of patients with schizophrenia. Neurosci. Lett. 309 (3), 185–188.

Joober, R., Zarate, J.M., Rouleau, G.A., Skamene, E., Boksa, P., 2002. Provisional mapping of quantitative trait loci modulating the acoustic startle response and prepulse inhibition of acoustic startle. Neuropsychopharmacology 27 (5), 765–781. doi:10.1016/S0893-133X(02)00333-0.

Kappenman, E.S., Luck, S.J., 2012. The ups and downs of brainwave recordings. In: Luck, S., Kappenman, E. (Eds.), Oxford Handbook of Event-Related Potential Components Oxford University Press, Oxford, pp. 3–30.

Kawakubo, Y., Kamio, S., Nose, T., Iwanami, A., Nakagome, K., Fukuda, M., et al., 2007. Phonetic mismatch negativity predicts social skills acquisition in schizophrenia. Psychiatry Res. 152 (2–3), 261–265. http://dx.doi.org/10.1016/j.psychres.2006.02.010.

Kircher, T.T., Rapp, A., Grodd, W., Buchkremer, G., Weiskopf, N., Lutzenberger, W., et al., 2004. Mismatch negativity responses in schizophrenia: a combined fMRI and whole-head MEG study. Am. J. Psychiatry 161, 294–304.

Kumari, V., Peters, E.R., Fannon, D., Premkumar, P., Aasen, I., Cooke, M.A., et al., 2008. Uncontrollable voices and their relationship to gating deficits in schizophrenia. Schizophr. Res. 101 (1–3), 185–194. http://dx.doi.org/10.1016/j.schres.2007.12.481.

Kumari, V., Premkumar, P., Fannon, D., Aasen, I., Raghuvanshi, S., Anilkumar, A.P., et al., 2012. Sensorimotor gating and clinical outcome following cognitive behaviour therapy for psychosis. Schizophr. Res. 134 (2–3), 232–238. http://dx.doi.org/10.1016/j.schres.2011.11.020.

Kwon, J.S., O'Donnell, B.F., Wallenstein, G.V., Greene, R.W., Hirayasu, Y., Nestor, P.G., et al., 1999. Gamma frequency-range abnormalities to auditory stimulation in schizophrenia. Arch. Gen. Psychiatry 56 (11), 1001–1005.

Lewis, D.A., Hashimoto, T., Volk, D.W., 2005. Cortical inhibitory neurons and schizophrenia. Nat. Rev. Neurosci. 6 (4), 312–324. http://dx.doi.org/10.1038/nrn1648.

Light, G.A., Braff, D.L., 2005. Stability of mismatch negativity deficits and their relationship to functional impairments in chronic schizophrenia. Am. J. Psychiatry 162 (9), 1741–1743. http://dx.doi.org/10.1176/appi.ajp.162.9.1741.

Light, G.A., Swerdlow, N.R., 2014. Neurophysiological biomarkers informing the clinical neuroscience of schizophrenia: mismatch negativity and prepulse inhibition of startle. Curr. Top. Behav. Neurosci. 21, 293–314. http://dx.doi.org/10.1007/7854_2014_316.

Light, G.A., Swerdlow, N.R., Rissling, A.J., Radant, A., Sugar, C.A., Sprock, J., et al., 2012. Characterization of neurophysiologic and neurocognitive biomarkers for use in genomic and clinical outcome studies of schizophrenia. PLoS One 7 (7), e39434. http://dx.doi.org/10.1371/journal.pone.0039434.

Light, G.A., Swerdlow, N.R., Thomas, M.L., Calkins, M.E., Green, M.F., Greenwood, T.A., et al., 2015. Validation of mismatch negativity and P3a for use in multi-site studies of schizophrenia: characterization of demographic, clinical, cognitive, and functional correlates in COGS-2. Schizophr. Res. 163 (1–3), 63–72. http://dx.doi.org/10.1016/j.schres.2014.09.042.

Linden, D.E., 2005. The p300: where in the brain is it produced and what does it tell us? Neuroscientist 11 (6), 563–576. http://dx.doi.org/10.1177/1073858405280524.

Luntz-Leybman, V., Bickford, P.C., Freedman, R., 1992. Cholinergic gating of response to auditory stimuli in rat hippocampus. Brain Res. 587, 130–136.

Mathalon, D.H., Ford, J.M., Pfefferbaum, A., 2000. Trait and state aspects of P300 amplitude reduction in schizophrenia: a retrospective longitudinal study. Biol. Psychiatry 47 (5), 434–449.

Mathiak, K., Ackermann, H., Rapp, A., Mathiak, K.A., Shergill, S., Riecker, A., et al., 2011. Neuromagnetic oscillations and hemodynamic correlates of P50 suppression in schizophrenia. Psychiatry Res. 94 (1), 95–104. http://dx.doi.org/10.1016/j.pscychresns.2011.01.001.

McCarley, R.W., Salisbury, D.F., Hirayasu, Y., Yurgelun-Todd, D.A., Tohen, M., Zarate, C., et al., 2002. Association between smaller left posterior superior temporal gyrus volume on magnetic resonance imaging and smaller left temporal P300 amplitude in first-episode schizophrenia. Arch. Gen. Psychiatry 59 (4), 321–331.

Murray, J.D., Anticevic, A., Gancsos, M., Ichinose, M., Corlett, P.R., Krystal, J.H., et al., 2014. Linking microcircuit dysfunction to cognitive impairment: effects of disinhibition associated with schizophrenia in a cortical working memory model. Cereb. Cortex 24 (4), 859–872.

Näätänen, R., Kujala, T., Escera, C., Baldeweg, T., Kreegipuu, K., Carlson, S., et al., 2012. The mismatch negativity (MMN) – a unique window to disturbed central auditory processing in aging and different clinical conditions. Clin. Neurophysiol. 123 (3), 424–458. http://dx.doi.org/10.1016/j.clinph.2011.09.020.

Nieman, D.H., Ruhrmann, S., Dragt, S., Soen, F., van Tricht, M.J., Koelman, J.H., et al., 2014. Psychosis prediction: stratification of risk estimation with information-processing and premorbid functioning variables. Schizophr. Bull. 40 (6), 1482–1490. http://dx.doi.org/10.1093/schbul/sbt145.

Nieuwenhuis, S., Aston-Jones, G., Cohen, J.D., 2005. Decision making, the P3, and the locus coeruleus-norepinephrine system. Psychol. Bull. 131 (4), 510–532. http://dx.doi.org/10.1037/0033-2909.131.4.510.

O'Connor, S., Morzorati, S., Christian, J.C., Li, T.-K., 1994. Heritable features of the auditory oddball event-related potential: peaks, latencies, morphology and topography. Electroencephalogr. Clin. Neurophysiol. 92, 115–125.

Olincy, A., Harris, J.G., Johnson, L.L., Pender, V., Kongs, S., Allensworth, D., et al., 2006. Proof-of-concept trial of an alpha7 nicotinic agonist in schizophrenia. Arch. Gen. Psychiatry 63 (6), 630–638. http://dx.doi.org/10.1001/archpsyc.63.6.630.

Olincy, A., Braff, D.L., Adler, L.E., Cadenhead, K.S., Calkins, M.E., Dobie, D.J., et al., 2010. Inhibition of the P50 cerebral evoked response to repeated auditory stimuli: results from the Consortium on Genetics of Schizophrenia. Schizophr. Res. 119 (1–3), 175–182. http://dx.doi.org/10.1016/j.schres.2010.03.004.

Perez, V.B., Braff, D.L., Pianka, S., Swerdlow, N.R., Light, G.A., 2014a. Mismatch negativity predicts and corresponds to behavioral improvements following an initial dose of cognitive training in schizophrenia patients. In: Presented at Society of Biological Psychiatry, New York, NY.

Perez, V.B., Woods, S.W., Roach, B.J., Ford, J.M., McGlashan, T.H., Srihari, V.H., et al., 2014b. Automatic auditory processing deficits in schizophrenia and clinical high-risk patients: forecasting psychosis risk with mismatch negativity. Biol. Psychiatry 75 (6), 459–469. http://dx.doi.org/10.1016/j.biopsych.2013.07.038.

Phillips, K.G., Uhlhaas, P.J., 2015. Neural oscillations as a translational tool in schizophrenia research: rationale, paradigms and challenges. J. Psychopharmacol. 29 (2), 155–168. http://dx.doi.org/10.1177/0269881114562093.

Pinault, D., 2008. N-methyl D-aspartate receptor antagonists ketamine and MK-801 induce wake-related aberrant gamma oscillations in the rat neocortex. Biol. Psychiatry 63 (8), 730–735. http://dx.doi.org/10.1016/j.biopsych.2007.10.006.

Polich, J., 2007. Updating P300: an integrative theory of P3a and P3b. Clin. Neurophysiol. 118 (10), 2128–2148. http://dx.doi.org/10.1016/j.clinph.2007.04.019.

Polich, J., 2012. Neuropsychology of P300. In: Luck, S., Kappenman, E. (Eds.), Oxford Handbook of Event-Related Potential Components Oxford University Press, Oxford, pp. 159–188.

Polich, J., Kok, A., 1995. Cognitive and biological determinants of P300: an integrative review. Biol. Psychol. 41 (2), 103–146.

Popov, T., Jordanov, T., Rockstroh, B., Elbert, T., Merzenich, M.M., Miller, G.A., 2011a. Specific cognitive training normalizes auditory sensory gating in schizophrenia: a randomized trial. Biol. Psychiatry 69 (5), 465–471. http://dx.doi.org/10.1016/j.biopsych.2010.09.028.

Popov, T., Jordanov, T., Weisz, N., Elbert, T., Rockstroh, B., Miller, G.A., 2011b. Evoked and induced oscillatory activity contributes to abnormal auditory sensory gating in schizophrenia. Neuroimage 56 (1), 307–314. http://dx.doi.org/10.1016/j.neuroimage.2011.02.016.

Potter, D., Summerfelt, A., Gold, J., Buchanan, R.W., 2006. Review of clinical correlates of P50 sensory gating abnormalities in patients with schizophrenia. Schizophr. Bull. 32 (4), 692–700. http://dx.doi.org/10.1093/schbul/sbj050.

Reite, M., Teale, P.D., Neumann, R., Davis, K., 1987. Localization of a 50 msec latency auditory evoked field component. Electroencephalogr. Clin. Neurophysiol. Suppl. 40, 487–492.

Roach, B.J., Mathalon, D.H., 2008. Event-related EEG time-frequency analysis: an overview of measures and an analysis of early gamma band phase locking in schizophrenia. Schizophr. Bull. 34 (5), 907–926. http://dx.doi.org/10.1093/schbul/sbn093.

Schreiber, H., Stolz-Born, G., Kornhuber, H.H., Born, J., 1992. Event-related potential correlates of impaired selective attention in children at high risk for schizophrenia. Biol. Psychiatry 32 (8), 634–651.

Shelley, A.M., Ward, P.B., Catts, S.V., Michie, P.T., Andrews, S., McConaghy, N., 1991. Mismatch negativity: an index of a preattentive processing deficit in schizophrenia. Biol. Psychiatry 30 (10), 1059–1062.

Siegel, C., Waldo, M., Mizner, G., Adler, L.E., Freedman, R., 1984. Deficits in sensory gating in schizophrenic patients and their relatives. Evidence obtained with auditory evoked responses. Arch. Gen. Psychiatry 41 (6), 607–612.

Sivarao, D.V., Frenkel, M., Chen, P., Healy, F.L., Lodge, N.J., Zaczek, R., 2013. MK-801 disrupts and nicotine augments 40 Hz auditory steady state responses in the auditory cortex of the urethane-anesthetized rat. Neuropharmacology 73, 1–9. http://dx.doi.org/10.1016/j.neuropharm.2013.05.006.

Smith, A.K., Edgar, J.C., Huang, M., Lu, B.Y., Thoma, R.J., Hanlon, F.M., et al., 2010. Cognitive abilities and 50- and 100-msec paired-click processes in schizophrenia. Am. J. Psychiatry 167 (10), 1264–1275. http://dx.doi.org/10.1176/appi.ajp.2010.09071059.

Smucny, J., Olincy, A., Eichman, L.C., Lyons, E., Tregellas, J.R., 2013. Early sensory processing deficits predict sensitivity to distraction in schizophrenia. Schizophr. Res. 147 (1), 196–200. http://dx.doi.org/10.1016/j.schres.2013.03.025.

Sohal, V.S., Zhang, F., Yizhar, O., Deisseroth, K., 2009. Parvalbumin neurons and gamma rhythms enhance cortical circuit performance. Nature 459 (7247), 698–702. http://dx.doi.org/10.1038/nature07991.

Spencer, K.M., Nestor, P.G., Niznikiewicz, M.A., Salisbury, D.F., Shenton, M.E., McCarley, R.W., 2003. Abnormal neural synchrony in schizophrenia. J. Neurosci. 23 (19), 7407–7411.

Spencer, K.M., Niznikiewicz, M.A., Nestor, P.G., Shenton, M.E., McCarley, R.W., 2009. Left auditory cortex gamma synchronization and auditory hallucination symptoms in schizophrenia. BMC Neurosci. 10, 85. http://dx.doi.org/10.1186/1471-2202-10-85.

Squires, N.K., Squires, K.C., Hillyard, S.A., 1975. Two varieties of long-latency positive waves evoked by unpredictable auditory stimuli in man. Electroencephalogr. Clin. Neurophysiol. 38 (4), 387–401.

Stanzione, P., Fattapposta, F., Giunti, P., D'Alessio, C., Tagliati, M., Affricano, C., et al., 1991. P300 variations in parkinsonian patients before and during dopaminergic monotherapy: a suggested dopamine component in P300. Electroencephalogr. Clin. Neurophysiol. 80 (5), 446–453.

Stevens, K.E., Kem, W.R., Mahnir, V.M., Freedman, R., 1998. Selective alpha7-nicotinic agonists normalize inhibition of auditory response in DBA mice. Psychopharmacology (Berl) 136 (4), 320–327.

Swerdlow, N.R., Geyer, M.A., Braff, D.L., 2001. Neural circuit regulation of prepulse inhibition of startle in the rat: current knowledge and future challenges. Psychopharmacology (Berl) 156 (2–3), 194–215.

Swerdlow, N.R., Light, G.A., Cadenhead, K.S., Sprock, J., Hsieh, M.H., Braff, D.L., 2006. Startle gating deficits in a large cohort of patients with schizophrenia: relationship to medications, symptoms, neurocognition, and level of function. Arch. Gen. Psychiatry 63 (12), 1325–1335. http://dx.doi.org/10.1001/archpsyc.63.12.1325.

Thoma, R.J., Hanlon, F.M., Moses, S.N., Edgar, J.C., Huang, M., Weisend, M.P., et al., 2003. Lateralization of auditory sensory gating and neuropsychological dysfunction in schizophrenia. Am. J. Psychiatry 160 (9), 1595–1605.

Thönnessen, H., Zvyagintsev, M., Harke, K.C., Boers, F., Dammers, J., Norra, C., et al., 2008. Optimized mismatch negativity paradigm reflects deficits in schizophrenia patients. A combined EEG and MEG study. Biol. Psychol. 77, 205–216.

Turetsky, B.I., Colbath, E.A., Gur, R.E., 2000. P300 subcomponent abnormalities in schizophrenia: III. Deficits in unaffected siblings of schizophrenia probands. Biol. Psychiatry 47, 380–390.

Turetsky, B.I., Calkins, M.E., Light, G.A., Olincy, A., Radant, A.D., Swerdlow, N.R., 2007. Neurophysiological endophenotypes of schizophrenia: the viability of selected candidate measures. Schizophr. Bull. 33 (1), 69–94. http://dx.doi.org/10.1093/schbul/sbl060.

Turetsky, B.I., Greenwood, T.A., Braff, D.L., Cadenhead, K.S., Calkins, M.E., Dobie, D.J., et al., 2011. Spontaneous and auditory evoked gamma and theta oscillations in schizophrenia probands and unaffected first-degree relatives: data from the COGS study. Biol. Psychiatry 69, 216S–217S.

Turetsky, B., Frishberg, N., Neustadter, E., Wolf, D., Kohler, C., March, M., et al., 2013. Auditory steady state evoked potential abnormalities in schizophrenia are normalized by an mGluR2 positive allosteric modulator. Neuropsychopharmacology 38, S382–S383.

Turetsky, B.I., Dress, E.M., Braff, D.L., Calkins, M.E., Green, M.F., Greenwood, T.A., et al., 2015. The utility of P300 as a schizophrenia endophenotype and predictive biomarker: Clinical and socio-demographic modulators in COGS-2. Schizophr. Res. 163 (1–3), 53–62. http://dx.doi.org/10.1016/j.schres.2014.09.024.

Uhlhaas, P.J., Singer, W., 2010. Abnormal neural oscillations and synchrony in schizophrenia. Nat. Rev. Neurosci. 11 (2), 100–113. http://dx.doi.org/10.1038/nrn2774.

Umbricht, D., Krljes, S., 2005. Mismatch negativity in schizophrenia: a meta-analysis. Schizophr. Res. 76 (1), 1–23. http://dx.doi.org/10.1016/j.schres.2004.12.002.

Umbricht, D., Schmid, L., Koller, R., Vollenweider, F.X., Hell, D., Javitt, D.C., 2000. Ketamine-induced deficits in auditory and visual context-dependent processing in healthy volunteers: implications for models of cognitive deficits in schizophrenia. Arch. Gen. Psychiatry 57 (12), 1139–1147.

Umbricht, D., Koller, R., Vollenweider, F.X., Schmid, L., 2002. Mismatch negativity predicts psychotic experiences induced by NMDA receptor antagonist in healthy volunteers. Biol. Psychiatry 51 (5), 400–406.

Umbricht, D., Vyssotki, D., Latanov, A., Nitsch, R., Lipp, H.P., 2005. Deviance-related electrophysiological activity in mice: is there mismatch negativity in mice? Clin. Neurophysiol. 116 (2), 353–363. http://dx.doi.org/10.1016/j.clinph.2004.08.015.

van Tricht, M.J., Nieman, D.H., Koelman, J.H., van der Meer, J.N., Bour, L.J., de Haan, L., et al., 2010. Reduced parietal P300 amplitude is associated with an increased risk for a first psychotic episode. Biol. Psychiatry 68 (7), 642–648. http://dx.doi.org/10.1016/j.biopsych.2010.04.022.

Watson, T.D., Petrakis, I.L., Edgecombe, J., Perrino, A., Krystal, J.H., Mathalon, D.H., 2009. Modulation of the cortical processing of novel and target stimuli by drugs affecting glutamate and GABA neurotransmission. Int. J. Neuropsychopharmacol. 12 (3), 357–370. http://dx.doi.org/10.1017/S1461145708009334.

Whittington, M.A., Traub, R.D., Faulkner, H.J., Jefferys, J.G., Chettiar, K., 1998. Morphine disrupts long-range synchrony of gamma oscillations in hippocampal slices. Proc. Natl. Acad. Sci. USA 95 (10), 5807–5811.

Winterer, G., McCarley, R.W., 2011. Electrophysiology of schizophrenia. In: Weinberger, D., Harrison, P. (Eds.), Schizophrenia, third ed. Wiley-Blackwell, Chichester; Hoboken, NJ, pp. 311–333.

Winterer, G., Mulert, C., Mientus, S., Gallinat, J., Schlattmann, P., Dorn, H., et al., 2001. P300 and LORETA: comparison of normal subjects and schizophrenic patients. Brain Topogr. 13 (4), 299–313.

Yingling, C.D., Hosobuchi, Y., 1984. A subcortical correlate of P300 in man. Electroencephalogr. Clin. Neurophysiol. 59 (1), 72–76.

CHAPTER 14

CELLULAR AND CIRCUIT MODELS OF INCREASED RESTING STATE NETWORK GAMMA ACTIVITY IN SCHIZOPHRENIA

R.S. White and S.J. Siegel
Department of Psychiatry, University of Pennsylvania, Philadelphia, PA, United States

CHAPTER OUTLINE

Introduction	238
Studies in Schizophrenia	238
Resting State Brain Activity	238
Electroencephalography	238
Magnetoencephalography	239
Functional Magnetic Resonance Imaging	239
Findings in Disease	239
Resting State Network and Gamma Oscillations in Schizophrenia	239
Power and Coherence of Brain Activity	241
Inherent Noise and Relationship to Symptoms in Schizophrenia	242
Circuit Models and Evidence in Humans	242
Interneuron Network Gamma Generation	243
The Pyramidal and Interneuron Model of Generating Gamma Oscillations	244
Animal Models of a Noisy Brain	244
Defining Baseline Gamma in Animal Models	244
Animal Models of Schizophrenia With Relation to Resting Gamma Activity	245
In Vitro Examination of Animal Models of Increased Resting Gamma	247
Relating Behavior to EEG/LFP Validation of Models	248
Summary and Conclusions	250
References	251

INTRODUCTION

Gamma oscillations are correlated with cognitive processes such as perception, attention, memory, and consciousness. In disease states, these fast-frequency brain oscillations are often perturbed when compared to those of control subjects (Uhlhaas and Singer, 2013). Furthermore, different aspects of gamma oscillatory activity are thought to be pertinent to abnormal brain activity and perhaps the pathophysiology of various disorders. For example, resting activity, event-related/event-evoked activity, and task-related/task-induced activity are all aspects of electroencephalography (EEG) data that have been compared between controls and disease populations. This review focuses on the specific role of increased resting state activity. Specifically, we present and discuss data, suggesting that this measure may represent the primary underlying cause of each of the other EEG measures as well as a variety of behavioral abnormalities in schizophrenia. Additional measures of sensory processing and inhibition using behavior (eg, prepulse inhibition of startle) and event-related potentials (eg, P50) are discussed in chapter: "EEG and MEG Probes of Schizophrenia Pathophysiology."

STUDIES IN SCHIZOPHRENIA
RESTING STATE BRAIN ACTIVITY

Resting state brain activity is defined as activity in the brain when a subject is awake but not performing a specific cognitive task or responding to sensory stimuli. This activity has been recorded using a multitude of techniques in humans and animal models. For example, brain oscillations at certain frequencies can be investigated using a variety of techniques. These different frequencies of activity are thought to underlie the coordinated firing of different brain regions that are associated with cognition. Hans Berger first described a dominant oscillation of approximately 10 Hz, which he termed alpha (Berger, 1929; Buzsáki and Draguhn, 2004; Buzsáki, 2006). Berger and others coined terms still used today to designate brain activity within specific frequency bands: delta (0–4 Hz), theta (4–8 Hz), alpha (8–12 Hz), beta (12–30 Hz), and gamma (>30 Hz). Distinct frequency bands have been associated with unique cognitive processes and behavioral states (Basar et al., 2001). This review focuses on gamma band activity because it has been seen to be perturbed in the pathophysiology of many psychiatric disorders (Gandal et al., 2012b; Herrmann et al., 2010; Port et al., 2014). Furthermore, gamma band activity underlies cognitive processes and is found in virtually all mammalian brain structures at both cortical and subcortical locations. Specific structures (eg, thalamus, hippocampus, and cortex) contribute prominently to scalp recorded activity (Basar and Bullock, 1992). Methods to record brain activity in humans include EEG, magnetoencephalography (MEG), and functional magnetic resonance imaging (fMRI). All three of these approaches have distinct advantages and disadvantages.

ELECTROENCEPHALOGRAPHY

EEG is a noninvasive method to record brain activity transcranially to investigate how coordinated brain activity changes during different states and/or performance of various activities. EEG recordings use a series of electrodes placed on the outside of the scalp that record the global changes in current in the brain. Although direct localization of specific sources of the brain activity is not possible, reverse solutions are used to infer potential sources based on underlying assumptions about brain organization. However, coordinated oscillatory activity at all frequency ranges of the brain as a whole can be recorded and analyzed (Gross, 2014; Lopes da Silva, 2013). As noted, resting state EEG activity refers to the power within any frequency range in the absence of any tasks or stimuli given to the subjects. Resting state oscillations are thought to underlie, or at least reflect, the relative level of consciousness of the individual. The total power and coherence in various frequency ranges are therefore thought to reflect the mental state of an individual.

MAGNETOENCEPHALOGRAPHY

MEG source localization measures brain activity by recording magnetic fields produced by electrical currents occurring naturally in the brain using very sensitive magnetometers to noninvasively record the summation of local activity in the brain (Lopes da Silva, 2013; Malmivuo, 2012). MEG is not sensitive to radial components of dipolar sources, but rather to the tangential components. The amplitude of oscillations is affected by the size of brain regions activated, however, synchronous they are, and spatial resolution of the areas is better than that of EEG (Kim et al., 2014; Mantini et al., 2007; Ramyead et al., 2014; Rutter et al., 2009; Schnitzler and Gross, 2005). Both event-related paradigms and continuous or baseline paradigms for recording brain activity are used for humans using MEG (Gross, 2014). However, these baseline recordings of the "resting state" of the brain are difficult to interpret because of the need to fit the patterns of activity to a model that is predicated on normal brain organization (Gross et al., 2013). Please see chapter "EEG and MEG Probes of Schizophrenia Pathophysiology" for an additional discussion of MEG findings in schizophrenia.

FUNCTIONAL MAGNETIC RESONANCE IMAGING

fMRI is a technique primarily used to study which areas of the brain are active at rest or during a cognitive task. Using this approach, a great deal of attention has been devoted to the changes in functional connectivity in the brain in healthy controls and in patients with psychiatric disorders. fMRI provides an indirect measure of brain activity by assessing blood oxygenation during a time interval of approximately 10s. Therefore, the temporal sampling rate precludes any assessment related to the frequency of neuronal activity with this technique. However, fMRI does offer a high degree of spatial resolution, allowing for assessment of the degree to which the different brain regions are active at the same time, which is interpreted as being functionally connected during tasks, and of whether or not this regional connectivity changes in disease states. Some groups have also used fMRI in concert with EEG to examine both frequency and spatial domains of brain activity (Laufs, 2008; Neuner et al., 2014; Ritter and Villringer, 2006). These groups have been trying to correlate the connectivity seen in fMRI studies and the temporal resolution that is possible with EEG studies.

FINDINGS IN DISEASE
RESTING STATE NETWORK AND GAMMA OSCILLATIONS IN SCHIZOPHRENIA

Baseline oscillatory activity can be investigated in two different contexts. First, in "default mode" or "resting state" paradigms, EEG/MEG is acquired while the subject lies still without engaging in a task. Second, during tasks with a repetitive time-locked event, "prestimulus" baseline oscillatory activity can be extracted to examine the receptiveness of neural networks to the next stimulus. In resting state paradigms, several studies reported elevated high-frequency EEG activity in schizophrenia (Davis, 1942; Fenton et al., 1980; Finley, 1944; Giannitrapani and Kayton, 1974; Itil et al., 1972; Itil et al., 1974; Kennard and Schwartzman, 1957; Rodin et al., 1968; Winterer et al., 2004), although they tended to investigate activity less than 40 Hz. For example, in a large (n = 100/group) clinical study, schizophrenia patients demonstrated increased activity of 24–33 Hz (Itil et al., 1972) that was stable over the course of 3 months (Itil et al., 1974). However, three studies did report elevations in high beta (20–30 Hz) power, interpreted as reflecting "cortical noise" (Brockhaus-Dumke et al., 2008; Kissler et al., 2000; Krishnan et al., 2005), but another did not (Miyauchi et al., 1990). A larger EEG study observed increased power of 20–50 Hz in schizophrenia subjects and their relatives (Venables et al., 2009), and another group found broadband increases of baseline activity at all frequencies (Winterer et al., 2004; Fig. 14.1). A similar MEG study using source-space projections found opposite results (Rutter et al., 2009). These mixed results may reflect differences in sample size, imaging modality (EEG/MEG), or recording site (scalp vs source-space projections). Several groups have examined prestimulus baseline gamma activity in schizophrenia. The issue of baseline gamma differences is important, given that most studies examining group differences in poststimulus activity use some form of baseline correction (Urbach and Kutas, 2006). Two very large studies reported elevated prestimulus gamma power in schizophrenia patients during auditory paradigms (Hong et al., 2008; Winterer et al., 2004), in accordance with our previous data (Turetsky and Siegel, 2007; Fig. 14.1). Two smaller studies found no group

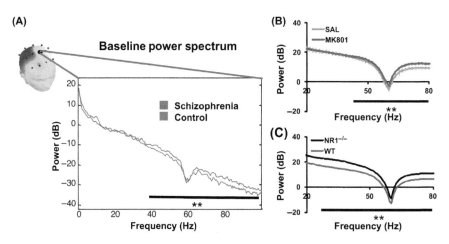

FIGURE 14.1

Brain activity in the gamma frequency range is increased at rest in schizophrenia. (A) Human resting state (baseline) EEG data from subjects with schizophrenia (red) and controls (blue), (B) Mice treated with MK801 (red) have an increase in baseline gamma power relative to control mice (blue) (Saunders et al., 2012), (C) Constitutive reduction of NMDA-R expression in mice also increases baseline gamma power, similar to the pattern in schizophrenia (Gandal et al., 2012c).

For (A): courtesy of Bruce Turetsky.

differences in prestimulus gamma band responses, but they did report elevated baseline beta power in schizophrenia (Brockhaus-Dumke et al., 2008).

The default mode network (DMN) is functional brain activity across different regions at rest, when a person is being introspective or not interacting with the world (Buckner et al., 2008; Damoiseaux, 2006). It is thought that the DMN is an interconnected series of brain regions that are active at rest but that deactivate during performance of a wide range of cognitive tasks. The medial frontal cortex is one of the principal constituents of this network, along with the posterior cingulate cortex/precuneus, parts of the parietal and temporal lobe cortex, and the hippocampus (Whitfield-Gabrieli and Ford, 2012). The salient network (SN) is the anterior insular SN that contains the fronto-insular operculum and dorsal anterior cingulate cortex (Manoliu et al., 2014). The network responds to behaviorally salient events and seems to interact with the DMN. Several studies have evaluated the role of resting state activity and connectivity in the performance deficits demonstrated among people with schizophrenia. For example, data suggest that there is a failure of deactivation of the DMN during the working memory task in schizophrenia patients relative to healthy controls (Landin-Romero et al., 2014). Similar deficits were noted in schizophrenia during an auditory oddball task, which requires that subjects respond to a target stimulus that is presented among a background of nontarget and distractor stimuli (Garrity et al., 2007). Furthermore, some investigators have attempted to relate the increase in resting state connectivity and/or failure to deactivate the DMN to the presence of delusions in schizophrenia patients. Studies using fMRI suggest that there is an increase in connectivity between the frontal cortex and the DMN early during the course of schizophrenia (Damoiseaux, 2006; Karbasforoushan and Woodward, 2012; Manoliu et al., 2014; Ongür et al., 2010; Orliac et al., 2013; Tu et al., 2013). Data indicate that there is an increase in connectivity between the frontal cortex and default motor network among patients with early-onset schizophrenia (Tang et al., 2013). Additional studies suggest that there are also differences in the connectivity among five neural networks in schizophrenia, including the DMN, fronto-parietal (FP), cingulo-opercular (CO), cerebellar (CER), and the SN (Mamah et al., 2013). Consistent with these studies, combining diffusion tensor imaging, which is a measure of structural connectivity, and fMRI also indicates that there is a perturbation in functional connectivity in schizophrenia (Skudlarski et al., 2010). This is consistent with resting state fMRI studies that show patients with schizophrenia have significantly lower global connectivity compared with healthy controls (Argyelan et al., 2014). Thus, combining different imaging techniques supports the hypothesis that there are differences in long-range functional connectivity in schizophrenia (Spellman and Gordon, 2014).

POWER AND COHERENCE OF BRAIN ACTIVITY

Power and coherence in various frequency ranges are thought to reflect the cross-sectional mental state of an individual, as evidenced by the level of symptoms at a given time. For example, there is a significant correlation between measures of EEG activity and the degree of hallucinations and delusions among people with schizophrenia (Herrmann and Demiralp, 2005). Some studies suggest that measures of EEG may also reflect traits of individuals with schizophrenia. A study involving drug-naive patients with schizophrenia during sleep found a decrease in the coherence of activity between the right central and right frontal areas in both beta and gamma frequencies (Yeragani et al., 2006). The investigators examined the coherence between the right central and the right frontal areas of the EEG during sleep states. Coherence involves the relationship of activity in two areas of the EEG in a specific frequency

(Yeragani et al., 2006). Similarly, Andreou et al. (2014a) examined resting state EEG analysis during different microstates of brain activity in three different populations of people: high-risk individuals, clinically stable first-episode patients with schizophrenia, and healthy controls. EEG microstates are periods of coordinated brain activity that are postulated to represent interactions between neural networks and local states. Their duration is approximately 100 ms, which is considered to be congruent with the timeframe in which spontaneous thought takes place. These microstates capture the moment-to-moment interactions among brain networks within certain activity frequencies.

INHERENT NOISE AND RELATIONSHIP TO SYMPTOMS IN SCHIZOPHRENIA

We propose that this increased autonomous gamma power throughout the brain leads to the misinterpretation of relationships among random events. Gamma activity is used for feature binding, that is, the mechanism of relating multimodal sensory stimuli and the context in which they occur into a coherent story with meaning. These misinterpretations of relatedness may then be the basis for referential delusions. Studies suggest that there is a link between deficits in the SN and default motor network in schizophrenia, especially as these relate to resting state EEG studies. EEG studies indicate that schizophrenia patients showed an increase in functional connectivity in the DMN at low frequencies (0.06 Hz) that is correlated with psychotic symptoms (Rolland et al., 2014; Rotarska-Jagiela et al., 2010). Aberrant salience, or lack of being able to identify relevant features during cognitive tasks, has been linked to the presence of positive symptoms in schizophrenia (Manoliu et al., 2014; White et al., 2010). Specifically, connectivity scores during a cognitive task indicate that there is a decrease in activity of the right anterior insula, which is part of the SN, among patients with hallucinations. Other studies have linked alterations in resting state gamma band connectivity and the core symptoms of schizophrenia with EEG (Andreou et al., 2014b; Sun et al., 2014). Studies using MEG have found that there is an increase in resting gamma activity in schizophrenia (Kim et al., 2014). Others have found a decrease in gamma power in schizophrenia patients and their siblings as measured with MEG source localization analysis (Rutter et al., 2009).

Similar to their role in abnormal beliefs, we propose that increased gamma power at rest may contribute to hallucinations. Hallucinations are an activation of sensory networks in the absence of external stimuli. Because there is an increase in activity throughout the brain, this would also include sensory pathways. Therefore, increased activity could be the basis for the active organization and formation of sensory experiences based on internally generated noise. Northoff has hypothesized that an increase in the resting state of the brain might contribute to auditory hallucinations (Northoff et al., 2010; Northoff, 2014). Andreou et al. examined resting state EEG analysis during different microstates of brain activity in three different populations of people: high-risk individuals, clinically stable first-episode patients with schizophrenia, and healthy controls.

CIRCUIT MODELS AND EVIDENCE IN HUMANS

Cortical function and neuronal oscillations are thought to be generated by a balance between excitation and inhibition throughout the brain. The two primary types of neurons are the excitatory pyramidal neurons and the diverse group of GABAergic inhibitory neurons (Tiesinga and Sejnowski, 2009). Studies have shown that gamma band activity is generated from the interplay between inhibitory neurons and the excitatory drive from the pyramidal neurons (Bartos et al., 2007; Buzsáki and Wang, 2012; Whittington et al., 1995). Mechanisms underlying gamma oscillations include: (1) feedback loops from pyramidal neurons onto interneurons (PING) and (2) oscillations in mutual interneuronal networks via chemical or electrical transmission (ING) (Nakazawa et al., 2012).

INTERNEURON NETWORK GAMMA GENERATION

According to the ING model, the primary deficit resides within interneurons. As a result, gamma oscillations are disrupted because their generation is dependent on mutual inhibition between reciprocally connected GABA interneurons. For example, the ING theory postulates that abnormalities in parvalbumin (PV) basket cells in dorsal lateral prefrontal cortex mediate gamma synchronization deficits via loss of perisomatic GABA-A receptor activity (Lewis et al., 2011). Specifically, several groups have found that there is a decrease in the number of these interneurons in postmortem tissue in schizophrenia (Benes et al., 2001). Other groups have suggested that there is a decrease in immunoreactivity for a specific subset of interneuron cell types without affecting the total number of interneurons (Konradi et al., 2011). Alternatively, there is some evidence that there is a decrease in the expression of a variety of more general interneuron markers (eg, GAD67) in schizophrenia (Fung et al., 2014).

One of the predominant theories for increased blood flow and background/resting gamma activity in schizophrenia has focused on the loss of inhibitory tone. Several lines of research support this idea. For example, there are synaptic abnormalities in postmortem samples from schizophrenia patients that are consistent with a decrease in GABA synthesis (Lewis et al., 2012; Lewis et al., 2011; Perry et al., 1979). There is also evidence of a decrease in GAD-67 expression in schizophrenia (Fung et al., 2014; Guidotti et al., 2000; Volk et al., 2000). Specifically, several studies have described a reduction in PV immunoreactive GABAergic interneurons, which is thought to reflect a relative reduction in glutamate N-methyl-D-aspartate receptors (NMDA-R) function in this population (Zhang and Reynolds, 2002). Basic and clinical studies also provide complementary support for this possibility. For example, decreasing GABA activity in the prefrontal cortex with bicuculline causes schizophrenia-like abnormalities (Enomoto et al., 2011). Similarly, NMDA-R hypofunction in PV interneurons alter gamma oscillations (Billingslea et al., 2014; Cunningham et al., 2006; Gonzalez-Burgos and Lewis, 2012).

There is also evidence of abnormalities in ion channels (termed channelopathies) that can lead to circuit dysfunction in schizophrenia and disinhibition. Genome-wide association studies suggest a genetic link between the pathophysiology of schizophrenia and specific gene loci that contain relevant ion channels (Consortium, 2013; Ohi et al., 2014). For example, the *CACNA1C* gene (α-1C subunit of the L-type voltage-gated calcium channel) has been identified as a risk gene for bipolar disorder and schizophrenia (Bigos et al., 2013; Gargus, 2006; Green et al., 2010). Similarly, the KCNS3 potassium channel subunit has lower expression in PV neurons in the prefrontal cortex of humans (Georgiev et al., 2014). Investigators in this study hypothesize that the loss of this potassium channel causes a longer repolarization of excitatory post synaptic potentials (EPSPs) of neurons, causing a greater likelihood of EPSP depolarization to summate to an action potential. Other groups have also shown that there is an increase in the ether-a-go-go-related K^+ channel mRNA in schizophrenic individuals. In vitro studies have shown that overexpression of KCNH2-3.1 in primary cortical neurons induces a rapidly deactivating K(+) current and a high-frequency, nonadapting firing pattern (Huffaker et al., 2009).

THE PYRAMIDAL AND INTERNEURON MODEL OF GENERATING GAMMA OSCILLATIONS

The PING model is a modification of the theory regarding how ensembles of neurons can form a functional circuit to generate gamma activity. This model incorporates a wider network of pyramidal neurons and interneurons that are reciprocally connected and generate a firing rhythm. In favor of this theory, studies have demonstrated that CA3 pyramidal cells drive feed-forward inhibition of CA1 pyramidal neurons. The oscillatory input from CA3 is imposed on CA1 via excitation of inhibitory neurons (Zemankovics et al., 2013).

In contrast to the disinhibition theory, many groups have found that there is a loss of excitatory drive on interneurons, supporting the PING model (Lisman et al., 2008; Marín, 2012; Whittington and Traub, 2003). In a review published in 2012, Gonzalez-Burgos and collaborators proposed a new circuit model of inhibition-based gamma oscillations relevant to schizophrenia in which pyramidal neuron dysfunction could be the primary source of reduced interneuron activation (Gonzalez-Burgos and Lewis, 2012). In this model, alterations in pyramidal neurons would lead to disrupted efferent drive onto interneurons, yielding abnormal synchronization of feedback inhibition. These data are particularly compelling because NMDA-R signaling is one of the major regulators of interneuron and pyramidal neuron excitability (Homayoun and Moghaddam, 2007; Xue et al., 2011). Preclinical and clinical studies focusing on pharmacology and genomics support the hypothesis that hypofunction of NMDA-R signaling contributes to the pathophysiology of schizophrenia (Javitt and Zukin, 1991; Jentsch et al., 1997; Kirihara et al., 2012; Krystal et al., 1994; Lahti et al., 1995; Lahti et al., 2001; Stone et al., 2008). Additionally, unaffected individuals can be induced to have schizophrenia-like symptoms by administration of NMDA-R antagonists (Coyle et al., 2003; Coyle, 2004, 2006). These clinical studies were performed with global administration of drugs and therefore cannot address whether NMDA-R hypofunction is present in both inhibitory and excitatory cells.

ANIMAL MODELS OF A NOISY BRAIN
DEFINING BASELINE GAMMA IN ANIMAL MODELS

The model systems and methods described here are discussed in relation to how they inform the "noisy brain" hypothesis. Specifically, we address the extent to which the data are consistent with the idea that increased resting state activity contributes to the underlying pathology of deficits in schizophrenia. We include in vivo as well as in vitro methods that assess the extent to which manipulation in animals leads to increased excitability. In vivo methods focus on the use of EEG, which provides one of the most directly translatable measures between preclinical and clinical populations. Of note, there is often confusion regarding the terminology when recording in vivo, particularly when distinguishing local field potentials (LFP) and EEG. The former refers to the use of high-impedance electrodes that are sensitive to only the local area (eg, hundreds of microns) in which the electrode tip is placed. Alternatively, EEG refers to the use of low-impedance electrodes that are sensitive to electrical activity (ie, vectors, generated throughout the brain) from a particular perspective. One also needs to be aware of how electrode configuration impacts the area from which electrical activity is sampled. Placing the positive and negative tips in close apposition yields a configuration that is relatively insensitive to distant electrical sources because the vector would be similar at both points. This would therefore favor accentuating local activity (Frankel et al., 2005). Alternatively, placing the electrodes at distant points from each other in the brain allows for differential activity from every vector, thus providing for sampling a wider spatial representation (Gandal et al., 2012c; Tatard-Leitman et al., 2015). A third approach uses either very high-impedance glass pipets or tetrode electrodes to isolate a single unit (ie, spikes) from a single cell. Among these options, the low-impedance electrodes placed far apart most accurately model the human EEG; therefore, they are the main focus of our discussion of in vivo studies. In vitro approaches include slice recordings of both LFP (eg, long-term potentiation), whole-cell intracellular recordings (patch studies), as well as special techniques that assess circuit dynamics in real time such as voltage-sensitive dye (Carlson et al., 2011). For in vitro studies, we review and discuss the extent to which changes in local circuits and cells are consistent with theories of how they may relate to the observed in vivo phenomena in humans.

ANIMAL MODELS OF SCHIZOPHRENIA WITH RELATION TO RESTING GAMMA ACTIVITY

As in humans, gamma activity has been studied in preclinical animal models to use schizophrenia-like electrophysiological phenotypes to help expedite drug discovery. Considerable evidence has been reported to support the "noisy brain" hypothesis as a relevant feature of abnormal behavior and deficit states. Many groups have modeled schizophrenia endophenotypes in mouse and rat models by modifying the amount of functional NMDA-R in the brain by genetic perturbation of gene expression and/or environmental/developmental manipulations. Many of these models include changing the expression of the NMDA-R subunit 1 (NR1), because this subunit is required to form functional receptors alone or in combination with NMDA-R2 subunits, of which there are several further subtypes (ie, NR2 A-D). For example, mice with a global reduction in the amount of NR1 protein, called NR1 hypomorphic mice, display increased LFP/EEG baseline gamma power (Gandal et al., 2012c; Fig. 14.1). Subsequently, mouse models that selectively knock-out NR1 in different neuronal cell types have also been investigated. Mice with a selective pyramidal cell knock-out of NR1 have increased gamma baseline power as measured by EEG (Tatard-Leitman et al., 2015). Similarly, mice with a selective knock-out of NR1 in a subset of interneurons that express the calcium-binding protein PV also have a significant increase in baseline EEG gamma power (Billingslea et al., 2014; Carlén et al., 2012; Fig. 14.2). Additionally, animals with a developmental knock-out of NR1 in a subset of interneurons (40–50% of cortical interneurons) exhibit high spontaneous LFP activity in the primary auditory cortex (Nakao and Nakazawa, 2014). Interestingly, genetic manipulations of genes in signaling pathways that modify NMDA-R activity also support the role of these receptors in mediating increased resting state activity. For example, previous studies indicate that neuregulin-1 activity at ErbB4 reduces ion currents through NMDA-Rs. Importantly, postmortem studies demonstrate that schizophrenia subjects have increased neuregulin-1-mediated activation of ErbB4, resulting in increased suppression of NMDA-R-mediated glutamate transmission. Additionally, developmental loss of ErbB4 from fast-spiking basket and chandelier interneurons causes an increase in LFP gamma power in awake freely moving mice (Del Pino et al., 2013). Dysbindin (*DTNBP1*) is another example of a gene that has received a great deal of attention due to its genetic association with schizophrenia. Dysbindin knockout mice ($Dys1^{-/-}$) show an increase in "late gamma," which the authors suggest is the manifestation of an inability to inhibit gamma activity (Carlson et al., 2011; Gandal et al., 2012b). As an alternative to genetic approaches, schizophrenia has also been modeled developmentally by administration of a mitotoxin methylazoxymethanol (MAM) on gestational day 17 (GD17) to rats. This model is thought to recapitulate disruption of the developmental process and leads to schizophrenia-like phenotypes in rodent

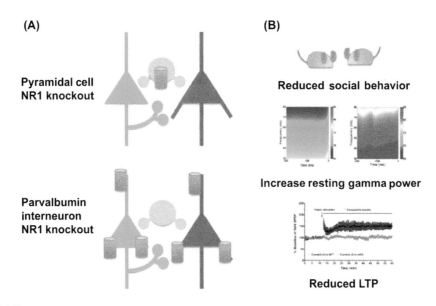

FIGURE 14.2

(A) A simple circuit is portrayed in which there is both excitatory connectivity and feedforward and feedback inhibition. The upper left figure depicts the pyramidal cell (triangles) NR1 knockouts, which lack NMDA-R (orange cylinders) on cortical pyramidal neurons. The lower left figure illustrates the parvalbumin neuron NR1 knockouts, which lack NMDA-R in this population of GABA interneurons (green circles), (B) The right figures portray the common findings in the two groups of animals: increases in resting gamma oscillations associated with reduced evoked gamma oscillations (top), reductions in long-term potentiation (middle), and reductions in social behavior (bottom). *KO*, knockout; *LTP*, long-term potentiation.

Adapted from: Krystal, J.H., 2015. Deconstructing N-Methyl-D-Aspartate glutamate receptor contributions to cortical circuit functions to construct better hypotheses about the pathophysiology of schizophrenia. Biol. Psychiatry 77, 508–510, (Krystal, 2015).

offspring, which include altered neuronal processing (Goto and Grace, 2006; Lavin et al., 2005; Lodge and Grace, 2007). The MAM gestational model of schizophrenia also shows an increase in baseline gamma activity in the prefrontal cortex in awake animals (Kocsis et al., 2013). However, another group found that there was no increase in gamma activity at rest (Phillips et al., 2012), and yet another (Lodge et al., 2009) found that there was no difference between groups for gamma power before auditory-evoked responses (note: in some studies, prestimulus baseline is used as a surrogate for resting state). Finally, groups have used another developmental model of schizophrenia in which injections of ibotenic acid are performed bilaterally in the ventral hippocampus to create lesions on postnatal day 6 or 7 (Vohs et al., 2009). Although the authors did not test for differences in spontaneous gamma activity, they note in the discussion that neonatal ventral hippocampus lesions (NVHL) rats take significantly longer to return to prestimulus levels of neuronal activity, suggestive of baseline elevations. All of these findings are consistent with pharmacological studies using NMDA-R antagonists because ketamine and MK801, which are NMDA-R antagonists, cause an increase in background EEG gamma activity (Ehrlichman et al., 2009; Lazarewicz et al., 2010; Saunders et al., 2012; Fig. 14.1).

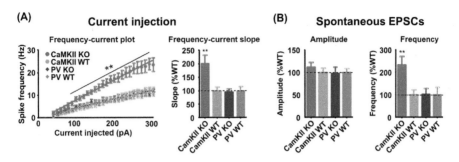

FIGURE 14.3

Membrane properties of conditional NR1 knockout mice. (A) Selective knockout of NR1 in pyramidal cells yields an increase in membrane excitability, as evidenced by increased spike frequency for any level of current injected, (B) There is an increase in the frequency of excitatory postsynaptic currents (EPSCs) on hippocampal pyramidal neurons in pyramidal cell-selective NR1 knockouts. Alternatively, the EPSC amplitude in these cells is unchanged (Billingslea et al., 2014; Tatard-Leitman et al., 2015). Red, pyramidal cell-selective NR1 KO; pink, pyramidal cell wild-type control mice; dark blue, PV interneuron selective NR1 KO mice; light blue, PV NR1 controls.

IN VITRO EXAMINATION OF ANIMAL MODELS OF INCREASED RESTING GAMMA

As noted in the previous section, a variety of studies have focused on proteins that modify glutamate signal transduction through NMDA-Rs. Hippocampal pyramidal cells from NR1 hypomorphic mice (Gandal et al., 2012a) have an increase in inherent membrane excitability, as indicated by both a reduction in the amount of current needed to elicit an action potential (lower rheobase value) and an increase in the current–firing frequency relationship. This is similar to the pattern of changes found among mice with a pyramidal cell-selective ablation of NR1. Pyramidal neurons in these mice show an increase in current frequency, spontaneous excitatory postsynaptic current (EPSC) frequency, and a decrease in resting membrane potential (Tatard-Leitman et al., 2015). Alternatively, mice with a selective reduction of NR1 limited to PV-containing interneurons do not display increased pyramidal cell excitability in vitro (Billingslea et al., 2014; Korotkova et al., 2010; Fig. 14.3).

Consistent with in vivo studies noted, ErbB4 conditional knockout mice also have more increased membrane excitability (lower rheobase value) values than their wild-type controls, which was observed in interneurons but not pyramidal cells (Del Pino et al., 2013). However, pyramidal cells in these animals have an increase in spontaneous EPSCs, consistent with the pattern of changes in pyramidal cell-selective reduction in NR1 expression. Importantly, these data suggest that altering the modulation of NMDA-R function either by increased ErbB4 activity, as in schizophrenia, or by decreased *ErbB4 Neuroligan-1* expression, as in animal models, yields similar outcomes with respect to baseline gamma power.

Although some animal models discussed have not been extensively evaluated using in vitro electrophysiological methods, perturbation of genes associated with functional interneurons and/or cell numbers has been used a surrogate for abnormal brain circuitry. MAM-treated rats have a decrease in expression of PV and glutamic acid decarboxylase 67 (GAD_{67}) and a decrease in the number of PV interneurons in the frontal cortex (Gill and Grace, 2014; Lodge and Grace, 2009). GAD_{67} is an enzyme that is crucial to the synthesis of GABA. The NVHL animals also have a decrease in the mRNA levels of GAD_{67} in the medial prefrontal cortex (Francois et al., 2009; Lipska, 2004). Similarly, reductions in PV immunoreactive cells are also seen in dysbindin knockout mice (Carlson et al., 2011).

RELATING BEHAVIOR TO EEG/LFP VALIDATION OF MODELS

Models that recreate increased background activity can also be assessed with regard to the extent to which they result in a broader behavioral phenotype similar to schizophrenia (Papaleo et al., 2012). Therefore, in the subsequent section we discuss the extent to which each model system meets this criterion.

Cognitive deficits are at the core of schizophrenia, including difficulties in problem solving, social cognition, and working memory (Powell, 2010). Therefore, rodent models have focused on tasks associated with learning and memory, working memory, and cognitive flexibility. Animals with global reduction of NR1 or in which NR1 is selectively knocked-out in pyramidal cells have deficits in spatial working memory when using the T-maze performance task (Gandal et al., 2012a; Tatard-Leitman et al., 2015; Fig. 14.4). *ErbB4* conditional mutants also have a decrease in correct alternation using

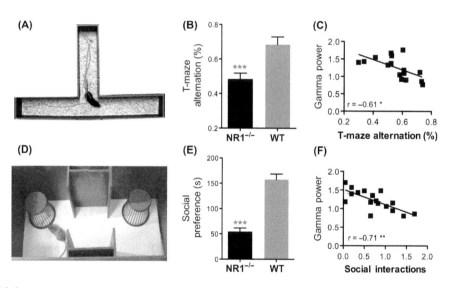

FIGURE 14.4

NR1 hypomorphic animals, an example of an animal model of schizophrenia, display both cognitive and social impairments. (A) Diagram of T-maze paradigm, (B) NR1 hypomorphs perform poorly on the T-maze alternation task, (C) There is an inverse correlation with increasing baseline gamma and reduced T-maze alternation task performance (ie, the higher the baseline gamma power, the more the animal is cognitively impaired), (D) Diagram of social behavior preference paradigm, (E) NR1 hypomorphs have a decrease in sociability, (F) The decrease in sociability has an inverse correlation with baseline gamma power such that social behavior reduces as a function of increasing baseline gamma power (Gandal et al., 2012c).

this task, suggesting a similar cognitive defect (Del Pino et al., 2013). Baclofen, a $GABA_B$ receptor agonist, rescued the T-maze phenotype in NR1 hypomorphs, coincident with its ability to restore normal background EEG gamma power and reduce the abnormal inherent membrane properties in pyramidal cells. The role of NMDA-R in interneurons in mediating cognitive performance has been less consistent. For example, developmental knockdown of NR1 in a broad range of interneurons causes a decrease in short-term memory (Belforte et al., 2010). However, animals with a selective knock-out of NR1 in PV interneurons have reduced working memory in some studies (Korotkova et al., 2010) but significantly improved T-maze performance in others (Billingslea et al., 2014). Thus, the role of impaired interneuron function in mediating functional deficits that are associated with increased resting gamma power remains unclear. Similar to data in the PV-selective NR1 knock-out mice, studies in dysbindin knockout mice are also mixed. Some studies indicate that these mice have increased T-maze performance, whereas others find the converse (Hattori et al., 2008; Karlsgodt et al., 2011; Papaleo et al., 2012; Takao et al., 2008). Alternatively, dysbindin knockout mice display impaired spatial reference memory and novel object recognition performance (Karlsgodt et al., 2011; Papaleo et al., 2012). Freely moving rats treated prenatally with MAM have a diminished gamma band response during task performance and display working memory deficits (Flagstad et al., 2005; Le Pen et al., 2006; Moore et al., 2006). Similarly, rats with NVHL exhibit working memory deficits (Lee et al., 2012; McDannald et al., 2011).

Hyperactivity is also used as a behavioral surrogate to model the psychotic symptoms of schizophrenia in animals (Lipska and Weinberger, 2000). Although increased locomotor activity has relatively poor construct validity for psychosis, it does have good predictive validity with regard to pharmacological agents. Specifically, agents that make rodents increase their locomotor activity tend to have psychotomimetic properties in humans (Jones et al., 2011; Swerdlow et al., 2000). Among genetic models with increased baseline gamma activity, increased locomotor activity has been noted in NR1 hypomorphs (Gandal et al., 2012c), pyramidal cell-selective NR1 knockout mice (Tatard-Leitman et al., 2015), mice with developmental knock-out of NR1 in a subset of interneurons (Belforte et al., 2010), ErbB4 conditional knockout mice (Del Pino et al., 2013), and dysbindin knockout mice (Karlsgodt et al., 2011; Papaleo et al., 2012). Similarly, developmental lesion models including both MAM-treated and NVHL rats exhibit hyperactivity (Lipska et al., 1993; Lipska and Weinberger, 2000). Thus, there is a large degree of overlap among manipulations that cause animal behaviors that are predictive of psychosis and that coincidentally cause an increase in background gamma activity.

Changes in social drive and social withdrawal are also symptoms of schizophrenia that can be modeled in animals (Miller et al., 2002; Fig. 14.4). Preference for social interactions is typically tested with a three-chambered social choice test. In this test, a test mouse's propensity to choose social interaction with another mouse as compared to an inanimate object is quantified. As with cognitive deficits and locomotion, there is a high degree of overlap between increased baseline gamma activity and reduced social interactions. For example, there is a reduction in social behavior in mice with knock-out of dysbindin, ErbB4, PV-selective NR1, general interneuron NR1, pyramidal cell-selective NR1, and NR1 hypomorphs (Belforte et al., 2010; Billingslea et al., 2014; Gandal et al., 2012c; Hattori et al., 2008; Korotkova et al., 2010; Tatard-Leitman et al., 2015). Similarly, rats treated prenatally with MAM (Flagstad et al., 2004; Le Pen et al., 2006) and NVHL rats also have social deficits, suggesting that both genetic and developmental lesion models that cause increased baseline gamma power also cause social impairments (O'Donnell, 2011; Sams-Dodd et al., 1997).

SUMMARY AND CONCLUSIONS

The increase in baseline resting gamma brain activity may underlie many of the symptoms of schizophrenia. For example, in humans one group has postulated that the increase in resting state gamma activity is a predictor for auditory verbal hallucinations (Northoff, 2014). Although such a link to subjective experiences may limit the capacity to model the relevant aspects of schizophrenia in rodents, it may also explain why mice with increased EEG resting gamma power are unable to respond to auditory stimuli in the same fashion as wild-type animals. Interestingly, alterations in gamma power have been identified among other disorders that share some, but not all, features of schizophrenia. Specifically, people with autism spectrum disorders (ASD) also display increased background gamma power but do not experience hallucinations and delusions (Van Diessen et al., 2014; Wang et al., 2013). We propose that the developmental timing during which increased background activity emerges determines the types of subjective and functional abnormalities that result. Specifically, the postadolescent emergence of increased background gamma power may lead to interference with previously developed sensory (hallucinations) and higher-order logical processing (eg, delusions). However, if this increased gamma power is present from birth, as in ASD, we propose that sensory and higher-order cognitive systems are able to incorporate the "noise" during development and therefore do not introduce abnormal sensory or logical interference. As such, increased gamma band noise at rest not only may underlie some of the common features among schizophrenia and ASD, like impaired social function and expression of language, but also may help explain some of the unique features of each disorder based on the temporal nature of its development.

In addition to informing the pathophysiology of schizophrenia, the ability to model basic alterations in EEG power spectra at rest will facilitate the discovery of new therapeutic approaches. Specifically, future research to find compounds that decrease baseline gamma power is likely to be among the most powerful tools for treating the basic underlying pathophysiology of schizophrenia. Furthermore, therapeutic agents that increase gamma signal-to-noise ratio by reducing background prestimulus "noise" may improve perceptual abnormalities in schizophrenia. Likewise, compounds that reduce baseline and elevate evoked or induced gamma power during cognitive paradigms may help alleviate these deficits in schizophrenia (Fig. 14.5).

In addition to pharmacological treatments noted, neuromodulatory approaches may also hold promise for addressing fundamental abnormalities in brain rhythms that we believe are the root cause of symptoms and functional deficits in schizophrenia. For example, deep brain stimulation (DBS) might be able to ameliorate the reduction in cross-frequency coupling and coordination between different brain regions. Previous studies in our group investigated a mouse model of DBS in auditory cortex of mice and found that the technique can modify slow wave activity (De Rojas et al., 2013). Similarly, DBS can reverse hippocampal information processing deficits among animals with a variety of developmental manipulations thought to model schizophrenia (Ewing and Grace, 2013; Klein et al., 2013; Perez et al., 2013). In humans, transcranial direct current stimulation has been used to induce changes in brain activity, suggesting that nonpharmacological approaches may also have promise in treating the abnormal underlying brain rhythms in the disorder (Jacobson et al., 2012; Keeser et al., 2011; Zaehle et al., 2011a; Zaehle et al., 2010; Zaehle et al., 2011b).

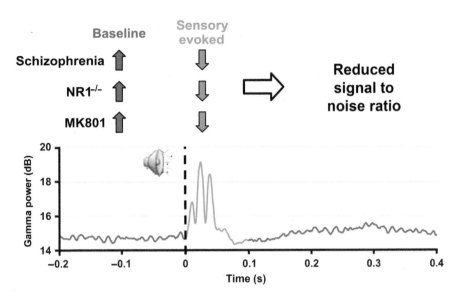

FIGURE 14.5

Schematic representation of similarities among schizophrenia, constitutive reduction in NMDA-R expression in mice, and NMDA-R antagonists in mice. All conditions are characterized by an increase in baseline coupled with a reduction of evoked gamma power, yielding a reduction in signal-to-noise ratio (Gandal et al., 2012b).

REFERENCES

Andreou, C., Faber, P.L., Leicht, G., Schoettle, D., Polomac, N., Hanganu-Opatz, I.L., et al., 2014a. Resting-state connectivity in the prodromal phase of schizophrenia: insights from EEG microstates. Schizophr. Res. 152, 513–520.

Andreou, C., Nolte, G., Leicht, G., Polomac, N., Hanganu-Opatz, I.L., Lambert, M., et al., 2014b. Increased resting-state gamma-band connectivity in first-episode schizophrenia. Schizophr. Bull. 41, 930–939.

Argyelan, M., Ikuta, T., Derosse, P., Braga, R.J., Burdick, K.E., John, M., et al., 2014. Resting-state fMRI connectivity impairment in schizophrenia and bipolar disorder. Schizophr. Bull. 40, 100–110.

Bartos, M., Vida, I., Jonas, P., 2007. Synaptic mechanisms of synchronized gamma oscillations in inhibitory interneuron networks. Nat. Rev. Neurosci. 8, 45–56.

Basar, E., Basar-Eroglu, C., Karakas, S., Schurmann, M., Başar, E., Başar-Eroglu, C., et al., 2001. Gamma, alpha, delta, and theta oscillations govern cognitive processes. Int. J. Psychophysiol. 39, 241–248.

Basar, E., Bullock, T.H., 1992. *Induced Rhythms in the Brain. Brain Dynamics Series*. Birkhäuser, Boston, MA.

Belforte, J.E., Zsiros, V., Sklar, E.R., Jiang, Z., Yu, G., Li, Y., et al., 2010. Postnatal NMDA receptor ablation in corticolimbic interneurons confers schizophrenia-like phenotypes. Nat. Neurosci. 13, 76–83.

Benes, F.M., Vincent, S.L., Todtenkopf, M., 2001. The density of pyramidal and nonpyramidal neurons in anterior cingulate cortex of schizophrenic and bipolar subjects. Biol. Psychiatry 50, 395–406.

Berger, H., 1929. Ueber das Electrocephalogramm des Menschen. Arch. Psychiatr. Nervenkr. 87, 527–570.
Bigos, K.L., Mattay, V.S., Callicott, J.H., Straub, R.D., Vakkalanka, R., Kolachana, B., et al., 2013. Genetic variation in CACNA1C affects brain circuitries related to mental illness. Arch. Gen. Psychiatry 67, 939–945.
Billingslea, E.N., Tatard-Leitman, V.M., Anguiano, J., Jutzeler, C.R., Suh, J., Saunders, J.A., et al., 2014. Parvalbumin cell ablation of NMDA-R1 causes increased resting network excitability with associated social and self-care deficits. Neuropsychopharmacology 39, 1603–1613.
Brockhaus-Dumke, A., Mueller, R., Faigle, U., Klosterkoetter, J., 2008. Sensory gating revisited: relation between brain oscillations and auditory evoked potentials in schizophrenia. Schizophr. Res. 99, 238–249.
Buckner, R.L., Andrews-Hanna, J.R., Schacter, D.L., 2008. The brain's default network: anatomy, function, and relevance to disease. Ann. N.Y. Acad. Sci. 1124, 1–38.
Buzsáki, G., 2006. Rhythms of the Brain. Oxford University Press, Oxford; NY.
Buzsáki, G., Draguhn, A., 2004. Neuronal oscillations in cortical networks. Science (New York, N.Y.) 304, 1926–1929.
Buzsáki, G., Wang, X.-J., 2012. Mechanisms of gamma oscillations. Annu. Rev. Neurosci. 35, 203–225.
Carlén, M., Meletis, K., Siegle, J.H., Cardin, J.A., Futai, K., Vierling-Claassen, D., et al., 2012. A critical role for NMDA receptors in parvalbumin interneurons for gamma rhythm induction and behavior. Mol. Psychiatry 17, 537–548.
Carlson, G.C., Talbot, K., Halene, T.B., Gandal, M.J., Kazi, H.A., Schlosser, L., et al., 2011. Dysbindin-1 mutant mice implicate reduced fast-phasic inhibition as a final common disease mechanism in schizophrenia. Proc. Natl. Acad. Sci. U.S.A. 108, 962–970.
Coyle, J.T., 2004. The GABA-glutamate connection in schizophrenia: which is the proximate cause? Biochem. Pharmacol. 68, 1507–1514.
Coyle, J.T., 2006. Glutamate and schizophrenia: beyond the dopamine hypothesis. Cell. Mol. Neurobiol. 26, 365–384.
Coyle, J.T., Tsai, G., Goff, D., 2003. Converging evidence of NMDA receptor hypofunction in the pathophysiology of schizophrenia. Ann. N.Y. Acad. Sci. 1003, 318–327.
Cross-Disorder Group of the Psychiatric Genomics Consortium, 2013. Identification of risk loci with shared effects on five major psychiatric disorders: a genome-wide analysis. Lancet 381, 1371–1379.
Cunningham, M.O., Hunt, J., Middleton, S., LeBeau, F.E., Gillies, M.J., Davies, C.H., et al., 2006. Region-specific reduction in entorhinal gamma oscillations and parvalbumin-immunoreactive neurons in animal models of psychiatric illness. J. Neurosci. 26, 2767–2776.
Damoiseaux, J., 2006. Consistent resting-state networks across healthy subjects. Proc. Natl. Acad. Sci. U.S.A. 103, 13848–13853.
Davis, P.A., 1942. Comparative study of the EEGs of schizophrenic and manic-depressive patients. Am. J. Psychiatry 99, 210–217.
Del Pino, I., García-Frigola, C., Dehorter, N., Brotons-Mas, J.R., Alvarez-Salvado, E., Martínez deLagrán, M., et al., 2013. ErbB4 deletion from fast-spiking interneurons causes schizophrenia-like phenotypes. Neuron 79, 1152–1168.
De Rojas, J.O., Saunders, J.A., Luminais, C., Hamilton, R.H., Siegel, S.J., 2013. Electroencephalographic changes following direct current deep brain stimulation of auditory cortex: a new model for investigating neuromodulation. Neurosurgery 72, 267–275. (discussion 275).
Ehrlichman, R.S., Gandal, M.J., Maxwell, C.R., Lazarewicz, M.T., Finkel, L.H., Contreras, D., et al., 2009. N-methyl-D-aspartic acid receptor antagonist-induced frequency oscillations in mice recreate pattern of electrophysiological deficits in schizophrenia. Neuroscience 158, 705–712.
Enomoto, T., Tse, M.T., Floresco, S.B., 2011. Reducing prefrontal gamma-aminobutyric acid activity induces cognitive, behavioral, and dopaminergic abnormalities that resemble schizophrenia. Biol. Psychiatry 69, 432–441.

Ewing, S.G., Grace, A.A., 2013. Deep brain stimulation of the ventral hippocampus restores deficits in processing of auditory evoked potentials in a rodent developmental disruption model of schizophrenia. Schizophr. Res. 143, 377–383.

Fenton, G.W., Fenwick, P.B., Dollimore, J., Dunn, T.L., Hirsch, S.R., 1980. EEG spectral analysis in schizophrenia. Br. J. Psychiatry 136, 445–455.

Finley, K.H., 1944. On the occurrence of rapid frequency potential changes in the human electroencephalogram. Am. J. Psychiatry 101, 194–200.

Flagstad, P., Glenthøj, B.Y., Didriksen, M., 2005. Cognitive deficits caused by late gestational disruption of neurogenesis in rats: a preclinical model of schizophrenia. Neuropsychopharmacology 30, 250–260.

Flagstad, P., Mørk, A., Glenthøj, B.Y., van Beek, J., Michael-Titus, A.T., Didriksen, M., 2004. Disruption of neurogenesis on gestational day 17 in the rat causes behavioral changes relevant to positive and negative schizophrenia symptoms and alters amphetamine-induced dopamine release in nucleus accumbens. Neuropsychopharmacology 29, 2052–2064.

Francois, J., Ferrandon, A., Koning, E., Angst, M.-J.J., Sandner, G., Nehlig, A., et al., 2009. Selective reorganization of GABAergic transmission in neonatal ventral hippocampal-lesioned rats. Int. J. Neuropsychopharmacol. 12, 1097–1110.

Frankel, W.N., Beyer, B., Maxwell, C.R., Pretel, S., Letts, V.A., Siegel, S.J., 2005. Development of a new genetic model for absence epilepsy: spike-wave seizures in C3H/He and backcross mice. J. Neurosci. 25, 3452–3458.

Fung, S.J., Fillman, S.G., Webster, M.J., Shannon Weickert, C., 2014. Schizophrenia and bipolar disorder show both common and distinct changes in cortical interneuron markers. Schizophr. Res. 155, 26–30.

Gandal, M.J., Anderson, R.L., Billingslea, E.N., Carlson, G.C., Roberts, T.P.L., Siegel, S.J., 2012a. Mice with reduced NMDA receptor expression: more consistent with autism than schizophrenia? Genes Brain Behav. 11, 740–750.

Gandal, M.J., Edgar, J.C., Klook, K., Siegel, S.J., 2012b. Gamma synchrony: towards a translational biomarker for the treatment-resistant symptoms of schizophrenia. Neuropharmacology 62, 1504–1518.

Gandal, M.J., Sisti, J., Klook, K., Ortinski, P.I., Leitman, V., Liang, Y., et al., 2012c. GABAB-mediated rescue of altered excitatory-inhibitory balance, gamma synchrony and behavioral deficits following constitutive NMDAR-hypofunction. Transl. Psychiatry 2, e142.

Gargus, J.J., 2006. Ion channel functional candidate genes in multigenic neuropsychiatric disease. Biol. Psychiatry 60, 177–185.

Garrity, A.G., Pearlson, G.D., Mckiernan, K., Lloyd, D., Kiehl, K.A., Calhoun, V.D., 2007. Aberrant "default mode" functional connectivity in schizophrenia. Am. J. Psychiatry 164, 450–457.

Georgiev, D., Arion, D., Enwright, J.F., Kikuchi, M., Minabe, Y., Corradi, J.P., et al., 2014. Lower gene expression for KCNS3 potassium channel subunit in parvalbumin-containing neurons in the prefrontal cortex in schizophrenia. Am. J. Psychiatry 171, 62–71.

Giannitrapani, D., Kayton, L., 1974. Schizophrenia and EEG spectral analysis. Electroencephalogr. Clin. Neurophysiol. 36, 377–386.

Gill, K.M., Grace, A.A., 2014. Corresponding decrease in neuronal markers signals progressive parvalbumin neuron loss in MAM schizophrenia model. Int. J. Neuropsychopharmacol. 17, 1609–1619.

Gonzalez-Burgos, G., Lewis, D.A., 2012. NMDA receptor hypofunction, parvalbumin-positive neurons, and cortical gamma oscillations in schizophrenia. Schizophr. Bull. 38, 950–957.

Goto, Y., Grace, A.A., 2006. Alterations in medial prefrontal cortical activity and plasticity in rats with disruption of cortical development. Biol. Psychiatry 60, 1259–1267.

Green, E.K., Grozeva, D., Jones, I., Jones, L., Kirov, G., Caesar, S., et al., 2010. The bipolar disorder risk allele at CACNA1C also confers risk of recurrent major depression and of schizophrenia. Mol. Psychiatry 15, 1016–1022.

Gross, J., 2014. Analytical methods and experimental approaches for electrophysiological studies of brain oscillations. J. Neurosci. Methods 228, 57–66.

Gross, J., Baillet, S., Barnes, G.R., Henson, R.N., Hillebrand, A., Jensen, O., et al., 2013. Good practice for conducting and reporting MEG research. Neuroimage 65, 349–363.

Guidotti, A., Auta, J., Davis, J.M., Gerevini, V.D., Dwivedi, Y., Grayson, D.R., et al., 2000. Decrease in reelin and glutamic acid decarboxylase 67 (GAD_{67}) expression in schizophrenia and bipolar disorder. Arch. Gen. Psychiatry 57, 1061–1069.

Hattori, S., Murotani, T., Matsuzaki, S., Ishizuka, T., Kumamoto, N., Takeda, M., et al., 2008. Behavioral abnormalities and dopamine reductions in sdy mutant mice with a deletion in Dtnbp1, a susceptibility gene for schizophrenia. Biochem. Biophys. Res. Commun. 373, 298–302.

Herrmann, C.S., Demiralp, T., 2005. Human EEG gamma oscillations in neuropsychiatric disorders. Clin. Neurophysiol. 116, 2719–2733.

Herrmann, C.S., Frund, I., Lenz, D., Fründ, I., Lenz, D., 2010. Human gamma-band activity: a review on cognitive and behavioral correlates and network models. Neurosci. Biobehav. Rev. 34, 981–992.

Homayoun, H., Moghaddam, B., 2007. NMDA receptor hypofunction produces opposite effects on prefrontal cortex interneurons and pyramidal neurons. J. Neurosci. 27, 11496–11500.

Hong, L.E., Summerfelt, A., Mitchell, B.D., McMahon, R.P., Wonodi, I., Buchanan, R.W., et al., 2008. Sensory gating endophenotype based on its neural oscillatory pattern and heritability estimate. Arch. Gen. Psychiatry 65, 1008–1016.

Huffaker, S.J., Chen, J., Nicodemus, K.K., Sambataro, F., Yang, F., Mattay, V., et al., 2009. A primate-specific, brain isoform of KCNH2 affects cortical physiology, cognition, neuronal repolarization and risk of schizophrenia. Nat. Med. 15, 509–518.

Itil, T.M., Saletu, B., Davis, S., 1972. EEG findings in chronic schizophrenics based on digital computer period analysis and analog power spectra. Biol. Psychiatry 5, 1–13.

Itil, T.M., Saletu, B., Davis, S., Allen, M., 1974. Stability studies in schizophrenics and normals using computer-analyzed EEG. Biol. Psychiatry 8, 321–335.

Jacobson, L., Ezra, A., Berger, U., Lavidor, M., 2012. Modulating oscillatory brain activity correlates of behavioral inhibition using transcranial direct current stimulation. Clin. Neurophysiol. 123, 979–984.

Javitt, D.C., Zukin, S.R., 1991. Recent advances in the phencyclidine model of schizophrenia. Am. J. Psychiatry 148, 1301–1308.

Jentsch, J.D., Andrusiak, E., Tran, A., Bowers, M.B., Roth, R.H., 1997. THC increases prefrontal cortical catecholaminergic utilization and impairs spatial working memory in the rat.pdf. Neuropsychopharmacology 16, 426–432.

Jones, C., Watson, D.J., Fone, K., 2011. Animal models of schizophrenia. Br. J. Pharmacol. 164, 1162–1194.

Karbasforoushan, H., Woodward, N., 2012. Resting-state networks in schizophrenia. Curr. Top. Med. Chem. 21, 2404–2414.

Karlsgodt, K.H., Robleto, K., Trantham-Davidson, H., Jairl, C., Cannon, T.D., Lavin, A., et al., 2011. Reduced dysbindin expression mediates N-methyl-D-aspartate receptor hypofunction and impaired working memory performance. Biol. Psychiatry 69, 28–34.

Keeser, D., Padberg, F., Reisinger, E., Pogarell, O., Kirsch, V., Palm, U., et al., 2011. Prefrontal direct current stimulation modulates resting EEG and event-related potentials in healthy subjects: a standardized low resolution tomography (sLORETA) study. Neuroimage 55, 644–657.

Kennard, M.A., Schwartzman, A.E., 1957. A longitudinal study of electroencephalographic frequency patterns in mental hospital patients and normal controls. Electroencephalogr. Clin. Neurophysiol. 9, 263–274.

Kim, J.S., Shin, K.S., Jung, W.H., Kim, S.N., Kwon, J.S., Chung, C.K., 2014. Power spectral aspects of the default mode network in schizophrenia: an MEG study. BMC Neurosci. 15, 104.

Kirihara, K., Rissling, A.J., Swerdlow, N.R., Braff, D.L., Light, G.A., 2012. Hierarchical organization of gamma and theta oscillatory dynamics in schizophrenia. Biol. Psychiatry 71, 873–880.

Kissler, J., Muller, M.M., Fehr, T., Rockstroh, B., Elbert, T., 2000. MEG gamma band activity in schizophrenia patients and healthy subjects in a mental arithmetic task and at rest. Clin. Neurophysiol. 111, 2079–2087.

Klein, J., Hadar, R., Götz, T., Männer, A., Eberhardt, C., Baldassarri, J., et al., 2013. Mapping brain regions in which deep brain stimulation affects schizophrenia-like behavior in two rat models of schizophrenia. Brain Stimulation 6, 490–499.

Kocsis, B., Lee, P., Deth, R., 2013. Enhancement of gamma activity after selective activation of dopamine D4 receptors in freely moving rats and in a neurodevelopmental model of schizophrenia. Brain Struct. Funct. 219, 2173–2180.

Konradi, C., Yang, C.K., Zimmerman, E.I., Lohmann, K.M., Gresch, P., Pantazopoulos, H., et al., 2011. Hippocampal interneurons are abnormal in schizophrenia. Schizophr. Res. 131, 165–173.

Korotkova, T.M., Fuchs, E.C., Ponomarenko, A., von Engelhardt, J., Monyer, H., 2010. NMDA receptor ablation on parvalbumin-positive interneurons impairs hippocampal synchrony, spatial representations, and working memory. Neuron 68, 557–569.

Krishnan, G.P., Vohs, J.L., Hetrick, W.P., Carroll, C.A., Shekhar, A., Bockbrader, M.A., et al., 2005. Steady state visual evoked potential abnormalities in schizophrenia. Clin. Neurophysiol. 116, 614–624.

Krystal, J.H., 2015. Deconstructing N-Methyl-D-Aspartate glutamate receptor contributions to cortical circuit functions to construct better hypotheses about the pathophysiology of schizophrenia. Biol. Psychiatry 77, 508–510.

Krystal, J.H., Karper, L.P., Seibyl, J.P., Freeman, G.K., Delaney, R., Bremner, J.D., et al., 1994. Subanesthetic effects of noncompetitive NMDA antagonist, ketamine, in humans: psychometric, perceptual, cognitive and neuroendocrine responses. Arch. Gen. Psychiatry 51, 199–214.

Lahti, A.C., Koffel, B., LaPorte, D., Tamminga, C.A., 1995. Subanesthetic doses of ketamine stimulate psychosis in schizophrenia. Neuropsychopharmacology 13, 9–19.

Lahti, A.C., Weiler, M. a, Michaelidis, T., Parwani, A., Tamminga, C.A., 2001. Effects of ketamine in normal and schizophrenic volunteers. Neuropsychopharmacology 25, 455–467.

Landin-Romero, R., McKenna, P.J., Salgado-Pineda, P., Sarró, S., Aguirre, C., Sarri, C., et al., 2014. Failure of deactivation in the default mode network: a trait marker for schizophrenia? Psychol. Med. 45, 1315–1325.

Laufs, H., 2008. Endogenous brain oscillations and related networks detected by surface EEG-combined fMRI. Hum. Brain Mapp. 29, 762–769.

Lavin, A., Moore, H.M., Grace, A.A., 2005. Prenatal disruption of neocortical development alters prefrontal cortical neuron responses to dopamine in adult rats. Neuropsychopharmacology 30, 1426–1435.

Lazarewicz, M.T., Ehrlichman, R.S., Maxwell, C.R., Gandal, M.J., Finkel, L.H., Siegel, S.J., 2010. Ketamine modulates theta and gamma oscillations. J. Cogn. Neurosci. 22, 1452–1464.

Lee, H., Dvorak, D., Kao, H.-Y., Duffy, Á.M., Scharfman, H.E., Fenton, A.A., 2012. Early cognitive experience prevents adult deficits in a neurodevelopmental schizophrenia model. Neuron 75, 714–724.

Le Pen, G., Gourevitch, R., Hazane, F., Hoareau, C., Jay, T.M., Krebs, M.O., 2006. Peri-pubertal maturation after developmental disturbance: a model for psychosis onset in the rat. Neuroscience 143, 395–405.

Lewis, D.A., Fish, K.N., Arion, D., Gonzalez-Burgos, G., 2011. Perisomatic inhibition and cortical circuit dysfunction in schizophrenia. Curr. Opin. Neurobiol. 21, 866–872.

Lewis, D.A., Curley, A.A., Glausier, J.R., Volk, D.W., 2012. Cortical parvalbumin interneurons and cognitive dysfunction in schizophrenia. Trends Neurosci. 35, 57–67.

Lipska, B.K., 2004. Using animal models to test a neurodevelopmental hypothesis of schizophrenia. J. Psychiatry Neurosci. 29, 282–286.

Lipska, B.K., Weinberger, D.R., 2000. To model a psychiatric disorder in animals: schizophrenia as a reality test. Neuropsychopharmacology 23, 223–239.

Lipska, B.K., Jaskiw, G.E., Weinberger, D.R., 1993. Postpubertal emergence of hyperresponsiveness to stress and to amphetamine after neonatal excitotoxic hippocampal damage: a potential animal model of schizophrenia. Neurosychopharmacology 9, 67–75.

Lisman, J.E., Coyle, J.T., Green, R.W., Javitt, D.C., Benes, F.M., Heckers, S., et al., 2008. Circuit-based framework for understanding neurotransmitter and risk gene interactions in schizophrenia. Trends Neurosci. 31, 234–242.

Lodge, D.J., Behrens, M.M., Grace, A.A., 2009. A loss of parvalbumin-containing interneurons is associated with diminished oscillatory activity in an animal model of schizophrenia. J. Neurosci. 29, 2344–2354.

Lodge, D.J., Grace, A.A., 2007. Aberrant hippocampal activity underlies the dopamine dysregulation in an animal model of schizophrenia. J. Neurosci. 27, 11424–11430.

Lodge, D.J., Grace, A.A., 2009. Gestational methylazoxymethanol acetate administration: a developmental disruption model of schizophrenia. Behav. Brain Res. 204, 306–312.

Lopes da Silva, F., 2013. EEG and MEG: relevance to neuroscience. Neuron 80, 1112–1128.

Malmivuo, J., 2012. Comparison of the properties of EEG and MEG in detecting the electric activity of the brain. Brain Topogr. 25, 1–19.

Mamah, D., Barch, D.M., Repovš, G., 2013. Resting state functional connectivity of five neural networks in bipolar disorder and schizophrenia. J. Affect. Disord. 150, 601–609.

Manoliu, A., Riedl, V., Zherdin, A., Mühlau, M., Schwerthöffer, D., Scherr, M., et al., 2014. Aberrant dependence of default mode/central executive network interactions on anterior insular salience network activity in schizophrenia. Schizophr. Bull. 40, 428–437.

Mantini, D., Perrucci, M.G., Del Gratta, C., Romani, G.L., Corbetta, M., 2007. Electrophysiological signatures of resting state networks in the human brain. Proc. Natl. Acad. Sci. U.S.A. 104, 13170–13175.

Marín, O., 2012. Interneuron dysfunction in psychiatric disorders. Nat. Rev. Neurosci. 13, 107–120.

McDannald, M.A., Whitt, J.P., Calhoon, G.G., Piantadosi, P.T., Karlsson, R.-M., O'Donnell, P., et al., 2011. Impaired reality testing in an animal model of schizophrenia. Biol. Psychiatry 70, 1122–1126.

Miller, P., Byrne, M., Hodges, A., Lawrie, S.M., Cunningham-Owens, D.G., Johnstone, E.C., 2002. Schizotypal components in people at high risk of developing schizophrenia: early findings from the Edinburgh High-Risk Study. Br. J. Psychiatry 180, 179–184.

Miyauchi, T., Tanaka, K., Hagimoto, H., Miura, T., Kishimoto, H., Matsushita, M., 1990. Computerized EEG in schizophrenic patients. Biol. Psychiatry 28, 488–494.

Moore, H., Jentsch, J.D., Ghajarnia, M., Geyer, M.A., Grace, A.A., 2006. A neurobehavioral systems analysis of adult rats exposed to methylazoxymethanol acetate on E17: implications for the neuropathology of schizophrenia. Biol. Psychiatry 60, 253–264.

Nakao, K., Nakazawa, K., 2014. Brain state-dependent abnormal LFP activity in the auditory cortex of a schizophrenia mouse model. Front. Neurosci. 8, 168.

Nakazawa, K., Zsiros, V., Jiang, Z., Nakao, K., Kolata, S., Zhang, S., et al., 2012. GABAergic interneuron origin of schizophrenia pathophysiology. Neuropharmacology 62, 1574–1583.

Neuner, I., Arrubla, J., Werner, C.J., Hitz, K., Boers, F., Kawohl, W., et al., 2014. The default mode network and EEG regional spectral power: a simultaneous fMRI-EEG study. PLoS ONE 9, 1–8.

Northoff, G., 2014. Are auditory hallucinations related to the brain's resting state activity? A "Neurophenomenal resting state hypothesis". Clin. Psychopharmacol. Neurosci. 12, 189–195.

Northoff, G., Qin, P., Nakao, T., 2010. Rest-stimulus interaction in the brain: a review. Trends Neurosci. 33, 277–284.

O'Donnell, P., 2011. Adolescent onset of cortical disinhibition in schizophrenia: insights from animal models. Schizophr. Bull. 37, 484–492.

Ohi, K., Hashimoto, R., Ikeda, M., Yamamori, H., Yasuda, Y., Fujimoto, M., et al., 2014. Glutamate networks implicate cognitive impairments in schizophrenia: genome-wide association studies of 52 cognitive phenotypes. Schizophr. Bull. 41, 909–918.

Ongür, D., Lundy, M., Greenhouse, I., Shinn, A.K., Menon, V., Cohen, B.M., et al., 2010. Default mode network abnormalities in bipolar disorder and schizophrenia. Psychiatry Res. 183, 59–68.

Orliac, F., Naveau, M., Joliot, M., Delcroix, N., Razafimandimby, A., Brazo, P., et al., 2013. Links among resting-state default-mode network, salience network, and symptomatology in schizophrenia. Schizophr. Res. 148, 74–80.

Papaleo, F., Lipska, B.K., Weinberger, D.R., 2012. Mouse models of genetic effects on cognition: relevance to schizophrenia. Neuropharmacology 62, 1204–1220.

Perez, S.M., Shah, A., Asher, A., Lodge, D.J., 2013. Hippocampal deep brain stimulation reverses physiological and behavioural deficits in a rodent model of schizophrenia. Int. J. Neuropsychopharmacol. 16, 1331–1339.

Perry, T.L., Kish, S.J., Buchanan, J., Hansen, S., 1979. Gamma-aminobutyric acid deficiency in brain of schizophrenic patients. Lancet 313, 237–239.

Phillips, K.G., Cotel, M.C., McCarthy, A.P., Edgar, D.M., Tricklebank, M., O'Neill, M.J., et al., 2012. Differential effects of NMDA antagonists on high frequency and gamma EEG oscillations in a neurodevelopmental model of schizophrenia. Neuropharmacology 62, 1359–1370.

Port, R.G., Gandal, M.J., Timothy, P., Roberts, L., Siegel, S.J., Carlson, G.C., 2014. Convergence of circuit dysfunction in ASD: a common bridge between diverse genetic and environmental risk factors and common clinical electrophysiology. Front. Cell. Neurosci. 8, 1–14.

Powell, S.B., 2010. Models of neurodevelopmental abnormalities in schizophrenia. Curr. Top. Behav. Neurosci. 4, 435–481.

Ramyead, A., Kometer, M., Studerus, E., Koranyi, S., Ittig, S., Gschwandtner, U., et al., 2014. Aberrant current source-density and lagged phase synchronization of neural oscillations as markers for emerging psychosis. Schizophr. Bull. 41, 919–929.

Ritter, P., Villringer, A., 2006. Simultaneous EEG-fMRI. Neurosci. Biobehav. Rev. 30, 823–838.

Rodin, E., Grisell, J., Gottlieb, J., 1968. Some electrographic differences between chronic schizophrenic patients and normal subjects. Recent Adv. Biol. Psychiatry 10, 194–204.

Rolland, B., Amad, A., Poulet, E., Bordet, R., Vignaud, A., Bation, R., et al., 2014. Resting-state functional connectivity of the nucleus accumbens in auditory and visual hallucinations in schizophrenia. Schizophr. Bull. 41, 291–299.

Rotarska-Jagiela, A., van de Ven, V., Oertel-Knöchel, V., Uhlhaas, P.J., Vogeley, K., Linden, D.E.J., 2010. Resting-state functional network correlates of psychotic symptoms in schizophrenia. Schizophr. Res. 117, 21–30.

Rutter, L., Carver, F.W., Holroyd, T., Nadar, S.R., Mitchell-Francis, J., Apud, J., et al., 2009. Magnetoencephalographic gamma power reduction in patients with schizophrenia during resting condition. Hum. Brain Mapp. 30, 3254–3264.

Sams-Dodd, F., Lipska, B.K., Weinberger, D.R., 1997. Neonatal lesions of the rat ventral hippocampus result in hyperlocomotion and deficits in social behaviour in adulthood. Psychopharmacology 132, 303–310.

Saunders, J.A., Gandal, M.J., Siegel, S.J., 2012. NMDA antagonists recreate signal-to-noise ratio and timing perturbations present in schizophrenia. Neurobiol. Dis. 46, 93–100.

Schnitzler, A., Gross, J., 2005. Normal and pathological oscillatory communication in the brain. Nat. Rev. Neurosci. 6, 285–296.

Skudlarski, P., Jagannathan, K., Anderson, K., Stevens, M.C., Calhoun, V.D., Skudlarska, B.A., et al., 2010. Brain connectivity is not only lower but different in schizophrenia: a combined anatomical and functional approach. Biol. Psychiatry 68, 61–69.

Spellman, T.J., Gordon, J.A., 2014. Synchrony in schizophrenia: a window into circuit-level pathophysiology. Curr. Opin. Neurobiol. 30C, 17–23.

Stone, J.M., Erlandsson, K., Arstad, E., Squassante, L., Teneggi, V., Bressan, R.A., et al., 2008. Relationship between ketamine-induced psychotic symptoms and NMDA receptor occupancy—A [123I]CNS-1261 SPET study. Psychopharmacology 197, 401–408.

Sun, J., Tang, Y., Lim, K.O., Wang, J., Tong, S., Li, H., et al., 2014. Abnormal dynamics of EEG oscillations in schizophrenia patients on multiple time scales. IEEE Trans. Biomed. Eng. 61, 1756–1764.

Swerdlow, N.R., Braff, D.L., Geyer, M.A., 2000. Animal models of deficient sensorimotor gating: what we know, what we think we know, and what we hope to know soon. Behav. Pharmacol. 11, 185–204.

Takao, K., Toyama, K., Nakanishi, K., Hattori, S., Takamura, H., Takeda, M., et al., 2008. Impaired long-term memory retention and working memory in sdy mutant mice with a deletion in Dtnbp1, a susceptibility gene for schizophrenia. Mol. Brain 1, 1–12.

Tang, J., Liao, Y., Song, M., Gao, J.-H., Zhou, B., Tan, C., et al., 2013. Aberrant default mode functional connectivity in early onset schizophrenia. PLoS ONE 8, e71061.

Tatard-Leitman, V.M., Jutzeler, C.R., Suh, J., Saunders, J.A., Billingslea, E.N., Morita, S., et al., 2015. Pyramidal cell selective ablation of N-Methyl-D-Aspartate receptor 1 causes increase in cellular and network excitability. Biol. Psychiatry 77, 556–568.

Tiesinga, P., Sejnowski, T.J., 2009. Cortical enlightenment: are attentional gamma oscillations driven by ING or PING? Neuron 63, 727–732.

Tu, P.-C., Lee, Y.-C., Chen, Y.-S., Li, C.-T., Su, T.-P., 2013. Schizophrenia and the brain's control network: aberrant within- and between-network connectivity of the frontoparietal network in schizophrenia. Schizophr. Res. 147, 339–347.

Turetsky, B.I., Siegel, S.J., 2007. Persistent auditory-evoked gamma band oscillations in schizophrenia. In: American College of Neuropsychopharmacology, Boca Raton, FL.

Uhlhaas, P.J., Singer, W., 2013. High-frequency oscillations and the neurobiology of schizophrenia. Dialogues. Clin. Neurosci. 15, 301–313.

Urbach, T.P., Kutas, M., 2006. Interpreting event-related brain potential (ERP) distributions: implications of baseline potentials and variability with application to amplitude normalization by vector scaling. Biol. Psychol. 72, 333–343.

Van Diessen, E., Senders, J., Jansen, F.E., Boersma, M., Bruining, H., 2014. Increased power of resting-state gamma oscillations in autism spectrum disorder detected by routine electroencephalography. Eur. Arch. Psychiatry Clin. Neurosci 265, 537–540.

Venables, N.C., Bernat, E.M., Sponheim, S.R., 2009. Genetic and disorder-specific aspects of resting state EEG abnormalities in schizophrenia. Schizophr. Bull. 35, 826–839.

Vohs, J.L., Chambers, R.A., Krishnan, G.P., O'Donnell, B.F., Hetrick, W.P., Kaiser, S.T., et al., 2009. Auditory sensory gating in the neonatal ventral hippocampal lesion model of schizophrenia. Neuropsychobiology 60, 12–22.

Volk, D.W., Austin, M.C., Pierri, J.N., Sampson, A.R., Lewis, D.A., 2000. Decreased glutamic acid decarboxylase 67 messenger RNA in a subset of prefrontal cortical gamma-aminobutyric acid neurons in subjects with schizophrenia. Arch. Gen. Psychiatry 57, 237–245.

Wang, J., Barstein, J., Ethridge, L.E., Mosconi, M.W., Takarae, Y., Sweeney, J.A., 2013. Resting state EEG abnormalities in autism spectrum disorders. J. Neurodevelopmental Disord. 5, 1–14.

White, T.P., Joseph, V., Francis, S.T., Liddle, P.F., 2010. Aberrant salience network (bilateral insula and anterior cingulate cortex) connectivity during information processing in schizophrenia. Schizophr. Res. 123, 105–115.

Whitfield-Gabrieli, S., Ford, J.M., 2012. Default mode network activity and connectivity in psychopathology. Annu. Rev. Clin. Psychol. 8, 49–76.

Whittington, M.A., Traub, R.D., 2003. Interneuron diversity series: inhibitory interneurons and network oscillations in vitro. Trends Neurosci. 26, 676–682.

Whittington, M.A., Traub, R.D., Jefferys, J.G., 1995. Synchronized oscillations in interneuron networks driven by metabotropic glutamate receptor activation. Nature 373, 612–615.

Winterer, G., Coppola, R., Goldberg, T.E., Egan, M.F., Jones, D.W., Sanchez, C.E., et al., 2004. Prefrontal broadband noise, working memory, and genetic risk for schizophrenia. Am. J. Psychiatry 161, 490–500.

Xue, J.-G., Masuoka, T., Gong, X.-D., Chen, K.-S., Yanagawa, Y., Law, S.K.A., et al., 2011. NMDA receptor activation enhances inhibitory GABAergic transmission onto hippocampal pyramidal neurons via presynaptic and postsynaptic mechanisms. J. Neurophysiol. 105, 2897–2906.

Yeragani, V.K., Cashmere, D., Miewald, J., Tancer, M., Keshavan, M.S., 2006. Decreased coherence in higher frequency ranges (beta and gamma) between central and frontal EEG in patients with schizophrenia: a preliminary report. Psychiatry Res. 141, 53–60.

Zaehle, T., Rach, S., Herrmann, C.S., 2010. Transcranial alternating current stimulation enhances individual alpha activity in human EEG. PLoS ONE 5, 1–7.

Zaehle, T., Beretta, M., Jäncke, L., Herrmann, C.S., Sandmann, P., 2011a. Excitability changes induced in the human auditory cortex by transcranial direct current stimulation: direct electrophysiological evidence. Exp. Brain Res. 215, 135–140.

Zaehle, T., Sandmann, P., Thorne, J.D., Jäncke, L., Herrmann, C.S., 2011b. Transcranial direct current stimulation of the prefrontal cortex modulates working memory performance: combined behavioural and electrophysiological evidence. BMC Neurosci. 12, 2.

Zemankovics, R., Veres, J.M., Oren, I., Ha, N., 2013. Feedforward inhibition underlies the propagation of cholinergically induced gamma oscillations from hippocampal CA3 to CA1. J. Neurosci. 33, 12337–12351.

Zhang, Z.J., Reynolds, G.P., 2002. A selective decrease in the relative density of parvalbumin-immunoreactive neurons in the hippocampus in schizophrenia. Schizophr. Res. 55, 1–10.

PART VI

THE STRUCTURE AND FUNCTION OF NEURAL CIRCUITS IN SCHIZOPHRENIA

CHAPTER 15

COMPUTATIONAL NEUROANATOMY OF SCHIZOPHRENIA

C. Davatzikos[1] and N. Koutsouleris[2]

[1]Center for Biomedical Image Computing and Analytics, University of Pennsylvania, Philadelphia, PA, United States
[2]Department of Psychiatry and Psychotherapy, Ludwig-Maximilians-University, Munich, Germany

CHAPTER OUTLINE

Introduction 263
Methods for Computational Neuroanatomy 266
　Region of Interest Versus Voxel-Based Analysis 266
　Regional Tissue Volumetrics 267
　Optimally Discriminant Voxel-Based Analysis 269
　Multivariate Pattern Analysis (MVPA) and Machine Learning 269
Topography of Reduced Brain Volumes in Schizophrenia 270
　Single-Subject Classification: From Computational Neuroanatomy of Schizophrenia to
　Individualized Diagnostic Tests 274
　Single-Subject Differentiation Between the Psychosis Prodrome and the At-Risk for Mental
　Psychosis Without Subsequent Disease Transition 275
　Are SCZ-Like Neuroanatomical Patterns Possibly Endophenotypes of Disease? 276
References 278

INTRODUCTION

Over the past two decades, a large body of scientific literature has provided accumulating and convincing evidence for a neuroanatomical surrogate of schizophrenia. Due to technological constraints, the first phase of this MRI-based research focused on specific hypothesis-driven brain regions (regions of interest (ROI)), such as the medial temporal, superior temporal, and frontal brain structures that (1) significantly differ in volume and shape between patients and healthy controls (Keshavan et al., 1998; Shenton et al., 2001; Sun et al., 2009); (2) were correlated with different core phenotypes of the disorder, including psychosis, disturbed cognition, and anhedonia/avolition (Antonova et al., 2004; Sanfilipo et al., 2000; Turetsky et al., 1995); (3) were present early in the course of the disease; and (4) accumulated in patients with residual courses of the disorder (Gur et al., 2007; Sigmundsson et al., 2001; Velakoulis et al., 1999, 2006). These findings indicate that these spatially confined brain imaging phenotypes of the disorder may parallel the clinical heterogeneity of the *group of schizophrenias*, as originally termed by Eugen Bleuer (1911). Furthermore, the advent of comparative neuroanatomical studies provided insight into the significant amount of overlap in these regional brain measures between schizophrenia and other neuropsychiatric diseases, such as bipolar disorder and unipolar depression. This shared structural brain variance suggested that the neuroanatomical substrates of these disorders may not fully align with the nosological boundaries drawn by the current disease taxonomy. Hence, these findings have stirred the debate whether the schizophrenia *umbrella construct* may be better deconstructed into new neurobiologically validated disease entities that may transcend the purely clinically defined psychiatric disease classification. This hypothesis received further support from the next phase of research into the neuroanatomical underpinnings of schizophrenia, which used whole-brain analytical methods to trace the brain structural patterns in schizophrenia (Sowell et al., 2000; Wolkin et al., 1998). Instead of confining researchers to specific hypothesis-defined structures, these voxel-based methods now allowed for a largely automatic, rater-unbiased analysis of structural brain variation operating at the whole-brain level (Criss et al., 1998; Davatzikos et al., 2001; Gaser et al., 2001). Thus, these methods dramatically increased our understanding of the complexity and distributed nature of structural brain anomalies in schizophrenia, which were not only limited to the previously studied ROI foci but also involved in a multitude of different brain structures, ranging from the prefrontal cortices, through the perisylvian structures, to the parieto-occipital and cerebellar areas (Honea et al., 2005). This vast array of affected higher-order cortical, limbic, subcortical, and cerebellar brain areas strengthened the notion of schizophrenia being a "disconnection syndrome." This syndrome implied that the pathophysiological core of the disorder consisted of deficient information processing within and across the neural systems spanned by these structures. Hence, this still-debated theory posited that the multifaceted phenotypes of schizophrenia may be subserved by the single or unitary pathophysiology of "cognitive dysmetria" (Andreasen et al., 1998).

Beyond the impact on hypothesis formation, the automated voxel-based methods enabled researchers to study significantly larger patient populations and, therefore, to draw much more robust statistical inferences in terms of localization, extent, and intensity of morphometric brain abnormalities. Despite the initial "teething problems" of these techniques (low spatial resolution of stereotactic registration), the new era of high-dimensional voxel-based morphometric studies (Ashburner and Friston, 2000; Davatzikos et al., 2001), which enabled a more detailed evaluation of regional morphological characteristics, provided sound evidence for a prefronto-temporo-limbic-subcortical pattern of gray matter volume reductions in patients with schizophrenia versus healthy controls (Davatzikos et al., 2005a). These studies were complemented by structural cortex modeling techniques (Fischl et al., 1999; Fischl and Dale, 2000), which enabled researchers to dissociate gray matter volume into the more specific measures of cortical thickness, area, and gyrification (Palaniyappan and Liddle, 2012b), with potentially different underlying genetic mediators (Winkler et al., 2010). Surface-based morphometry revealed patterns of perisylvian and parieto-occipital hypogyria as well as prefrontal hypergyria in patients with schizophrenia (Palaniyappan and Liddle, 2012a) and widespread cortical hypogyria in patients with psychotic disorders and their first-degree relatives (Nanda et al., 2014). Importantly, cross-sectional and longitudinal studies demonstrated that the extent and intensity of prefrontal, temporal/perisylvian, and subcortical neuroanatomical abnormalities were already present in the first episode of schizophrenia and, to a lesser degree, were detectable not only in clinically and genetically defined high-risk populations (Fusar-Poli et al., 2011; Thermenos et al., 2013) but also in healthy persons with moderate levels of schizotypical personality features (Ettinger et al., 2012). These findings may point to a *longitudinal continuum* of accumulating brain changes, which evolve on temporal trajectories from the clinical high-risk (CHR) state to the full-blown manifestation of schizophrenia (Cannon et al., 2015; Pantelis et al., 2005; Wood et al., 2008) and to an *epidemiological continuum*, in which the psychosis-related brain patterns intensify with increasing clinical proximity to the cross-sectional and longitudinal dimensions of the schizophrenia phenotype (Ettinger et al., 2014).

The current neuroimaging literature indicates that the neuroanatomical trajectories of those CHR individuals who subsequently transit to a frank psychotic disorder compared to those without disease transition separate early during the course of the illness. Again, this hypothesis was substantiated by voxel-based imaging studies, which revealed complex patterns of gray matter volume reductions in prefrontal, temporo-limbic, and cerebellar brain region abnormalities of truly prodromal CHRs (Koutsouleris et al., 2009b; Pantelis et al., 2003). Importantly, these findings implicated the possibility of using neuroimaging markers to predict the later onset of a frank psychotic illness in vulnerable CHR subjects—a cardinal clinical objective given the low predictive value of prodromal symptoms that typically remit in the majority of ~70% affected persons within a 3-year follow-up period (Fusar-Poli et al., 2012). However, the VBM methodology used so far to describe the "morphometric gap" between the true prodrome and the CHR state without later conversion did not provide the adequate statistical apparatus to extract a clinically useful signature of the emerging illness. Two major methodological properties account for the limitation of VBM in generating neuroanatomical markers for individualized disease prediction (Davatzikos, 2004). First, the mass univariate statistical approach, instead of modeling group differences in the neuroanatomical space of potentially disease-affected brain systems, projected these differences onto overlapping and unrelated voxel-by-voxel measurements. Second, the main outcome measure of the VBM approach—voxel-level significance values corrected for multiple comparisons—was mainly useless in terms of individualized inference and personalized medicine, thus benchmarking diagnostic and prognostic tests in terms of sensitivity and specificity as well as positive and negative likelihood ratios. Hence, the methodological drawbacks of the mass univariate approach pushed the psychosis field toward multivariate voxel pattern analysis (MVPA) methods.

These methods originated in the machine learning community (Vapnik, 1998) and were subsequently successfully translated into clinical neuroscience and computational neuroanatomy applications in both neuropsychiatric (Chaim et al., 2012; Davatzikos et al., 2005b; Koutsouleris et al., 2009a; Zanetti et al., 2013) and dementia research (Davatzikos et al., 2008; Fan et al., 2008a; Kloppel et al., 2008; Lao et al., 2004; Misra et al., 2009; Vemuri et al., 2008). In particular, the superiority of these methods in modeling the healthy brain and the diseased brain as a *system* of interconnected measurements met the field's need for statistical tools that would drive the identification of biological markers for the differential diagnostic classification and early recognition of dementias. However, what made the difference between MVPA and the aforementioned univariate techniques? MVPA algorithms enable the flexible extraction of those variance patterns from the high-dimensional imaging data that provide superior prediction or classification accuracy, even if their voxel-level constituents show highly overlapping class distributions. This property makes these pattern recognition methods particularly interesting for the psychosis field because the underlying pathophysiology is assumed to involve the system-level disconnectivity of brain regions spanning structural and functional networks at multiple scales. The first analyses conducted involving patients with schizophrenia by means of MVPA (Davatzikos et al., 2005b) suggested that complex patterns of brain abnormalities exist that separate healthy volunteers from the patients affected by the disorder with an accuracy ranging from 81% to 85% (see Fig. 15.8). These initial results have been replicated in larger and independent samples within an accuracy range of 70% to 85%, thus providing accumulating evidence for a neuroanatomical signature of schizophrenia underlying this clinically heterogeneous disorder (Kambeitz et al., 2015). Furthermore, prospective studies conducted in CHR cohorts over the course of 4–5 years indicated that neuroanatomical MVPA may provide markers for individualized prediction and risk stratification, thereby potentially constituting the prognostic grounds required for targeted early intervention. These findings suggest that a predictive neuroanatomical biomarker of psychosis may be detectable years before the onset of frank illness. Nevertheless, it remains unclear whether this marker scales to larger, sociodemographically heterogeneous CHR populations, specifically predicts schizophrenic psychosis or, alternatively, is generally associated with psychiatric illness and constitutes a "mixture signature" of more homogeneous, and thus diagnostically more specific, neuroanatomical entities.

METHODS FOR COMPUTATIONAL NEUROANATOMY
REGION OF INTEREST VERSUS VOXEL-BASED ANALYSIS

Traditionally, volumetric analyses involved the outlining of a number of anatomical regions of interest (ROIs), such as brain lobes, deep structures, or specific gyri, based on a subject's MRI scan (Andreasen et al., 1994; Cannon et al., 1998; Shenton et al., 1992). Although this approach is still important for addressing *a priori* hypotheses (eg, the volume of the hippocampus), it is insufficient in two important ways. First, an *a priori* hypothesis pertains only to certain structures and does not allow us to evaluate the entire brain. Given that MRI measures brain structure in its entirety, it is of great interest to be able to evaluate all brain regions, not only selected ones. Second, predefined ROIs might or might not coincide with the region affected by schizophrenia. For example, Fig. 15.1 shows typical brain

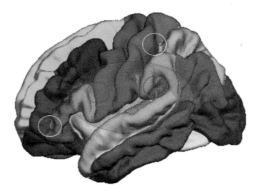

FIGURE 15.1

Schematic demonstrating limitations of ROI-based analyses. If, for the sake of the argument, schizophrenia happens to affect the shaded oval region, then all of the predefined anatomical ROIs would have only partial overlap with that region of pathology. As a result, our ability to detect this region would be substantially diminished. VBMA approaches visit each and every voxel in the brain and investigate whether a small region around the voxel displays significant differences between patients and controls. The regions that do show significant differences (such as the regions enclosed by the red, but not the yellow, circles) collectively form the spatial pattern of disease effect.

parcellation into ROIs corresponding to known anatomical regions. If, for the sake of the argument, brain structure in schizophrenia is affected in the shaded region, then many of these ROIs overlap only partially with the region of pathology. Consequently, volumetric analyses in those ROIs would mix affected and unaffected brain regions and therefore would significantly reduce our ability to detect disease effects.

An alternative to ROI analyses is the family of voxel-based morphometric analyses (VBMA) methods, which have appeared in various different types in the literature, as described in the previous section. These methods make no *a priori* hypotheses about the region of abnormality. Instead, they search throughout the entire brain MRI, voxel-by-voxel, to find regions that display disease effects (eg, differences between healthy controls and schizophrenia patients). Fig. 15.1 shows a schematic of this type of approach in which a local searching filter is applied to the data, aiming to find voxels around which brain volumes are relatively lower in patients compared with controls. The entire region of anatomical difference will finally emerge as the collection of voxels showing significant group differences.

REGIONAL TISSUE VOLUMETRICS

The brains of different individuals differ by size and shape. Hence, a point of spatial coordinates (x,y,z) in one person's MRI scan might correspond to the hippocampus, whereas the same spatial coordinates in another person's scan might correspond to the motor cortex. This issue must be resolved prior to embarking on any type of VBMA. This is achieved via a fundamental process in the voxel-by-voxel investigation of anatomical differences, namely that of deformable registration of subjects' MRI scans to a standardized coordinate system, usually called the atlas space. This process spatially normalizes brain scans, so that a given location (x,y,z) in the atlas space corresponds to the same anatomical region in all subjects. A number of methods for deformable registration have been developed over the past two decades. A comprehensive summary can be found in the work by Sotiras et al. (2013).

Because deformable registration changes the morphology of the brain being measured, by making it similar to that of the atlas, most computational neuroanatomy methods consider this effect by measuring the exact deformation that takes a subject's MRI to the atlas space, subsequently accounting for it. A frequently derived entity is called tissue density maps, an image whose value at a voxel with coordinates (x,y,z) is proportional to the amount of tissue (GM, WM, or CSF) present in the vicinity of that location for each individual.

The most commonly used VBMA method is the one that falls under the name VBM (Ashburner and Friston, 2000). In its original form, this method first warped all MRI scans to a common template, thereby removing relatively global shape differences. The residual tissue segmentation maps were then filtered by a fixed Gaussian filter with sizes selected by the user, usually in the range of 6 to 15 mm. This filtering step Gaussianizes the data, thereby preparing them for subsequent linear statistical tests, and somewhat mitigates registration inaccuracies. The general linear model was then applied on each and every voxel, allowing investigators to determine group differences after removing the effects of covariates, such as age and sex. Effectively, this approach places a spherical ROI around each and every voxel, as in Fig. 15.1, and investigates group differences in each of these ROIs. This approach was later modified to account for the deformation that takes each individual's scan to the template space and resulted in the "modulated VBM" method (Good et al., 2001).

All results described herein have been obtained using the RAVENS approach, which was described in detail (Davatzikos et al., 2001; Goldszal et al., 1998; Shen and Davatzikos, 2003) and is highly related to the modulated VBM approach. A schematic representation of this process is shown in Fig. 15.2.

Fig. 15.3 shows a representative RAVENS (tissue density) map for GM, reflecting the regional variations of the amount of GM present in that individual's brain, but after his brain has been spatially coregistered to a standardized atlas (template) space.

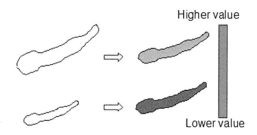

FIGURE 15.2

Schematic of two hippocampi of different sizes (left). Respective RAVENS maps are on the right. Both hippocampi were deformed into the shape of the atlas hippocampus on the right. However, more gray matter was forced into this same hippocampus template for the top (compared to the bottom) hippocampus. Respective intensity (gray value) of the RAVENS maps reflects this and allows us to perform voxel-by-voxel quantitative analyses of regional tissue volume.

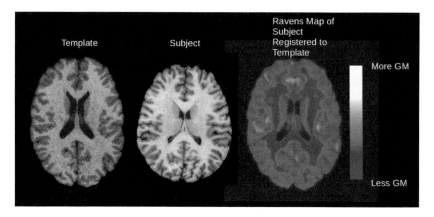

FIGURE 15.3

Representative slice from a 3D RAVENS map, whose (color-coded) intensity is directly proportional to the gray matter volume present around each voxel. Voxel-by-voxel analysis of these RAVENS maps allows us to examine regional patterns of reduced or increased gray matter volume.

OPTIMALLY DISCRIMINANT VOXEL-BASED ANALYSIS

VBMA methods perform a voxel-by-voxel analysis of the amount of brain tissue present in the vicinity of each voxel. However, the size of this "vicinity" is very important. If it is too small, then too much noise, registration inaccuracies, and anatomical variability across individuals will dramatically reduce our ability to detect statistically significant group effects. However, if this search neighborhood is too large, then the analysis is no longer localized around each voxel, and the same problems as the ones found in ROI analyses (see Fig. 15.2) hinder our ability to optimally detect a disease effect. In essence, the optimal way to detect a structural abnormality is to filter the RAVENS maps with a filter that exactly matches the spatial extent and shape of this abnormality (eg, shaded region of Fig. 15.2). However this region is not known in advance. Recent developments in the field of computational neuroanatomy have allowed us to mathematically determine the best way of filtering the data to tease out group effects. A method called optimally discriminant voxel-based analysis (ODVBA), which performs such an optimal estimation (Zhang and Davatzikos, 2011, 2013), has been used to obtain many of the results presented here.

MULTIVARIATE PATTERN ANALYSIS (MVPA) AND MACHINE LEARNING

Both ROI-based and voxel-based analysis methods and, to a lesser extent, ODVBA have a key limitation: they are mass univariate methods. In other words, they examine one measurement after the other, in sequence, thereby largely ignoring correlations in the data that might be able to better elucidate group effects. From the perspective of understanding the neuroanatomical underpinnings of schizophrenia, because the focus is on answering the question, "which brain regions are affected by schizophrenia?," this limitation is not as important. However, from the perspective of constructing personalized diagnostic and predictive markers of sufficiently high sensitivity and specificity *on an individual basis*, which is critical for these methods to be adopted in the clinical practice, these methods are insufficient. For example, even though certain frontal and temporal regions display relatively reduced brain volumes in schizophrenia, volumetrics of these regions do not lead to any clinically meaningful diagnostic quantity due to large interindividual variability and due to lack of spatial specificity of these measures (see Fig. 15.4).

During the past decade, the computational neuroanatomy literature has seen a rapid increase in the use of MVPA methods, increasingly leveraging strengths of the machine learning literature to overcome the aforementioned limitation. The main premise of these methods is that by combining many different types of measurements, they form increasingly more distinctive anatomical phenotypes of schizophrenia. Multiple measurements here originate from multiple brain regions, thereby forming a spatial pattern of abnormal brain structure. In other words, even though no individual brain measurement of regional brain volume is close to being a sufficiently sensitive and specific marker of schizophrenia, the brain changes in a number of brain regions jointly form a spatial pattern that is quite distinctive and detectable ion individual patients.

Several methods for choosing the best set of measurements have been developed in the literature. Herein, we use two general approaches to this problem (Fan et al., 2007; Koutsouleris et al., 2009a). We integrate these spatial patterns into indices via a machine learning method called support vector machines (SVM), which theoretically and experimentally possess many desired properties for

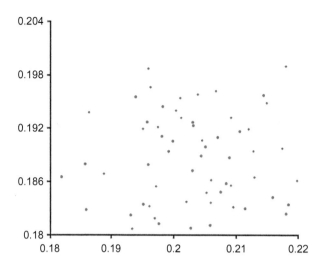

FIGURE 15.4

Scatterplot of (normalized) frontal versus temporal lobe volumes for normal controls (red dots) and patients with schizophrenia (blue diamonds) from the University of Pennsylvania (UPenn) study (Davatzikos et al., 2005b). These volumetric measurements are almost entirely overlapping between the two groups, potentially giving the false impression that these regions are structurally spared by the disease. However, advanced ODVBA analysis of Fig. 15.5 clearly shows that certain parts of them are significantly affected by the disease. Moreover, this spatial pattern can be used to construct highly sensitive and specific imaging biomarkers, as in Fig. 15.8.

classifying high-dimensionality patterns. Results are shown from application of these machine learning methods to the problems of detecting schizophrenia-like anatomical signatures in individual patients, in unaffected family members, and in at-risk individuals.

TOPOGRAPHY OF REDUCED BRAIN VOLUMES IN SCHIZOPHRENIA

In this section, we draw from results previously reported by two large studies of schizophrenia using the aforementioned techniques for computational neuroanatomy that shed light on spatially complex and subtle, but highly distinctive, neuroanatomical patterns of schizophrenia that have not been entirely appreciated before the use of these methods. In particular, in Davatzikos et al., 69 (46 men and 23 women) patients with chronic schizophrenia (SCZ) and 79 (41 men and 38 women) healthy controls (HC) participated in the study. Participants underwent medical, neurological, and psychiatric evaluations to exclude for history of illness affecting brain function, including substance abuse, hypertension, metabolic disorders, neurological disorders, and head trauma with loss of consciousness. For patients, the comprehensive intake evaluation included a structured diagnostic interview and a review of records and information available from family and care providers that contributed to a consensus diagnosis of schizophrenia. T1-weighted structural MRI scans were used to extract regional volumetric RAVENS

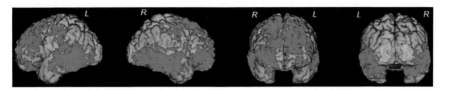

FIGURE 15.5

Optimally discriminative analysis (ODVBA) of 148 scans from the UPenn study revealed a widespread pattern of relatively reduced GM tissue in SCZ patients, spanning large frontal, temporal, and occipito-temporal regions.

maps, which were analyzed in a common stereotaxic coordinate space associated with a brain atlas. Original reports (Davatzikos et al., 2005a) showed clear hints toward a relatively widespread pattern of lower GM tissue in frontal, temporal, and occipito-temporal regions. Recent reanalysis of those datasets using the more powerful aforementioned ODVBA methodology further elucidated the extent of this pattern, as shown in Fig. 15.5.

These patterns of reduced GM volumes were consistent with ROI analyses. For example, volumetric reductions in superior and inferior temporal gyri and in hippocampus were also reported in previous ROI analyses (Gur et al., 2000). More generally, in frontotemporal regions, the ODVBA analysis yielded findings consistent with previous studies and with the literature on ROI-based approaches (Lawrie and Abukmeil, 1998; Nelson et al., 1998; Shenton et al., 2001; Wright et al., 2000). However, ODVBA also pointed to a more precise spatial localization of neuroanatomical effects of schizophrenia by not relying on *a priori* partitioning of the brain into standard anatomical regions. For example, a notable new area of abnormality is in the inferior occipito-temporal region, where substantial volume reduction was observed. These regions are involved in visual association, face recognition, and orientation (Cohen et al., 1996; Kaas, 1996). Although the visual system has been considered generally intact in schizophrenia, our anatomic findings suggest further investigation, especially of occipito-temporal connections. Deficits in the visual processing cascade may underlie downstream dysfunction in more complex operations requiring visual input.

These results overlapped with the findings obtained in the Munich schizophrenia cohort, which consisted of 175 (130 males and 45 females) patients and 177 (123 males and 54 females) healthy volunteers (Koutsouleris et al., 2008). This representative MRI database included structural brain scans of both first-episode and recurrently ill patients with schizophrenia who had a mean (SD) age of 31.5 (9.2) years and a mean illness duration of 1.6 (2.5) years and who were psychometrically assessed using the positive and negative symptom scale (PANSS) (Kay et al., 1987). The patients' medical and psychiatric histories were carefully assessed using a standardized clinical interview and examination. Patients meeting diagnostic criteria for somatic/neurological disorders potentially affecting the human brain structure were excluded from the study. Furthermore, potential comorbidities were evaluated by trained clinical raters using the standardized clinical interview for DSM-IV axis I disorders (First et al., 1995). Positive diagnostic criteria for other axis I disorders or lack of diagnostic consistency across two independent and experienced psychiatrists led to exclusion from the study.

The MRI data were preprocessed using a refined version of the optimized VBM protocol described in Good et al. (2001), which involved the postprocessing of the computed GM tissue segmentations to further increase segmentation quality by applying a spatial smoothing approach, known as the hidden Markov field (HMRF) model. In contrast to the UPenn dataset, nonlinear registration to a common stereotactic space was performed using less deformable normalization algorithms of statistical parametric mapping (version 2; SPM2) and did not involve the computation of RAVENS maps, thus ensuring that a precise quantitative measurement of regional volume is obtained. Thus, the subsequent analysis step targeted residual contrast differences between the HC and SCZ groups after smoothing the data with a 12-mm Gaussian kernel, whereas the UPenn study reported findings on voxel-based volumetric differences. Beyond these methodological differences, the cohorts also were distinct in terms of disease stage, with the Munich sample including patients from first episode to chronic states of the disorder, whereas the UPenn cohort represented rather chronic disease stages. Despite these differences, striking spatial overlaps were observed between the prefronto-temporal volumetric reductions in the UPenn patients and the pattern of extended prefrontal, perisylvian, and temporo-limbic gray matter density reductions in the Munich schizophrenia cohort (Figs. 15.5 and 15.6). This fronto-temporo-limbic pattern independently and consistently detected in both patient cohorts may point to a robust structural brain signature of a complex disorder that primarily affects higher-order human brain functions like cognitive control, language, thought, and mnemonic processing as well as social cognition, theory of mind, and affective reasoning. Therefore, to further dissect the structural correlates of these domains, clinically driven subgroup analyses were performed in the Munich cohort that provided a rich set of psychopathological information. Using exploratory factor analysis, three psychopathological dimensions were detected in the Munich patients' PANSS data that, in keeping with the previous literature, mapped to factors for negative, positive, and disorganized symptoms. Then, the patients' loading on these factors was used to create three patient subgroups that corresponded to predominant negative, positive, and disorganized symptomatology. The normalized gray matter density data of each of these subgroups were compared to those of the HC group, revealing overlapping and distinct patterns of structural brain alterations. The negative symptom sample showed the most extended set of gray matter reductions covering primarily the medial, orbitofrontal, and lateral prefrontal cortices, and the medial temporal lobe and perisylvian areas, the temporal poles, and the inferior temporal cortices. Alterations in the other two subgroups were more circumscribed and involved gray matter density reductions in the thalamus, the superior temporal gyri, and the lateral prefrontal cortices (paranoid-hallucinatory subgroup), and the orbitofrontal, medial prefrontal, and the superior temporal cortices (disorganized subgroup). A structural substrate shared by all subgroups was evident in the anterior insula, the medial prefrontal, and superior temporal cortices, in line with previous findings from the UPenn sample and the more recent results highlighting the potentially crucial role of these regions in the pathophysiology of the disorder (Palaniyappan et al., 2014, 2013, 2012). Alterations in these regions may impair the capacity of the salience network to switch between dynamic brain states that gives rise to a disruption of reward processing and the promotion of false, context-independent associations, which subserve the phenomenology of both negative and positive symptoms in schizophrenia (Gradin et al., 2013).

However, this initial analysis did not directly explore the more subtle morphometric differences between these three schizophrenia subgroups. These differences were revealed by means of the more sensitive ODVBA methodology, described previously (results are displayed in Fig. 15.7) (Zhang et al., 2015), and consisted of a specific cerebellar involvement in patients with predominant negative symptoms, which distinguished this group from the other two subgroups, and specific occipital and medial orbitofrontal volume reductions in patients with predominant positive symptoms. Interestingly, in keeping with the initial study (Koutsouleris et al., 2008), patients with primarily disorganized

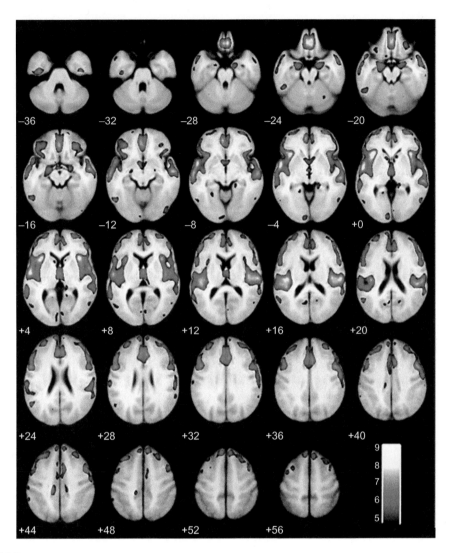

FIGURE 15.6

VBM comparison of gray matter density reductions in 175 patients with SCZ compared to 177 healthy controls (HC). The resulting T map was thresholded at $p < 0.05$ (FWE-corrected). Most significant areas of gray matter density reductions were identified in the perisylvian and superior temporal regions.

symptoms did not show specific brain alterations compared to the other samples, suggesting that the alterations in this sample compared to HC represent a core structural pattern of the disorder shared by patients with diverse schizophrenic phenotypes. This analysis also revealed significant correlations of the brain patterns with longitudinal disease variables, such as age of disease onset and illness duration. These cross-sectional findings point to possible disease-related brain trajectories that may determine when and to which extent these patterns occur across the patients' lifespan. Further prospective MRI

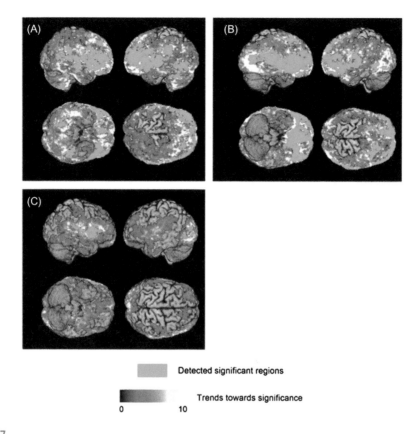

FIGURE 15.7

Reduced gray matter volumes in several brain regions. In the Munich dataset using ODVBA for three subtypes of schizophrenia patients, the following were found: (A) negative symptoms; (B) positive symptoms; and (C) disorganized symptoms.

studies may elucidate the neuroanatomical evolution and the predictive value of the patterns in determining the therapeutic responsivity as well as the clinical and functional outcome of a given patient across time.

SINGLE-SUBJECT CLASSIFICATION: FROM COMPUTATIONAL NEUROANATOMY OF SCHIZOPHRENIA TO INDIVIDUALIZED DIAGNOSTIC TESTS

One of the first studies using MVPA machine learning methods is the one already presented in the previous section (Davatzikos et al., 2005a). That study derived individualized neuroanatomical markers of schizophrenia, demonstrating that neuroanatomical signatures of the disease are recognizable on an individual basis using advanced analytical methods. This work paved the way toward the development of neuroanatomical markers of high sensitivity and specificity that could ultimately lead to clinically useful markers of schizophrenia. In that study, the previously described spatial pattern of reduced GM

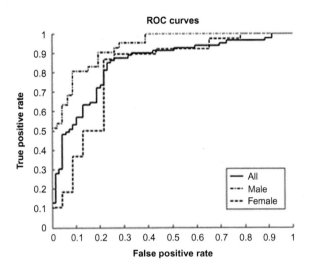

FIGURE 15.8

Relatively high sensitivity and specificity achieved by SVM-based classification of structural MRI-based on spatial patterns of regional GM volumes.

Adapted from Davatzikos, C., Shen, D.G., Wu, X., Lao, Z., Hughett, P., Turetsky, B.I., et al., 2005b. Whole-brain morphometric study of schizophrenia reveals a spatially complex set of focal abnormalities. JAMA Arch. Gen. Psychiatry. 62, 1218–1227.

volumes (Fig. 15.5) was synthesized into an index of schizophrenia-like phenotypes via a machine learning method (Fan et al., 2007). Using this approach, a multivariate high-dimensional model, or pattern, was determined from patients and controls. This model was subsequently applied to new scans, seeking the schizophrenia-like neuroanatomical pattern of each individual. Accuracy of classification of individual scans was reported to be 81.1%. The same study showed that the neuroanatomical pattern of schizophrenia was slightly different between men and women. Therefore, when these techniques were applied separately to each sex, they achieved classification accuracy of 85% and 82% in men and women, respectively, despite the reduced sample size in each subgroup relative to the combined data (see Fig. 15.8). These classification rates began to approach those of clinically useful diagnostic tests, and they clearly showed that these distinctive neuroanatomical signatures are spatially complex and subtle and cannot be appreciated via standard radiologic readings of MRI.

Subsequent work (Koutsouleris et al., 2009a) further supported these conclusions.

SINGLE-SUBJECT DIFFERENTIATION BETWEEN THE PSYCHOSIS PRODROME AND THE AT-RISK FOR MENTAL PSYCHOSIS WITHOUT SUBSEQUENT DISEASE TRANSITION

Since the beginning of this century, the field increasingly focused on identifying brain phenotypes of early and even prodromal phases of schizophrenia with the ultimate goal of implementing biomarker-driven individualized early recognition of illness in persons with a clinically defined risk for psychosis. A landmark study (Pantelis et al., 2003) reported subtle and distributed gray matter reductions in CHR individuals with or without subsequent disease transition that were located in the right medial temporal, lateral temporal, and inferior frontal cortex, as well as the cingulate cortex, bilaterally. Overlapping findings were reported in subsequent VBM studies (Borgwardt et al., 2007; Fusar-Poli et al., 2011; Koutsouleris et al., 2009b) that corroborated evidence for accumulating prefronto-temporo-limbic abnormalities preceding the onset of frank psychosis in these at-risk populations. However, due to the aforementioned limitations of the univariate VBM methodology, it remained unclear whether these subtle structural brain differences could provide any added predictive value in terms of an improved risk stratification of persons with an approximately 30% risk of developing psychosis over the course of 3 years (Fusar-Poli et al., 2012).

The first evidence for a potentially high predictive value of structural brain imaging was reported by Koutsouleris et al. (2009a,b). The authors extracted a multivariate pattern of distributed brain abnormalities that distinguished subsequent converters from nonconverters with an accuracy of 88% by training nonlinear support vector machines on neuroanatomical features obtained through principal component analysis. Again, the main foci of brain regions involved in the decision boundary of the SVM was in the prefrontal, temporal, and cerebellar areas, suggesting that a disruption of cortico-subcortical systems spanning these regions not only may characterize the established disease but also may predate the onset of schizophrenia by several years. These initial findings were first replicated in a second and completely independent population of at-risk individuals (Koutsouleris et al., 2011) and then further substantiated in a third SVM analysis performed in the pooled Basel and Munich CHR cohort. The latter analysis demonstrated that it may be possible to identify predictive neuroanatomical disease signatures across heterogeneous CHR populations examined by means of different MRI scanner hardware and data acquisition protocols and that the individual expression of these brain signatures may be strongly linked with different "survival" curves in terms of disease transition risk. Based on these data, several large-scale CHR projects have started to probe the predictive value of MVPA-based neuroanatomical risk stratification models in the multicenter study settings across the United States and Europe.

ARE SCZ-LIKE NEUROANATOMICAL PATTERNS POSSIBLY ENDOPHENOTYPES OF DISEASE?

A number of studies have found evidence for genetic modulation of the brain structure in unaffected family members of patients with schizophrenia. However, most studies were based on measurements of a priori sets of selected brain regions of interest (ROI), thereby potentially biasing the results toward these ROIs. The ROI approach might also miss effects in brain regions where the preselected ROIs are not optimally defined to capture phenotypic characteristics of family members. Moreover, although most studies have reported group differences in some brain structures, there has been substantial overlap of brain volumes among family members, healthy controls, and patients, thereby rendering it difficult to evaluate structural phenotypes in individuals.

Results from the UPenn study (Fan et al., 2008b) indicated that schizophrenia-like neuroanatomical patterns of family members were mostly in the same range as that of the patient, thereby indicating that the structural phenotype that distinguishes between SCZ and healthy controls was present in most family members. Male family members displayed relatively higher similarity to schizophrenia patients, thereby indicating a potential sex-specific relationship between genotype and phenotype in schizophrenia. To the extent that brain structural phenotypes portend risk for SCZ, the results also indicate that male family members might be at higher risk for developing the disease, although female family members also had patterns similar to those of the patients, as Fig. 15.9 shows.

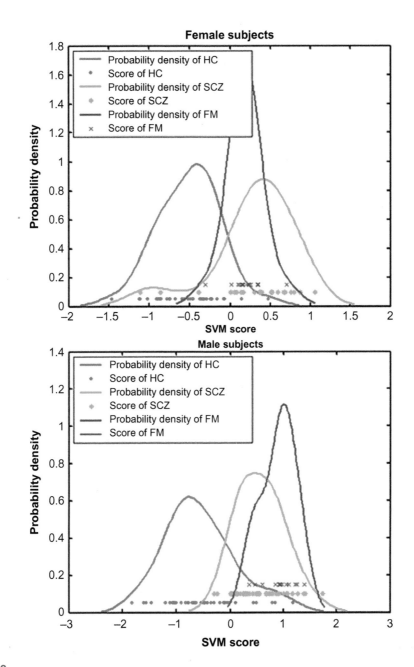

FIGURE 15.9

Schizophrenia-like neuroanatomical patterns were present in unaffected family members (blue) of schizophrenia patients (green), whereas healthy controls (red) had negative (normal) neuroanatomical patterns.

REFERENCES

Andreasen, N.C., Flashman, L., Flaum, M., Arndt, S., Swayze 2nd, V., O'Leary, D.S., et al., 1994. Regional brain abnormalities in schizophrenia measured with magnetic resonance imaging. J. Am. Med. Assoc. 272 (22), 1763–1769.

Andreasen, N.C., Paradiso, S., O'Leary, D.S., 1998. "Cognitive dysmetria" as an integrative theory of schizophrenia: a dysfunction in cortical-subcortical-cerebellar circuitry? Schizophr. Bull. 24 (2), 203–218.

Antonova, E., Sharma, T., Morris, R., Kumari, V., 2004. The relationship between brain structure and neurocognition in schizophrenia: a selective review. Schizophr. Res. 70 (2-3), 117–145.

Ashburner, J., Friston, K.J., 2000. Voxel-based morphometry – the methods. Neuroimage 11 (6 Pt 1), 805–821. http://dx.doi.org/10.1006/nimg.2000.0582 S1053-8119(00)90582-2 [pii].

Bleuler, E., 1911. Dementia Praeox oder die Gruppe der Schizophrenien, Deuticke, Leipzig.

Borgwardt, S.J., Riecher-Rossler, A., Dazzan, P., Chitnis, X., Aston, J., Drewe, M., et al., 2007. Regional gray matter volume abnormalities in the at risk mental state. Biol. Psychiatry. 61 (10), 1148–1156. http://dx.doi.org/10.1016/j.biopsych.2006.08.009.

Cannon, T.D., van Erp, T.G.M., Huttunen, M., Lonnqvist, J., Salonen, O., Valanne, L., et al., 1998. Regional gray matter, white matter, and cerebrospinal fluid distributions in schizophrenic patients, their siblings, and controls. Arch. Gen. Psychiatry. 55 (12), 1084–1091.

Cannon, T.D., Chung, Y., He, G., Sun, D., Jacobson, A., van Erp, T.G., et al., 2015. Progressive reduction in cortical thickness as psychosis develops: a multisite longitudinal neuroimaging study of youth at elevated clinical risk. Biol. Psychiatry. 77 (2), 147–157. http://dx.doi.org/10.1016/j.biopsych.2014.05.023.

Chaim, T.M., Silva, M.A., Varol, E., Doshi, J., Zanetti, M.V., Gaonkar, B., et al., 2012. High-dimensional pattern classification of brain morphometric and DTI data of adult ADHD. Biol. Psychiatry. 71 (8), 190s.

Cohen, M., Kosslyn, S., Breiter, H., DiGirolamo, G., Thompson, W., Anderson, A., et al., 1996. Changes in cortical activity during mental rotation. A mapping study using functional MRI. Brain 119 (1), 89–100.

Criss, T.B., Davatzikos, C., Liao, R., Williams, J.A., 1998. Constrained optimization for stereotaxic radiosurgery using deformable neuroanatomical models. Int. J. Radiat. Oncol. Biol. Phys. 42 (1), 361. (Suppl.).

Davatzikos, C., 2004. Why voxel-based morphometric analysis should be used with great caution when characterizing group differences. Neuroimage 23 (1), 17–20.

Davatzikos, C., Genc, A., Xu, D., Resnick, S.M., 2001. Voxel-based morphometry using the RAVENS maps: methods and validation using simulated longitudinal atrophy. Neuroimage 14 (6), 1361–1369. http://dx.doi.org/10.1006/nimg.2001.0937.

Davatzikos, C., Fan, Y., Wu, X., Shen, D., Resnick, S.M., 2008. Detection of prodromal Alzheimer's disease via pattern classification of magnetic resonance imaging. Neurobiol. Aging. 29 (4), 514–523. http://dx.doi.org/10.1016/j.neurobiolaging.2006.11.010.

Davatzikos, C., Shen, D.G., Gur, R.C., Wu, X.Y., Liu, D.F., Fan, Y., et al., 2005a. Whole-brain morphometric study of schizophrenia revealing a spatially complex set of focal abnormalities. Arch. Gen. Psychiatry. 62 (11), 1218–1227.

Davatzikos, C., Shen, D.G., Wu, X., Lao, Z., Hughett, P., Turetsky, B.I., et al., 2005b. Whole-brain morphometric study of schizophrenia reveals a spatially complex set of focal abnormalities. JAMA Arch. Gen. Psychiatry. 62, 1218–1227.

Ettinger, U., Williams, S.C., Meisenzahl, E.M., Moller, H.J., Kumari, V., Koutsouleris, N., 2012. Association between brain structure and psychometric schizotypy in healthy individuals. World. J. Biol. Psychiatry. 13 (7), 544–549. http://dx.doi.org/10.3109/15622975.2011.559269.

Ettinger, U., Meyhofer, I., Steffens, M., Wagner, M., Koutsouleris, N., 2014. Genetics, cognition, and neurobiology of schizotypal personality: a review of the overlap with schizophrenia. Front Psychiatry 5, 18. http://dx.doi.org/10.3389/fpsyt.2014.00018.

Fan, Y., Shen, D., Gur, R.C., Gur, R.E., Davatzikos, C., 2007. COMPARE: classification of morphological patterns using adaptive regional elements. IEEE. Trans. Med. Imaging. 26 (1), 93–105.

Fan, Y., Batmanghelich, N., Clark, C.M., Davatzikos, C., the Alzheimer's Disease Neuroimaging Initiative, 2008a. Spatial patterns of brain atrophy in MCI patients, identified via high-dimensional pattern classification, predict subsequent cognitive decline. Neuroimage (one of the top 10 cited papers of 2008) 39 (4), 1731–1743.

Fan, Y., Gur, R.E., Gur, R.C., Wu, X., Shen, D., Calkins, M.E., et al., 2008b. Unaffected family members and schizophrenia patients share brain structure patterns: a high-dimensional pattern classification study. Biol. Psychiatry. 63 (1), 118–124. http://dx.doi.org/10.1016/j.biopsych.2007.03.015.

First, M., Spitzer, R., Gibbon, M., Williams, J., 1995. Structured Clinical Interview for DSM-IV Axis I Disorders, Non-Patient Edition (SCID-NP).

Fischl, B., Dale, A.M., 2000. Measuring the thickness of the human cerebral cortex from magnetic resonance images. Proc. Natl. Acad. Sci. U.S.A. 97 (20), 11050–11055. http://dx.doi.org/10.1073/pnas.200033797.

Fischl, B., Sereno, M.I., Dale, A.M., 1999. Cortical surface-based analysis: II: inflation, flattening, and a surface-based coordinate system. Neuroimage 9 (2), 195–207. http://dx.doi.org/10.1006/nimg.1998.0396.

Fusar-Poli, P., Borgwardt, S., Crescini, A., Deste, G., Kempton, M.J., Lawrie, S., et al., 2011. Neuroanatomy of vulnerability to psychosis: a voxel-based meta-analysis. Neurosci. Biobehav. Rev. 35 (5), 1175–1185. http://dx.doi.org/10.1016/j.neubiorev.2010.12.005.

Fusar-Poli, P., Bonoldi, I., Yung, A.R., Borgwardt, S., Kempton, M.J., Valmaggia, L., et al., 2012. Predicting psychosis: meta-analysis of transition outcomes in individuals at high clinical risk. Arch. Gen. Psychiatry. 69 (3), 220–229. http://dx.doi.org/10.1001/archgenpsychiatry.2011.1472.

Gaser, C., Nenadica, I., Buchsbaumb, B.R., Hazlettc, E.A., Buchsbaumc, M.S., 2001. Deformation-based morphometry and its relation to conventional volumetry of brain lateral ventricles in MRI. Neuroimage 13 (6), 1140–1145.

Goldszal, A.F., Davatzikos, C., Pham, D.L., Yan, M.X., Bryan, R.N., Resnick, S.M., 1998. An image-processing system for qualitative and quantitative volumetric analysis of brain images. J. Comput. Assist. Tomogr. 22 (5), 827–837.

Good, C.D., Johnsrude, I.S., Ashburner, J., Henson, R.N., Friston, K.J., Frackowiak, R.S., 2001. A voxel-based morphometric study of ageing in 465 normal adult human brains. Neuroimage 14 (1 Pt 1), 21–36. http://dx.doi.org/10.1006/nimg.2001.0786.

Gradin, V.B., Waiter, G., O'Connor, A., Romaniuk, L., Stickle, C., Matthews, K., et al., 2013. Salience network-midbrain dysconnectivity and blunted reward signals in schizophrenia. Psychiatry. Res. 211 (2), 104–111. http://dx.doi.org/10.1016/j.pscychresns.2012.06.003.

Gur, R., Turetsky, B., Cowell, P., Finkelman, C., Maany, V., Grossman, R., et al., 2000. Temporolimbic volume reductions in schizophrenia. Arch. Gen. Psychiatry. 57 (8), 769–775.

Gur, R.E., Keshavan, M.S., Lawrie, S.M., 2007. Deconstructing psychosis with human brain imaging. Schizophr. Bull. 33 (4), 921–931. http://dx.doi.org/10.1093/schbul/sbm045.

Honea, R., Crow, T.J., Passingham, D., Mackay, C.E., 2005. Regional deficits in brain volume in schizophrenia: a meta-analysis of voxel-based morphometry studies. Am. J. Psychiatry 162 (12), 2233–2245. http://dx.doi.org/10.1176/appi.ajp.162.12.2233.

Kaas, J., 1996. Theories of visual cortex organization in primates: areas of the third level. Prog. Brain. Res. 112, 213–221.

Kambeitz, J., Kambeitz-Ilankovic, L., Leucht, S., Wood, S., Davatzikos, C., Malchow, B., et al., 2015. Detecting neuroimaging biomarkers for schizophrenia: a meta-analysis of multivariate pattern recognition studies. Neuropsychopharmacology 40 (7), 1742–1751. http://dx.doi.org/10.1038/npp.2015.22.

Kay, S.R., Fiszbein, A., Opler, L.A., 1987. The positive and negative syndrome scale (PANSS) for schizophrenia. Schizophr. Bull. 13 (2), 261–276.

Keshavan, M.S., Haas, G.L., Kahn, C.E., Aguilar, E., Dick, E.L., Schooler, N.R., et al., 1998. Superior temporal gyrus and the course of early schizophrenia: progressive, static, or reversible? J. Psychiatr. Res. 32 (3–4), 161–167.

Kloppel, S., Stonnington, C.M., Chu, C., Draganski, B., Scahill, R.I., Rohrer, J.D., et al., 2008. Automatic classification of MR scans in Alzheimer's disease. Brain 131 (Pt 3), 681–689.

Koutsouleris, N., Gaser, C., Jäger, M., Bottlender, R., Frodl, T., Holzinger, S., et al., 2008. Structural correlates of psychopathological symptom dimensions in schizophrenia: a voxel-based morphometric study. Neuroimage 39 (4), 1600–1612. http://dx.doi.org/10.1016/j.neuroimage.2007.10.029.

Koutsouleris, N., Meisenzahl, E.M., Davatzikos, C., Bottlender, R., Frodl, T., Scheuerecker, J., et al., 2009a. Use of neuroanatomical pattern classification to identify subjects in at-risk mental States of psychosis and predict disease transition. Arch. Gen. Psychiatry. 66 (7), 700–712.

Koutsouleris, N., Schmitt, G.J., Gaser, C., Bottlender, R., Scheuerecker, J., McGuire, P., et al., 2009b. Neuroanatomical correlates of different vulnerability states for psychosis and their clinical outcomes. Br. J. Psychiatry 195 (3), 218–226. http://dx.doi.org/10.1192/bjp.bp.108.052068.

Koutsouleris, N., Borgwardt, S., Meisenzahl, E.M., Bottlender, R., Möller, H.-J., Riecher-Rössler, A., 2011. Disease prediction in the at-risk mental state for psychosis using neuroanatomical biomarkers: results from the FePsy Study. Schizophr. Bull. http://dx.doi.org/10.1093/schbul/sbr145.

Lao, Z., Shen, D., Xue, Z., Karacali, B., Resnick, S.M., Davatzikos, C., 2004. Morphological classification of brains via high-dimensional shape transformations and machine learning methods. Neuroimage 21 (1), 46–57. doi:S1053811903005731 [pii].

Lawrie, S., Abukmeil, S., 1998. Brain abnormality in schizophrenia. A systematic and quantitative review of volumetric magnetic resonance imaging studies. Br. J. Psychiatry 172 (2), 110–120.

Misra, C., Fan, Y., Davatzikos, C., 2009. Baseline and longitudinal patterns of brain atrophy in MCI patients, and their use in prediction of short-term conversion to AD: results from ADNI. Neuroimage 44 (4), 1415–1422.

Nanda, P., Tandon, N., Mathew, I.T., Giakoumatos, C.I., Abhishekh, H.A., Clementz, B.A., et al., 2014. Local gyrification index in probands with psychotic disorders and their first-degree relatives. Biol. Psychiatry. 76 (6), 447–455. http://dx.doi.org/10.1016/j.biopsych.2013.11.018.

Nelson, M.D., Saykin, A.J., Flashman, L.A., Riordan, H.J., 1998. Hippocampal volume reduction in schizophrenia as assessed by magnetic resonance imaging. Arch. Gen. Psychiatry. 55 (5), 433–440.

Palaniyappan, L., Liddle, P.F., 2012a. Aberrant cortical gyrification in schizophrenia: a surface-based morphometry study. J. Psychiatry. Neurosci. 37 (6), 399–406. http://dx.doi.org/10.1503/jpn.110119.

Palaniyappan, L., Liddle, P.F., 2012b. Differential effects of surface area, gyrification and cortical thickness on voxel based morphometric deficits in schizophrenia. Neuroimage 60 (1), 693–699. http://dx.doi.org/10.1016/j.neuroimage.2011.12.058.

Palaniyappan, L., White, T.P., Liddle, P.F., 2012. The concept of salience network dysfunction in schizophrenia: from neuroimaging observations to therapeutic opportunities. Curr. Top. Med. Chem. 12 (21), 2324–2338.

Palaniyappan, L., Simmonite, M., White, T.P., Liddle, E.B., Liddle, P.F., 2013. Neural primacy of the salience processing system in schizophrenia. Neuron 79 (4), 814–828. http://dx.doi.org/10.1016/j.neuron.2013.06.027.

Palaniyappan, L., Park, B., Balain, V., Dangi, R., Liddle, P., 2014. Abnormalities in structural covariance of cortical gyrification in schizophrenia. Brain. Struct. Funct. http://dx.doi.org/10.1007/s00429-014-0772-2.

Pantelis, C., Velakoulis, D., McGorry, P.D., Wood, S.J., Suckling, J., Phillips, L.J., et al., 2003. Neuroanatomical abnormalities before and after onset of psychosis: a cross-sectional and longitudinal MRI comparison. Lancet 361 (9354), 281–288. doi:10.1016/s0140-6736(03)12323-9.

Pantelis, C., Yucel, M., Wood, S.J., Velakoulis, D., Sun, D., Berger, G., et al., 2005. Structural brain imaging evidence for multiple pathological processes at different stages of brain development in schizophrenia. Schizophr. Bull. 31 (3), 672–696. http://dx.doi.org/10.1093/schbul/sbi034.

Sanfilipo, M., Lafargue, T., Rusinek, H., Arena, L., Loneragan, C., Lautin, A., et al., 2000. Volumetric measure of the frontal and temporal lobe regions in schizophrenia: relationship to negative symptoms. Arch. Gen. Psychiatry. 57 (5), 471–480.

Shen, D.G., Davatzikos, C., 2003. Very high resolution morphometry using mass-preserving deformations and HAMMER elastic registration. Neuroimage 18 (1), 28–41.

Shenton, M.E., Kikinis, R., Jolesz, F.A., Pollak, S.D., LeMay, M., Wible, C.G., et al., 1992. Abnormalities of the left temporal lobe and thought disorder in schizophrenia: a quantitative magnetic resonance imaging study. New Engl. J. Med. 327, 604–612.

Shenton, M.E., Dickey, C.C., Frumin, M., McCarley, R.W., 2001. A review of MRI findings in schizophrenia. Schizophr. Res. 49 (1–2), 1–52.

Sigmundsson, T., Suckling, J., Maier, M., Williams, S.C.R., Bullmore, E.T., Greenwood, K.E., et al., 2001. Structural abnormalities in frontal, temporal, and limbic regions and interconnecting white matter tracts in schizophrenic patients with prominent negative symptoms. Am. J. Psychiatry 158 (2), 234–243.

Sotiras, A., Davatzikos, C., Paragios, N., 2013. Deformable medical image registration: a survey. IEEE. Trans. Med. Imaging. 32 (7), 1153–1190. http://dx.doi.org/10.1109/TMI.2013.2265603.

Sowell, E.R., Levitt, J., Thompson, P.M., Holmes, C.J., Blanton, R.E., Kornsand, D.S., et al., 2000. Brain abnormalities in early-onset schizophrenia spectrum disorder observed with statistical parametric mapping of structural magnetic resonance images. Am. J. Psychiatry 157 (9), 1475–1484.

Sun, J., Maller, J.J., Guo, L., Fitzgerald, P.B., 2009. Superior temporal gyrus volume change in schizophrenia: a review on region of interest volumetric studies. Brain. Res. Rev. 61 (1), 14–32. http://dx.doi.org/10.1016/j.brainresrev.2009.03.004.

Thermenos, H.W., Keshavan, M.S., Juelich, R.J., Molokotos, E., Whitfield-Gabrieli, S., Brent, B.K., et al., 2013. A review of neuroimaging studies of young relatives of individuals with schizophrenia: a developmental perspective from schizotaxia to schizophrenia. Am. J. Med. Genet. B. Neuropsychiatr. Genet. 162B (7), 604–635. http://dx.doi.org/10.1002/ajmg.b.32170.

Turetsky, B., Cowell, P., Gur, R., Grossman, R., Shtasel, D., Gur, R., 1995. Frontal and temporal lobe brain volumes in schizophrenia: relationship to symptomatology and clinical subtype. Arch. Gen. Psychiatry. 52, 1061–1070.

Vapnik, V.N., 1998. Statistical Learning Theory. Wiley, New York.

Velakoulis, D., Pantelis, C., McGorry, P.D., Dudgeon, P., Brewer, W., Cook, M., et al., 1999. Hippocampal volume in first-episode psychoses and chronic schizophrenia: a high-resolution magnetic resonance imaging study. Arch. Gen. Psychiatry. 56 (2), 133–141.

Velakoulis, D., Wood, S.J., Wong, M.T., McGorry, P.D., Yung, A., Phillips, L., et al., 2006. Hippocampal and amygdala volumes according to psychosis stage and diagnosis: a magnetic resonance imaging study of chronic schizophrenia, first-episode psychosis, and ultra-high-risk individuals. Arch. Gen. Psychiatry. 63 (2), 139–149. http://dx.doi.org/10.1001/archpsyc.63.2.139.

Vemuri, P., Gunter, J.L., Senjem, M.L., Whitwell, J.L., Kantarci, K., Knopman, D.S., et al., 2008. Alzheimer's disease diagnosis in individual subjects using structural MR images: validation studies. Neuroimage 39 (3), 1186–1197.

Winkler, A.M., Kochunov, P., Blangero, J., Almasy, L., Zilles, K., Fox, P.T., et al., 2010. Cortical thickness or grey matter volume? The importance of selecting the phenotype for imaging genetics studies. Neuroimage 53 (3), 1135–1146. http://dx.doi.org/10.1016/j.neuroimage.2009.12.028.

Wolkin, A., Rusinek, H., Vaid, G., Arena, L., Lafargue, T., Sanfilipo, M., et al., 1998. Structural magnetic resonance image averaging in schizophrenia. Am. J. Psychiatry 155 (8), 1064–1073.

Wood, S.J., Pantelis, C., Velakoulis, D., Yucel, M., Fornito, A., McGorry, P.D., 2008. Progressive changes in the development toward schizophrenia: studies in subjects at increased symptomatic risk. Schizophr. Bull. 34 (2), 322–329. http://dx.doi.org/10.1093/schbul/sbm149.

Wright, I.C., Rabe-Hesketh, S., Woodruff, P.W.R., David, A.S., Murray, R.M., Bullmore, E.T., 2000. Meta-analysis of regional brain volumes in schizophrenia. Am. J. Psychiatry 157 (1), 16–25.

Zanetti, M.V., Schaufelberger, M.S., Doshi, J., Ou, Y., Ferreira, L.K., Menezes, P.R., et al., 2013. Neuroanatomical pattern classification in a population-based sample of first-episode schizophrenia. Prog. Neuropsychopharmacol. Biol. Psychiatry 3 (43), 116–125. http://dx.doi.org/10.1016/j.pnpbp.2012.12.005.

Zhang, T.H., Davatzikos, C., 2011. ODVBA: optimally-discriminative voxel-based analysis. IEEE. Trans. Med. Imaging. 30 (8), 1441–1454. http://dx.doi.org/10.1109/Tmi.2011.2114362.

Zhang, T.H., Davatzikos, C., 2013. Optimally-discriminative voxel-based morphometry significantly increases the ability to detect group differences in schizophrenia, mild cognitive impairment, and Alzheimer's disease. Neuroimage 79, 94–110. http://dx.doi.org/10.1016/j.neuroimage.2013.04.063.

Zhang, T., Koutsouleris, N., Meisenzahl, E., Davatzikos, C., 2015. Heterogeneity of structural brain changes in subtypes of schizophrenia revealed using magnetic resonance imaging pattern analysis. Schizophr. Bull. 41 (1), 74–84. http://dx.doi.org/10.1093/schbul/sbu136.

CHAPTER 16

BRAIN COMPUTATIONS IN SCHIZOPHRENIA

R.A. Adams[1,2] and K.J. Friston[3]

[1]*Institute of Cognitive Neuroscience, University College London, London, United Kingdom* [2]*Division of Psychiatry, University College London, Charles Bell House, London, United Kingdom* [3]*The Wellcome Trust Centre for Neuroimaging, University College London, London, United Kingdom*

CHAPTER OUTLINE

The Bayesian Brain, Precision, and Hierarchical Models 283
Psychosis and Synaptic Gain 285
Computational Implications of Decreased High-Level Synaptic Gain..287
Computational Implications of Striatal Presynaptic Dopamine Elevation 288
    Tonic Dopamine Signaling...289
    Phasic Dopamine Signaling...289
    Incentive (and Aberrant) Salience 290
    Jumping to Conclusions: Overweighting Evidence versus Lowered Decision Threshold 290
Conclusions and Further Questions 291
Acknowledgments...292
References...292

THE BAYESIAN BRAIN, PRECISION, AND HIERARCHICAL MODELS

In this chapter, we assume that the brain performs Bayesian inference for the causes of its sensory data using a hierarchical generative model and predictive coding. Under these assumptions, many disparate findings and symptoms in schizophrenia can be understood as abnormalities in the hierarchical encoding of precision (inverse variance) in the brain (Adams et al., 2013a, 2016).

We refer to beliefs, inference, priors, and precision in a Bayesian sense, in which a belief (conscious or unconscious) is a probability distribution over some unknown state. The precision of a belief corresponds to its confidence or certainty: high precision means that the probability distribution is concentrated over the most likely value—the mean or expectation. Beliefs prior to observing data are called prior beliefs, which are updated to posterior beliefs after seeing the data.

What does it mean to say the brain performs Bayesian inference? To infer the states of its environment, the brain combines its prior beliefs with sensory evidence (or the likelihood of the data), and these distributions are *weighted according to their relative precision* (Fig. 16.1). For simplicity, these

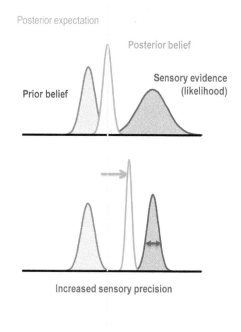

FIGURE 16.1

This schematic illustrates the importance of precision when forming posterior beliefs. The graphs show Gaussian probability distributions that represent prior beliefs, posterior beliefs, and the likelihood of some data or sensory evidence as functions of some hidden (unknown) parameter. The dotted line corresponds to the posterior expectation, whereas the width of the distributions corresponds to their variance. Precision is the inverse of this variance and can have a profound effect on posterior beliefs, because the posterior belief is biased toward the prior or sensory evidence in proportion to their relative precision. This means that the posterior expectation can be biased toward sensory evidence by either increasing sensory precision (or failing to attenuate it) or decreasing prior precision.

probability distributions are often assumed to be those that can be represented by a few "sufficient statistics" (eg, the mean and precision of a normal distribution); in this case, both prior and likelihood can be weighted by their (scalar) precision.

Despite its inherent uncertainty and huge complexity, sensory data contain patterns due to the hierarchical causal structure of the environment. The brain's prior beliefs can respect the hierarchical structure in its sensory data if they take the form of a hierarchical model. Hierarchical models explain complex patterns of low-level data features in terms of more abstract causes, such as the configuration of facial features prescribed by a person's identity.

Hierarchical models can use predictive coding (Rao and Ballard, 1999) to predict low-level data by exploiting their high-level descriptions, such as reconstructing the missing part of an image. In predictive coding, a unit at a given hierarchical level sends messages to one or more units at lower levels, which predict their activity; discrepancies between these predictions and the actual input are then passed back up the hierarchy in the form of prediction errors (PEs). These PEs revise the higher-level predictions, and this hierarchical message passing continues iteratively. There are theoretical reasons to suppose that the brain's fundamental task is simply to minimize its PEs (or "free energy") through both perception and action (Friston et al., 2012).

Exactly which predictions should be changed to explain away a given PE is a crucial question for hierarchical models. The Bayesian solution is that if you are very uncertain about your beliefs, but your source is very reliable, then you should change your beliefs a lot. Put more formally, the uncertainty (inverse precision) at each level helps determine the learning rate (update size) at that level (Mathys et al., 2011).

A classic psychology paradigm illustrates uncertainty at different levels (see Fig. 16.2). Imagine you are shown two jars of beads, one containing 85% green beads and 15% red beads and the other containing 85% red beads and 15% green beads. The jars are then hidden and a sequence of beads is drawn (with replacement) as GGRGG. You are asked to guess the color of the next bead. Even if you are quite certain of the identity of the jar (say, green), you will still be only 85% certain that the next bead will be green: this is "outcome uncertainty" or risk. Imagine you see more beads—the total sequence is GGRGGRR. Now you are very uncertain about the identity of the jar: this is "state uncertainty" or ambiguity. Imagine you see a much longer sequence—GGRGGRRRRRGGGRGGGGGGRGGGG. From this, it seems that the jar changes from green (five draws), to red (five draws), to green (remaining sequence). Such temporal changes in hidden causes (eg, someone surreptitiously switching the jar during the experiment) give rise to "volatility" (Mathys et al., 2011; Yu and Dayan, 2005).

Now, suppose that although the real proportions are 85% and 15%, a malicious experimenter told you that they are 99.9% and 0.1%. From the 25-draw sequence, you might conclude that the jars had changed eight times—whenever the color changed. This is what happens when the precision at the bottom of a hierarchical model is too high relative to the top. Following a (overly precise) sensory PE, the model concludes that there must have been a change in the environment (eg, the jar) rather than attributing it to chance.

This effect is reminiscent of the formation of so-called delusional perceptions, in which the perception of a chance event leads to a delusional idea; for example, a stranger touching his face is perceived as conveying a threatening message from the CIA. How might such an imbalance in the encoding of precision arise in schizophrenia? To answer this question, we must first explore the key neurobiological abnormalities in schizophrenia and then assess how they might impact the encoding of precision.

PSYCHOSIS AND SYNAPTIC GAIN

We now explore the commonalities among various neurobiological abnormalities in schizophrenia. We discuss reductions in synaptic gain in higher hierarchical (cortical) areas and increased presynaptic dopamine availability in the striatum.

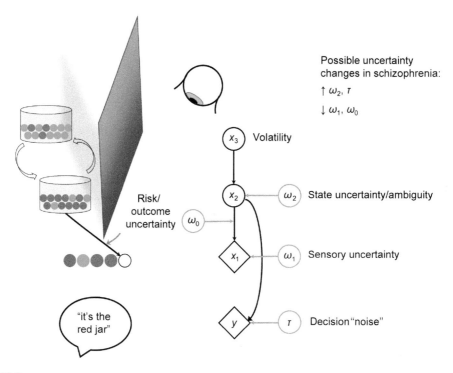

FIGURE 16.2

A hierarchical probabilistic model, illustrated using the beads or urn task. On the left, two jars are hidden behind a screen, one containing mostly green balls (eg, 70–85%) and the rest containing red balls, the other containing the converse. A sequence of balls is being drawn from one of these jars, in view of an observer, who is asked to guess from which jar they are coming. We have illustrated a simple hierarchical model of this process on the right: the observer is using such a model to make his/her guess. Variables in diamonds are observed, and variables in circles are hidden (ie, part of the model only). x_1 denotes the color of the currently observed bead, and its uncertainty (eg, if the light is low or if the subject is color blind) is denoted as ω_1. x_2 denotes the belief about the identity of the current jar, and its associated uncertainty ω_2 is known as state uncertainty or ambiguity. Another form of uncertainty, risk, or outcome uncertainty (ω_0) governs the relationship between the identity of the jar and the next outcome. Even if we are sure of the jar's identity, we cannot be certain of the color of the next bead. x_3 is the belief about the probability that the jars could be swapped at any time, known as volatility (further levels and uncertainties are not shown). The mapping between the belief about the jar x_2 and the response y is affected by a degree of stochasticity or decision "noise," τ. In schizophrenia, there may be too much uncertainty (ie, lower precision) in higher hierarchical areas that encode states or make decisions and an underestimation of uncertainty in lower (sensory) areas.

Abnormalities of N-methyl-D-aspartate receptors (NMDARs), γ-amino butyric acid (GABA) transmission, and dopamine receptors have all been proposed to be causal factors of schizophrenia. NMDARs have been thought to contribute to schizophrenia because NMDAR antagonists were found to produce psychotic-like symptoms (Krystal et al., 1994), and NMDAR and GABAergic systems are implicated by both genetic (Greenwood et al., 2012) and neuropathological evidence (reviewed by Harrison et al., 2011), particularly in hippocampus (HC) and prefrontal cortex (PFC).

A widely replicated finding in acute psychosis is elevated striatal dopamine availability, which Howes and Kapur (2009) propose is more closely linked to the state of psychosis than the trait of schizophrenia. This makes sense because antipsychotics work by blocking dopamine 2 receptors (D_2Rs), although an important caveat is that patients with symptoms resistant to D_2R blockade do not have elevated striatal dopamine synthesis (Demjaha et al., 2012).

What might dysfunction in NMDAR, GABA, and dopaminergic transmission have in common? All have profound effects on synaptic gain (or "short-term" synaptic plasticity) (Stephan et al., 2006), which is a multiplicative change in the influence of presynaptic input on postsynaptic responses.

Of all the receptors that determine synaptic gain, the most ubiquitous is the glutamatergic NMDA receptor (NMDAR). NMDARs can drive (ie, induce an excitatory postsynaptic potential) postsynaptic cells like other ionotropic glutamatergic (AMPA and kainate) receptors, but their time constants can be much longer. This enables integration of synaptic inputs over tens to hundreds of milliseconds, increasing the gain of synaptic inputs to distal dendrites. NMDARs also have a major impact on the short-term plasticity of glutamatergic synapses because they regulate the functional state and number of AMPARs (as well as mediate calcium-induced long-term synaptic depression or potentiation). Together, these properties make a significant contribution to the dynamics of neural networks, especially to oscillatory behavior and sustained firing patterns (Durstewitz, 2009).

Other key determinants of synaptic gain are the classical neuromodulator receptors (eg, dopamine and serotonin) and the metabotropic glutamate receptor (mGluR). These are metabotropic receptors. They do not activate ion channels; instead, they alter neuronal excitability via intracellular second messengers, such as cyclic adenosine monophosphate (cAMP), that change the production, surface expression, or activity of ion channels. Dopamine receptor subtypes have opposite effects on synaptic gain: D_1R activation stimulates cAMP production and increases gain, whereas D_2R activation inhibits cAMP production and reduces gain (reviewed in Frank, 2005).

Synaptic gain is determined not only by receptor activity but also by network dynamics, like the synchronization of fast oscillations, especially in 40–100 Hz or gamma frequencies. The fast-acting inhibitory $GABA_A$ receptor is instrumental in this synchronization process. In the cortex, a GABAergic (parvalbumin-positive basket cell) interneuron contacts many pyramidal cells, which it transiently hyperpolarizes. When this hyperpolarization wears off, all the cortical pyramidal cells can then fire together, leading to synchronous firing and oscillations across the network (Gonzalez-Burgos and Lewis, 2008). Synchronous firing has a facilitatory effect (ie, gain) on neural communication (Fries, 2005).

COMPUTATIONAL IMPLICATIONS OF DECREASED HIGH-LEVEL SYNAPTIC GAIN

What could be the computational effect of altered synaptic gain? Because precision determines the influence that one piece of information has over another in Bayesian inference, the neurobiological substrate of precision could be synaptic gain (Feldman and Friston, 2010). So, if the brain represents only the mean and precision of its probability distributions, then a loss of synaptic gain at a given level could cause a loss of precision (increase in variance) at that level.

A loss of precision encoding in PFC or HC would cause a loss of influence of the model's priors over the sensory data, making the world look less predictable and more surprising. This simple computational change can describe many phenomena in schizophrenia (more references and some simulations are in Adams et al., 2013a):

- At a neurophysiological level, responses to predictable stimuli resemble responses to unpredictable stimuli, and vice versa, in both perceptual electrophysiology experiments (eg, the P50 or P300 responses to tones) (Turetsky et al., 2007) and cognitive fMRI paradigms;
- At a network level, higher regions of cortex (eg, PFC and HC) have diminished connectivity to the thalamus relative to controls, whereas primary sensory areas are coupled more strongly with this region (Anticevic et al., 2014);
- At a perceptual level, there is greater resistance to visual illusions (Silverstein and Keane, 2011) that exploit the effects of visual priors on ambiguous images and a failure to attenuate the sensory consequences of one's own actions, which could diminish one's sense of agency (Shergill et al., 2005);
- At a motor level, there is impaired smooth visual pursuit of a predictably moving target but better tracking of a sudden unpredictable change in a target's motion (Hong et al., 2008);
- At a cognitive level, there are diffuse generalized cognitive problems (Dickinson et al., 2007).

How do these ideas relate to the positive symptoms of psychosis, such as delusions and hallucinations? A loss of high-level precision could precipitate the formation of delusional ideas because the former permits updates to beliefs that are larger and less constrained. However, decreased high-level precision alone should make unusual beliefs themselves vulnerable to rapid updating, unlike delusions. This account raises two important (and possibly related) questions that are addressed in the next section. First, if high-level precision is generally low, then why do delusions, which appear to exist at a reasonably high (conceptual) level in the hierarchy, become so fixed? Second, what is the computational impact of the best-established neurobiological abnormality in schizophrenia—an elevation in presynaptic dopamine?

COMPUTATIONAL IMPLICATIONS OF STRIATAL PRESYNAPTIC DOPAMINE ELEVATION

Given the evidence for striatal D_2R hyperactivity in psychosis, one might conclude that this is a sufficient cause of psychosis, but the exact nature of this stimulation and how it causes psychotic symptoms remain unclear.

In electrophysiological studies, dopamine neurons show both tonic and phasic firing patterns (Goto and Grace, 2005), to which D_1Rs and D_2Rs are more sensitive (respectively); however, these patterns cannot be distinguished using (human) brain imaging. It is unclear how the increased presynaptic availability of dopamine alters these patterns in schizophrenia; an increase in tonic release could reduce phasic release, such as via inhibitory presynaptic receptors (Beaulieu and Gainetdinov, 2011), and these two modes of release have also been argued to be at least partially independent (Goto and Grace, 2005). We explore how these patterns may be disrupted in schizophrenia and how this might affect computations.

TONIC DOPAMINE SIGNALING

How might (tonic) striatal D_2R hyperstimulation cause psychotic symptoms? The cortico-striato-thalamo-cortical loops take one of two routes through the striatum. Dopamine facilitates the "direct" pathway via phasic responses at excitatory D_1Rs and inhibits the "indirect" pathway through tonic activity at inhibitory D_2Rs (Dreyer et al., 2010). Dopamine has thus been proposed to facilitate the currently selected action (Maia and Frank, 2011) by stepping on the (direct) "accelerator" and/or releasing the (indirect) "brake."

The indirect pathway may not be just a brake, however. It contains two inhibitory pathways. One (via the subthalamic nucleus) causes blanket inhibition of action and acts as a brake, but the other is channelized (Smith et al., 1998) such that it can help switching to alternative actions. If the indirect pathway enables switching, then increased tonic D_2R activity in the striatum should oppose this (interestingly, D_2Rs also suppress alternative task "rules" in PFC; Ott et al., 2014). Indeed, reversal learning performance decreases with increasing occupancy of D_2Rs in the dorsal (Clatworthy et al., 2009; Groman et al., 2011) and ventral (Goto and Grace, 2005) striatum and with genetic variants in the dopamine transporter that might increase tonic dopamine (Den Ouden et al., 2013). Those with schizophrenia are impaired at reversal learning over and above their generalized cognitive impairment (Leeson et al., 2009); therefore, perhaps D_2R-mediated inflexibility makes delusions so resistant to change.

Importantly, this suggests the existence of two parallel deficits: although cortical synaptic gain problems would impair precision and lead to priors from the PFC losing influence, striatal D_2R hyperstimulation would separately explain perseveration by directly cementing in action (and belief) tendencies. There are theoretical suggestions that tonic dopamine (in the dorsal striatum and PFC) might encode the precision of action sequences (Friston et al., 2013). It is conceivable that this hyperdopaminergic increase in (dorsal striatal) precision of policies might occur as an adaptation to a loss of (prefrontal) high-level precision; for example, the excessive dorsal striatal dopamine release found in prefrontal dysfunction (Fusar-Poli et al., 2011; Sesack and Carr, 2002) is an attempt to stabilize beliefs and action selection in the face of cognitive instability.

Of note, other models of tonic dopamine signaling have proposed that it encodes (in the ventral striatum) average reward rate (Niv et al., 2007). Those with schizophrenia do show a small bias toward enacting rather than withholding actions (Waltz et al., 2011)—as the "average reward rate" model would predict—and acutely psychotic (putatively very hyperdopaminergic) states are certainly characterized by high vigor. Although mPFC and ventral striatal dopamine both correlate with average reward rate, ventral striatal dopamine is more closely linked to actual choice behavior and uncertainty of outcome (St. Onge et al., 2012), as the precision of policies would be.

PHASIC DOPAMINE SIGNALING

The phasic responses of dopamine neurons comprise bursts (activating D_1Rs) and pauses (inhibiting D_2Rs) that have been proposed to reflect positive and negative reward PEs, respectively (Schultz et al., 1997), and also aversive PEs (Lammel et al., 2011) and the precision or "salience" of PEs (Friston et al., 2013) in ventral and dorsal striatum. Actor–critic models (Maia and Frank, 2011) propose that dopaminergic reward PEs teach the ventral striatum (critic) the values of states and teach the dorsal striatum (actor) to associate states with optimal actions, and empirical evidence concurs (Steinberg et al., 2013).

In fMRI paradigms designed to elicit reward PE and reward prediction signals, subjects with schizophrenia show diminished appropriate activations (Winton-Brown et al., 2014) but greater inappropriate activations in ventral striatum (Murray et al., 2007) compared with controls. Similar patterns were observed in an associative learning task (Corlett et al., 2007); however, fMRI cannot tell whether these abnormalities are due to abnormal phasic dopamine signaling. Although behavioral responses in such tasks are often less abnormal than the underlying neural activity (Murray et al., 2010), computational modeling of behavior in both the beads task (Averbeck et al., 2011) and reward learning tasks (Gold et al., 2012) suggests that the impact of phasic positive feedback is diminished in schizophrenia. Elevated tonic dopamine levels may be reducing phasic bursts (reward learning) but not phasic pauses (punishment learning), and PET studies suggest an inverse relationship between tonic DA and phasic BOLD signals (Deserno et al., 2015). Alternatively, this apparent reward learning deficit may actually be caused by reduced working memory capacity (Collins et al., 2014).

INCENTIVE (AND ABERRANT) SALIENCE

The theory of incentive salience proposes that ventral striatal dopamine signaling (phasic or tonic) gives motivational impetus to act on stimuli whose values have already been learned (Berridge, 2007). Incentive salience is closely related to model-free learning of values (Flagel et al., 2011), or the precision or confidence of beliefs that actions will have preferred outcomes (Friston et al., 2013), but distinct from "informational" salience, or the degree to which sensory information can minimize uncertainty (Friston et al., 2012). In the "aberrant salience" hypothesis, Kapur (2003) proposed that there is aberrant (ie, increased inappropriate) signaling of incentive salience in psychosis. Unmedicated prodromal psychotic subjects do experience—in proportion to their positive symptoms—irrelevant features of stimuli as "aberrantly salient" (although this is not obviously reflected in their reaction times), but their striatal activations are harder to interpret (Roiser et al., 2013).

A weakness of the aberrant salience hypothesis is that the connection between aberrant motivational signaling and abnormal inference (hallucinations) and abnormal learning (delusions) is not intuitive (although this presupposes that delusions do involve abnormal learning). Conversely, diminished appropriate salience signaling could underlie negative symptoms, and a loss of ventral striatal activation of rewards has been shown to be proportional to negative symptoms in unmedicated subjects with schizophrenia (Juckel et al., 2006). Finally, the aberrant salience hypothesis does not explain why positive symptoms are typically aversive rather than appetitive; it seems to be a better explanation for manic rather than paranoid psychosis.

Aside from a loss of appropriate motivational signaling (or optimal encoding of precision), there are many other potential explanations for negative symptoms, such as pronounced asymmetry in learning (ie, a failure to learn stimulus–reward associations but intact learning of stimulus–punishment associations), a failure to infer the values of actions, and a loss of uncertainty-driven exploration such that valuable states are never discovered; these are reviewed extensively elsewhere (Strauss et al., 2014).

JUMPING TO CONCLUSIONS: OVERWEIGHTING EVIDENCE VERSUS LOWERED DECISION THRESHOLD

The "jumping to conclusions" bias is a well-established finding in schizophrenia (Fine et al., 2007) and is manifest—when performing the beads task (Fig. 16.2)—as a tendency to decide on the color of the jar after seeing only one or two beads (found in approximately 50% of patients). One might hypothesize that a loss of high-level precision (ω_2 in Fig. 16.2) causes larger belief updating (ie, overweighting of evidence) and, hence, premature decisions; however, in fact, it seems that underlying this effect is a lowered decision threshold (Averbeck et al., 2011) or increased stochasticity of decision-making (τ in Fig. 16.2; Moutoussis et al., 2011). The latter two problems may still be due to PFC rather than striatal pathology, however, because attempts to recreate this bias in healthy subjects by boosting phasic and tonic DAergic release have failed (Andreou et al., 2014; Ermakova et al., 2014), and it may simply reflect changes in the cortical precision of states or decisions.

CONCLUSIONS AND FURTHER QUESTIONS

In this brief overview, we described how some neurobiological risk factors for schizophrenia (NMDAR and GABAergic interneuron hypofunction) reduce both synaptic and oscillatory gain at high hierarchical areas. This could impair the encoding of precision at higher levels of the brain's hierarchical model and increase expected precision at lower levels. This imbalance can account for many neurobiological and phenomenological findings in schizophrenia.

We also described the potential computational consequences of striatal D_2R hyperactivity. This may increase the precision of current policies by inhibiting indirect pathway-mediated behavioral or cognitive switching. This could be a (dysfunctional) consequence of or even an attempt to compensate for prefrontal or hippocampal pathology. This D_2R hyperactivity may also reduce learning from positive outcomes and affect the encoding of motivational (or informational) salience.

We conclude with some brief speculations about some other important aspects of psychosis. First, precision imbalance in a predictive coding network could explain auditory verbal hallucinations (AVH). Although many kinds of hallucinations are likely due to an *increase* in the relative precision of prior beliefs over contradictory—but imprecise—sensory evidence (Friston, 2005), AVH may be better understood as a failure to attenuate the sensory consequences (corollary discharge) of inner speech. Previous versions of this hypothesis (Frith et al., 1998) proposed that a dysfunctional forward model causes a mismatch between desired inner speech and its imagined sensory consequences (which the forward model generates from efference copy of speech commands) and that this mismatch makes the inner speech seem as if it were externally caused. This account has been criticized because efference copy of inner speech seems redundant for thinking (Gallagher, 2004). Predictive coding in sensorimotor systems does not require efference copy, however (Adams et al., 2013b). Here, motor predictions of auditory input (ie, inner speech) may themselves be perceived as externally caused because of an imbalance between high-level and sensory-level precision encoding.

Second, many delusional beliefs seem to stem not from chance environmental events but rather from individuals' goals or fantasies (ie, from the top of the hierarchical model, not the bottom). Computational models of the HC could illuminate this issue. The HC may instantiate a complex model of the current (and past/future) state(s) of the environment, which can be updated by sensory information or generate sensory (ie, imagined) data by receiving or sending (respectively) messages to sensory hierarchies and enable goal-driven planning through its interactions with limbic and prefrontal cortices (Penny et al., 2013). Critically, sensory input must be switched off during planning, and goals must be switched off during sensory state estimation. If the latter fails to happen, then goals themselves may contribute to the estimation of the current state. Patterns of HC–PFC disconnectivity in schizophrenia have some intriguing parallels with this model. In schizophrenia, connectivity is increased relative to controls when it should be downmodulated during a working memory task (Meyer-Lindenberg et al., 2005) but, conversely, during an episodic memory task, when it should be upmodulated, PFC fails to stimulate HC activation (Heckers et al., 1998).

ACKNOWLEDGMENTS

Dr. Adams thanks Dr. Quentin Huys, Dr. Jon Roiser, Dr. Klaas Stephan, Dr. Harriet Brown, and Prof. Chris Frith for their contributions to this work. Dr. Adams is supported by the National Institutes of Health Research. Prof. Friston's work was funded by the Wellcome Trust Grant 088130/Z/09/Z. The Wellcome Trust Centre for Neuroimaging is supported by core funding from Wellcome Trust Grant 091593/Z/10/Z.

REFERENCES

Adams, R.A., Shipp, S., Friston, K.J., 2013a. Predictions not commands: active inference in the motor system. Brain. Struct. Funct. 218 (3), 611–643.

Adams, R.A., Stephan, K.E., Brown, H.R., Frith, C.D., Friston, K.J., 2013b. The computational anatomy of psychosis. Front. Psychiatry 4, 47.

Adams, R.A., Huys, Q.J.M., Roiser, J.P., 2016. Computational psychiatry – towards a mathematically informed understanding of depression and schizophrenia. J. Neurol. Neurosurg. Psychiatry 87 (1), 53–63. http://dx.doi.org/10.1136/jnnp-2015-310737.

Andreou, C., Moritz, S., Veith, K., Veckenstedt, R., Naber, D., 2014. Dopaminergic modulation of probabilistic reasoning and overconfidence in errors: a Double-Blind Study. Schizophr. Bull. 40 (3), 558–565.

Anticevic, A., Cole, M.W., Repovs, G., Murray, J.D., Brumbaugh, M.S., Winkler, A.M., et al., 2014. Characterizing thalamo-cortical disturbances in schizophrenia and bipolar illness. Cereb Cortex (New York, NY: 1991) 24 (12), 3116–3130.

Averbeck, B.B., Evans, S., Chouhan, V., Bristow, E., Shergill, S.S., 2011. Probabilistic learning and inference in schizophrenia. Schizophr. Res. 127 (1–3), 115–122.

Beaulieu, J.-M., Gainetdinov, R.R., 2011. The physiology, signaling, and pharmacology of dopamine receptors. Pharmacol. Rev. 63 (1), 182–217.

Berridge, K.C., 2007. The debate over dopamine's role in reward: the case for incentive salience. Psychopharmacology 191 (3), 391–431.

Clatworthy, P.L., Lewis, S.J.G., Brichard, L., Hong, Y.T., Izquierdo, D., Clark, L., et al., 2009. Dopamine release in dissociable striatal subregions predicts the different effects of oral methylphenidate on reversal learning and spatial working memory. J. Neurosci. 29 (15), 4690–4696.

Collins, A.G.E., Brown, J.K., Gold, J.M., Waltz, J.A., Frank, M.J., 2014. Working memory contributions to reinforcement learning impairments in schizophrenia. J. Neurosci. 34 (41), 13747–13756.

Corlett, P.R., Murray, G.K., Honey, G.D., Aitken, M.R.F., Shanks, D.R., Robbins, T.W., et al., 2007. Disrupted prediction-error signal in psychosis: evidence for an associative account of delusions. Brain 130 (9), 2387–2400.

Demjaha, A., Murray, R.M., McGuire, P.K., Kapur, S., Howes, O.D., 2012. Dopamine synthesis capacity in patients with treatment-resistant schizophrenia. Am. J. Psychiatry 169 (11), 1203–1210.

Den Ouden, H.E.M., Daw, N.D., Fernandez, G., Elshout, J.A., Rijpkema, M., Hoogman, M., et al., 2013. Dissociable effects of dopamine and serotonin on reversal learning. Neuron 80 (4), 1090–1100.

Deserno, L., Huys, Q.J.M., Boehme, R., Buchert, R., Heinze, H.-J., Grace, A.A., et al., 2015. Ventral striatal dopamine reflects behavioral and neural signatures of model-based control during sequential decision making. Proc. Natl. Acad. Sci. U.S.A. 112 (5), 1595–1600.

Dickinson, D., Ramsey, M.E., Gold, J.M., 2007. Overlooking the obvious: a meta-analytic comparison of digit symbol coding tasks and other cognitive measures in schizophrenia. Arch. Gen. Psychiatry 64 (5), 532–542.

Dreyer, J.K., Herrik, K.F., Berg, R.W., Hounsgaard, J.D., 2010. Influence of phasic and tonic dopamine release on receptor activation. J. Neurosci. 30 (42), 14273–14283.

Durstewitz, D., 2009. Implications of synaptic biophysics for recurrent network dynamics and active memory. Neural Netw. 22 (8), 1189–1200.

Ermakova, A.O., Ramachandra, P., Corlett, P.R., Fletcher, P.C., Murray, G.K., 2014. Effects of methamphetamine administration on information gathering during probabilistic reasoning in healthy humans. PLoS One. 9 (7), e102683.

Feldman, H., Friston, K.J., 2010. Attention, uncertainty, and free-energy. Front. Human Neurosci. 4, 215.

Fine, C., Gardner, M., Craigie, J., Gold, I., 2007. Hopping, skipping or jumping to conclusions? Clarifying the role of the JTC bias in delusions. Cognit. Neuropsychiatry 12 (1), 46–77.

Flagel, S.B., Clark, J.J., Robinson, T.E., Mayo, L., Czuj, A., Willuhn, I., et al., 2011. A selective role for dopamine in stimulus-reward learning. Nature 469 (7328), 53–57.

Frank, M.J., 2005. Dynamic dopamine modulation in the basal ganglia: a neurocomputational account of cognitive deficits in medicated and nonmedicated Parkinsonism. J. Cogn. Neurosci. 17 (1), 51–72.

Fries, P., 2005. A mechanism for cognitive dynamics: neuronal communication through neuronal coherence. Trends Cogn. Sci. 9 (10), 474–480.

Friston, K., Adams, R.A., Perrinet, L., Breakspear, M., 2012. Perceptions as hypotheses: saccades as experiments. Front. Psychol. 3, 151.

Friston, K., Schwartenbeck, P., Fitzgerald, T., Moutoussis, M., Behrens, T., Dolan, R.J., 2013. The anatomy of choice: active inference and agency. Front. Human Neurosci. 7, 598.

Friston, K.J., 2005. Hallucinations and perceptual inference. Behav. Brain Sci. 28 (06), 764–766.

Frith, C., Rees, G., Friston, K., 1998. Psychosis and the experience of self. Brain systems underlying self-monitoring. Ann. N.Y. Acad. Sci. 843, 170–178.

Fusar-Poli, P., Howes, O.D., Allen, P., Broome, M., Valli, I., Asselin, M.-C., et al., 2011. Abnormal prefrontal activation directly related to pre-synaptic striatal dopamine dysfunction in people at clinical high risk for psychosis. Mol. Psychiatry 16 (1), 67–75.

Gallagher, S., 2004. Neurocognitive models of schizophrenia: a neurophenomenological critique. Psychopathology 37 (1), 8–19.

Gold, J.M., Waltz, J.A., Matveeva, T.M., Kasanova, Z., Strauss, G.P., Herbener, E.S., et al., 2012. Negative symptoms and the failure to represent the expected reward value of actions: behavioral and computational modeling evidence. Arch. Gen. Psychiatry 69 (2), 129–138.

Gonzalez-Burgos, G., Lewis, D.A., 2008. GABA neurons and the mechanisms of network oscillations: implications for understanding cortical dysfunction in schizophrenia. Schizophr Bull. 34 (5), 944–961.

Goto, Y., Grace, A.A., 2005. Dopaminergic modulation of limbic and cortical drive of nucleus accumbens in goal-directed behavior. Nat. Neurosci. 8 (6), 805–812.

Greenwood, T.A., Light, G.A., Swerdlow, N.R., Radant, A.D., Braff, D.L., 2012. Association analysis of 94 candidate genes and schizophrenia-related endophenotypes. PLoS One 7 (1), e29630.

Groman, S.M., Lee, B., London, E.D., Mandelkern, M.A., James, A.S., Feiler, K., et al., 2011. Dorsal striatal D2-like receptor availability covaries with sensitivity to positive reinforcement during discrimination learning. J. Neurosci. 31 (20), 7291–7299.

Harrison, P.J., Lewis, D.A., Kleinman, J.E., 2011. Neuropathology of schizophrenia. In: Weinberger, D.R., Harrison, P.J. (Eds.), Schizophrenia Wiley-Blackwell, pp. 372–392.

Heckers, S., Rauch, S.L., Goff, D., Savage, C.R., Schacter, D.L., Fischman, A.J., et al., 1998. Impaired recruitment of the hippocampus during conscious recollection in schizophrenia. Nat. Neurosci. 1 (4), 318–323.

Hong, L.E., Turano, K.A., O'Neill, H., Hao, L., Wonodi, I., McMahon, R.P., et al., 2008. Refining the predictive pursuit endophenotype in schizophrenia. Biol. Psychiatry 63 (5), 458–464.

Howes, O.D., Kapur, S., 2009. The dopamine hypothesis of schizophrenia: version III – the final common pathway. Schizophr. Bull. 35 (3), 549–562.

Juckel, G., Schlagenhauf, F., Koslowski, M., Wüstenberg, T., Villringer, A., Knutson, B., et al., 2006. Dysfunction of ventral striatal reward prediction in schizophrenia. NeuroImage 29 (2), 409–416.

Kapur, S., 2003. Psychosis as a state of aberrant salience: a framework linking biology, phenomenology, and pharmacology in schizophrenia. Am. J. Psychiatry 160 (1), 13–23.

Krystal, J.H., Karper, L.P., Seibyl, J.P., Freeman, G.K., Delaney, R., Bremner, J.D., et al., 1994. Subanesthetic effects of the noncompetitive NMDA antagonist, ketamine, in humans. Psychotomimetic, perceptual, cognitive, and neuroendocrine responses. Arch. Gen. Psychiatry 51 (3), 199–214.

Lammel, S., Ion, D.I., Roeper, J., Malenka, R.C., 2011. Projection-specific modulation of dopamine neuron synapses by aversive and rewarding stimuli. Neuron 70 (5), 855–862.

Leeson, V.C., Robbins, T.W., Matheson, E., Hutton, S.B., Ron, M.A., Barnes, T.R.E., et al., 2009. Discrimination learning, reversal, and set-shifting in first-episode schizophrenia: stability over six years and specific associations with medication type and disorganization syndrome. Biol. Psychiatry 66 (6), 586–593.

Maia, T.V., Frank, M.J., 2011. From reinforcement learning models to psychiatric and neurological disorders. Nat. Neurosci. 14 (2), 154–162.

Mathys, C., Daunizeau, J., Friston, K.J., Stephan, K.E., 2011. A Bayesian foundation for individual learning under uncertainty. Front. Human Neurosci. 5, 39.

Meyer-Lindenberg, A.S., Olsen, R.K., Kohn, P.D., Brown, T., Egan, M.F., Weinberger, D.R., et al., 2005. Regionally specific disturbance of dorsolateral prefrontal-hippocampal functional connectivity in schizophrenia. Arch. Gen. Psychiatry 62 (4), 379–386.

Moutoussis, M., Bentall, R.P., El-Deredy, W., Dayan, P., 2011. Bayesian modelling of jumping-to-conclusions bias in delusional patients. Cognit. Neuropsychiatry 16 (5), 422–447.

Murray, G.K., Corlett, P.R., Clark, L., Pessiglione, M., Blackwell, A.D., Honey, G., et al., 2007. Substantia nigra/ventral tegmental reward prediction error disruption in psychosis. Mol. Psychiatry 13 (3), 267–276.

Murray, G.K., Corlett, P.R., Fletcher, P.C., 2010. The neural underpinnings of associative learning in health and psychosis: how can performance be preserved when brain responses are abnormal? Schizophr. Bull. 36 (3), 465–471.

Niv, Y., Daw, N.D., Joel, D., Dayan, P., 2007. Tonic dopamine: opportunity costs and the control of response vigor. Psychopharmacology 191 (3), 507–520.

Ott, T., Jacob, S.N., Nieder, A., 2014. Dopamine receptors differentially enhance rule coding in primate prefrontal cortex neurons. Neuron 84 (6), 1317–1328.

Penny, W.D., Zeidman, P., Burgess, N., 2013. Forward and backward inference in spatial cognition. PLoS Comput. Biol. 9 (12), e1003383.

Rao, R.P., Ballard, D.H., 1999. Predictive coding in the visual cortex: a functional interpretation of some extraclassical receptive-field effects. Nat. Neurosci. 2 (1), 79–87.

Roiser, J.P., Howes, O.D., Chaddock, C.A., Joyce, E.M., McGuire, P., 2013. Neural and behavioral correlates of aberrant salience in individuals at risk for psychosis. Schizophr. Bull. 39 (6), 1328–1336.

Schultz, W., Dayan, P., Montague, P.R., 1997. A neural substrate of prediction and reward. Science (New York, NY) 275 (5306), 1593–1599.

Sesack, S.R., Carr, D.B., 2002. Selective prefrontal cortex inputs to dopamine cells: implications for schizophrenia. Physiol. Behav. 77 (4–5), 513–517.

Shergill, S.S., Samson, G., Bays, P.M., Frith, C.D., Wolpert, D.M., 2005. Evidence for sensory prediction deficits in schizophrenia. Am. J. Psychiatry 162 (12), 2384–2386.

Silverstein, S.M., Keane, B.P., 2011. Perceptual organization impairment in schizophrenia and associated brain mechanisms: review of research from 2005 to 2010. Schizophr. Bull. 37 (4), 690–699.

Smith, Y., Bevan, M.D., Shink, E., Bolam, J.P., 1998. Microcircuitry of the direct and indirect pathways of the basal ganglia. Neuroscience 86 (2), 353–387.

Steinberg, E.E., Keiflin, R., Boivin, J.R., Witten, I.B., Deisseroth, K., Janak, P.H., 2013. A causal link between prediction errors, dopamine neurons and learning. Nat. Neurosci. 16 (7), 966–973.

Stephan, K.E., Baldeweg, T., Friston, K.J., 2006. Synaptic plasticity and dysconnection in schizophrenia. Biol. Psychiatry 59 (10), 929–939.

St Onge, J.R., Ahn, S., Phillips, A.G., Floresco, S.B., 2012. Dynamic fluctuations in dopamine efflux in the prefrontal cortex and nucleus accumbens during risk-based decision making. J. Neurosci. 32 (47), 16880–16891.

Strauss, G.P., Waltz, J.A., Gold, J.M., 2014. A review of reward processing and motivational impairment in schizophrenia. Schizophr. Bull. 40 (Suppl. 2), S107–S116.

Turetsky, B.I., Calkins, M.E., Light, G.A., Olincy, A., Radant, A.D., Swerdlow, N.R., 2007. Neurophysiological endophenotypes of schizophrenia: the viability of selected candidate measures. Schizophr. Bull. 33 (1), 69–94.

Waltz, J.A., Frank, M.J., Wiecki, T.V., Gold, J.M., 2011. Altered probabilistic learning and response biases in schizophrenia: behavioral evidence and neurocomputational modeling. Neuropsychology 25 (1), 86–97.

Winton-Brown, T.T., Fusar-Poli, P., Ungless, M.A., Howes, O.D., 2014. Dopaminergic basis of salience dysregulation in psychosis. Trends Neurosci. 37 (2), 85–94.

Yu, A.J., Dayan, P., 2005. Uncertainty, neuromodulation, and attention. Neuron 46 (4), 681–692.

CHAPTER 17

SCHIZOPHRENIA AND FUNCTIONAL IMAGING

K. Pauly[1] and C. Moessnang[2]

[1]Department of Psychiatry, Psychotherapy and Psychosomatic Medicine, RWTH Aachen University, Aachen, Germany
[2]Department of Psychiatry and Psychotherapy, Central Institute of Mental Health, Medical Faculty Mannheim/Heidelberg University, Mannheim, Germany

CHAPTER OUTLINE

Functional Magnetic Resonance Imaging: Challenges and Perspectives for the Study of Schizophrenia 297
Functional Imaging of Schizophrenia Symptom Clusters 298
    Cerebral Correlates of Delusions 298
    Cerebral Correlates of Hallucinations 306
Summary and Outlook 308
References 308

FUNCTIONAL MAGNETIC RESONANCE IMAGING: CHALLENGES AND PERSPECTIVES FOR THE STUDY OF SCHIZOPHRENIA

Blood oxygenation level–dependent (BOLD) functional magnetic resonance imaging (fMRI) is a noninvasive imaging technique that indirectly measures brain activation by quantifying magnetic effects of the so-called hemodynamic response. It is based on the fact that neural activity requires oxygen. The brain overcompensates this need by increased levels of oxygenated hemoglobin, which even exceed actual oxygen consumption. The hemodynamic response thus leads to a relative increase in oxygenated hemoglobin in activated brain areas. Differences in the ratio of oxygenated to deoxygenated hemoglobin induce changes of the magnetic signal. These changes are detected by the MRI scanner and reconstructed in 3D space. The spatial resolution is given by the voxel size, which is usually in the range of 2 to 4 mm^3 and, therefore, much higher than the resolution offered by other functional imaging techniques. The temporal resolution, in contrast, is limited by the inertia of the hemodynamic response, which occurs approximately 4–8 s subsequent to stimulus delivery (eg, a picture or a tone in the context of an experimental task). Given the high spatial but relatively low temporal resolution, BOLD fMRI is the technique of choice for questions about the localization of processes related to cognitive or emotional functions in the brain. In addition, advanced analysis techniques are able to reconstruct the time course of brain activation in the range of seconds, which allows a solid description of network dynamics, but will not reach the temporal precision of electrophysiological techniques such as electroencephalography.

One main way of relating brain activation to certain mental processes is the *subtraction method*. Brain responses during a well-matched control condition (eg, presentation of neutral facial expressions) are subtracted from brain responses during the experimental condition of interest (eg, presentation of fearful facial expressions). The difference in brain activation should identify brain regions specifically involved in the function of interest (eg, perception of fearfulness in facial expressions) while controlling for unwanted effects of other processes (eg, perception of faces in general). Due to great interindividual differences, neurofunctional studies normally investigate groups of participants instead of single subjects to allow for greater generalizability to the entire population using inferential statistical analysis. Moreover, to investigate specific symptoms, such as delusions in schizophrenia, patient samples are compared to other populations, such as healthy participants or schizophrenia patients without delusions.

A major challenge of fMRI studies in schizophrenia is posed by the heterogeneity of the individual symptoms as well as of the experimental procedures that can be applied (eg, the parameters of experimental paradigms, the inferential statistics applied). Moreover, especially in previous studies, rather small sample sizes (oftentimes even less than 10 subjects) were investigated, resulting in high variability of findings, low statistical power, and, consequently, the application of very liberal statistical criteria. Notwithstanding the scientific contribution of those early studies, their generalizability has to be evaluated with caution. An important means to overcome this shortcoming is the pooling of single studies into *meta-analyses*, whereby results are tested for consistency using statistical methods. Findings derived from meta-analyses are principally more robust and are therefore prioritized in the following overview of fMRI findings in schizophrenia. It should be noted, however, that meta-analytic results are highly dependent on the contributing studies, the selection of which inevitably introduces a bias with regard to content and statistics. Finally, some current fMRI projects are aiming at big sample sizes, in the range of hundreds of data sets, which are mostly realized in nationally and internationally funded multi-center imaging consortia. Given a best possible standardization of related potential confounds (such as differences in scanner hardware or task instructions across sites), big data approaches are a promising endeavor in the search for neurofunctional correlates of schizophrenia.

FUNCTIONAL IMAGING OF SCHIZOPHRENIA SYMPTOM CLUSTERS

Symptom clusters of schizophrenia are manifold. In addition to delusional thinking and hallucinations, patients may exhibit disorganized speech, grossly disorganized or catatonic behavior, and/or negative symptoms, such as diminished emotional expressions or avolition (Tandon et al., 2013). Impairments of such a wide range of behavioral and mental domains suggest the involvement of multiple brain systems in the etiology and manifestation of schizophrenia.

In this chapter, we report functional imaging—mainly fMRI—studies describing the neural correlates and proposed pathomechanisms of two core symptom clusters of schizophrenia, namely, delusions and auditory hallucinations.

CEREBRAL CORRELATES OF DELUSIONS

> There were different things that made me realize they are after me. When I came back from the supermarket I saw the road sign. They had positioned it exactly there to let me know they planned to

kill me. Every time I leave or enter the house, the neighbors are waiting for me behind their curtains. I think they don't know that I realized they had been in my apartment.

Similar to the challenge of studying a wide range of symptom clusters to understand schizophrenia, many contributing symptoms need to be investigated to understand a single symptom cluster, such as delusional thinking. To date, only a few neuroimaging studies have explored the neural correlates of specific delusional themes. One example is *persecutory delusions* (also referred to as paranoia), which are quite common in schizophrenia. Neurofunctionally, persecutory delusions have been linked to the anterior and the posterior cingulate cortex, which are core regions of self-perception (eg, Blackwood et al., 2004; see below). For example, if affected patients were asked to read menacing (compared to neutral) sentences and then decide whether those were self-related or not, patients with persecutory symptoms revealed decreased activation in the anterior ventral cingulate cortex as well as activation increases in the posterior cingulate cortex in comparison to healthy subjects (Sabri et al., 1997). Moreover, the degree of delusion, mistrust, and hallucinations correlated negatively with the cerebral blood flow in frontal, temporal, cingulate, and thalamic areas.

Delusions of reference are another common theme that can be found in schizophrenia. Patients may believe that radio broadcasts, television news, or even other people's clothes are intended to address them personally to transfer a certain message. Thus, actually unrelated input is self-referred and endowed with a certain (often negative) meaning. One way to experimentally examine the underlying neurofunctional correlates is to present impersonal sentences as well as sentences specifically tailored to the participant and to compare the reactions of healthy subjects and patients with delusions of reference (Menon et al., 2011). In studies like this, deluded patients showed reduced ability to differentiate between self-related and self-unrelated information, as revealed on behavioral as well as neuronal levels. Dysfunctional brain activation was mainly located in cortical midline structures, such as the medial prefrontal cortex, the posterior cingulate cortex, and the precuneus, which are typically associated with self-referential processing. In addition, aberrant activation was observed in the insula and the striatum, which correlated positively with the degree of delusions of reference (Fig. 17.1; Menon et al., 2011). These findings point to a dysfunctional salience attribution in that neutral information is aberrantly processed as self-relevant.

Aberrant Salience Attribution

Delusional symptoms are closely tied to the concept of salience, which refers to the property of a stimulus to attract our attention (Seeley et al., 2007). We usually attribute salience to only a small fraction of input we perceive every day and ignore information irrelevant to our goals to avoid stimulus overload. However, if the ascription of salience is dysfunctional, then it can result in distorted perception, as is postulated by the *aberrant-salience hypothesis* of schizophrenia (Kapur, 2003). According to this neurobehavioral model, patients with schizophrenia suffer from dysfunctional attribution of salience to irrelevant aspects of the environment, resulting in input overflow, inability to differentiate between relevant and irrelevant information, and increased cognitive efforts to make sense of these experiences. It is therefore comprehensible that patients with schizophrenia develop delusions when attempting to establish order in an otherwise uncontrollable environment. Random events are perceived as important and apparently meaningful, and the patient's environment is turned into a world full of signs. Although the primary cause of aberrant salience attribution is still a matter of debate (eg, Barkus et al., 2014; Winton-Brown et al., 2014), a well-accepted finding is the dysregulation of the mesolimbic system,

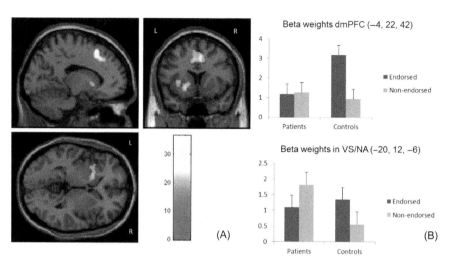

FIGURE 17.1

(A) When comparing functional responses to sentences endorsed as self-referential to those that were not endorsed as self-referential, significant differences between healthy subjects and patients with schizophrenia emerged within the dorsomedial prefrontal cortex (dmPFC), ventral striatum/nucleus accumbens (VS/NA), and insula. (B) Beta estimates extracted from dmPFC and VS/NA reveal greater responses to endorsed items in control subjects, whereas patients showed comparable responses or even greater responses to nonendorsed items, suggesting reduced ability to differentiate between self-referential and neutral information.

From Menon, M., Schmitz, T.W., Anderson, A.K., Graff, A., Korostil, M., Mamo, D., et al., 2011. Exploring the neural correlates of delusions of reference. Biol. Psychiatry 70 (12), 1127–1133.

which plays a key role in the attribution of salience (Bhattacharyya et al., 2015; Menon et al., 2007). In a series of fMRI studies using classical and instrumental conditioning (eg, Esslinger et al., 2012; Jensen et al., 2008; Juckel et al., 2006a; Schlagenhauf et al., 2009), schizophrenia patients revealed altered activation in the ventral striatum in response to conditioned and neutral stimuli (ie, stimuli that are predictive or not predictive of reward or punishment). For instance, patients demonstrated blunted responses in the ventral striatum toward reward-predicting cues, which was correlated with both positive (Nielsen et al., 2012) and negative symptoms (Juckel et al., 2006a,b). Activation partially normalized when atypical medication was administered (Juckel et al., 2006a). The finding of decreased striatal responses to conditioned stimuli has been interpreted as a failure to acknowledge salience in the environment. This assumption is further supported by complementary findings of elevated striatal responses to neutral (unconditioned) stimuli, which was interpreted as aberrant (over)attribution of salience (Fig. 17.2; Jensen et al., 2008; Murray et al., 2008). Taken together, these neurofunctional data suggest that patients with schizophrenia are impaired in successfully distinguishing between neutral stimuli and those that predict reward or punishment. Recent evidence even lends support to the hypothesis that aberrant striatal responses represent an endophenotype of schizophrenia (Grimm et al., 2014). The pattern of blunted striatal responses during reward anticipation could be not only replicated in healthy first-degree relatives of patients with schizophrenia but also linked to the schizophrenia candidate gene variants in NRG1 (Stefansson et al., 2002).

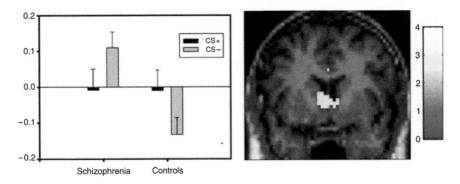

FIGURE 17.2

Aberrantly increased response of the ventral striatum to neutral cues (CS−) compared to cues that were predictive of punishment (CS+) in patients with schizophrenia during an aversive Pavlovian learning task.

From Jensen, J., Willeit, M., Zipursky, R.B., Savina, I., Smith, A.J., Menon, M., et al., 2008. The formation of abnormal associations in schizophrenia: neural and behavioral evidence. Neuropsychopharmacology 33 (3), 473–479.

Although mesolimbic dysfunctions and associated aberrant salience attribution can be postulated as some of the basic mechanisms underlying delusions, the wide variety of delusional symptoms as well as their persistence and pervasiveness point to the additional involvement of far-reaching cognition-related and emotion-related networks.

Neuronal Correlates of Cognitive and Emotional Biases

Cognitive biases support the generation and maintenance of delusions. Social-cognitive schemes hypothesized to underlie paranoia encompass inferential biases such as "jumping to conclusion" based on insufficient information and probabilistic reasoning biases, but also deficits in theory of mind (ie, the ability to put oneself in the position of another person to explain or predict his or her behavior) (Blackwood et al., 2001).

A classical task for the investigation of *jumping to conclusions* is the "beads in the bottle" task, which requires participants to determine from which of two jars a series of beads has been drawn. Importantly, participants know a priori that the jars contain beads of two colors at different ratios (eg, the ratio of green and red beads is 80:20 in jar A and 20:80 in jar B). Beads are drawn from one jar only, one bead at a time, and subjects are asked to indicate from which bottle they were taken as soon as they feel confident about the correct answer. Across repeated draws, subjects get more information about the origin of the bead, which in turn increases the probability for a correct response. Many studies have found that deluded patients tend to gather less information until they come up with a decision as compared to healthy subjects (ie, they jump to their conclusion) (Corcoran et al., 2008; Van Deal et al., 2006; see Ross et al., 2015 for a meta-analysis). Performing a similar "beads in the bottle" task in an fMRI study, paranoid patients showed reduced activation of the lateral and medial frontal gyri as well as parietal areas, which represent key nodes of the working memory network (Fig. 17.3; Krug et al., 2014). The central role of this network for reacting to unpredicted stimuli and acting in uncertain situations has been supported by findings in healthy subjects, for whom increasing uncertainty led to increasing activation in the dorsal prefrontal, posterior parietal, and insular cortices (Huettel et al.,

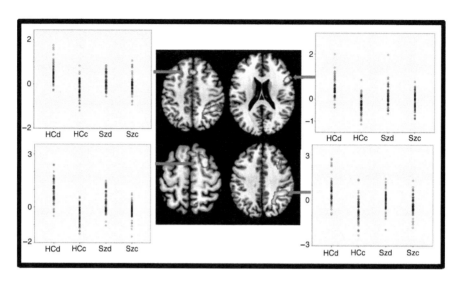

FIGURE 17.3

Interaction of group and condition in healthy controls (HC) and chronically deluded patients with schizophrenia (Sz) during decision-making under uncertainty (d) and a counting (c) control condition.

From Krug, A., Cabanis, M., Pyka, M., Pauly, K., Kellermann, T., Walter, H., et al., 2014. Attenuated prefrontal activation during decision-making under uncertainty in schizophrenia: a multi-center fMRI study. Schizophr. Res. 152, 176–183.

2005). Similar results were provided by another study about probabilistic reasoning involving patients with schizophrenia (Paulus et al., 2003). Although healthy participants showed increased activation in the superior parietal cortices and the precuneus with increasing uncertainty, no comparable modulation of activation was found in patients. These findings suggest that the observed difficulties in appropriately appraising and weighing information during decision-making under uncertainty arise from inefficient information processing within brain regions involved in working memory and executive functioning. Interestingly, the extent to which information serves goal-directed behavior is regarded as another form of salience (eg, Winton-Brown et al., 2014). The failure to acknowledge the goal-directedness of information could therefore be interpreted as failure to acknowledge salience, which is consistent with the aberrant salience account of psychosis.

Other cognitive biases in schizophrenia are more closely related to social interactions. In this context, many studies have focused on the investigation of theory of mind (ToM). One line of evidence suggests that patients with schizophrenia suffer from *hyper-ToM*, which is the tendency to perceive agency when there is none and which is assumed to result from an overactive ToM network (Abu-Akel and Bailey, 2000). This view was supported by an fMRI study that required subjects to infer the intention of a cartoon protagonist. Results revealed that activation in parts of the so-called social brain network, particularly the medial prefrontal cortex and temporo-parietal junction, was decreased during social conditions but increased during nonsocial control conditions in patients with schizophrenia (Walter et al., 2009). A similar activation pattern was reported in another fMRI experiment. Here, subjects watched short video clips depicting two actors who manipulated objects either with or without cooperation (Backasch et al., 2013). Authors reported elevated responses in the medial prefrontal cortex, temporo-parietal junction, and posterior superior temporal sulcus during the noncooperative (ie, less social) condition in patients with schizophrenia, which correlated with delusional symptoms. Authors reasoned that the observed increase in activation might be a result of greater ambiguity of the noncooperation condition for patients who tended to actively search for cues for cooperation and, consequently, to overattribute intention.

Moreover, a recent meta-analysis on social cognition (Sugranyes et al., 2011) points to increased recruitment of somatosensory areas in patients with schizophrenia, which might contribute to the overattribution of agency. In addition, the underrecruitment of brain areas related to social processing in response to social stimuli seems to underlie the reduced ability to discriminate between social and nonsocial events.

In mental disorders, symptoms do not coexist without interacting with each other. Cognitive biases are often accompanied and often aggravated by *negative emotional biases*. For example, if frightening music is presented, then deluded patients often jump to conclusions even more (compared to healthy participants). This is not the case if happy music is played (Moritz et al. 2009).

Patients commonly reveal negatively biased cognitive schemes (eg, Garety et al. 2001; Lee et al. 2004), and delusions usually have a negative connotation. Neutral or even positive events are misinterpreted as frightening or dangerous (Cohen and Minor, 2010). Moreover, negative cognitive schemes are not restricted to the environment. Schizophrenia patients also more often self-ascribe negative characteristics than healthy participants (Pauly et al., 2011).

This negative biasing seems to be related to activation changes in emotion-related areas, such as the amygdala (Heinz and Schlagenhauf, 2010). In an experiment involving reward processing, schizophrenia patients revealed blunted responses not only in the thalamo-striatal network but also in the right amygdalo-hippocampal complex during the processing of unexpected rewards. Moreover, amygdalo-hippocampal activation was negatively correlated with the degree of psychotic symptoms (Fig. 17.4; Gradin et al., 2011).

The pattern of aberrant salience processing therefore seems to exist in the limbic system as well. Even stronger support for this hypothesis is offered by converging findings derived from reviews and meta-analyses that point to general overactivity of the amygdala during the processing of emotionally neutral stimuli (eg, neutral facial expressions) but blunted responses to positively or negatively valenced stimuli (eg, facial expressions of fear, anger, or happiness) in patients with schizophrenia (Aleman and Kahn, 2005; Anticevic et al., 2012; Taylor et al., 2012).

Another line of evidence for the role of emotional dysfunction in delusional symptoms comes from Williams et al. (2004), who compared patients with and without delusions with healthy participants while they watched fearful and neutral faces. Healthy participants revealed activation in the amygdala and the medial prefrontal cortex during increased excitement. In contrast to nondeluded patients, patients with paranoid ideation showed increased physical stress responses but decreased activation of the amygdala and medial prefrontal cortex when exposed to fearful facial expressions. This pattern was interpreted as a functional disconnection of cortical and subcortical networks as well as a disturbed interaction between the amygdala and the autonomic system, likely resulting in "hypervigilance" and misattributions.

Taken together, the dysregulation of the amygdalo–hippocampal complex in particular and also dysfunctions in higher-order cortical networks involved in emotion regulation in general contribute to the well-documented abnormalities in emotion processing and the anticipation of a hedonic experience and therefore might represent a major neural correlate of emotional biases in schizophrenia.

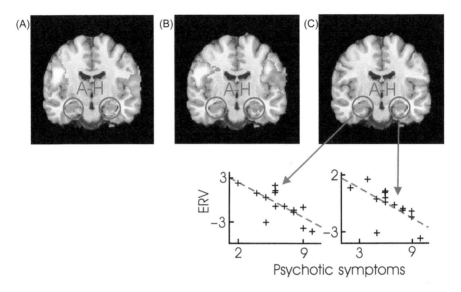

FIGURE 17.4

In contrast to healthy subjects (A) and patients with major depressive disorder (B), patients with schizophrenia (C) did not show a modulation of amygdalo-hippocampal activation by an expected reward value (ERV) in an instrumental reward learning task. The lack of modulation was correlated with the degree of psychotic symptomatology.

From Gradin, V.B., Kumar, P., Waiter, G., Ahearn, T., Stickle, C., Milders, M., et al., 2011. Expected value and prediction error abnormalities in depression and schizophrenia. Brain 134, 1751–1764.

Neural Correlates of Impaired Self-Monitoring and Self-Reflection

Cognitive and emotional biases can have an impact on self-evaluation, an ability that is closely related to social cognition. Both functions recruit meta-cognitive abilities and share overlapping neural networks, with key nodes located in cortical midline structures (eg, anterior and posterior cingulate cortex, medial prefrontal cortex) (Van der Meer et al., 2010). Impaired self-monitoring can result in reduced ability to distinguish between internally generated stimuli (eg, thoughts, memories) and externally generated stimuli (eg, events occurring in the surrounding environment). This failure has been postulated to contribute to the emergence of *passivity symptoms*, which describe the belief that one's own thoughts are withdrawn, inserted, or broadcasted, and the feeling that one's own actions are being controlled externally. In line with this hypothesis, functional imaging studies have repeatedly demonstrated abnormal activation of cortical midline structures during self-reflection in patients with schizophrenia. Several studies involving self-evaluation revealed decreased responses of anterior cortical midline structures, such as the medial prefrontal cortex (Bedford et al., 2012; Holt et al., 2011) or the anterior cingulate cortex (Blackwood et al., 2004), but increased responses of the posterior cingulate cortex, which point to an anterior-to-posterior shift of activation (Blackwood et al., 2004; Holt et al., 2011). In addition, reduced engagement of the medial prefrontal cortex during ToM tasks has been linked to an impaired decoupling of other people's perspectives from one's own (Brunet-Gouet and Decety, 2006).

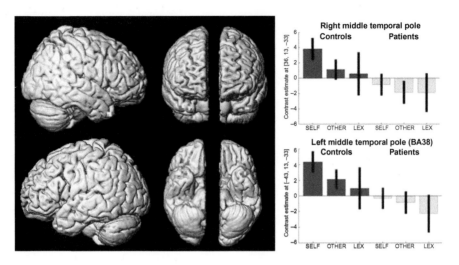

FIGURE 17.5

Left: Decreased activation in the temporal poles and the insular cortex in patients with schizophrenia as compared to healthy participants during self-ascription of traits (SELF) exclusively masked with activation during the evaluation of a close other person (OTHER). Right: Parameter estimates reveal continuous activation decrease from self-evaluation to a lexical baseline in healthy controls (blue) but disturbed modulation in patients with schizophrenia (gray).

Adapted from Pauly, K., Kircher, T.T., Schneider, F., Habel, U., 2014. Me, myself and I: temporal dysfunctions during self-evaluation in patients with schizophrenia. Soc. Cogn. Affect. Neurosci. 9, 1779–1788.

In addition to cortical midline structures, other brain regions involved in self-referential processing have been shown to be significantly altered in patients with schizophrenia. For instance, several temporal areas and the insula, all of which are implicated in mentalizing and empathy, showed a graded activation pattern with increasing self-relevance (ie, self-evaluation, evaluation of a close other person, lexical control task) in healthy subjects. However, no such modulation was found in patients with schizophrenia (Fig. 17.5; Pauly et al., 2014). Interestingly, even in healthy persons, decreased activation in the lateral temporal cortex during the perception of self and others has been associated with increased delusional thinking (Brent et al., 2014).

Another important brain region for self-related information processing is the parietal cortex, which seems to play a crucial role in the feeling of self-agency versus alien action control. An increase in activation in the inferior parietal lobe, especially in the right angular and supramarginal gyri, has been related to decreased experience of self-agency, which is a perceived discordance between executed and perceived action (Farrer et al., 2003). In patients with schizophrenia, no correlation was found between perceived action control and regional cerebral blood flow in the angular gyrus or the insular cortex. However, when comparing other agency (ie, action attribution to someone else) to the experience of self-agency, activation in the right angular gyrus correlated positively with degrees of delusions and passivity symptoms (Farrer et al., 2004). Interestingly, cerebral dysfunctions in cortical midline structures during self-reflection can already be found in early phases of the illness, namely in the so-called prodromal state (see Nelson et al., 2009 for an overview), which suggests abnormal self-reflective processes represent a trait marker of the disease.

CEREBRAL CORRELATES OF HALLUCINATIONS

> Mostly, they talk about me. Talk to each other how stupid it has been to act as I did. Sometimes they discuss how I could be punished if I go on like this. But they also talk directly to me, insult and frighten me.

Most hallucinations in schizophrenia are acoustic in nature. Hallucinating patients may hear not only one or more voices but also nonverbal sounds, such as humming, knocking, or whistling. Accordingly, taking the perspective of a brain researcher, dysfunctions of language and speech areas as well as the auditory system seem likely.

Neural Correlates of Acute Auditory Hallucinations

State studies of auditory hallucinations are aiming at identifying the neural correlates of *acute hallucinations*. In these studies, periods of active hallucinations are typically compared to periods without hallucinations within the same participant. The investigation of acute hallucinations holds the difficulty that hallucinations may come and go and may differ in content, length, and intensity. One approach to overcome this problem is to ask subjects to indicate the presence and absence of acute hallucinations by button press and to report their sensory experiences after the fMRI scan (see Jardri et al., 2011 for an overview). An example is the pioneering fMRI study by Dierks et al. (1999). A small sample of schizophrenia patients was asked to press a button as long as they experienced auditory verbal hallucinations. In comparison conditions, participants listened to a spoken text, to a text played backwards, or a modulating tone. Auditory verbal hallucinations could be related to activation in the transverse temporal gyrus, especially in the dominant hemisphere. This area of the primary auditory cortex was also active during the auditory stimulation session and may very well explain why patients perceive acoustic hallucinations as external voices or sounds. In addition, auditory hallucinations were related to activation in the posterior superior and middle temporal gyrus, the frontoparietal operculum, the amygdala, and the hippocampus. Activation was increased as long as patients indicated they were hallucinating, and decreased to the baseline level afterwards. The authors hypothesized that activation in the amygdala and hippocampus was related to emotional reactions or the retrieval of memory content in reaction to what was heard.

This study and most of the following early studies on acute hallucinations were based on very small sample sizes, sometimes even single case investigations. To overcome the statistical disadvantages of small sample sizes, several meta-analyses have attempted to extract converging evidence across studies. One meta-analysis of 10 imaging studies suggested that hallucinations result from aberrant activation in a bilateral fronto-temporal network, including regions of speech perception and generation as well as medial temporal regions involved in verbal memory (Jardri et al., 2011). Contrary to some single-study results, the primary auditory cortex (ie, Heschl's gyrus) was not identified in this meta-analysis, which suggests that auditory hallucinations do not rely on activation in this region. Authors speculated that the emergence of auditory hallucinations might result from involuntary activation of verbal memories in medial temporal areas, which are experienced as sensory when propagated to areas of the auditory and frontal cortices. Because these experiences are involuntary (ie, they lack a "self-tag"), auditory activations are perceived as external voices. This hypothesis was motivated by the finding of a phasic deactivation in a distributed network including the parahippocampal gyrus, a region involved in episodic memory, immediately before symptom occurrence, which might act as a trigger for fronto-temporal cortical activations during hallucinations (Allen et al., 2012; Diederen et al., 2010; Hoffman et al., 2008).

Partially overlapping findings were derived from another meta-analysis that was also based on 10 imaging studies (Kühn and Gallinat, 2012). Consistent activation in brain regions involved in speech production (eg, Broca's area and postcentral gyri) during auditory hallucinations was interpreted as defective monitoring of inner speech, whereas no consistent engagement of the primary auditory cortex was found. An additional activation cluster was identified in the inferior parietal lobule. As described, this region plays a role in the generation of the sense of agency and has been shown to be similarly hyperactivated during the experience of passivity symptoms (Spence et al., 1997). Its involvement might explain the lack of "ownership" accompanying the experience of hearing voices or the feeling that actions are controlled by external forces.

Neural Correlates of Hallucination Proneness

Another possible approach to the exploration of auditory hallucinations is the investigation of *hallucination proneness* as a trait (ie, neural alterations that might predispose to the emergence of hallucinatory events). These studies usually compare patients with a history of hallucinations during symptom-free periods to patients who have never experienced hallucinations and/or healthy subjects. Typically, participants are asked to listen to auditory verbal stimuli or to engage in inner speech. The aforementioned meta-analysis by Kühn and Gallinat (2012) identified consistent hypoactivation of the speech perception network (eg, left superior and middle temporal gyrus) in persons who had suffered from hallucinations. This might result from tonic changes (increase or decrease) in the level of activation within temporal areas, which in turn lead to blunted differences between conditions. Decreased activation was also detected in the anterior cingulate cortex, which suggests that more permanent deficits in self-monitoring contribute to the emergence of auditory hallucinations. In this context, it has been suggested that patients are impaired in differentiating inner speech from external input (see also Taber and Hurley, 2007). Another meta-analysis similarly suggested reduced activation of the left primary auditory cortex during processing of external auditory stimuli (Kompus et al., 2011). In contrast to the meta-analysis results reported in the previous section, however, additional analyses of state-related activation revealed increased responses of the same region, thereby suggesting the involvement of the primary auditory cortex in active hallucinations. This "paradoxical" activation (ie, decreased activation in response to external stimulation vs. increased activation in the absence of external stimulation) was interpreted as the result of a dysfunctional attentional bias, which favors the processing of endogenously generated information over external auditory input. This bias, in turn, might originate from impaired interplay between the default mode network (ie, task-unrelated activation at wakeful rest) and auditory processing network. More precisely, the authors reasoned that a failure to downregulate the default mode network leads to excessive spontaneous activation in the primary auditory cortex, which is subsequently "shut down" for external input.

The pattern of blunted responses toward (externally generated) auditory stimuli and the elevated activation during the absence thereof suggest a certain similarity to blunted responses toward salient events. Support is given by a recent study that suggests that hyperactivity in the auditory cortex during silence results from a deficient predictive coding mechanism (Horga et al., 2014). This impairment has also been postulated to underlie the emergence of delusional symptoms because predictive coding plays a central role in salience attribution. In general, predictive coding allows for the anticipation of events based on prior experience. Predictions are continuously updated to minimize prediction errors, resulting in an internal model about occurrences in the environment. Using verbal stimuli with varying predictability and model-based fMRI, Horga et al. (2014) quantified prediction error signals in patients with schizophrenia suffering from frequent hallucinations and in healthy participants. Patients revealed a significant reduction of prediction error signals in the right auditory cortex in response to auditory stimulation, which correlated with hyperactivity in auditory regions during silent periods. The excess activation during silence was interpreted as the result of disrupted prediction coding because successful prediction usually leads to attenuation of neural responses. These findings suggest that patients experiencing hallucinations fail to adequately use sensory evidence for updating their internal predictions and consequently rely on (false) expectations, all of which culminate in the formation of false beliefs and hallucinations.

SUMMARY AND OUTLOOK

In this chapter, we sought to provide an overview of neurofunctional correlates of the most common symptom clusters in schizophrenia. The presented findings suggest that alterations in neural responses in cortical and subcortical brain regions can be linked to delusional thinking and the experience of auditory (verbal) hallucinations. Complementary to using the symptom level as a "starting point" for studying brain activation changes in patients with schizophrenia, the focus can also be set on major functional domains, such as perception (Green et al., 2011), cognitive processes (eg, executive function; Minzenberg et al., 2009), episodic memory (Ragland et al., 2009), social cognition (Brunet-Gouet and Decety, 2006), and emotional (Pauly and Habel, 2011) or language processes (Strelnikov, 2010). Identified alterations are then translated to the symptom level and discussed with respect to their contribution to various symptom clusters. For instance, one may ask to what extent alterations in attention or problem solving, which are also reflected in brain activation changes, might be related to behavioral deficits, which in turn contribute to the emergence of positive or negative symptoms.

In this chapter, we focused on only some of the most common (so-called positive) symptoms in schizophrenia. Other symptoms, such as disorganized speech, grossly disorganized or catatonic behavior, or negative symptoms, might be related to other cognitive-emotional mechanisms and underlying brain activation changes. However, irrespective of the specific functional domain undergoing investigation, the ultimate aim of all neuroimaging studies in schizophrenia goes beyond the identification of pathomechanisms and is to inform the development of therapeutic approaches to alleviate the disease burden and eventually prevent and cure this devastating illness.

REFERENCES

Abu-Akel, A., Bailey, A.L., 2000. The possibility of different forms of theory of mind impairment in psychiatric and developmental disorders. Psychol. Med. 30 (3), 735–738.

Aleman, A., Kahn, R.S., 2005. Strange feelings: do amygdala abnormalities dysregulate the emotional brain in schizophrenia? Prog. Neurobiol. 77 (5), 283–298.

Allen, P., Modinos, G., Hubl, D., Shields, G., Cachia, A., Jardri, R., et al., 2012. Neuroimaging auditory hallucinations in schizophrenia: from neuroanatomy to neurochemistry and beyond. Schizophr. Bull. 38 (4), 695–703.

Anticevic, A., Van Snellenberg, J.X., Cohen, R.E., Repovs, G., Dowd, E.C., Barch, D.M., 2012. Amygdala recruitment in schizophrenia in response to aversive emotional material: a meta-analysis of neuroimaging studies. Schizophr. Bull. 38 (3), 608–621.

Backasch, B., Straube, B., Pyka, M., Klohn-Saghatolislam, F., Muller, M.J., Kircher, T.T., et al., 2013. Hyperintentionality during automatic perception of naturalistic cooperative behavior in patients with schizophrenia. Soc. Neurosci. 8 (5), 489–504.

Barkus, C., Sanderson, D.J., Rawlins, J.N., Walton, M.E., Harrison, P.J., Bannerman, D.M., 2014. What causes aberrant salience in schizophrenia? A role for impaired short-term habituation and the GRIA1 (GluA1) AMPA receptor subunit. Mol. Psychiatry 19 (10), 1060–1070.

Bedford, N.J., Surguladze, S., Giampietro, V., Brammer, M.J., David, A.S., 2012. Self- evaluation in schizophrenia: an fMRI study with implications for the understanding of insight. BMC Psychiatry 12, 106.

Bhattacharyya, S., Falkenberg, I., Martin-Santos, R., Atakan, Z., Crippa, J.A., Giampietro, V., et al., 2015. Cannabinoid modulation of functional connectivity within regions processing attentional salience. Neuropsychopharmacology 40 (6), 1343–1352.

Blackwood, N.J., Howard, R.J., Bentall, R.P., Murray, R.M., 2001. Cognitive neuropsychiatric models of persecutory delusions. Am. J. Psychiatry 158, 527–539.

Blackwood, N.J., Bentall, R.P., Ffytche, D.H., Simmons, A., Murray, R.M., Howard, R.J., 2004. Persecutory delusions and the determination of self-relevance: an fMRI investigation. Psychol. Med. 34, 591–596.

Brent, B.J., Coombs, G., Keshavan, M.S., Seidman, L.J., Moran, J.M., Holt, D.J., 2014. Subclinical delusional thinking predicts lateral temporal cortex responses during social reflection. Soc. Cogn. Affect. Neurosci. 9, 273–282.

Brunet-Gouet, E., Decety, J., 2006. Social brain dysfunctions in schizophrenia: a review of neuroimaging studies. Psychiatry Res. 148 (2-3), 75–92.

Cohen, A.S., Minor, K.S., 2010. Emotional experience in patients with schizophrenia revisited: meta-analysis of laboratory studies. Schizophr. Bull. 36, 143–150.

Corcoran, R., Rowse, G., Moore, R., Blackwood, N., Kinderman, P., Howard, R., et al., 2008. A transdiagnostic investigation of "theory of mind" and "jumping to conclusions" in patients with persecutory delusions. Psychol. Med. 38, 1577–1583.

Diederen, K.M., Neggers, S.F., Daalman, K., Blom, J.D., Goekoop, R., Kahn, R.S., et al., 2010. Deactivation of the parahippocampal gyrus preceding auditory hallucinations in schizophrenia. Am. J. Psychiatry 167 (4), 427–435.

Dierks, T., Linden, D.E.J., Jandl, M., Formisano, E., Goebel, R., Lanfermann, H., et al., 1999. Activation of Heschl's gyrus during auditory hallucinations. Neuron 22, 615–621.

Esslinger, C., Englisch, S., Inta, D., Rausch, F., Schirmbeck, F., Mier, D., et al., 2012. Ventral striatal activation during attribution of stimulus saliency and reward anticipation is correlated in unmedicated first episode schizophrenia patients. Schizophr. Res. 140 (1–3), 114–121.

Farrer, C., Franck, N., Georgieff, N., Frith, C.D., Decety, J., Jeannerod, M., 2003. Modulating the experience of agency: a positron emission tomography study. NeuroImage 18, 324–333.

Farrer, C., Franck, N., Frith, C.D., Decety, J., Georgieff, N., d'Amato, T., et al., 2004. Neural correlates of action attribution in schizophrenia. Psychiat. Res. Neuroim. 131, 31–44.

Garety, P.A., Kuipers, E., Fowler, D., Freeman, D., Bebbington, P.E., 2001. A cognitive model of the positive symptoms of psychosis. Psychol. Med. 31, 189–195.

Gradin, V.B., Kumar, P., Waiter, G., Ahearn, T., Stickle, C., Milders, M., et al., 2011. Expected value and prediction error abnormalities in depression and schizophrenia. Brain 134, 1751–1764.

Green, M.F., Lee, J., Wynn, J.K., Mathis, K.I., 2011. Visual masking in schizophrenia: overview and theoretical implications. Schizophr. Bull. 37 (4), 700–708.

Grimm, O., Heinz, A., Walter, H., Kirsch, P., Erk, S., Haddad, L., et al., 2014. Striatal response to reward anticipation: evidence for a systems-level intermediate phenotype for schizophrenia. JAMA Psychiatry 71 (5), 531–539.

Heinz, A., Schlagenhauf, F., 2010. Dopaminergic dysfunction in schizophrenia: salience attribution revisited. Schizophr. Bull. 36, 472–485.

Hoffman, R.E., Anderson, A.W., Varanko, M., Gore, J.C., Hampson, M., 2008. Time course of regional brain activation associated with onset of auditory/verbal hallucinations. Br. J. Psychiatry 193 (5), 424–425.

Holt, D.J., Cassidy, B.S., Andrews-Hanna, J.R., Lee, S.M., Coombs, G., Goff, D.C., et al., 2011. An anterior-to-posterior shift in midline cortical activity in schizophrenia during self-reflection. Biol. Psychiatry 69, 415–423.

Horga, G., Schatz, K.C., Abi-Dargham, A., Peterson, B.S., 2014. Deficits in predicitve coding underlie hallucinations in schizophrenia. J. Neurosci. 34, 8072–8082.

Huettel, S.A., Song, A.W., McCarthy, G., 2005. Decisions under uncertainty: probabilistic context influences activation of prefrontal and parietal cortices. J. Neurosci. 25, 3304–3311.

Jardri, R., Pins, D., Lafargue, G., Very, E., Ameller, A., Delmaire, C., et al., 2011. Increased overlap between the brain areas involved in self-other distinction in schizophrenia. PLoS One. 6, e17500.

Jensen, J., Willeit, M., Zipursky, R.B., Savina, I., Smith, A.J., Menon, M., et al., 2008. The formation of abnormal associations in schizophrenia: neural and behavioral evidence. Neuropsychopharmacology 33 (3), 473–479.

Juckel, G., Schlagenhauf, F., Koslowski, M., Filonov, D., Wustenberg, T., Villringer, A., et al., 2006a. Dysfunction of ventral striatal reward prediction in schizophrenic patients treated with typical, not atypical, neuroleptics. Psychopharmacology (Berl) 187 (2), 222–228.

Juckel, G., Schlagenhauf, F., Koslowski, M., Wustenberg, T., Villringer, A., Knutson, B., et al., 2006b. Dysfunction of ventral striatal reward prediction in schizophrenia. NeuroImage 29 (2), 409–416.

Kapur, S., 2003. Psychosis as a state of aberrant salience: a framework linking biology, phenomenology, and pharmacology in schizophrenia. Am. J. Psychiatry 160, 13–23.

Kompus, K., Westerhausen, R., Hugdahl, K., 2011. The "paradoxical" engagement of the primary auditory cortex in patients with auditory verbal hallucinations: a meta-analysis of functional neuroimaging studies. Neuropsychologia 49 (12), 3361–3369.

Krug, A., Cabanis, M., Pyka, M., Pauly, K., Kellermann, T., Walter, H., et al., 2014. Attenuated prefrontal activation during decision-making under uncertainty in schizophrenia: a multi-center fMRI study. Schizophr. Res. 152, 176–183.

Kühn, S., Gallinat, J., 2012. Quantitative meta-analysis on state and trait aspects of auditory verbal hallucinations in schizophrenia. Schizophr. Bull. 38 (4), 779–786.

Lee, D.A., Randall, F., Beattie, G., Bentall, R.P., 2004. Delusional discourse: an investigation comparing the spontaneous causal attributions of paranoid and non-paranoid individuals. Psychol. Psychother. Theor. Res. Pract. 77, 525–540.

Menon, M., Jensen, J., Vitcu, I., Graff-Guerrero, A., Crawley, A., Smith, M.A., et al., 2007. Temporal difference modeling of the blood-oxygen level dependent response during aversive conditioning in humans: effects of dopaminergic modulation. Biol. Psychiatry 62 (7), 765–772.

Menon, M., Schmitz, T.W., Anderson, A.K., Graff, A., Korostil, M., Mamo, D., et al., 2011. Exploring the neural correlates of delusions of reference. Biol. Psychiatry 70 (12), 1127–1133.

Minzenberg, M.J., Laird, A.R., Thelen, S., Carter, C.S., Glahn, D.C., 2009. Meta-analysis of 41 functional neuroimaging studies of executive function in schizophrenia. Arch. Gen. Psychiatry 66 (8), 811–822.

Moritz, S., Veckenstedt, R., Randjbar, S., Hottenrott, B., Woodward, T.S., von Eckstaedt, F.V., et al., 2009. Decision making under uncertainty and mood induction: further evidence for liberal acceptance in schizophrenia. Psychol. Med. 39, 1821–1829.

Murray, G.K., Corlett, P.R., Clark, L., Pessiglione, M., Blackwell, A.D., Honey, G., et al., 2008. Substantia nigra/ventral tegmental reward prediction error disruption in psychosis. Mol. Psychiatry 13 (3) 239, 267–276.

Nelson, B., Fornito, A., Harrison, B.J., Yücel, M., Sass, L.A., Yung, A.R., et al., 2009. A disturbed sense of self in the psychosis prodrome: linking phenomenology and neurobiology. Neurosci. Biobehav. Rev. 33, 807–817.

Nielsen, M.O., Rostrup, E., Wulff, S., Bak, N., Lublin, H., Kapur, S., et al., 2012. Alterations of the brain reward system in antipsychotic naive schizophrenia patients. Biol. Psychiatry 71 (10), 898–905.

Paulus, M.P., Frank, L., Brown, G.G., Braff, D.L., 2003. Schizophrenia subjects show intact success-related neural activation but impaired uncertainty processing during decision-making. Neuropsychopharmacology 28, 795–806.

Pauly, K., Habel, U., 2011. Neural substrates of emotion dysfunctions in patients with schizophrenia spectrum disorders. In: Ritsner, M.S. (Ed.), Textbook of Schizophrenia Spectrum and Related Disorders. Volume I: Conceptual Issues and Neurobiological Advances Springer, Dordrecht, pp. 405–429.

Pauly, K., Kircher, T., Weber, J., Schneider, F., Habel, U., 2011. Self-concept, emotion and memory performance in schizophrenia. Psychiatry Res. 186, 11–17.

Pauly, K., Kircher, T.T., Schneider, F., Habel, U., 2014. Me, myself and I: temporal dysfunctions during self-evaluation in patients with schizophrenia. Soc. Cogn. Affect. Neurosci. 9, 1779–1788.

Ragland, J.D., Laird, A.R., Ranganath, C., Blumenfeld, R.S., Gonzales, S.M., Glahn, D.C., 2009. Prefrontal activation deficits during episodic memory in schizophrenia. Am. J. Psychiatry 166 (8), 863–874.

Ross, R.M., McKay, R., Coltheart, M., Langdon, R., 2015. Jumping to conclusions about the beads task? A meta-analysis of delusional ideation and data-gathering. Schizophr. Bull. 41 (5), 1183–1191.

Sabri, O., Erkwoh, R., Schreckenberger, M., Owega, A., Sass, H., Buell, U., 1997. Correlation of positive symptoms exclusively to hyperperfusion or hypoperfusion of cerebral cortex in never-treated schizophrenics. Lancet 349, 1735–1739.

Schlagenhauf, F., Sterzer, P., Schmack, K., Ballmaier, M., Rapp, M., Wrase, J., et al., 2009. Reward feedback alterations in unmedicated schizophrenia patients: relevance for delusions. Biol. Psychiatry 65 (12), 1032–1039.

Seeley, W.W., Menon, V., Schatzberg, A.F., Keller, J., Glover, G.H., Kenna, H., et al., 2007. Dissociable intrinsic connectivity networks for salience processing and executive control. J. Neurosci. 27 (9), 2349–2356.

Spence, S.A., Brooks, D.J., Hirsch, S.R., Liddle, P.F., Meehan, J., Grasby, P.M., 1997. A PET study of voluntary movement in schizophrenic patients experiencing passivity phenomena (delusions of control). Brain 120, 1997–2011.

Stefansson, H., Sigurdsson, E., Steinthorsdottir, V., Bjornsdottir, S., Sigmundsson, T., Ghosh, S., et al., 2002. Neuregulin 1 and susceptibility to schizophrenia. Am. J. Hum. Genet. 71 (4), 877–892.

Strelnikov, K., 2010. Schizophrenia and language – shall we look for a deficit of deviance detection? Psychiatry Res. 178 (2), 225–229.

Sugranyes, G., Kyriakopoulos, M., Corrigall, R., Taylor, E., Frangou, S., 2011. Autism spectrum disorders and schizophrenia: meta-analysis of the neural correlates of social cognition. PLoS One 6 (10), e25322.

Taber, K.H., Hurley, R.A., 2007. Neuroimaging in schizophrenia: misattributions and religious delusions. J. Neuropsychiatry Clin. Neurosci. 19 iv–4.

Tandon, R., Gaebel, W., Barch, D.M., Bustillo, J., Gur, R.E., Heckers, S., et al., 2013. Definition and description of schizophrenia in DSM-5. Schizophr. Res. 150, 3–10.

Taylor, S.F., Kang, J., Brege, I.S., Tso, I.F., Hosanagar, A., Johnson, T.D., 2012. Meta-analysis of functional neuroimaging studies of emotion perception and experience in schizophrenia. Biol. Psychiatry 71 (2), 136–145.

Van Deal, F., Versmissen, D., Janssen, I., Myin-Germeys, I., van Os, J., Krabbendam, L., 2006. Data gathering: biased in psychosis? Schizophr. Bull. 32, 341–351.

Van der Meer, L., Costafreda, S., Aleman, A., David, A.S., 2010. Self-reflection and the brain: a theoretical review and meta-analysis of neuroimaging studies with implications for schizophrenia. Neurosci. Biobehav. Rev. 34 (6), 935–946.

Walter, H., Ciaramidaro, A., Adenzato, M., Vasic, N., Ardito, R.B., Erk, S., et al., 2009. Dysfunction of the social brain in schizophrenia is modulated by intention type: an fMRI study. Soc. Cogn. Affect. Neurosci. 4, 166–176.

Williams, L.M., Das, P., Harris, A.W., Liddell, B.B., Brammer, M.J., Olivieri, G., et al., 2004. Dysregulation of arousal and amygdala-prefrontal systems in paranoid schizophrenia. Am. J. Psychiatry 161, 480–489.

Winton-Brown, T.T., Fusar-Poli, P., Ungless, M.A., Howes, O.D., 2014. Dopaminergic basis of salience dysregulation in psychosis. Trends Neurosci. 37 (2), 85–94.

CHAPTER 18

ANATOMICAL AND FUNCTIONAL BRAIN NETWORK ARCHITECTURE IN SCHIZOPHRENIA

G. Collin[1,2] and M.P. van den Heuvel[1,2]

[1]*Department of Psychiatry, University Medical Center Utrecht, Utrecht, The Netherlands*
[2]*Brain Center Rudolf Magnus, Utrecht, The Netherlands*

CHAPTER OUTLINE

Introduction	313
What is the Brain Network?	314
Exploring the Brain's Connectional Anatomy	314
Chapter Structure	316
Part 1. Brain Connectivity in Schizophrenia	**317**
Structural Connectivity	317
Functional Connectivity	318
Structural-Functional Coupling	319
Part 2. Connectome Topology in Schizophrenia	**321**
Connectome Segregation	322
Connectome Integration	324
Relation to Clinical Symptoms and Outcome	327
Conclusion and Future Directions	**328**
Specificity of Hub Pathology to Schizophrenia	328
Relation Between Connectome Organization and Cognitive Deficits	329
Potential of Connectomics in Informing Long-Term Outcome and Clinical Practice	329
References	**330**

INTRODUCTION

From the first description of *dementia praecox* in 1887 by German physician Emil Kraepelin (Kraepelin, 1893) until today, the nosology and underlying mechanisms of what we now refer to as schizophrenia have been the topic of debate (Bernet, 2013; Burns, 2007). Around the turn of the 20th century, pioneering German psychiatrist and neuropathologist Carl Wernicke (1848–1905) suggested the term

Sejunktionspsychose, or "dissociation psychosis," to describe the syndrome (Wernicke, 1900). In following years, alternative proposals included *dissociationsprozess* (Stransky, 1903), *dementia sejunctiva* (Gross, 1904), and *dementia dissecans* (Zweig, 1908). The common denominator among these terms is that the patients' signs and symptoms are suggested to stem from some form of dissociation. Carl Wernicke was one of the first to argue that this dissociation was, in essence, structural; he assumed anatomical disruption of association fibers to be at the core of the disorder (Wernicke, 1900). Swiss psychiatrist Eugen Bleuler, who eventually coined the term "schizophrenia" in 1911, preferred psychological understanding to a biological explanation and argued that a disturbance of associations was the core disruption (Bleuler, 1911; Heckers, 2011). Nevertheless, Bleuler—although deeming it "not absolutely necessary"—also assumed that a physical disease process underlies the observed symptoms (Heckers, 2011). However, at that time—as Bleuler himself stated—there was no method to test this hypothesis (Burns, 2007). Today, advances in diffusion imaging have enabled the delineation of the white matter connections of the brain in vivo (Assaf and Pasternak, 2008; Mori and Zhang, 2006). In addition, using graph theory as a mathematical framework, the architectural organization of the brain's wiring pattern as a whole can be explored (Hagmann, 2005; Sporns et al., 2005). As a result, more than a century after Wernicke and Bleuler postulated their pioneering hypothesis, it has become possible to test the dissociation hypothesis and examine whether disruptions in neural integration are indeed a central feature of schizophrenia's neuropathology.

WHAT IS THE BRAIN NETWORK?

Healthy brain function requires effective communication and integration of neural information between distributed brain regions (Bullmore and Sporns, 2009). The anatomical infrastructure to support interregional neural interaction is the complex network of axonal projections that is known as the human connectome (Hagmann, 2005; Sporns et al., 2005). The connectome has been proposed to give rise to, and shape, the collective and coordinated neural phenomena underlying cognitive processes (Sporns, 2011). Studies have linked connectome organization to general intelligence (Li et al., 2009; Van den Heuvel et al., 2009a; Zalesky et al., 2011), working memory performance (Bassett et al., 2009; Cole et al., 2012), executive functioning and memory (Baggio et al., 2014; Reijmer et al., 2013), major personality traits (Adelstein et al., 2011; Gao et al., 2013), and creativity (Ryman et al., 2014). The notion that (complex) brain functions are not solely attributable to the properties of individual brain regions but emerge from their interplay within the connectome as a whole (Van den Heuvel et al., 2009b) implies that the pattern of brain wiring may be crucial to healthy brain function and, conversely, brain disorders (Fornito et al., 2015; Griffa et al., 2013). In this context, it has been proposed that schizophrenia may relate to a failure of proper functional integration between the distributed regions of the brain due to aberrant or 'dys'-connectivity (Friston and Frith, 1995; Friston, 1998; Stephan et al., 2009; Van den Heuvel and Fornito, 2014). The hypothesis that faulty communication between regions throughout the brain (ie, dissociation) is the core disturbance in schizophrenia's etiology provides an intuitive explanation for the vast phenomenological heterogeneity that characterizes the disorder (Collin et al., 2012; Derks et al., 2012; Picardi et al., 2012; Tandon et al., 2009).

EXPLORING THE BRAIN'S CONNECTIONAL ANATOMY

An important first step to understanding the brain's wiring pattern and how it relates to brain (dys)function is to map its elements and connections (Sporns, 2011). Brain connections are organized on multiple nested spatial scales, ranging from synaptic connections between individual neurons on the microscale, to mesoscale axonal projections linking neuronal populations, and large axonal bundles connecting spatially distributed and functionally specialized brain regions at the macroscale or systems level (Sporns, 2011). Synaptic connections can be visualized using light (Kim et al., 2013) and electron (Torrealba and Carrasco, 2004) microscopy and axonal connections through the application of viral tracing techniques (Oztas, 2003; Vercelli et al., 2000). The only current complete map of all interactions in a neural system is that of the roundworm *Caenorhabditis elegans* (Varshney et al., 2011; White et al., 1986). In addition, efforts to map the connectome at the mesoscale are ongoing in *Drosophila* (Shih et al., 2013), zebrafish (Hughes, 2013), and mouse (www.mouseconnectome.org). Due to its vast size and complexity (an estimated 80 billion neurons and 100 trillion connections), it is not currently possible to map the human whole-brain connectome at the scale of individual neuronal connections or even at the mesoscale of brain organization, although developing techniques are starting to offer a glimpse into brain connectivity complexity (Eberle et al., 2015; Kaynig et al., 2015; Lichtman et al., 2014). Despite current limitations, novel insights into human connectome organization are being provided by studies of whole-brain network reconstructions at the scale of brain regions based on data from various imaging modalities (Box 18.1). Importantly, using the analogy of a road map for which one does not need to know every byway to understand the road system's major organization, it has been argued that is not necessary for the connectome to be an exact replica of the connectional anatomy down to the finest ramifications of neuritis and individual synaptic boutons (Sporns, 2011). Rather, the connectome should provide a description of the connectional anatomy across multiple scales, including the macroscopic or systems-level network of functional cortical regions and their mutual connections (Sporns, 2010, 2012). In this chapter, we focus on the connectome at the macroscale, as derived from in vivo neuroimaging measurements of brain connectivity.

BOX 18.1 IMAGING BRAIN CONNECTIVITY

Structural Connectivity

Macroscale structural connections of the brain can be reconstructed in vivo using diffusion-weighted imaging (DWI) (Basser and Pierpaoli, 1996; Basser et al., 2000; Beaulieu, 2002; Mori and Zhang, 2006). This technique estimates the diffusion of water molecules in the brain, which is constrained by large-scale white matter fiber tracts, allowing these tracts to be delineated (Mori and van Zijl, 2002). In addition to the spatial localization of macroscopic white matter connections, DWI can be used to approximate the strength or quality of structural connections. Commonly used metrics of inter-regional connectivity include the number of reconstructed fibers between two regions as a proxy of connection strength (Cammoun et al., 2012; Hagmann et al., 2010; Van den Heuvel and Sporns, 2011; Zalesky et al., 2011) as well as measures of connection quality including fractional anisotropy (FA), mean, radial, and axial diffusivity (MD, RD, AD, respectively) (Hagmann et al., 2010; Van den Heuvel et al., 2010; Verstraete et al., 2014), and magnetic transfer ratio (MTR) (Mandl et al., 2010) as proxies of myelin integrity and local fiber organization. Of note, relying on water diffusion as an indirect marker of axon geometry, diffusion-weighted imaging is associated with a number of inherent limitations such as constraints in the detection of crossing, diverging, and converging fibers (for review, see Jbabdi and Johansen-Berg, 2011). Notwithstanding these limitations, a crucial strength of diffusion imaging is its ability to measure brain connections in vivo. In addition, validation studies in nonhuman primates have shown a considerable degree of correlation between diffusion-weighted indices of structural connectivity and anatomical connectivity as derived from viral tract tracing (Van den Heuvel et al., 2015).

Functional Connectivity

The functional collaboration between distributed brain regions can be studied using a variety of functional imaging techniques. Functional connectivity refers to the temporal coherence in activation patterns of anatomically separated brain regions (Friston et al., 1993; Friston, 1994; Van den Heuvel and Hulshoff Pol, 2010) and can be estimated using functional magnetic resonance imaging (fMRI), electro-encephalography (EEG), magneto-encephalography (MEG), and positron emission tomography (PET) during task or rest. Recordings of spontaneous fluctuations in activation patterns during rest (Biswal et al., 1997; Greicius et al., 2003; Lowe et al., 2000) have been used to identify distinct functional modules overlapping with specialized functional subsystems of the brain (Biswal et al., 1995; 1997; Cordes et al., 2000; Damoiseaux et al., 2006; Fox and Raichle, 2007; Greicius et al., 2003; Lowe et al., 2000; Van den Heuvel and Hulshoff Pol, 2010). Importantly, functional connectivity unfolding within the brain's structural network is significantly more variable over time (Sporns, 2011).

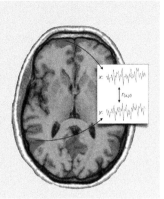

CHAPTER STRUCTURE

This chapter provides an overview of findings on alterations in brain connectivity and connectome organization in schizophrenia, focusing on findings from network theoretical studies. In part 1, we summarize contemporary evidence for disrupted structural and functional brain connectivity and discuss findings of studies on the intersection of structural and functional connectivity deficits in schizophrenia (Fig. 18.1). Then we go from single brain connections to a connectome perspective of neural systems in part 2, starting with a general overview of connectome organization in terms of its major architectural features. We discuss studies showing aberrant brain network organization in schizophrenia patients and how these changes have been reported to relate to phenomenology, analyze how these findings might converge, and highlight important open questions reflecting avenues for future connectome research in schizophrenia.

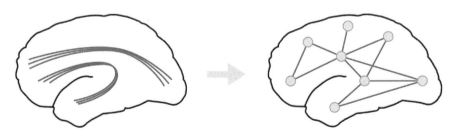

FIGURE 18.1

From connections to the connectome. The left plot illustrates part 1, in which alterations in (a collection of) individual brain connections in schizophrenia patients are discussed. In part 2, we move from connections to the topological organization of the connectome as a whole in schizophrenia.

PART 1. BRAIN CONNECTIVITY IN SCHIZOPHRENIA
STRUCTURAL CONNECTIVITY

Studies using diffusion MRI have demonstrated a range of structural brain connectivity impairments in schizophrenia (for review, see Ellison-Wright and Bullmore, 2009; Fitzsimmons et al., 2013; Fornito et al., 2012; Kubicki and Shenton, 2014; Kubicki et al., 2007; Wheeler and Voineskos, 2014). Reports of altered white matter microstructure of major white matter bundles include the corpus callosum, cingulum bundle, uncinate fasciculus, and internal capsule (Fig. 18.2), as well as superior and inferior longitudinal fasciculi, arcuate fasciculus, and fornix (Ellison-Wright and Bullmore, 2009; Whitford et al., 2011). In addition to tract-specific differences, diffusion abnormalities have been demonstrated throughout the entire white matter (Wheeler and Voineskos, 2014). Reductions in white matter volume, as shown by structural T1-weighted imaging studies, and alterations in DWI indices of anatomical connectivity may reflect decreased myelination and/or abnormal local fiber organization of white matter tracts (Ellison-Wright and Bullmore, 2009; Kubicki et al., 2007; Skudlarski et al., 2013). In addition, histological examinations of postmortem brain tissue indicate that the proportion of pathologically (as opposed to normally) myelinated fibers in prefrontal brain regions is higher in schizophrenia patients than in controls (Uranova et al., 2011). In line with this finding, MRI magnetic transfer studies are indicative of deficits in myelination and increases in free water and related glutamate concentrations in some of the major white matter tracts (Foong et al., 2001; Kubicki et al., 2005; Mandl et al., 2010). In short, structural connectivity studies over a range of spatial scales corroborate the existence of widespread impairments in white matter connectivity in schizophrenia.

FIGURE 18.2

Affected white matter tracts in schizophrenia. Findings suggest that schizophrenia involves disruptions in major white matter tracts including the uncinate fasciculus (red), cingulum bundle (purple), corpus callosum (blue), and internal capsule (green). Notably, the corpus callosum and internal capsule are only partly depicted. The association and projection fibers are shown unilaterally, but disruptions in these tracts have been found bilaterally.

Reproduced from Wheeler, A.L., Voineskos, A.N., 2014. A review of structural neuroimaging in schizophrenia: from connectivity to connectomics. Front. Hum. Neurosci. 8, 1–18, under the CC-BY license.

FUNCTIONAL CONNECTIVITY

In addition to deficits in structural connectivity, an array of functional systems in the brain has been implicated in schizophrenia, including fronto-parietal (Roiser et al., 2013; Tu et al., 2013), cingulo-opercular (Palaniyappan et al., 2013), default mode (Whitfield-Gabrieli et al., 2009), fronto-striatal (Dandash et al., 2014; Fornito et al., 2013; Hoffman et al., 2011), fronto-temporal (Hoffman et al., 2011), and cerebellar (Collin et al., 2011) circuits (Fig. 18.3A–C). Moreover, functional coupling between disparate functional systems has been shown to be reduced in schizophrenia

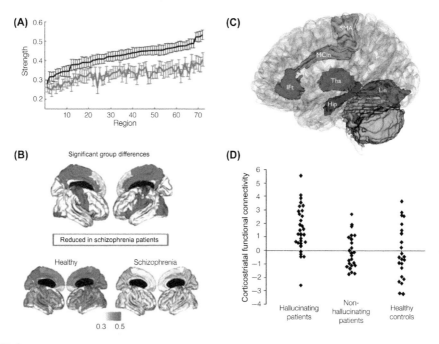

FIGURE 18.3

Functional connectivity deficits in schizophrenia. Schizophrenia patients show a general reduction in mean connectivity strength across brain regions (A) distributed throughout the cortex (B) and between cerebral and cerebellar cortex (C). Functional connectivity has been shown to be reduced in both patients and their relatives (here, in a corticostriatal loop) and to be related to symptomatology (D).

Reproduced from Lynall, M.-E., Bassett, D.S., Kerwin, R., McKenna, P.J., Kitzbichler, M., Muller, U., et al., 2010. Functional connectivity and brain networks in schizophrenia. J. Neurosci., 30 (28), 9477–9487: A, B; Collin, G., Hulshoff Pol, H.E., Haijma, S.V., Cahn, W., Kahn, R.S., van den Heuvel, M.P., 2011. Impaired cerebellar functional connectivity in schizophrenia patients and their healthy siblings. Front Psychiatry, 2, 73: C; and Hoffman, R.E., Fernandez, T., Pittman, B., Hampson, M., 2011. Elevated functional connectivity along a corticostriatal loop and the mechanism of auditory/verbal hallucinations in patients with schizophrenia. Biol. Psychiatry, 69 (5), 407–414: D, with permission.

(Repovs et al., 2011). Notably, as opposed to changes in structural connectivity, which is generally found to be reduced in schizophrenia, both decreases and increases of functional connectivity have been demonstrated in schizophrenia patients (Skudlarski et al., 2010; Zhou et al., 2008), although reduced functional connectivity is reported more commonly (Fornito et al., 2012; Lynall et al., 2010; Pettersson-Yeo et al., 2011; Van den Heuvel and Fornito, 2014; Fig. 18.3A). Considering that studies mainly report decreases in structural connectivity in schizophrenia patients while functional connectivity changes may include both reduced and increased connectivity, it appears that reduced structural connectivity does not necessarily need to lead to reduced functional coupling. Possible explanations for this observation include the notion that damage to structural connections in one location could lead to a compensatory upregulation of activity and/or connectivity in other areas (Johansen-Berg et al., 2002; Riecker et al., 2010; Van den Heuvel and Fornito, 2014) or that differential patterns of functional connectivity relate to heterogeneity in phenomenology or first episode versus chronic disease (Pettersson-Yeo et al., 2011). In addition, impairments of structural brain connections might render some (ie, fronto-temporal) subnetworks relatively isolated from the rest of the connectome, invoking increased communication within that subnetwork (Van den Heuvel et al., 2010). It has been proposed that mechanisms such as these might relate to the manifestation of psychotic symptoms (Van den Heuvel and Fornito, 2014; Fig. 18.3D).

STRUCTURAL-FUNCTIONAL COUPLING

The brain's functional dynamics are constrained by its anatomical connections (Honey et al., 2010; Nakagawa et al., 2013; Sporns, 2011). Specialized functional systems of the brain exhibit dense white matter connectivity among its elements (Greicius et al., 2009; Honey et al., 2009; Van den Heuvel et al., 2009b Fig. 18.4), and a limited number of long-range structural connections have been shown to cross-link distinct functional domains (Van den Heuvel and Sporns, 2013a). In addition, the presence of well-myelinated white matter pathways has been shown to accurately predict synchronization of neural activity (ie, functional connectivity) between corresponding brain regions (Honey et al., 2009; Fig. 18.4B–C), although the converse is not necessarily true; functional connectivity may be established in the absence of a direct structural pathway (eg, being mediated by a third brain region) (Cocchi et al., 2014; Stephan et al., 2009). If disruptions in structural connectivity in schizophrenia are related to the changes in functional connectivity, then there should be a degree of consistency between the observed alterations in connectivity. A number of neuroimaging studies have shown spatial overlap in terms of aberrant functional and structural connectivity in schizophrenia patients (Camchong et al., 2011; Cocchi et al., 2014; Yan et al., 2012; Zalesky et al., 2010). For example, a study by Zalesky et al. (2010) demonstrated overlapping anatomical and functional alterations in brain networks encompassing mediofrontal, temporal, parietal, and occipital cortices. Similarly, Cocchi et al. (2014) found reduced functional connectivity to converge with impaired white matter integrity in a network encompassing frontal, striatal, thalamic, and temporal regions. Moreover, studies examining the relationship between structural connectivity (SC) and functional connectivity (FC) show evidence for altered SC-FC coupling in schizophrenia patients relative to healthy controls (Cocchi et al., 2014; Skudlarski et al., 2010; Van den Heuvel et al., 2013), suggestive of affected structural constraint over functional interactions between brain regions in schizophrenia.

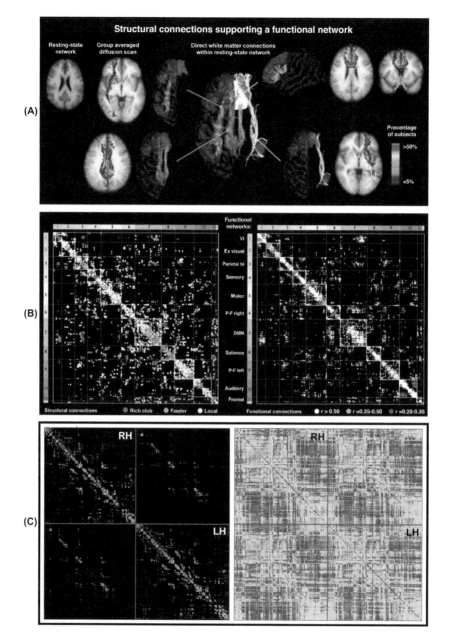

FIGURE 18.4

Structural connections underlying functional connectivity. Structural white matter connections—bilateral cingulum bundle and superior fronto-occipital fasciculus and genu—between regions of the functional default mode network (A) and side-by-side plots showing high compatibility between structural and functional connectivity matrices (B and C).

(A), (B), and (C) reproduced from Van den Heuvel, M.P., Mandl, R.C.W., Kahn, R.S., Hulshoff Pol, H.E., 2009a. Functionally linked resting-state networks reflect the underlying structural connectivity architecture of the human brain. Hum. Brain Mapp. 30 (10), 3127–3141; Van den Heuvel, M.P., Sporns, O., 2013a. An anatomical substrate for integration among functional networks in human cortex. J. Neurosci. 33 (36), 14489–14500; and Honey, C.J., Thivierge, J.P., Sporns, O., 2010. Can structure predict function in the human brain? NeuroImage 52 (3), 766–776, respectively, with permission.

PART 2. CONNECTOME TOPOLOGY IN SCHIZOPHRENIA

The field of connectomics has created a novel platform to gain a deeper understanding of the overall organization of the brain's wiring architecture, both in healthy individuals and in those with brain disorders. One of the main reasons for structural connectome reconstruction and analysis is the supposed importance of anatomical brain structure for understanding neural dynamics and, ultimately, cognition and behavior (Honey et al., 2010; Sporns, 2011). The basic principles of connectome examination using a graph theory are described in Box 18.2.

BOX 18.2 BRAIN NETWORK SCIENCE

The connectome may be described mathematically as a graph: a collection of nodes interconnected by a set of "edges" (Bullmore and Sporns, 2009; Van den Heuvel and Hulshoff Pol, 2010). Construction and analysis of structural and functional brain networks from empirical data involve a number of steps (Bullmore and Sporns, 2009). The first step of node definition—when examining neuroimaging-derived brain networks—generally involves parcellation of the cortex or whole brain into coherent brain regions (Sporns, 2011). Second, the association between the nodes is quantified in terms of structural or functional connectivity. Third, these pairwise associations are aggregated into a connection matrix representing a graph or network.

Once the connectome has been reconstructed, its organization is explored using a variety of graph measures. Based on the type of information the measure provides about the networks organization, graph measures can be approximately divided into three classes: measures of segregation, integration, and influence (Rubinov and Sporns, 2010; Sporns, 2011).

Network Segregation
Measures of segregation reflect the degree to which the network can be subdivided into local communities of clusters or modules that are strongly interconnected, with relatively sparse connectivity to the rest of the network. Network clustering can be quantified using the clustering coefficient, indicating the extent to which the neighbors of a node are also mutually connected. Modularity reflects the extent to which the network as a whole can be decomposed into modules.

Network Integration
Measures of integration reflect the efficiency of communication among all nodes in the overall network. Two commonly used and inversely related measures are path length and global efficiency. Path length indicates the average number of steps it takes to traverse the network. Global efficiency reflects the efficiency with which information can be distributed throughout the network.

Node and edge centrality reflect the number of shortest paths in the network traveling through that particular node or edge. Hubs are nodes with a high degree of centrality. Rich club organization refers to the tendency of high-degree nodes (ie, hubs) to be more densely connected mutually than to nodes of lower degree.

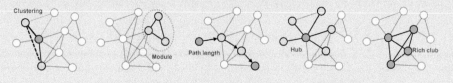

Graph theoretical studies of the human connectome are indicative of a nonrandom organization in the brain's wiring architecture, including a propensity for nodes (ie, brain regions) to cluster into structural communities (Sporns, 2011). The brain's tendency for local clustering is coupled with a high capacity for global information flow, as indicated by high efficiency and short path length (Bullmore and Sporns, 2009; Sporns, 2011). The combination of a high level of global integration and a high level of local information processing is a type of network organization that has been referred to as a small-world topology (Watts and Strogatz, 1998). This type of network architecture is thought to promote local information processing while at the same time permitting the integration of information throughout the system (Kaiser and Hilgetag, 2006). The connectome's modular community structure (He et al., 2009; Meunier et al., 2010; Sporns, 2011) is thought to be advantageous to brain function because high connectivity between nodes in the same module favors locally segregated processing of specialized functions such as primary sensory processing (Meunier et al., 2010) while reducing wiring cost (Bullmore and Sporns, 2012). Recent findings indicate a large degree of overlap between functional and structural modules of the connectome (Fig. 18.3B; Rudie et al., 2013; Van den Heuvel and Sporns, 2013a). In addition, it has been suggested that a modular organization enables functional dynamics with high complexity (Sporns, 2013) and supports the maintenance of dynamic balance, maintaining neural activity between the extremes of dying out and invading the whole network (Kaiser et al., 2007; Kaiser and Hilgetag, 2010; Senden et al., 2014). The structural communities and associated functional systems of the brain have been shown to be interconnected by a relatively small number of highly connected and central brain hubs (Sporns et al., 2007; Van den Heuvel and Sporns, 2011) localized in association regions of the frontal, parietal, and insular cortex (Van den Heuvel and Sporns, 2013b). The collective of white matter tracts linking these brain hubs over large spatial distances has been shown to comprise strong anatomical projections (Collin et al., 2013; De Reus and van den Heuvel, 2013; Zamora-López et al., 2010). Similar to interstate highways in the US road network, the connections between brain hubs have been argued to form a topologically central backbone for global "neural traffic"; this connectivity core has been referred to as the rich club system (Van den Heuvel and Sporns, 2011). Damage to *rich club* connections is believed to confer considerable disadvantages to the system as a whole (Van den Heuvel et al., 2012).

CONNECTOME SEGREGATION

As previously mentioned, the tendency of brain regions to cluster into segregated communities is captured by the graph theoretical metrics of clustering and modularity. In DWI structural connectome studies of schizophrenia, weighted overall clustering has been demonstrated to be reduced in patients as compared to controls (Collin et al., 2014; Griffa et al., 2015; Van den Heuvel et al., 2013; Fig. 18.5A), whereas findings on binary (unweighted) or normalized clustering (ie, dividing the clustering coefficient by that of a comparable random network) include unchanged or even increased structural clustering (Van den Heuvel et al., 2010; Zalesky et al., 2011, 2010; Fig. 18.5B). In terms of individual brain regions, findings of reduced clustering of frontal, temporal, parietal, and cerebellar regions (Van den Heuvel et al., 2010) are in support of a general tendency for reduced connectome clustering in schizophrenia patients (Fig. 18.5C). Moreover, findings from functional network studies in schizophrenia are also generally indicative of decreased connectome clustering in patients (Alexander-Bloch et al., 2010; Anderson and Cohen, 2013; Liu et al., 2008; Rubinov et al., 2009). Incorporating these findings on connectome clustering in schizophrenia across methodologies, the evidence appears to converge on a

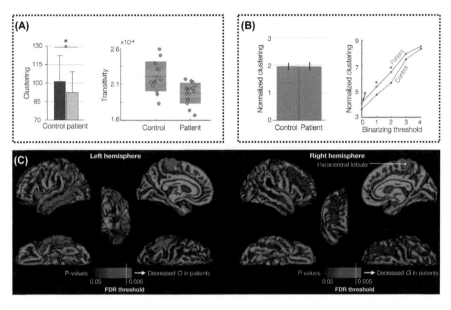

FIGURE 18.5

Altered connectome segregation in schizophrenia. Findings of altered global (A, B) and regional (C) connectome segregation in schizophrenia patients relative to healthy controls. Absolute global clustering and transitivity (related metric) of weighted structural connectome reconstructions have been reported to be reduced in patients (A). Normalized clustering of weighted networks similar to controls (B, left) and normalized clustering of binary networks increased in patients (B, right). Regional changes mainly involve reductions in clustering (C).

Graphs reproduced, with permission from: (A), Van den Heuvel, M.P., Sporns, O., Collin, G., Scheewe, T., Mandl, R.C.W., Cahn, W., et al., 2013. Abnormal rich club organization and functional brain dynamics in schizophrenia. JAMA Psychiatry, 70 (8), 783–792 and Griffa, A., Baumann, P., Ferrari, C., Do, K., Conus, P., Thiran, J., et al., 2015. Characterizing the dysconnectivity syndrome of schizophrenia with diffusion spectrum imaging. Hum. Brain Mapp., 366 (2015), 354–366; (B), Van den Heuvel, M.P., Mandl, R.C.W., Stam, C.J., Kahn, R.S., Hulshoff Pol, H.E., 2010. Aberrant frontal and temporal complex network structure in schizophrenia: a graph theoretical analysis. J. Neurosci., 30 (47), 15915–15926 and Zalesky, A., Fornito, A., Seal, M.L., Cocchi, L., Westin, C., Bullmore, E.T., et al., 2011. Disrupted Axonal Fiber Connectivity in Schizophrenia. Biol. Psychiatry 69 (1), 80–89; and (C), Van den Heuvel, M.P., Mandl, R.C.W., Stam, C.J., Kahn, R.S., Hulshoff Pol, H.E., 2010. Aberrant frontal and temporal complex network structure in schizophrenia: a graph theoretical analysis. J. Neurosci. 30 (47), 15915–15926.

subtle reduction or redistribution of connections weights, thereby reducing weighted clustering, rather than missing or abnormally connected binary edges.

In terms of the modular organization of the connectome, tentative evidence on structural connectome modularity suggests that patients may possess an elevated level of structural modularity (Van den Heuvel et al., 2013). This finding suggests that the connections between, as opposed to within, community modules may be preferentially affected in schizophrenia, effectively rendering the brain's community modules more isolated from one another. The implications of increased structural modularity on functional brain dynamics remain to be determined. Considering the previously discussed overlap in functional and structural connectome modules (Van den Heuvel and Sporns, 2013a), an increase in structural modularity may give rise to increased functional modularity. Contrary to this hypothesis, one resting-state fMRI study reported reduced modularity of the functional connectome in childhood-onset schizophrenia (Alexander-Bloch et al., 2010), but the interpretation of this finding is not straightforward considering the general reduction of functional connectivity in schizophrenia patients (eg, Fig. 18.2A–B). As a result, matching patient and control connectome reconstructions on connection density (ie, the absolute number of connections) may result in the inclusion of more low-value edges in the patients' networks (Fornito et al., 2012), which could also be an explanation of findings of reduced overall functional specialization and modularity.

CONNECTOME INTEGRATION

Measures of connectome integration are reflective of the ease with which neural information can be distributed throughout the brain's network, to enable spatially distributed brain regions to continuously share and integrate neural information (Sporns, 2011). Studies examining DWI-derived structural connectome organization in individuals with schizophrenia report reduced levels of global efficiency and/or increased path length (Collin et al., 2014; Griffa et al., 2015; Van den Heuvel et al., 2013; Wang et al., 2012; Zalesky et al., 2011, 2010; Fig. 18.6A), indicative of reduced capacity for communication between more segregated parts of the brain. Across DWI network studies, the most affected areas in terms of integration include frontal, temporal, and, to a lesser extent, parietal cortices (Collin et al., 2014; Griffa et al., 2015; Van den Heuvel et al., 2010; Zalesky et al., 2010) (Fig. 18.6B) and subcortical structures such as striatum (Griffa et al., 2015; Van den Heuvel et al., 2010). Functional brain network studies reporting on global integration paint a more diffuse picture, with some demonstrating no significant changes in measures of integration (Becerril et al., 2011; Fornito et al., 2011; Liu et al., 2008; Wang et al., 2010), and others reporting increased global efficiency (Alexander-Bloch et al., 2010; Lynall et al., 2010) and reduced path length (Rubinov et al., 2009). The apparent discrepancy between findings from structural and functional studies may relate to the temporal nature of functional findings (Rubinov and Sporns, 2010). In this context, a recent study demonstrating reduced variance over time in dynamic graph properties including global efficiency and clustering of resting fMRI functional networks in schizophrenia patients (Yu et al., 2015) is of particular interest, suggesting that there may also be a role for altered temporal connectivity patterns impacting brain functioning in schizophrenia (Narr and Leaver, 2015).

In addition to reduced levels of global network efficiency, DWI connectome studies in schizophrenia suggest a reduced centrality of mainly frontal cortical regions (Griffa et al., 2015; Van den Heuvel et al., 2010; Fig. 18.7A). Corroborating these findings, studies of structural co-variance networks (reconstructed from interregional correlation in grey matter volume or cortical thickness) report on a less prominent role of central hubs in frontal and parietal cortices, accompanied by the emergence of additional hubs (in schizophrenia patients relative to healthy controls) in other brain areas, including speech and (primary) sensorimotor cortices (Bassett et al., 2008; Shi et al., 2012; Zhang et al., 2012; Fig. 18.7B). Moreover, recent studies on hub connectivity indicate that the hubs' connections may not be at equal risk to be affected. Rather, studies examining so-called rich club connections among frontal, parietal, and insular hubs (as opposed to connections from hubs to nonhub brain regions) have suggested hub-to-hub rich club connections to be disproportionally affected in schizophrenia

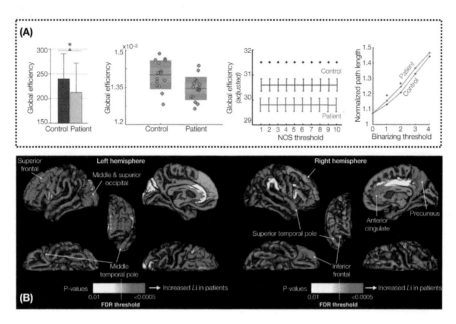

FIGURE 18.6

Disrupted connectome integration in schizophrenia. Findings of altered global (A) and regional (B) structural connectome efficiency. (A) Connectome efficiency is reduced in schizophrenia patients, as compared to control, irrespective of examining weighted (left three graphs) or binary (far right graph) connectome reconstructions. Regional findings corroborate with global effects, showing an increase in path length (ie, decreased efficiency) in a range of predominantly frontal and temporal cortices (B).

Graphs in (A) from left to right reproduced with permission from Van den Heuvel, M.P., Sporns, O., Collin, G., Scheewe, T., Mandl, R.C.W., Cahn, W., et al., 2013. Abnormal rich club organization and functional brain dynamics in schizophrenia. JAMA Psychiatry, 70(8), 783–792; Griffa, A., Baumann, P., Ferrari, C., Do, K., Conus, P., Thiran, J., et al., 2015. Characterizing the dysconnectivity syndrome of schizophrenia with diffusion spectrum imaging. Hum. Brain Mapp., 366 (2015), 354–366; Wang, Q., Su, T.P., Zhou, Y., Chou, K.H., Chen, I.Y., Jiang, T., et al., 2012. Anatomical insights into disrupted small-world networks in schizophrenia. NeuroImage 59 (2), 1085–1093; and Zalesky, A., Fornito, A., Seal, M.L., Cocchi, L., Westin, C., Bullmore, E.T., et al., 2011. Disrupted Axonal Fiber Connectivity in Schizophrenia. Biol. Psychiatry 69 (1), 80–89; (B) from Van den Heuvel, M.P., Mandl, R.C.W., Stam, C.J., Kahn, R.S., Hulshoff Pol, H.E., 2010. Aberrant frontal and temporal complex network structure in schizophrenia: a graph theoretical analysis. J. Neurosci. 30 (47), 15915–15926.

(Van den Heuvel et al., 2013; Fig. 18.7C). Moreover, unaffected siblings of schizophrenia patients have been shown to show similar, albeit more subtle, reductions in rich club connectivity (Collin et al., 2014). This findings suggests that the reductions in structural connectivity among key brain hubs are not (solely) attributable to a negative impact of psychosis or antipsychotic medication on the brain, but may be linked to familial (possibly reflecting genetic) factors involved in the illness. In addition, reduced functional coupling and disrupted low-frequency power of rich club hubs have been shown in schizophrenia patients (Yu et al., 2013), and stronger reductions in rich club strength have been

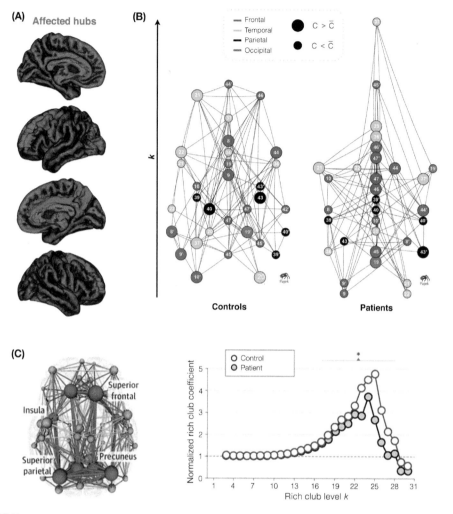

FIGURE 18.7

Affected hub connectivity. Hub regions showing reduced centrality or increased path length in schizophrenia patients (A). (B) Node ordering according to degree (y-axis) in controls and schizophrenia patients, in which the patient plot shows less prominent (frontal) hubs. (C) The rich club system, a central infrastructure comprising brain hubs (regions in the superior frontal and parietal cortex, precuneus, and insula) and their mutual connections, and a rich club curve for controls (white) and schizophrenia patients (blue), indicating that particular connections spanning high-degree hubs are affected in patients (C).

Reproduced from Van den Heuvel, M.P., Mandl, R.C.W., Stam, C.J., Kahn, R.S., Hulshoff Pol, H.E., 2010. Aberrant frontal and temporal complex network structure in schizophrenia: a graph theoretical analysis. J. Neurosci. 30 (47), 15915–15926. (A), Bassett, D.S., Bullmore, E., Verchinski, B.A., Mattay, V.S., Weinberger, D.R., Meyer-Lindenberg, A., 2008. Hierarchical organization of human cortical networks in health and schizophrenia. J. Neurosci. 28 (37), 9239–9248 (B), and Van den Heuvel, M.P., Sporns, O., Collin, G., Scheewe, T., Mandl, R.C.W., Cahn, W., et al., 2013. Abnormal rich club organization and functional brain dynamics in schizophrenia. JAMA Psychiatry, 70(8), 783–792 (C), with permission.

related to lower network efficiency and more pronounced alterations of functional network dynamics (Van den Heuvel et al., 2013). Taken together, it appears that hub pathology may be a core disturbance in schizophrenia (Rubinov and Bullmore, 2013; Van den Heuvel and Fornito, 2014; Van den Heuvel et al., 2013). Disrupting the brain's capacity for global information flow, such a disturbance is likely to impede synchronization of neural activity and is thus consistent with neural 'malintegration' as a key pathophysiological mechanism in the etiology of schizophrenia.

RELATION TO CLINICAL SYMPTOMS AND OUTCOME

Abnormalities in brain network organization such as reduced global network efficiency and clustering have been associated with both positive and negative symptom severity in schizophrenia patients (Wang et al., 2012; Yu et al., 2011; Fig. 18.8A). In addition, functional connectivity among key structural brain hubs including precuneus and medial prefrontal cortex (Whitfield-Gabrieli et al., 2009) and between frontal hubs and striatum (Fornito et al., 2013) has been associated with positive symptomatology (Fig. 18.8C). In terms of a possible relation between affected connectome organization and cognitive impairment in schizophrenia, one study has reported an absence of the normal relationship between structural network efficiency and intelligence in schizophrenia patients (Zalesky et al., 2011). In addition,

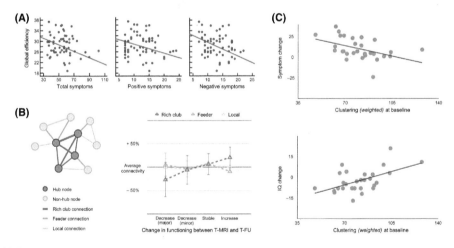

FIGURE 18.8

Clinical correlates of connectome alterations in schizophrenia. Graphs illustrating clinical correlates of altered connectome organization in schizophrenia patients. Global efficiency has been related to severity of positive, negative, and total (PANSS) symptoms (A). Rich club connectivity has been found to be predictive of subsequent changes in functional outcome (B) and connectome clustering of longitudinal changes in total symptoms and IQ (C).

Reproduced, with permission, from Wang, Q., Su, T.P., Zhou, Y., Chou, K.H., Chen, I.Y., Jiang, T., et al., 2012. Anatomical insights into disrupted small-world networks in schizophrenia. NeuroImage 59 (2), 1085–1093 (A), and Collin, G., de Nijs, J., Hulshoff Pol, H.E., Cahn, W., van den Heuvel, M.P. Connectome organization is related to longitudinal changes in general functioning, symptoms and IQ in chronic schizophrenia. Schizophr. Res., 2016 (B, C).

reduced functional network cost-efficiency has been reported to correlate with poorer working memory performance (Bassett et al., 2009). In terms of functional performance, impaired rich club organization has been linked to global functioning (Collin et al., 2014) and is predictive of subsequent changes in functional outcome (Fig. 18.8B). Finally, connectome clustering at baseline has been shown to predict longitudinal changes in symptoms and intellectual performance in patients (Collin et al., 2016 Fig. 18.8D). These findings suggest that topological alterations in brain wiring may relate to clinical symptoms, cognitive functioning, and functional outcome in schizophrenia.

CONCLUSION AND FUTURE DIRECTIONS

Dating back to Wernicke in the 19th century, an influential notion of schizophrenia is that its psychopathology relates to a failure of proper integration in the brain due to a disruption of association fiber tracts (Friston and Frith, 1995; Friston, 1998; Stephan et al., 2009). Recent studies on structural and functional connectivity confirm widespread connectivity impairments in schizophrenia patients (eg, Ellison-Wright and Bullmore, 2009; Fornito et al., 2012). With the advent of connectomics, mounting evidence suggests that in addition to a general connectivity impairment, the topological organization of the connectome as a whole may be disrupted in these patients (eg, Bullmore and Sporns, 2012; Rubinov and Bullmore, 2013; Van den Heuvel and Fornito, 2014). Across connectomic studies, converging evidence suggests that a disruption of central brain hubs and their mutual connections—leading to a reduced capacity for integrative processing between segregated communities in the brain—may have a central role in schizophrenia's etiology.

SPECIFICITY OF HUB PATHOLOGY TO SCHIZOPHRENIA

Notwithstanding the importance of these new insights into hub pathology as an etiological substrate for schizophrenia, a number of questions remain. Among these, it needs to be determined whether affected hub organization is specific to schizophrenia. It has been argued that the overload and failure of hubs might be central to the etiology of all brain disorders (Stam, 2014), for example due to excitotoxicity related to the high neuronal activity levels of brain hubs (De Haan et al., 2012). A recent meta-analysis of hub pathology across brain disorders indicates that brain hubs are also affected in neurodegenerative disorders (Crossley et al., 2014). However, the observation that hubs are affected is not necessarily synonymous with hub pathology being the central and/or primary disruption; changes in hub structure or connections may reflect primary effects in some, and be secondary in other brain disorders. For example, it has been suggested that disease propagation may be constrained by the topology of the anatomical brain network in neurodegenerative disorders (Raj et al., 2012; Schmidt et al., 2016; Zhou et al., 2012). One implication of such a mechanism is that brain hubs—considering their topological centrality—may be more vulnerable to becoming affected earlier than an "average" region, irrespective of where in the brain the central disruption occurs. Although highly tentative, there is some evidence to suggest that in schizophrenia, disruptions of central white matter connections (such as those spanning brain hubs) may constitute a primary effect of the disorder. One piece of evidence comes from insights into the developmental trajectory of large-scale white matter connections. In the time window during which schizophrenia commonly manifests, between late childhood and early adulthood, the ongoing maturation of long fiber pathways enables increasing integrative processing between the disparate communities of the brain (Hagmann et al., 2010), the functional influence of frontal hubs over other brain regions gradually increases (Uddin et al., 2011), and the rich club effect intensifies (Dennis et al., 2013a). In addition, the intermediate disruption (relative to schizophrenia patients and controls) of rich club connections in unaffected siblings of patients (Collin et al., 2014), suggests that rich club disconnectivity precedes—and is related to heritable factors associated with—the manifestation of the illness.

RELATION BETWEEN CONNECTOME ORGANIZATION AND COGNITIVE DEFICITS

Another open question is how changes in connectome organization relate to schizophrenia's phenomenology. In addition to further elucidating the mechanisms underlying the manifestation of psychotic symptoms, an important avenue for future research is to elucidate the relation between disease-related changes of the connectome and cognitive dysfunction in schizophrenia, which has been proposed to be at the core of its psychopathology (Kahn and Keefe, 2013). Connectome organization has been shown to relate to many aspects of higher-order cognitive functioning (Baggio et al., 2014; Bassett et al., 2009; Cole et al., 2012; Li et al., 2009; Reijmer et al., 2013; Van den Heuvel et al., 2009a; Zalesky et al., 2011) but to date, there have been limited studies of altered connectome topology as the substrate for cognitive deficits in schizophrenia. As a result, the full potential of connectomics for understanding cognition in brain disorders including schizophrenia remains to be realized.

POTENTIAL OF CONNECTOMICS IN INFORMING LONG-TERM OUTCOME AND CLINICAL PRACTICE

A third important avenue for future research relates to the potential value of connectome exploration in terms of informing long-term outcome and clinical decision-making. This is of specific interest in terms of early detection in individuals at increased risk for schizophrenia. In this context, it is important to consider that the connectome is a dynamic system that does not come into existence in a mature state; elaborate dynamic processes including the development, expansion, transformation, and elimination of the brain networks nodes and connections start in utero; and continue throughout life (eg, Collin and van den Heuvel, 2013; Dennis and Thompson, 2014; Dennis et al., 2013b; Hagmann et al., 2010; Van den Heuvel et al., 2014). These processes may be governed by a set of sequentially implemented rules or algorithms comparable to those dictating the routes of migrating neurons and the growth and guidance of axonal and dendritic projections (Innocenti and Price, 2005). Genetically and environmentally induced alterations of such putative biological rules may exert a profound effect on the development and maintenance of connectome organization, and thus brain function, throughout life and may form an important common pathway for the manifestation of neuropsychiatric disorders including schizophrenia. Elucidating how and when those at risk of schizophrenia deviate from the normal pattern of connectome formation and maturation may be one of the central directions in contemporary schizophrenia research because it may ultimately move us closer to targeted preventative interventions aimed to halt or mitigate this neuropathological process.

REFERENCES

Adelstein, J.S., Shehzad, Z., Mennes, M., Deyoung, C.G., Zuo, X.-N., Kelly, C., et al., 2011. Personality is reflected in the brain's intrinsic functional architecture. PLoS One 6 (11), e27633.

Alexander-Bloch, A.F., Gogtay, N., Meunier, D., Birn, R., Clasen, L., Lalonde, F., et al., 2010. Disrupted modularity and local connectivity of brain functional networks in childhood-onset schizophrenia. Front. Syst. Neurosci. 4, 147.

Anderson, A., Cohen, M.S., 2013. Decreased small-world functional network connectivity and clustering across resting state networks in schizophrenia: an fMRI classification tutorial. Front. Human Neurosci. 7, 520.

Assaf, Y., Pasternak, O., 2008. Diffusion tensor imaging (DTI)-based white matter mapping in brain research: a review. J. Mol. Neurosci. 34 (1), 51–61.

Baggio, H.-C., Sala-Llonch, R., Segura, B., Marti, M.-J., Valldeoriola, F., Compta, Y., et al., 2014. Functional brain networks and cognitive deficits in Parkinson's disease. Hum. Brain Mapp. 35 (9), 4620–4634.

Bassett, D.S., Bullmore, E., Verchinski, B.A., Mattay, V.S., Weinberger, D.R., Meyer-Lindenberg, A., 2008. Hierarchical organization of human cortical networks in health and schizophrenia. J. Neurosci. 28 (37), 9239–9248.

Bassett, D.S., Bullmore, E.T., Meyer-Lindenberg, A., Apud, J.A., Weinberger, D.R., Coppola, R., 2009. Cognitive fitness of cost-efficient brain functional networks. Proc. Natil. Acad. Sci. U.S.A. 106 (28), 11747–11752.

Basser, P.J., Pierpaoli, C., 1996. Microstructural and physiological features of tissues elucidated by quantitative-diffusion-tensor MRI. J. Magn. Reson. B. 111, 209–219.

Basser, P.J., Pajevic, S., Pierpaoli, C., Duda, J., Aldroubi, A., 2000. In vivo fiber tractography using DT-MRI data. Mag. Reson. Med. 44, 625–632.

Beaulieu, C., 2002. The basis of anisotropic water diffusion in the nervous system – a technical review. NMR Biomed. 15, 435–455.

Becerril, K.E., Repovs, G., Barch, D.M., 2011. Error processing network dynamics in schizophrenia. NeuroImage 54 (2), 1495–1505.

Bernet, B., 2013. Schizophrenie. Entstehung und Entwicklung eines psychiatrischen Krankheitsbilds um 1900. Chronos-Verlag, Zurich.

Biswal, B., Zerrin Yetkin, F., Haughton, V.M., Hyde, J.S., 1995. Functional connectivity in the motor cortex of resting human brain using echo-planar mri. Mag. Reson. Med. 34 (9), 537–541.

Biswal, B.B., Van Kylen, J., Hyde, J.S., 1997. Simultaneous assessment of flow and BOLD signals in resting-state functional connectivity maps. NMR Biomed. 10, 165–170.

Bleuler, E., 1911. Dementia Praecox oder die Gruppe der Schizophrenien. Aschaffenburgs Handbuch, Deutike, Leipzig.

Bullmore, E., Sporns, O., 2009. Complex brain networks: graph theoretical analysis of structural and functional systems. Nature Rev. Neurosci. 10 (3), 186–198.

Bullmore, E., Sporns, O., 2012. The economy of brain network organization. Nature Rev. Neurosci. 13 (5), 336–349.

Burns, J., 2007. The Descent of Madness. Evolutionary Origins of Psychosis and the Social Brain. Routledge, New York, NY.

Camchong, J., MacDonald, A.W., Bell, C., Mueller, B.A., Lim, K.O., 2011. Altered functional and anatomical connectivity in schizophrenia. Schizophr. Bull. 37 (3), 640–650.

Cammoun, L., Gigandet, X., Meskaldji, D., Thiran, J.P., Sporns, O., Do, K.Q., et al., 2012. Mapping the human connectome at multiple scales with diffusion spectrum MRI. J. Neurosci. Methods 203 (2), 386–397.

Cocchi, L., Harding, I.H., Lord, A., Pantelis, C., Yucel, M., Zalesky, A., 2014. Disruption of structure-function coupling in the schizophrenia connectome. Neuroimage Clin. 4, 779–787.

Cole, M.W., Yarkoni, T., Repovs, G., Anticevic, A., Braver, T.S., 2012. Global connectivity of prefrontal cortex predicts cognitive control and intelligence. J. Neurosci. 32 (26), 8988–8999.

Collin, G., van den Heuvel, M.P., 2013. The ontogeny of the human connectome: development and dynamic changes of brain connectivity across the life span. Neuroscientist 19 (6), 616–628.

Collin, G., Hulshoff Pol, H.E., Haijma, S.V., Cahn, W., Kahn, R.S., van den Heuvel, M.P., 2011. Impaired cerebellar functional connectivity in schizophrenia patients and their healthy siblings. Front. Psychiatry 2, 73.

Collin, G., Derks, E.M., van Haren, N.E.M., Schnack, H.G., Hulshoff Pol, H.E., Kahn, R.S., et al., 2012. Symptom dimensions are associated with progressive brain volume changes in schizophrenia. Schizophr. Res. 138 (2–3), 171–176.

Collin, G., Sporns, O., Mandl, R.C.W., van den Heuvel, M.P., 2013. Structural and functional aspects relating to cost and benefit of rich club organization in the human cerebral cortex. Cereb. Cortex, 2258–2267.

Collin, G., Kahn, R., de Reus, M., Cahn, W., van den Heuvel, M., 2014. Impaired rich club connectivity in unaffected siblings of schizophrenia patients. Schizophr. Bull. 40 (2), 438–448.

Collin, G., de Nijs, J., Hulshoff Pol, H.E., Cahn, W., van den Heuvel, M.P., 2016. Connectome organization is related to longitudinal changes in general functioning, symptoms and IQ in chronic schizophrenia. Schizophr. Res. 173 (3), 166–173.

Cordes, D., Haughton, V.M., Arfanakis, K., Wendt, G.J., Turski, P.A., Moritz, C.H., et al., 2000. Mapping functionally related regions of brain with functional connectivity MR imaging. Am. J. Neuroradiol. 21, 1636–1644.

Crossley, N.A., Mechelli, A., Scott, J., Carletti, F., Fox, P.T., McGuire, P., et al., 2014. The hubs of the human connectome are generally implicated in the anatomy of brain disorders. Brain 137, 2382–2395.

Damoiseaux, J., Rombouts, S., Barkhof, F., Scheltens, P., Stam, C., Smith, S., et al., 2006. Consistent resting-state networks across healthy subjects. Proc. Natil. Acad. Sci. U.S.A. 103 (37), 13848–13853.

Dandash, O., Fornito, A., Lee, J., Keefe, R.S.E., Chee, M.W.L., Adcock, R.A., et al., 2014. Altered striatal functional connectivity in subjects with an at-risk mental state for psychosis. Schizophr. Bull. 40, 904–913.

De Haan, W., Mott, K., van Straaten, E.C.W., Scheltens, P., Stam, C.J., 2012. Activity dependent degeneration explains hub vulnerability in Alzheimer's disease. PLoS Comput. Biol. 8 (8), e1002582.

De Reus, M.A., van den Heuvel, M.P., 2013. Rich club organization and intermodule communication in the cat connectome. J. Neurosci. 33 (32), 12929–12939.

Dennis, E.L., Thompson, P.M., 2014. Reprint of: mapping connectivity in the developing brain. Int. J. Develop. Neurosci. 32 (7), 41–57.

Dennis, E.L., Jahanshad, N., Toga, A.W., McMahon, K.L., De Zubicaray, G.I., Hickie, I., et al., 2013a. Development of the "rich club" in brain connectivity networks from 438 adolescents & adults aged 12 to 30. Proc IEEE Int Symp. Biomed. Imaging, 624–627.

Dennis, E.L., Jahanshad, N., McMahon, K.L., De Zubicaray, G.I., Martin, N.G., Hickie, I.B., et al., 2013b. Development of brain structural connectivity between ages 12 and 30: a 4-Tesla diffusion imaging study in 439 adolescents and adults. NeuroImage 64 (1), 161–684.

Derks, E.M., Allardyce, J., Boks, M.P., Vermunt, J.K., Hijman, R., Ophoff, R.A., 2012. Kraepelin was right: a latent class analysis of symptom dimensions in patients and controls. Schizophr. Bull. 38 (3), 495–505.

Eberle, A.L., Mikula, S., Schalek, R., Lichtman, J.W., Knothe Tate, M.L., Zeidler, D., 2015. High-resolution, high-throughput imaging with a multibeam scanning electron microscope. J. Microsc., 1–7.

Ellison-Wright, I., Bullmore, E., 2009. Meta-analysis of diffusion tensor imaging studies in schizophrenia. Schizophr. Res. 108 (1-3), 3–10.

Fitzsimmons, J., Kubicki, M., Shenton, M.E., 2013. Review of functional and anatomical brain connectivity findings in schizophrenia. Curr. Opin. Psychiatry 26 (2), 172–187.

Foong, J., Symms, M.R., Barker, G.J., Maier, M., Woermann, F.G., Miller, D.H., et al., 2001. Neuropathological abnormalities in schizophrenia: evidence from magnetization transfer imaging. Brain 124, 882–892.

Fornito, A., Zalesky, A., Bassett, D.S., Meunier, D., Ellison-Wright, I., Yücel, M., et al., 2011. Genetic influences on cost-efficient organization of human cortical functional networks. J. Neurosci. 31 (9), 3261–3270.

Fornito, A., Zalesky, A., Pantelis, C., Bullmore, E.T., 2012. Schizophrenia, neuroimaging and connectomics. NeuroImage 62 (4), 2296–2314.

Fornito, A., Harrison, B.J., Goodby, E., Dean, A., Ooi, C., Nathan, P.J., et al., 2013. Functional dysconnectivity of corticostriatal circuitry as a risk phenotype for psychosis. JAMA Psychiatry 70 (11), 1143–1151.

Fornito, A., Zalesky, A., Breakspear, M., 2015. The connectomics of brain disorders. Nature Publishing Group 16 (3), 159–172.

Fox, M.D., Raichle, M.E., 2007. Spontaneous fluctuations in brain activity observed with functional magnetic resonance imaging. Nat. Rev. Neurosci. 8, 700–711.

Friston, K.J., 1994. Functional and effective connectivity in neuroimaging: a synthesis. Hum. Brain Mapp. 2, 56–78.

Friston, K.J., 1998. The disconnection hypothesis. Schizophr. Res. 30 (2), 115–125.

Friston, K.J., Frith, C.D., 1995. Schizophrenia: a disconnection syndrome? Clin. Neurosci. 3, 89–97.

Friston, K.J., Frith, C.D., Liddle, P.F., Frackowiak, R.S., 1993. Functional connectivity: the principal-component analysis of large (PET) data sets. J. Cereb. Blood Flow Metab. 13, 5–14.

Gao, Q., Xu, Q., Duan, X., Liao, W., Ding, J., Zhang, Z., et al., 2013. Extraversion and neuroticism relate to topological properties of resting-state brain networks. Front. Hum. Neurosci. 7, 257.

Greicius, M.D., Krasnow, B., Reiss, A.L., Menon, V., 2003. Functional connectivity in the resting brain: a network analysis of the default mode hypothesis. Proc. Natl. Acad. Sci. U.S.A. 100 (1), 253–258.

Greicius, M.D., Supekar, K., Menon, V., Dougherty, R.F., 2009. Resting-state functional connectivity reflects structural connectivity in the default mode network. Cereb. Cortex 19 (1), 72–78.

Griffa, A., Baumann, P.S., Thiran, J.P., Hagmann, P., 2013. Structural connectomics in brain diseases. NeuroImage 80, 515–526.

Griffa, A., Baumann, P., Ferrari, C., Do, K., Conus, P., Thiran, J., et al., 2015. Characterizing the dysconnectivity syndrome of schizophrenia with diffusion spectrum imaging. Hum. Brain Mapp. 366 (2015), 354–366.

Gross, O., 1904. Dementia Sejunctiva. Neurologisches Centralblatt 23, 1144–1166.

Hagmann, P., 2005. From Diffusion MRI to Brain Connectomics (PhD thesis). Ecole Polytechnique Fédérale de Lausanne (EPFL), Lausanne, 127 p.

Hagmann, P., Sporns, O., Madan, N., Cammoun, L., Pienaar, R., Wedeen, V.J., et al., 2010. White matter maturation reshapes structural connectivity in the late developing human brain. Proc. Natl. Acad. Sci. U.S.A. 107 (44), 19067–19072.

He, Y., Wang, J., Wang, L., Chen, Z.J., Yan, C., Yang, H., et al., 2009. Uncovering intrinsic modular organization of spontaneous brain activity in humans. PLoS One 4 (4), 23–25.

Heckers, S., 2011. Bleuler and the neurobiology of schizophrenia. Schizophr. Bull. 37 (6), 1131–1135.

Hoffman, R.E., Fernandez, T., Pittman, B., Hampson, M., 2011. Elevated functional connectivity along a corticostriatal loop and the mechanism of auditory/verbal hallucinations in patients with schizophrenia. Biol. Psychiatry 69 (5), 407–414.

Honey, C.J., Honey, C.J., Sporns, O., Sporns, O., Cammoun, L., Cammoun, L., et al., 2009. Predicting human resting-state functional connectivity from structural connectivity. Proc. Natl. Acad. Sci. U.S.A. 106 (6), 2035–2040.

Honey, C.J., Thivierge, J.P., Sporns, O., 2010. Can structure predict function in the human brain? NeuroImage 52 (3), 766–776.

Hughes, V., 2013. Fish-bowl neuroscience: tiny fish trapped in a virtual world provide a window into complex brain connections. Nature 493, 466–468.

Innocenti, G.M., Price, D.J., 2005. Exuberance in the development of cortical networks. Nature Rev. Neurosci. 6 (12), 955–965.

Jbabdi, S., Johansen-Berg, H., 2011. Tractography: where do we go from here? Brain Connect. 1 (3), 169–183.

Johansen-Berg, H., Rushworth, M.F.S., Bogdanovic, M.D., Kischka, U., Wimalaratna, S., Matthews, P.M., 2002. The role of ipsilateral premotor cortex in hand movement after stroke. Proc. Natl. Acad. Sci. U.S.A. 99, 14518–14523.

Kahn, R.S., Keefe, R.S.E., 2013. Schizophrenia Is a Cognitive Illness Time for a Change in Focus. JAMA Psychiatry 70 (10), 1107–1112.

Kaiser, M., Hilgetag, C.C., 2006. Nonoptimal component placement, but short processing paths, due to long-distance projections in neural systems. PLoS Comput. Biol. 2 (7), 0805–0815.

Kaiser, M., Hilgetag, C.C., 2010. Optimal hierarchical modular topologies for producing limited sustained activation of neural networks. Front. Neuroinform. 4, 8.

Kaiser, M., Görner, M., Hilgetag, C.C., 2007. Criticality of spreading dynamics in hierarchical cluster networks without inhibition. New J. Phys. 9 (5) 110–110.

Kaynig, V., Vazquez-Reina, A., Knowles-Barley, S., Roberts, M., Jones, T.R., Kasthuri, N., et al., 2015. Large-scale automatic reconstruction of neuronal processes from electron microscopy images. Med. Image. Anal. 22, 77–88.

Kim, S.Y., Chung, K., Deisseroth, K., 2013. Light microscopy mapping of connections in the intact brain. Trends Cogn. Sci. 17 (12), 596–599.

Kraepelin, E., 1893. Ein Lerbuch für Studirende und Aertze (4. Aufl.). Leipzig.

Kubicki, M., Shenton, M.E., 2014. Diffusion Tensor Imaging findings and their implications in schizophrenia. Curr. Opin. Psychiatry 27 (3), 179–184.

Kubicki, M., Park, H., Westin, C.F., Nestor, P.G., Mulkern, R.V., Maier, S.E., et al., 2005. DTI and MTR abnormalities in schizophrenia: analysis of white matter integrity. NeuroImage 26 (4), 1109–1118.

Kubicki, M., McCarley, R., Westin, C.-F., Park, H.-J., Maier, S., Kikinis, R., et al., 2007. A review of diffusion tensor imaging studies in schizophrenia. J. Psychiatr. Res. 41 (1-2), 15–30.

Li, Y., Liu, Y., Li, J., Qin, W., Li, K., Yu, C., et al., 2009. Brain anatomical network and intelligence. PLoS. Comput. Biol. 5 (5), e1000395.

Lichtman, J.W., Pfister, H., Shavit, N., 2014. The big data challenges of connectomics. Nat. Neurosci. 17 (11), 1448–1454.

Liu, Y., Liang, M., Zhou, Y., He, Y., Hao, Y., Song, M., et al., 2008. Disrupted small-world networks in schizophrenia. Brain 131, 945–961.

Lowe, M.J., Dzemidzic, M., Lurito, J.T., Mathews, V.P., Philips, M.D., 2000. Correlations in low-frequency BOLD fluctuations reflect cortico-cortical connections. Neuroimage 12 (5), 582–587.

Lynall, M.-E., Bassett, D.S., Kerwin, R., McKenna, P.J., Kitzbichler, M., Muller, U., et al., 2010. Functional connectivity and brain networks in schizophrenia. J. Neurosci. 30 (28), 9477–9487.

Mandl, R.C.W., Schnack, H.G., Luigjes, J., van den Heuvel, M.P., Cahn, W., Kahn, R.S., et al., 2010. Tract-based analysis of magnetization transfer ratio and diffusion tensor imaging of the frontal and frontotemporal connections in schizophrenia. Schizophr. Bull. 36 (4), 778–787.

Meunier, D., Lambiotte, R., Bullmore, E.T., 2010. Modular and hierarchically modular organization of brain networks. Front. Neurosci. 4, 1–11.

Mori, S., van Zijl, P.C.M., 2002. Fiber tracking: principles and strategies - a technical review. NMR. Biomed. 15 (7-8), 468–480.

Mori, S., Zhang, J., 2006. Principles of diffusion tensor imaging and its applications to basic neuroscience research. Neuron 51, 527–539.

Nakagawa, T.T., Jirsa, V.K., Spiegler, A., McIntosh, A.R., Deco, G., 2013. Bottom up modeling of the connectome: linking structure and function in the resting brain and their changes in aging. NeuroImage 80, 318–329.

Narr, K.L., Leaver, A.M., 2015. Connectome and schizophrenia. Curr. Opin. Psychiatry 28 (3), 229–235.

Oztas, E., 2003. Neuronal tracing. Neuroanatomy 2, 2–5.

Palaniyappan, L., Simmonite, M., White, T., Liddle, E., Liddle, P., 2013. Neural primacy of the salience processing system in schizophrenia. Neuron 79 (4), 814–828.

Pettersson-Yeo, W., Allen, P., Benetti, S., McGuire, P., Mcchelli, A., 2011. Dysconnectivity in schizophrenia: where are we now? Neurosci. Biobehav. Rev. 35 (5), 1110–1124.

Picardi, A., Viroli, C., Tarsitani, L., Miglio, R., de Girolamo, G., Dell'Acqua, G., et al., 2012. Heterogeneity and symptom structure of schizophrenia. Psychiatry Res. 198 (3), 386–394.

Raj, A., Kuceyeski, A., Weiner, M., 2012. A network diffusion model of disease progression in dementia. Neuron 73 (6), 1204–1215.

Reijmer, Y.D., Leemans, A., Caeyenberghs, K., Heringa, S.M., Koek, H.L., Biessels, G.J., 2013. Disruption of cerebral networks and cognitive impairment in Alzheimer disease. Neurology 80 (15), 1370–1377.

Repovs, G., Csernansky, J.G., Barch, D.M., 2011. Brain network connectivity in individuals with schizophrenia and their siblings. Biol. Psychiatry 69 (10), 967–973.

Riecker, A., Gröschel, K., Ackermann, H., Schnaudigel, S., Kassubek, J., Kastrup, A., 2010. The role of the unaffected hemisphere in motor recovery after stroke. Hum. Brain Mapp. 31, 1017–1029.

Roiser, J.P., Wigton, R., Kilner, J.M., Mendez, M.A., Hon, N., Friston, K.J., et al., 2013. Dysconnectivity in the frontoparietal attention network in schizophrenia. Front. Psychiatry 4, 1–13.

Rubinov, M., Bullmore, E., 2013. Schizophrenia and abnormal brain network hubs. Dialogues Clin. Neurosci. 15, 339–349.

Rubinov, M., Sporns, O., 2010. Complex network measures of brain connectivity: uses and interpretations. NeuroImage 52 (3), 1059–1069.

Rubinov, M., Knock, S. a, Stam, C.J., Micheloyannis, S., Harris, A.W.F., Williams, L.M., et al., 2009. Small-world properties of nonlinear brain activity in schizophrenia. Hum. Brain Mapp. 30, 403–416.

Rudie, J.D., Brown, J.A., Beck-Pancer, D., Hernandez, L.M., Dennis, E.L., Thompson, P.M., et al., 2013. Altered functional and structural brain network organization in autism. Neuroimage Clin. 2 (1), 79–94.

Ryman, S.G., van den Heuvel, M.P., Yeo, R.A., Caprihan, A., Carrasco, J., Vakhtin, A.A., et al., 2014. Sex differences in the relationship between white matter connectivity and creativity. NeuroImage 101, 380–389.

Schmidt, R., de Reus, M.A., Scholtens, L.H., van den Berg, L.H., van den Heuvel, M.P., 2016. Simulating disease propagation across white matter connectome reveals anatomical substrate for neuropathology staging in amyotrophic lateral sclerosis. NeuroImage 124, 762–769.

Senden, M., Deco, G., de Reus, M.A., Goebel, R., van den Heuvel, M.P., 2014. Rich club organization supports a diverse set of functional network configurations. NeuroImage 96, 174–182.

Shi, F., Yap, P.-T., Gao, W., Lin, W., Gilmore, J.H., Shen, D., 2012. Altered structural connectivity in neonates at genetic risk for schizophrenia: a combined study using morphological and white matter networks. NeuroImage 62 (3), 1622–1633.

Shih, C.-T., Sporns, O., Chiang, A.-S., 2013. Toward the Drosophila connectome: structural analysis of the brain network. BMC Neurosci. 14 (Suppl. 1), P63.

Skudlarski, P., Jagannathan, K., Anderson, K., Stevens, M.C., Calhoun, V.D., Skudlarska, B.A., et al., 2010. Brain connectivity is not only lower but different in schizophrenia: a combined anatomical and functional approach. Biol. Psychiatry 68 (1), 61–69.

Skudlarski, P., Schretlen, D.J., Thaker, G.K., Stevens, M.C., Keshavan, M.S., Sweeney, J.A., et al., 2013. Diffusion tensor imaging white matter endophenotypes in patients with schizophrenia or psychotic bipolar disorder and their relatives. Am. J. Psych. 170 (8), 886–898.

Sporns, O., 2010. Networks of the Brain. MIT Press Ltd.

Sporns, O., 2011. The human connectome: a complex network. Ann. N.Y. Acad. Sci. 1224, 109–125.

Sporns, O., 2012. Discovering the Human Connectome. MIT Press Ltd.

Sporns, O., 2013. Network attributes for segregation and integration in the human brain. Curr. Opin. Neurobiol. 23 (2), 162–171.

Sporns, O., Tononi, G., Kötter, R., 2005. The human connectome: a structural description of the human brain. PLoS Comput. Biol. 1 (4), e42.

Sporns, O., Honey, C.J., Ko, R., 2007. Identification and classification of hubs in brain networks. PLoS One 2 (10), e1049.

Stam, C.J., 2014. Modern network science of neurological disorders. Nature Rev. Neurosci. 15 (10), 683–695.

Stephan, K.E., Friston, K.J., Frith, C.D., 2009. Dysconnection in schizophrenia: from abnormal synaptic plasticity to failures of self-monitoring. Schizophr. Bull. 35 (3), 509–527.

Stransky, E., 1903. Zur kenntnis gewisser erworbener blödsinssformen (zugleich ein beitrag zur lehre von der dementia praecox). Jahrbuch Der Psychiatrie Und Neurology 24, 1–149.

Tandon, R., Nasrallah, H.A., Keshavan, M.S., 2009. Schizophrenia, "just the facts" 4. Clinical features and conceptualization. Schizophr. Res. 110 (1-3), 1–23.

Torrealba, F., Carrasco, M.A., 2004. A review on electron microscopy and neurotransmitter systems. Brain Res. Rev. 47 (1–3), 5–17.

Tu, P.C., Lee, Y.C., Chen, Y.S., Li, C.T., Su, T.P., 2013. Schizophrenia and the brain's control network: aberrant within- and between-network connectivity of the frontoparietal network in schizophrenia. Schizophr. Res. 147 (2-3), 339–347.

Uddin, L.Q., Supekar, K.S., Ryali, S., Menon, V., 2011. Dynamic reconfiguration of structural and functional connectivity across core neurocognitive brain networks with development. J. Neurosci. 31 (50), 18578–18589.

Uranova, N.A., Vikhreva, O.V., Rachmanova, V.I., Orlovskaya, D.D., 2011. Ultrastructural alterations of myelinated fibers and oligodendrocytes in the prefrontal cortex in schizophrenia: a postmortem morphometric study. Schizophr. Res. Treatment 2011, 1–13.

Van den Heuvel, M.P., Hulshoff Pol, H.E., 2010. Exploring the brain network: a review on resting-state fMRI functional connectivity. Eur. Neuropsychopharmacol. 20 (8), 519–534.

Van den Heuvel, M.P., Sporns, O., 2011. Rich-club organization of the human connectome. J. Neurosci. 31 (44), 15775–15786.

Van den Heuvel, M.P., Sporns, O., 2013a. An anatomical substrate for integration among functional networks in human cortex. J. Neurosci. 33 (36), 14489–14500.

Van den Heuvel, M.P., Sporns, O., 2013b. Network hubs in the human brain. Trends. Cogn. Sci. 17 (12), 683–696.

Van den Heuvel, M.P., Fornito, A., 2014. Brain networks in schizophrenia. Neuropsychol. Rev. 24 (1), 32–48.

Van den Heuvel, M.P., Stam, C.J., Kahn, R.S., Hulshoff Pol, H.E., 2009a. Efficiency of functional brain networks and intellectual performance. J. Neurosci. 29 (23), 7619–7624.

Van den Heuvel, M.P., Mandl, R.C.W., Kahn, R.S., Hulshoff Pol, H.E., 2009b. Functionally linked resting-state networks reflect the underlying structural connectivity architecture of the human brain. Hum. Brain Mapp. 30 (10), 3127–3141.

Van den Heuvel, M.P., Mandl, R.C.W., Stam, C.J., Kahn, R.S., Hulshoff Pol, H.E., 2010. Aberrant frontal and temporal complex network structure in schizophrenia: a graph theoretical analysis. J. Neurosci. 30 (47), 15915–15926.

Van den Heuvel, M.P., Kahn, R.S., Goñi, J., Sporns, O., 2012. High-cost, high-capacity backbone for global brain communication. Proc. Natl. Acad. Sci. U.S.A. 109 (28), 11372–11377.

Van den Heuvel, M.P., Sporns, O., Collin, G., Scheewe, T., Mandl, R.C.W., Cahn, W., et al., 2013. Abnormal rich club organization and functional brain dynamics in schizophrenia. JAMA Psychiatry 70 (8), 783–792.

Van den Heuvel, M.P., Kersbergen, K.J., de Reus, M.A., Keunen, K., Kahn, R.S., Groenendaal, F., et al., 2014. The neonatal connectome during preterm brain development. Cereb. Cortex, 1–14.

Van den Heuvel, M.P., De Reus, M.A., Feldman Barrett, L., Scholtens, L.H., Coopmans, F.M.T., Preuss, T.M., et al. 2015. Comparison of diffusion tractography and tract-tracing measures of connectivity strength in rhesus macaque connectome. Hum. Brain Mapp. 36 (8), 3064–3075.

Varshney, L.R., Chen, B.L., Paniagua, E., Hall, D.H., Chklovskii, D.B., 2011. Structural properties of the Caenorhabditis elegans neuronal network. PLoS Comput. Biol. 7 (2), e1001066.

Vercelli, A., Repici, M., Garbossa, D., Grimaldi, A., 2000. Recent techniques for tracing pathways in the central nervous system of developing and adult mammals. Brain Res. Bull. 51 (1), 11–28.

Verstraete, E., Veldink, J.H., van den Berg, L.H., Van den Heuvel, M.P., 2014. Structural brain network imaging shows expanding disconnection of the motor system in amyotrophic lateral sclerosis. Hum. Brain. Mapp. 35, 1351–1361.

Wang, L., Metzak, P.D., Honer, W.G., Woodward, T.S., 2010. Impaired efficiency of functional networks underlying episodic memory-for-context in schizophrenia. J. Neurosci. 30 (39), 13171–13179.

Wang, Q., Su, T.P., Zhou, Y., Chou, K.H., Chen, I.Y., Jiang, T., et al., 2012. Anatomical insights into disrupted small-world networks in schizophrenia. NeuroImage 59 (2), 1085–1093.

Watts, D.J., Strogatz, S.H., 1998. Collective dynamics of "small-world" networks. Nature 393, 440–442.

Wernicke, C., 1900. Grundriss der Psychiatrie in klinische Vorlesungen. Thieme, Leipzig.

Wheeler, A.L., Voineskos, A.N., 2014. A review of structural neuroimaging in schizophrenia: from connectivity to connectomics. Front. Hum. Neurosci. 8, 1–18.

White, J.G., Southgate, E., Thomson, J.N., Brenner, S., 1986. The structure of the nervous system of the nematode Caenorhabditis elegans. Phil. Trans. R Soc. Lond. B Biol. Sci. 12 (314 (1165)), 1–340.

Whitfield-Gabrieli, S., Thermenos, H.W., Milanovic, S., Tsuang, M.T., Faraone, S.V., McCarley, R.W., et al., 2009. Hyperactivity and hyperconnectivity of the default network in schizophrenia and in first-degree relatives of persons with schizophrenia. Proc. Natl. Acad. Sci. U.S.A. 106 (4), 1279–1284.

Whitford, T.J., Kubicki, M., Shenton, M.E., 2011. Diffusion tensor imaging, structural connectivity, and schizophrenia. Schizophr. Res. Treatment ID 709523, 1–7.

Yan, H., Tian, L., Yan, J., Sun, W., Liu, Q., Zhang, Y.B., et al., 2012. Functional and anatomical connectivity abnormalities in cognitive division of anterior cingulate cortex in schizophrenia. PLoS One 7, 9.

Yu, Q., Sui, J., Rachakonda, S., He, H., Gruner, W., Pearlson, G., et al., 2011. Altered topological properties of functional network connectivity in schizophrenia during resting state: a small-world brain Network study. PLoS One 6 (9).

Yu, Q., Sui, J., Liu, J., Plis, S.M., Kiehl, K.A., Pearlson, G., et al., 2013. Disrupted correlation between low frequency power and connectivity strength of resting state brain networks in schizophrenia. Schizophr. Res. 143 (1), 165–171.

Yu, Q., Erhardt, E.B., Sui, J., Du, Y., He, H., Hjelm, D., et al., 2015. Assessing dynamic brain graphs of time-varying connectivity in fMRI data: application to healthy controls and patients with schizophrenia. NeuroImage 107, 345–355.

Zalesky, A., Fornito, A., Bullmore, E.T., 2010. Network-based statistic: identifying differences in brain networks. NeuroImage 53 (4), 1197–1207.

Zalesky, A., Fornito, A., Seal, M.L., Cocchi, L., Westin, C., Bullmore, E.T., et al., 2011. Disrupted Axonal Fiber Connectivity in Schizophrenia. Biol. Psychiatry 69 (1), 80–89.

Zamora-López, G., Zhou, C., Kurths, J., 2010. Cortical hubs form a module for multisensory integration on top of the hierarchy of cortical networks. Front. Neuroinform. 4, 1–13.

Zhang, Y., Lin, L., Lin, C.-P., Zhou, Y., Chou, K.-H., Lo, C.-Y., et al., 2012. Abnormal topological organization of structural brain networks in schizophrenia. Schizophr. Res. 141 (2–3), 109–118.

Zhou, J., Gennatas, E.D., Kramer, J.H., Miller, B.L., Seeley, W.W., 2012. Predicting regional neurodegeneration from the healthy brain functional connectome. Neuron 73 (6), 1216–1227.

Zhou, Y., Shu, N., Liu, Y., Song, M., Hao, Y., Liu, H., et al., 2008. Altered resting-state functional connectivity and anatomical connectivity of hippocampus in schizophrenia. Schizophr. Res. 100 (1–3), 120–132.

Zweig, A., 1908. Dementia praecox jenseits des 30. Lebensjahres. Archiv Für Psychiatrie Und Nervenkrankheiten 44, 1015–1035.

CHAPTER 19

STATISTICAL LEARNING OF THE NEUROBIOLOGY OF SCHIZOPHRENIA: NEUROIMAGING META-ANALYSES AND MACHINE-LEARNING IN SCHIZOPHRENIA RESEARCH

D. Bzdok[1,2,3] and S.B. Eickhoff[1,2]

[1]*Institut für Neurowissenschaften und Medizin (INM-1), Forschungszentrum Jülich GmbH, Jülich, Germany*
[2]*Institut für klinische Neurowissenschaften und Medizinische Psychologie, Heinrich-Heine Universität Düsseldorf, Düsseldorf, Germany* [3]*Parietal Team, INRIA, Neurospin, Gif-sur-Yvette, France*

CHAPTER OUTLINE

Introduction	337
Quantitative Meta-Analysis	338
Structural Meta-Analyses in Schizophrenia	339
Functional Meta-Analyses of Schizophrenia	341
Discussion	342
Machine Learning	343
Structural ML Studies in Schizophrenia	344
Hypothesis-Guided ML Studies in Schizophrenia	345
More Advanced ML Studies in Schizophrenia	346
Discussion	346
Conclusion	347
References	348

INTRODUCTION

Schizophrenia is a major burden on human societies. It causes substantial distress to the affected individuals as well as to their relatives and acquaintances. This major psychiatric disorder is also an important financial challenge for our health systems. Clinical descriptions of key schizophrenia symptoms have remained virtually identical over the past 100 years (Insel, 2010), thus basing schizophrenia diagnosis on century-old diagnostic categories. The pathophysiology of schizophrenia is generally acknowledged to evade mechanistic understanding. The risk of schizophrenia onset is augmented in a probabilistic fashion by different environmental factors, substance abuse, and other etiological categories. Genetic findings indicate that heterogeneous genetic vulnerabilities can lead to a set of psychopathologies that is uniformly labeled "schizophrenia" by psychiatrists. In fact, the genetic foundation of schizophrenia also appears to be inextricably linked with an array of a priori unrelated developmental disorders, including (but not exclusive to) autism, schizoaffective, and bipolar disorder. As a consequence of the unsatisfactory understanding of schizophrenia's pathophysiology, this major psychiatric disorder is still treated heuristically by focusing on alleviating symptoms instead of abolishing causes.

For decades, histopathology, lesion studies, and invasive research involving nonhuman primates have been the main neuroscientific workhorses. The advent of neuroimaging methods then leveraged unprecedented insight into brain biology. Task-based neuroimaging studies typically attempt to solicit target mental operations in participants by means of sensory, cognitive, or affective paradigms. This research greatly increased our understanding of the healthy brain and the diseased human brain. Yet, such work rests on a set of strict assumptions, including: historically inherited psychological categories are useful description systems for neurobiological phenomena; target and nontarget mental operations can be statistically distinguished by so-called cognitive subtraction; and a few dozen participants are sufficient to reliably establish fundamental (patho-)physiological mechanisms in the brain. All these assumptions have repeatedly been questioned. Apart from that, structural neuroimaging approaches, such as voxel-based morphometry, may be used to identify locations in which two groups of subjects, such as patients versus controls, differ from each other in local gray matter (GM) volume. Although not affected by the choice of a particular task, these approaches are also frequently criticized for insufficient sample sizes. In addition, they only allow characterizing presence, absence, or a group-mean effect at a particular location. Finally, methods for individual structural analyses are manifold and have repeatedly been in disagreement.

QUANTITATIVE META-ANALYSIS

How can the discrepancy between (much) data and (little) insight in the pathophysiology of schizophrenia be explained? First, sample sizes between 15 and 30 participants in typical neuroimaging studies on schizophrenia are orders of magnitude smaller than those for population genetics or clinical trials. Second, considerable variability is introduced by differences in inclusion criteria, inconsistent noncompliance, and incongruent treatment regimes. Third, such clinical neuroimaging studies incur substantial logistic costs and challenges. Fourth, a same functional neuroimaging dataset can be analyzed in hundreds of different ways (Carp, 2012). Nevertheless, by combining different paradigms, stimuli, instructions, and comparisons, there is a myriad of possible ways to, for instance, delineate "dysfunctional working memory" in schizophrenia.

It is in this setting that the necessity for unbiased synthesis of neuroimaging research (ie, quantitative meta-analysis) emerged (Turkeltaub et al., 2002). The activation likelihood estimation (ALE; Eickhoff et al., 2012) has now become the de facto standard for quantitative meta-analysis. ALE allows statistical integration of hundreds of neuroimaging studies purely based on the activation coordinates routinely reported in neuroimaging publications in a standardized fashion. ALE provides statistically defensible answers to the question, "where in the brain do the included activation foci cluster more tightly than it would be expected if they were randomly distributed?" ALE models individual activation peaks as centers of 3D Gaussian probability distributions. This allows for local averaging across coordinates from various studies to formally summarized "activation maps." A random-effects approach then results in significant activation convergence results that extend to the general population. By judging concordance and noise across a high number of studies, the limitation of individual neuroimaging studies may be overcome and a synoptic view of the (previously isolated) published findings can be acquired. Importantly, ALE is objective in algorithmically weighing all results equally. This lack of subjectivity precludes overinterpretation of expected and easily interpretable findings. Because algorithms for quantitative meta-analyses rest on the model-based integration of published activation coordinates, they can potentially be applied to the entirety of the available literature.

STRUCTURAL META-ANALYSES IN SCHIZOPHRENIA

Despite repeated controversy, most psychiatrists now believe that schizophrenia is associated with global brain atrophy and ventricular enlargement (Fusar-Poli et al., 2013; Johnstone et al., 1976). To quantify and detail morphometric studies, one of the first coordinate-based meta-analyses using ALE on schizophrenia statistically summarized 31 voxel-based morphometry (VBM) reports (Glahn et al., 2008). These investigators found statistically significant convergence of GM volume reduction in the bilateral insula, dorsal anterior cingulate cortex, subgenual anterior cingulate, and ventral anterior cingulate, thalamus, as well as the parahippocampal gyrus and middle frontal gyrus on the left (Fig. 19.1).

It is noteworthy that these findings involved both the "affective" and "cognitive" portions of the anterior cingulate. Additionally, the meta-analysis reported increased GM density in striatum regions. These investigators emphasized that the striatal GM increase might be read as a consequence of antipsychotic drugs, but that this drug–brain–volume association has been found only inconsistently. These findings are in line with a more recent coordinate-based meta-analysis of VBM studies (Nickl-Jockschat et al., 2011). ALE meta-analysis congruently identified GM reduction in the left peri-insular region, left thalamus, as well as the ventral striatum. However, additional GM reductions were found in the left amygdala and superior temporal gyrus. These findings have been interpreted as reflecting potential impairment of reward, emotional, and language processes in patients diagnosed with schizophrenia.

Complementing these research endeavors, other investigators (Olabi et al., 2011) tested whether the morphological aberrations are more likely due to a neurodevelopmental or neurodegenerative pathophysiology, that is, whether the structural brain changes in schizophrenia are static or progressive. In a quantitative review, in contrast to the aforementioned ALE meta-analysis, these investigators tracked the percentage of GM volume change in 32 predefined brain regions between scans of same individuals. The time interval between consecutive brain scans ranged from 1 to 10 years. Data from 27 studies thus were entered in a random-effects meta-analysis. The results showed that patients with schizophrenia have a *significantly greater decrease* of brain volume over time. This was noted in the combined GM and white matter (WM) volume and the GM volume, as well as in the frontal GM and WM, parietal WM, and temporal WM. Additionally, this report confirmed increased volume in the lateral ventricles. The differences between patients and control subjects in the "annualized percentage volume change" were shown between patients and healthy controls as 59% (entire GM) and 7% (whole-brain volume),

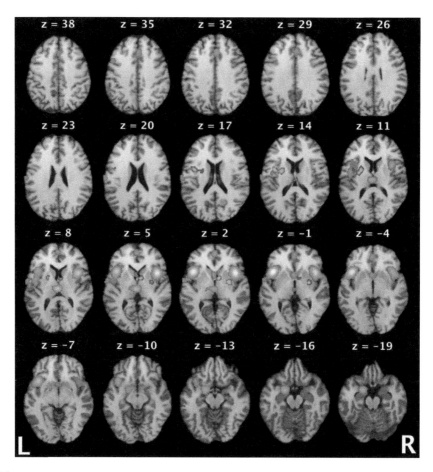

FIGURE 19.1

Significant gray matter volume alterations in schizophrenia according to ALE meta-analysis of 31 voxel-based morphometry studies (Glahn et al., 2008).

respectively. In contrast to other reports, these investigators did not observe volumetric changes in the medial temporal lobe (ie, hippocampus, amygdala). A more recent (non-ALE) report (Fusar-Poli et al., 2013) involving 30 original longitudinal studies confirmed the overall decrease in GM volume and increase of volume in the lateral ventricles in patients that were not observed in healthy controls. Similarly, Haijma et al. (2013) addressed the question of whether brain volume alterations are a pathophysiological feature already present before treatment onset or whether it is largely a result of treatment. Using a quantitative (non-ALE) meta-analysis of 317 MRI studies, they demonstrated that antipsychotics-naive patients show brain changes very similar to those observed in antipsychotics-treated schizophrenics, albeit to a lesser extent. Interestingly, the thalamus and caudate nucleus volume were bigger in nonmedicated patients. These investigators therefore concluded that brain volume aberrations in schizophrenia mostly reflect neurodevelopmental reduction in brain volume that is aggravated by illness progression and antipsychotic treatment regimes.

Finally, a study focusing on white matter alterations compared graph-analytical measures of fiber-tract connectivity estimations in controls with ALE results from VBM studies involving psychiatrically ill individuals (Crossley et al., 2014). In several psychiatric disorders, including schizophrenia, GM involvement is located to "connectome hubs." WM alterations can also be framed as structural brain features and have been meta-analyzed in schizophrenia. ALE of diffusion MRI (dMRI) results has thus emphasized different fiber connections in the left frontal and temporal lobe (Ellison-Wright and Bullmore, 2009). Other meta-analytic dMRI studies have confirmed these spots (Yao et al., 2013) and further detailed, for instance, the left arcuate fasciculus (Geoffroy et al., 2014) as well as the splenium (Patel et al., 2011). An important consequence of these data-driven approaches combining different modalities is that a small set of brain regions emerged as playing a key role across all considered psychiatric disorders.

FUNCTIONAL META-ANALYSES OF SCHIZOPHRENIA

Schizophrenia patients are known to be impaired in perception of emotions. Although many functional MRI studies reported abnormal activation patterns during facial emotion processing, there is quite some disagreement between their findings. Li and colleagues (2010) therefore set out to reach a consensus on aberrant activation during emotion processing using ALE meta-analysis of 17 functional neuroimaging studies. This report identified the bilateral amygdala and the right fusiform gyrus as the most important correlates of emotional disturbance in these patients. In fact, the pattern of neural activation increase during exposure to facial stimuli was highly similar between patients and controls, yet the increase was more prominent in controls.

In the same vain, another ALE meta-analysis (Kühn et al., 2013) delineated consistent alternations of resting-state neuroimaging findings in schizophrenic patients. A significant metabolic decrease was identified in the ventromedial prefrontal cortex, (left) hippocampus, posterior cingulate cortex, and the precuneus of schizophrenic patients. A significant metabolic increase, however, was identified in the bilateral lingual gyrus. These investigators concluded that the observed metabolic patterns reflect impaired self-attribution and self-reflection. This would be in line with dysfunction of the self as an important feature of psychopathology. In addition to disrupted self-processing, audio-visual hallucinations (AVH) are an important, if not specific, clinical symptom of schizophrenia. An ALE meta-analysis of AVH-related findings was conducted for 10 functional neuroimaging studies (Jardri et al., 2011) to investigate the robust underlying neurophysiology (Fig. 19.2).

The brain locations most consistently associated with AVHs comprise Broca's area, anterior insula, precentral gyrus, frontal operculum, middle/superior temporal gyri, inferior parietal lobule, as well as the (para-)hippocampal region. Jardri and collaborators concluded that experiencing AVHs is essentially associated with disrupted processing in brain areas involved in speech generation, speech perception, and verbal memory. Concentrating on the morphological rather than functional basis of AVHs, Modinos and collaborators corroborated these findings by a random-effects (non-ALE) meta-analysis of nine VBM studies of AVHs (Modinos et al., 2013), showing that the severity of AVHs was related to GM volume reductions in the left and also the right superior temporal gyri. These investigators discussed right temporal lobe involvement in AVH severity as being potentially related to processing of prosody and emotional salience.

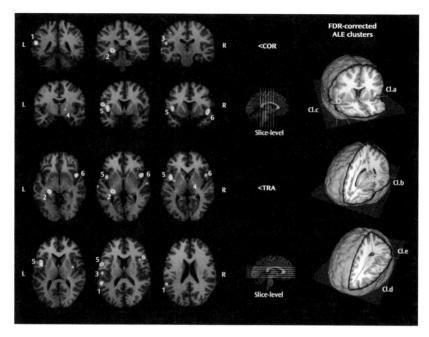

FIGURE 19.2

Significant functional alterations in schizophrenia according to ALE meta-analysis of 10 PET/fMRI studies depicted as coronal/axial slices (left) and brain renderings (right) (Jardri et al., 2011).

DISCUSSION

Descriptive review articles and opinion papers on the pathophysiology of schizophrenia are commonly and eagerly used means to informally integrate distributed neuroimaging findings. However, such critical verbal analyses tend to focus on preselected aspects and tend to be biased by the authors' own research agenda. Coordinate-based meta-analyses by ALE have particular potential in schizophrenia research given that they can be largely free of hypotheses regarding a particular pathophysiological mechanism and data-driven and, hence, allow for algorithmically weighing all results at an identical importance. This discourages overinterpretation of expected and easily interpretable findings and neglect of unexpected and barely reconcilable findings in schizophrenia research.

Quantitative meta-analyses thus represent powerful ways to gain a synoptic view of the heterogeneous neuroimaging findings in schizophrenia in a quantitative and impartial fashion. This should then reconcile conflicting views that arose from subsampling among neurobiologically distinct schizophrenia types, the difficulty to control for heterogeneous medication regimes, and lack of task compliance, as well as from subjectively biased interpretation. Prospectively, regarding the goal of clinical translation, meta-analysis-derived regions of interest can be used in a variety of neuroimaging methods, including structural, functional, and effective connectivity, in new groups of (individual) patients (Schilbach et al., 2014). Quantitative meta-analyses thus permit identification of the quintessence of emergent knowledge across various types of neuroimaging research on schizophrenia that can be used as a foundation for future examinations.

MACHINE LEARNING

As a complementary methodological family, machine-learning (ML) approaches are characterized by making the least assumptions possible, being more motivated by mathematical models rather than cognitive theory, and automatically mining structured knowledge from massive data resources. Given the widely acknowledged intricacies of and slow progress in schizophrenia research, ML methods lend themselves particularly well. They have the potential to decipher and subsequently render explicit biological indices that might assist in diagnosis, treatment, and clinical outcome prediction.

ML has developed into a set of statistical methods that focus on the detection and prediction of robust patterns in data (Hastie et al., 2011; Domingos, 2012). The different methods subsumed under ML have a number of common properties (Danilo, 2016). In an ML framework, the "truth" is believed to be in the data. This is why ML tries to explore and discover structure in typically large datasets. Technically, ML is inherently multivariate (ie, simultaneously operating on the entirety of input variables rather than in a mass-univariate regime) and nonparametric (making least assumptions on the data) and exploits computer algorithms (instead of being based on one-shot calculations, such as for Student *t* tests or tests of Gaussianity).

ML can be divided into *unsupervised* and *supervised* methods. Unsupervised ML prominently includes clustering algorithms, such as k-means, hierarchical, or ward clustering. Neuroscientific research has, for instance, capitalized on clustering approaches for the automatic grouping of observations into coherent subgroups of behavioral performance profiles (eg, patient groups based on clinical questionnaires), brain voxels (eg, parcellation of brain regions), fiber bundles (eg, parcellation of macroscopic pathways), or networks of brain regions (ie, functional brain architecture). The main advantage of unsupervised methods relies on the possible discovery of neurobiologically meaningful relationships while making few or no subjective decisions during the investigation. In supervised ML, an algorithm is automatically learning to detect a certain pattern in the *training phase*, and the success of which is then quantified in the subsequent *testing phase*. That is, a classification task can be learned by feeding data point examples with known class labels (ie, training set) into the algorithm. Importantly, this precludes the need to explicitly know and program how the distinction is performed. Once the learning algorithm has been trained, it is used to predict the class of a previously unused part of the data (ie, test set). This second part of the dataset provides quantification of the out-of-sample performance of the learned classification, that is, how likely the classification will succeed in future data. The popularity of such supervised methods in neuroimaging has increased dramatically in the attempt to "mind-read" or "decode" human thought from neural activity patterns (Haynes and Rees, 2005; Kamitani and Tong, 2005).

Diversity of off-the-shelf ML algorithms exists, with each permitting many flexible ways of application. Neuroimaging studies using ML in schizophrenia (Veronese et al., 2013) have so far mainly been applied to structural MRI measurements, with particular emphasis on the brain's GM. This is typically motivated by the wish to automatically distinguish patients from healthy controls by finding discriminative morphological patterns. As a byproduct, the neuroanatomical locations of the identified discriminative patterns are issued and reported. However, similar classification algorithms can readily be used for other neuroimaging modalities. This includes resting-state correlations that reflect the coupling strength of metabolic fluctuations between brain regions (Biswal et al., 1995) or diffusion MRI measurements that reflect properties of white-matter fibers (Jones, 2008). All three can be conducted without requiring the active participation of the patient. Particular interest lies in the combination of one or several of these neuroimaging modalities with behavioral measures, including clinical questionnaires (eg, PANSS), demographic information (eg, educational level), and treatment outcomes (eg, drug response). We detail and discuss such applications of machine-learning algorithms in clinical neuroimaging on schizophrenia in the following sections.

STRUCTURAL ML STUDIES IN SCHIZOPHRENIA

Present knowledge on the location of *discriminative patterns* of volume differences or morphology has mostly been derived from rather recent VBM studies. In one of the first ML applications for schizophrenia, Kawasaki and colleagues (2007) trained a multivariate linear model involving 30 male patients with schizophrenia and 30 healthy controls. The trained classifier was then tested for accuracy in an independent test population with 16 male patients and 16 matched healthy controls; 90% of the participants were correctly classified as schizophrenic/control based on characteristic GM distributions in the medial/lateral prefrontal and medial/lateral temporal cortex, as well as insula, thalamus, putamen, and cerebellum. Another ML study from the same year assessed cortical thickness measures in schizophrenia as a potentially more genuine quantification of cortical morphology (Yoon et al., 2007). The quantified cortical thickness of 53 patients and 52 healthy controls was spatially normalized and subsequently reduced to its most important sources of variability by lobe-wise principal component analysis (PCA). The ensuing information was fed into support vector machine (SVM) algorithms. SVMs are currently the most frequently used off-the-shelf classification algorithms in bioinformatics in general and in neuroimaging in particular. As a key property, they find the most representative data points (ie, brain images) for each class (ie, "support vectors") to define a decision boundary (eg, between healthy and diagnosed individuals). This allowed for classification accuracy of 91% based on differences in a dispersed set of temporal, limbic, and pre-/postcentral regions. It is noteworthy that these discriminative locations concurred with a "traditional" univariate statistical analysis performed in parallel. Additionally, the principal components with the highest explained variance did not consistently feature the highest diagnostic discriminability. As a rare example of targeting WM instead of GM, a diffusion MRI study (Ardekani et al., 2011) extracted measures of fractional anisotropy (FA) and mean diffusivity (MD) in 50 patients and 50 healthy controls. Applying Fisher's linear discriminant analysis (LDA) to the participants' FA and MD maps discriminated 94% and 98% of the cases, respectively. Interestingly, however, a combination of FA and MD did not enhance classification performance. Another VBM study (Ulas et al., 2012) concentrated on a few preselected target brain regions comprising the amygdala, entorhinal cortex, superior temporal gyrus, and thalamus bilaterally. These regions of interest were segmented by expert knowledge. Fifty schizophrenia patients and 50 matched controls were then examined by three quantitative descriptors: GM volume distribution by histograms; geometric features reflected by the shape index; and geometric features reflected by the curvature from triangle mesh vertices. The three obtained descriptors were used as input for the SVM. In the single-kernel setting, the left amygdala and right thalamus reached the highest accuracies of 77% and 76%, respectively, whereas multikernel results yielded success in up to 84%. Similarly, when concentrating only on the dorsolateral prefrontal cortex (dlPFC) in an advanced SVM pipeline involving bag-of-words techniques, 54 patients and 54 healthy controls were correctly categorized in 75% (left side) and 66% (right side) (Castellani et al., 2012), with slightly higher scores in female (84% vs 77%) and senior (81% vs 71%) subsamples.

HYPOTHESIS-GUIDED ML STUDIES IN SCHIZOPHRENIA

Rather than focusing on the mere localization of discriminative patterns, a number of MRI reports have been motivated by targeted a priori hypotheses. "At-risk mental state" (ARMS) refers to a preclinical state indicating a likely conversion to psychosis/schizophrenia without formally meeting the corresponding diagnosis criteria (yet). Individuals with ARMS motivated a series of reports by Koutsouleris et al. (2009). These investigators first compared 45 (neuroleptic-naïve) individuals with ARMS belonging to either early (ARMS-E) or late (ARMS-L) subpopulations with matched healthy controls in a 4-year follow-up setting. The VBM data were submitted to PCA. The ensuing principal components (PC) reflecting the main variability sources were subsequently used as features for a radial-based kernel SVM. This allowed automatic searching for the most useful number of PCs (a known, often difficult, problem) by peak accuracy in an independent analysis. The ensuing number of PCs provided the features for SVM classification. In 81% of the cases, SVM was able to tell ARMS-E and ARMS-L apart from healthy controls. That is, both early and late high-risk populations (ARMS-E or ARMS-L) featured patterns of neurocognitive deficits on a behavioral level. More specifically, individuals in the ARMS-E (ie, low conversion rate) group were classified with an accuracy of 91%, whereas individuals in the ARMS-L group (ie, high conversion rate) were classified with 86% accuracy. This unexpected finding was interpreted by the investigators to possibly reflect bigger clinical heterogeneity of the ARMS-L group. High diagnostic accuracy was related to structural patterns in the prefrontal/orbitofrontal cortex, and structural patterns in the perisylvian, limbic, and cerebellar regions were potential correlates of pending psychosis risk. The same investigators later published another report purely based on a neuropsychological test battery from 48 ARMS and 30 healthy individuals (Koutsouleris et al., 2012). Converters were correctly distinguished from individuals not converting to psychosis (in 90.8%). Behavioral items that differed between ARMS-E/ARMS-L and controls related to verbal learning/memory capacities, whereas executive performance and verbal IQ related more specifically to ARMS-L. These ML analyses identified impairments in verbal and executive learning domains during disease transition. Thus, discriminative patterns can be extracted from an unspecific set of neuropsychological tests. In yet another hypothesis-driven ML study by these investigators (Koutsouleris et al., 2014), a model of brain aging was learned from 800 healthy controls. In addition, such an aging model was also learned from patients with schizophrenia ($n = 141$), ARMS ($n = 89$), bipolar personality disorder (BPD; $n = 57$), and major depression (MD; $n = 104$) in healthy controls. This confirmed a previously proposed accelerated aging hypothesis in schizophrenia, an effect that was not observed in the BPD and MD. This might indicate trans-nosologically specific mechanisms of brain maturation and aging. In sum, morphological and neuropsychological aberrations predate psychosis and might be potentially important in the trend toward early detection and treatment of schizophrenia.

Furthermore, individuals diagnosed with schizophrenia can be clinically separated into cognitive deficit (CD) and cognitively spared (CS) subpopulations. A VBM-ML study (Gould et al., 2014) attempted to distinguish these from healthy controls ($n = 134$) based on multivariate patterns of volumetric brain data. CS and CD could be automatically distinguished from each other at an accuracy of 71%. It is noteworthy that this score increased to 83% when restricting the samples to females. That is, clinically distinct schizophrenia subpopulations can be distinguished based on subtle neuroanatomical differences, especially in females. Finally, the usefulness of ML based on VBM data was validated in a recent study by learning a discriminative model in a first sample with 128 patients and 111 healthy controls (Nieuwenhuis et al., 2012). It also tested the same unaltered model in a second sample with

155 patients and 122 healthy controls scanned with a different scanner. The SVM-provided model generalized to the general population at 70%.

MORE ADVANCED ML STUDIES IN SCHIZOPHRENIA

The advantage of ML approaches to seamlessly integrate with diverse imaging and analysis methods has been exploited on several recent occasions. In an ML-neurogenetics approach (Yang et al., 2010), single-nucleotide polymorphisms (SNP) data were combined with fMRI data from an auditory oddball task. First, the most discriminative SNPs capable of distinguishing patients with schizophrenia from healthy controls were selected to train a group of SVMs (ie, "ensemble of SVMs"). Second, individual voxels in the fMRI task data were analogously selected to build another group of trained SVMs. Third, dimensionality reduction of fMRI activation was performed by means of ICA to train one single SVM. The three obtained models were then combined in a majority voting approach on joint SNP-fMRI information with an overall classification accuracy of 87%. This prediction score was better than when using SNP (74%), voxel (82%), or ICA (83%) information alone. This indicates complementary information, which provides further evidence for an intertwined relationship between genetic predisposition and the neurobiology in schizophrenia. Rather than fusing diagnostically relevant information from different modalities, other investigators (Demirci et al., 2008) have performed successive compression of functional neuroimaging data from an oddball task. After application of a (rather conventional) spatiotemporal reduction of the input data into "independent components," these condensed data were further reduced by so-called projection pursuit algorithms to find interesting low-dimensional projections. In so doing, 70 individuals could be classified as schizophrenic and healthy with accuracies between 80% and 90%. Moreover, when using yet another classification algorithm, namely sparse multinomial logistic regression on GM density maps, 36 schizophrenia spectrum individuals were correctly distinguished from 36 controls in 86% (Sun et al., 2009).

DISCUSSION

The number of ML applications in schizophrenia is still surprisingly small. However, existing reports have demonstrated the potential to find diagnostically relevant information in diverse types of brain images (eg, of structure, function, connectivity) and behavioral indices (eg, from clinical questionnaires). These sources of information have been studied typically in isolation but are increasingly used in combination. The reported accuracies ranged from 70% to 98%.

The biggest accuracy scores are typically reported by studies using ingenious means to combine, simplify, and represent the preprocessed imaging data (ie, feature engineering). Such data transformation compresses the imaging information into a (often much) lower-dimensional space. Relevant information should thus mostly persist in this new representation, whereas irrelevant information is hopefully mostly discarded. This greatly facilitates the learning of a distinctive pattern for classification by avoiding the curse of dimensionality (Domingos, 2012). In this aspect of feature engineering, published ML reports on schizophrenia have frequently used PCA and ICA.

Moreover, undue emphasis is put on *how well* schizophrenics can be distinguished from neurotypicals rather than inquiring the *why*. That is, schizophrenia investigators have been particularly devoted to perfecting classification accuracy scores. Yet, this does not teach us exactly what elements of neurobiology have led to this discrimination success. Even more so, the discriminative features (eg, brain regions) might be explained by differences in medication regime, sampling effects, or site differences, if not further investigated. Rather, the focus should be on informed interpretation of statistically learned patterns to further our understanding of the pathophysiological foundation of schizophrenia. It is also important to appreciate that the same statistical learning approach is generally not optimal for both best classificatory performance and biggest insight by easy interpretability, which is a general property of the statistical learning theory (Hastie et al., 2011). Inexistence of any optimal ML approach for all research questions will necessitate in-depth knowledge of learning algorithms to optimally suit appropriate methods for clinical research questions.

Apart from that, selection bias might have unavoidably been introduced by conducting imaging research on participants who reside in or have resided in psychiatric institutions (rather than psychiatrically ill individuals who have never been diagnosed as such). Schizophrenia research aims at describing the group of individuals meeting diagnosis criteria for this disorder in the *general population*. Yet, clinical neuroscientists only have access to imaging data from the subgroup of schizophrenic individuals who have been hospitalized. The subgroup of highly functioning or subclinical schizophrenics who have never been in contact with a psychiatrist therefore systematically evades our research efforts. At the moment, ML techniques are unlikely to replace medical doctors, yet they are likely to increasingly assist in the diagnosis process in the very near future (Bostrom, 2014).

In sum, statistical learning of the brain–behavior architecture of schizophrenia (including psychopathological, social, and genetic indices) might have the highest potential in the detection of high-risk or first-episode schizophrenics for early therapeutic intervention as well as the prediction of treatment response for reduction of the substantial drug costs. Additionally, ML approaches might tackle the challenging obstacle of widely assumed but not well-understood biological subtypes of the *schizophrenia spectrum* (Hyman et al., 2007; Stephan et al., 2009).

CONCLUSION

Schizophrenia disrupts thinking, speech, and behavior. There is a general consensus that this psychiatric disorder emerges from interplay of genetic, environmental, and neurobiological factors. Furthermore, there is an increasing suspicion that this results in a spectrum of distinct brain pathophysiological trajectories with partly identical clinical phenotypes. This poses a particular challenge to studies using neuroimaging techniques. As an attractive alternative to experimental approaches motivated by cognitive theory, statistical learning approaches for neuroimaging data suggest themselves because they leave the phenomenon undergoing study—the pathoneurobiology of schizophrenia—as unaltered as possible by directly learning mechanisms from data. Quantitative meta-analyses allow for formalized synthesis of neuroimaging findings consistent across laboratories and patient populations. This remedies unavoidable treatment heterogeneities, high dropout rates, and usually small sample sizes. Machine learning, however, allows for automatically revealing and formalizing complex relationships between clinical behavior and brain data. This should permit creation of a data-driven definition of clinical subgroups and diagnosis thereof for a novel, biologically founded taxonomy of *schizophrenia spectrum disorders*. Taken together, the methods introduced here should foster our understanding of schizophrenia disease endophenotypes for stratification, diagnosis, and prediction of treatment response.

REFERENCES

Ardekani, B.A., Tabesh, A., Sevy, S., Robinson, D.G., Bilder, R.M., Szeszko, P.R., 2011. Diffusion tensor imaging reliably differentiates patients with schizophrenia from healthy volunteers. Hum. Brain Mapp. 32, 1–9.

Biswal, B., Yetkin, F.Z., Haughton, V.M., Hyde, J.S., 1995. Functional connectivity in the motor cortex of resting human brain using echo-planar MRI. Magn. Reson. Med. 34, 537–541.

Bostrom, N., 2014. Superintelligence: Paths, Dangers, Strategies. Oxford University Press.

Carp, J., 2012. On the plurality of (methodological) worlds: estimating the analytic flexibility of FMRI experiments. Front. Neurosci. 6, 149.

Castellani, U., Rossato, E., Murino, V., Bellani, M., Rambaldelli, G., Perlini, C., et al., 2012. Classification of schizophrenia using feature-based morphometry. J. Neural. Transm. 119, 395–404.

Crossley, N.A., Mechelli, A., Scott, J., Carletti, F., Fox, P.T., McGuire, P., et al., 2014. The hubs of the human connectome are generally implicated in the anatomy of brain disorders. Brain 137, 2382–2395.

Danilo B., 2016. Classical statistics and statistical learning in imaging neuroscience. arXiv preprint arXiv:1603.01857.

Demirci, O., Clark, V.P., Calhoun, V.D., 2008. A projection pursuit algorithm to classify individuals using fMRI data: application to schizophrenia. NeuroImage 39, 1774–1782.

Domingos, P., 2012. A few useful things to know about machine learning. Commun. ACM 55, 78–87.

Eickhoff, S.B., Bzdok, D., Laird, A.R., Kurth, F., Fox, P.T., 2012. Activation likelihood estimation meta-analysis revisited. NeuroImage 59, 2349–2361.

Ellison-Wright, I., Bullmore, E., 2009. Meta-analysis of diffusion tensor imaging studies in schizophrenia. Schizophr. Res. 108, 3–10.

Fusar-Poli, P., Smieskova, R., Kempton, M.J., Ho, B.C., Andreasen, N.C., Borgwardt, S., 2013. Progressive brain changes in schizophrenia related to antipsychotic treatment? A meta-analysis of longitudinal MRI studies. Neurosci. Biobehav. Rev. 37, 1680–1691.

Geoffroy, P.A., Houenou, J., Duhamel, A., Amad, A., De Weijer, A.D., Curcic-Blake, B., et al., 2014. The Arcuate Fasciculus in auditory-verbal hallucinations: a meta-analysis of diffusion-tensor-imaging studies. Schizophr. Res. 159, 234–237.

Glahn, D.C., Laird, A.R., Ellison-Wright, I., Thelen, S.M., Robinson, J.L., Lancaster, J.L., et al., 2008. Meta-analysis of gray matter anomalies in schizophrenia: application of anatomic likelihood estimation and network analysis. Biol. Psychiatry 64, 774–781.

Gould, I.C., Shepherd, A.M., Laurens, K.R., Cairns, M.J., Carr, V.J., Green, M.J., 2014. Multivariate neuroanatomical classification of cognitive subtypes in schizophrenia: a support vector machine learning approach. Neuroimage Clin. 6, 229–236.

Haijma, S.V., Van Haren, N., Cahn, W., Koolschijn, P.C., Hulshoff Pol, H.E., Kahn, R.S., 2013. Brain volumes in schizophrenia: a meta-analysis in over 18 000 subjects. Schizophr. Bull. 39, 1129–1138.

Hastie, T., Tibshirani, R., Friedman, J., 2011. The Elements of Statistical Learning. Springer Series in Statistics, Heidelberg, Germany.

Haynes, J.D., Rees, G., 2005. Predicting the orientation of invisible stimuli from acitvity in human primary visual cortex. Nat. Neurosci. 8, 686–691.

Hegarty, J.D., Baldessarini, R.J., Tohen, M., Waternaux, C., Oepen, G., 1994. One hundred years of schizophrenia: a meta-analysis of the outcome literature. Am. J. Psychiatry 151, 1409–1416.

Hyman, S.E., 2007. Can neuroscience be integrated into the DSM-V? Nat. Rev. Neurosci. 8, 725–732.

Insel, T.R., 2010. Rethinking schizophrenia. Nature 468, 187–193.

Jardri, R., Pouchet, A., Pins, D., Thomas, P., 2011. Cortical activations during auditory verbal hallucinations in schizophrenia: a coordinate-based meta-analysis. Am. J. Psychiatry 168, 73–81.

Johnstone, E.C., Crow, T.J., Frith, C.D., Husband, J., Kreel, L., 1976. Cerebral ventricular size and cognitive impairment in chronic schizophrenia. Lancet 2, 924–926.

Jones, D.K., 2008. Studying connections in the living human brain with diffusion MRI. Cortex 44, 936–952.
Kamitani, Y., Tong, F., 2005. Decoding the visual and subjective contents of the human brain. Nat. Neurosci. 8, 679–685.
Kawasaki, Y., Suzuki, M., Kherif, F., Takahashi, T., Zhou, S.Y., Nakamura, K., et al., 2007. Multivariate voxel-based morphometry successfully differentiates schizophrenia patients from healthy controls. NeuroImage 34, 235–242.
Koutsouleris, N., Meisenzahl, E.M., Davatzikos, C., Bottlender, R., Frodl, T., Scheuerecker, J., et al., 2009. Use of neuroanatomical pattern classification to identify subjects in at-risk mental states of psychosis and predict disease transition. Arch. Gen. Psychiatry 66 (7), 700–712.
Koutsouleris, N., Davatzikos, C., Bottlender, R., Patschurek-Kliche, K., Scheuerecker, J., Decker, P., et al., 2012. Early recognition and disease prediction in the at-risk mental states for psychosis using neurocognitive pattern classification. Schizophr. Bull. 38, 1200–1215.
Koutsouleris, N., Davatzikos, C., Borgwardt, S., Gaser, C., Bottlender, R., Frodl, T., et al., 2014. Accelerated brain aging in schizophrenia and beyond: a neuroanatomical marker of psychiatric disorders. Schizophr. Bull. 40, 1140–1153.
Kühn, S., Gallinat, J., 2013. Resting-state brain activity in schizophrenia and major depression: a quantitative meta-analysis. Schizophr Bull. 39 (2), 358–365.
Li, H., Chan, R.C., McAlonan, G.M., Gong, Q.Y., 2010. Facial emotion processing in schizophrenia: a meta-analysis of functional neuroimaging data. Schizophr. Bull. 36, 1029–1039.
Modinos, G., Costafreda, S.G., van Tol, M.J., McGuire, P.K., Aleman, A., Allen, P., 2013. Neuroanatomy of auditory verbal hallucinations in schizophrenia: a quantitative meta-analysis of voxel-based morphometry studies. Cortex 49, 1046–1055.
Nickl-Jockschat, T., Schneider, F., Pagel, A.D., Laird, A.R., Fox, P.T., Eickhoff, S.B., 2011. Progressive pathology is functionally linked to the domains of language and emotion: meta-analysis of brain structure changes in schizophrenia patients. Eur. Arch. Psychiatry Clin. Neurosci 261 (suppl 2), 166–171.
Nieuwenhuis, M., van Haren, N.E., Hulshoff Pol, H.E., Cahn, W., Kahn, R.S., Schnack, H.G., 2012. Classification of schizophrenia patients and healthy controls from structural MRI scans in two large independent samples. NeuroImage 61, 606–612.
Olabi, B., Ellison-Wright, I., McIntosh, A.M., Wood, S.J., Bullmore, E., Lawrie, S.M., 2011. Are there progressive brain changes in schizophrenia? A meta-analysis of structural magnetic resonance imaging studies. Biol. Psychiatry 70, 88–96.
Patel, S., Mahon, K., Wellington, R., Zhang, J., Chaplin, W., Szeszko, P.R., 2011. A meta-analysis of diffusion tensor imaging studies of the corpus callosum in schizophrenia. Schizophr. Res. 129, 149–155.
Schilbach, L., Muller, V.I., Hoffstaedter, F., Clos, M., Goya-Maldonado, R., Gruber, O., et al., 2014. Meta-analytically informed network analysis of resting state FMRI reveals hyperconnectivity in an introspective socio-affective network in depression. PLoS One 9, e94973.
Stephan, K.E., Friston, K.J., Frith, C.D., 2009. Dysconnection in schizophrenia: from abnormal synaptic plasticity to failures of self-monitoring. Schizophr. Bull. 35, 509–527.
Sun, D., van Erp, T.G., Thompson, P.M., Bearden, C.E., Daley, M., Kushan, L., et al., 2009. Elucidating a magnetic resonance imaging-based neuroanatomic biomarker for psychosis: classification analysis using probabilistic brain atlas and machine learning algorithms. Biol. Psychiatry 66, 1055–1060.
Turkeltaub, P.E., Eden, G.F., Jones, K.M., Zeffiro, T.A., 2002. Meta-analysis of the functional neuroanatomy of single-word reading: method and validation. NeuroImage 16, 765–780.
Ulas, A., Castellani, U., Murino, V., Bellani, M., Tansella, M., Brambilla, P., 2012. Biomarker evaluation by multiple Kernel learning for schizophrenia detection. In: Proceedings of the International Workshop on Pattern Recognition in NeuroImaging.
Veronese, E., Castellani, U., Peruzzo, D., Bellani, M., Brambilla, P., 2013. Machine learning approaches: from theory to application in schizophrenia. Comput. Math. Methods Med. 2013, 867924.

Yang, H., Liu, J., Sui, J., Pearlson, G., Calhoun, V.D., 2010. A hybrid machine learning method for fusing fMRI and genetic data: combining both improves classification of schizophrenia. Front. Hum. Neurosci. 4, 192.

Yao, L., Lui, S., Liao, Y., Du, M.Y., Hu, N., Thomas, J.A., et al., 2013. White matter deficits in first episode schizophrenia: an activation likelihood estimation meta-analysis. Prog. Neuropsychopharmacol. Biol. Psychiatry 45, 100–106.

Yoon, U., Lee, J.M., Im, K., Shin, Y.W., Cho, B.H., Kim, I.Y., et al., 2007. Pattern classification using principal components of cortical thickness and its discriminative pattern in schizophrenia. NeuroImage 34, 1405–1415.

PART VII

MODELING SCHIZOPHRENIA IN ANIMALS

CHAPTER 20

MODELING SCHIZOPHRENIA IN ANIMALS: OLD CHALLENGES AND NEW OPPORTUNITIES

Y. Ayhan[1,2], C.E. Terrillion[2] and M.V. Pletnikov[2,3,4,5]

[1]Department of Psychiatry, Faculty of Medicine, Hacettepe University, Ankara, Turkey
[2]Department of Psychiatry and Behavioral Sciences, School of Medicine, Johns Hopkins University, Baltimore, MD, United States [3]Department of Molecular and Comparative Pathobiology, School of Medicine, Johns Hopkins University, Baltimore, MD, United States [4]Solomon H. Snyder Department of Neuroscience, School of Medicine, Johns Hopkins University, Baltimore, MD, United States [5]Department of Molecular Microbiology and Immunology, Bloomberg School of Public Health, Johns Hopkins University, Baltimore, MD, United States

CHAPTER OUTLINE

Introduction 353
Genetic Animal Models of Schizophrenia 355
    Pathophysiological Models 356
    Models Based on Etiological Findings 358
Challenges in Modeling Schizophrenia 366
Conclusions 367
Acknowledgments 370
References 370

INTRODUCTION

Schizophrenia is a major psychiatric disorder that is characterized by positive, negative, disorganization, cognitive, psychomotor, and mood symptoms (Tandon et al., 2009). The onset of schizophrenia occurs in late adolescence or early adulthood, with subtle motor, cognitive, and social abnormalities being frequently observed prior to the full diagnosis (Insel, 2010). Despite extensive in vivo imaging and postmortem studies, no specific morphological, biochemical, or electrophysiological biomarkers have been identified for schizophrenia (Insel, 2010). Although schizophrenia is a multifactorial disorder, it has a strong genetic component (Tsuang et al., 2000), as has been recently reinforced by genome-wide association studies (GWAS) (Sullivan, 2010).

As with any complex human disorder, the main goal of animal models for schizophrenia is to reproduce major features of the disorder (eg, an etiological factor or a behavioral phenotype), because recreating the entire spectrum of psychiatric phenomenological manifestations is impossible and counterproductive (Peleg-Raibstein et al., 2012; Swerdlow and Geyer, 1998; Weiner and Arad, 2009). For the past 30 plus years, numerous animal models relevant to aspects of schizophrenia have been introduced (Lipska and Weinberger, 1993, 2000). Behavioral models mimic behavioral symptoms and signs observed in patients, and several cross-species translational tests have been proposed to evaluate schizophrenia-like behaviors in laboratory animals (Barch et al., 2009). Tests related to the positive symptoms include spontaneous or psychostimulant-induced locomotor activity (O'Tuathaigh and Waddington, 2010) and prepulse inhibition (PPI) of the acoustic startle reflex. PPI is regarded as a highly translational test for the ability to gate sensory stimuli, although the PPI test is not specific to psychoses (Geyer et al., 2002). Tests for cognitive impairment, a major treatment-resistant feature of schizophrenia, include those for learning and memory and sustained attention (Kellendonk et al., 2009). Although modeling the negative symptoms of schizophrenia is a daunting task, assays of social behaviors, anhedonia, blunted affect, and avolition represent a promising direction (O'Tuathaigh and Waddington, 2010).

Pharmacological challenge models manipulate the neurotransmitter systems thought to underlie the pathophysiological mechanisms of behavioral or cognitive symptoms in patients. Exposure to psychostimulants, cannabis, phencyclidine (PCP), or other psychotropic drugs may increase susceptibility to psychiatric disorders (Ellenbroek, 1995). Thus, these pharmacologic models mimic the behavioral effects of treatments (pharmacological isomorphism) and may help screen for experimental therapeutics (predictive validity) (Matthysse, 1986; Ellenbroek, 1995).

Models that exhibit construct validity reproduce the pathogenic mechanisms of the disease (Ellenbroek, 1995; Pletnikov, 2002). Among these models, abnormal neurodevelopment models are poised to advance our understanding of the disease's development because abnormalities in brain maturation have been strongly implicated in the pathogenesis of schizophrenia (Jaaro-Peled et al., 2010). One prominent model is based on a lesion of the ventral hippocampus (VHP) of rats. Both bilateral and unilateral lesions of the VHP result in increased responsivity to psychostimulants, impaired PPI, and latent inhibition (LI) of attenuated social interaction and spatial memory (Tseng et al., 2009). This model highlights several important points relevant to the development of schizophrenia. First, the model shows that earlier lesions produce greater neurobehavioral abnormalities (Wood et al., 1997). Second, even though the lesion itself does not mimic the brain pathology of schizophrenia, some of the behavioral alterations following the lesion respond to antipsychotics. This suggests that the model recreates some of the key pathophysiological mechanisms of psychotic disorders. Third, the model demonstrates that even if the lesion is limited to a specific brain area, the ensuing structural alterations elsewhere in the brain, such as reduction of GAD67 immunoreactivity in the cortex and hippocampus, are relevant to schizophrenia pathology (Francois et al., 2009). Still, nongenetic models do not completely address etiologies of the disease, and thus they may limit the discovery of new etiopathogenic treatments. The readers are also referred to excellent reviews on this topic (Lillrank et al., 1995; Lipska et al., 1993; Lipska and Weinberger, 1993). Fig. 20.1 summarizes the main approaches to modeling schizophrenia. Here, we turn our attention to the genetic models for this disorder.

Human genetics studies have provided the strongest rationale for genetic animal models that can better illuminate the etiology and pathogenesis of the disease. Because schizophrenia is increasingly considered a disorder of brain development, mouse models of susceptibility genes involved in

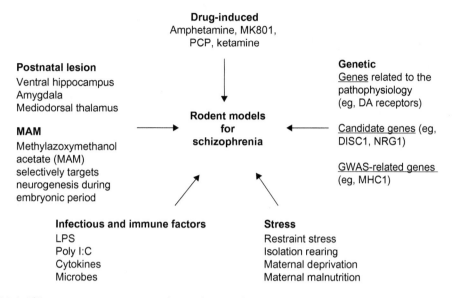

FIGURE 20.1 Different approaches to modeling schizophrenia.

Various approaches to model aspects of schizophrenia ranging (counterclockwise) from postnatal lesions and prenatal neurotoxins to etiologically relevant environmental factors (eg, immune or stress) and genetic models are presented. Notably, many of the same nongenetic manipulations have been used to model other psychiatric disorders (eg, depression or autism). This notion supports the main idea of the chapter that diagnostically distinct diseases may be caused by overlapping etiologies, both genetic and nongenetic, and that animal models based on reproducing different dimensions or domains of psychopathology rather than the entire disease will provide more efficient experimental systems for mechanistic studies.

neurodevelopment will likely be the most informative (Abakay et al., 2012; Lu et al., 2011). In this chapter, we focus on some of the genetic mouse models available, evaluate their advantages and limitations, and propose new prospects for the field.

GENETIC ANIMAL MODELS OF SCHIZOPHRENIA

Although the goal of genetic models is to identify the genetic underpinnings of a behavioral disease, each model has unique features that address key aspects of the disease. Although some models have been generated using genetic information from linkage or association studies, others have tried to mimic the pathophysiology of the disorder even if particular genetic manipulation had little support from human studies. Yet another group of genetic models is based on both approaches, with one example being mouse models of the 22q11.2 deletion syndrome, resulting from microdeletions of 1.5–3 MB on human chromosome 22. This syndrome, among other features, is characterized by a high incidence of schizophrenia (Karayiorgou and Gogos, 2004; Meechan et al., 2015).

These genetic models not only provide useful tools for our understanding of the pathophysiology of the disease and serve as potential tools for pharmacological studies but also allow researchers to model etiologically relevant situations, especially the interaction of genetic and environmental factors that emerge from human epidemiological studies (Ayhan et al., 2015; Kannan et al., 2013).

PATHOPHYSIOLOGICAL MODELS

Some genetic models of schizophrenia were built with mutations that were not necessarily derived from genetic epidemiologic studies performed in humans. Instead, the mutations in these models were produced in the genes that play a role in the pathophysiology of schizophrenia, such as dopamine or glutamate receptors. These models have been useful in pharmacological studies, as well as for understanding the molecular mechanisms of schizophrenia (Ross et al., 2006). In the following sections, we provide an overview of some major models in this group. We also refer readers to several excellent recent reviews for more comprehensive information about these models (Dawe et al., 2009; O'Tuathaigh and Waddington, 2010).

Models targeting dopaminergic transmission

Dopamine (DA) dysfunction in schizophrenia is a popular hypothesis (Abi-Dargham, 2014), with D2 receptor antagonism linked to the antipsychotic effects of neuroleptics and effects of DA psychostimulants providing the main rationale for this hypothesis (Meisenzahl et al., 2007). It has been recently proposed that increased DA transmission in the striatum contributes to positive symptoms, whereas diminished DA signaling in the prefrontal cortex (PFC) contributes to cognitive impairment and negative symptoms (Goldman-Rakic et al., 2004; Kuepper et al., 2012).

Kellendonk and colleagues (2006) generated a Tet-off model of inducible overexpression of **D2 receptors** in the striatum using the CAMKII promoter. In this model, when the animals were fed with a regular diet, D2 receptors were overexpressed; overexpression was halted by administration of doxycycline (DOX)-containing food.

Overexpression of D2 receptors resulted in behavioral and physiological effects that were present even after turning off the transgene. Transgenic D2 mice (mutants) were tested in several behavioral tasks, including cognitive tests. In the eight-arm radial maze delayed nonmatch to sample (DNMTS) test, the mice were tested during two sessions daily for 10 consecutive days on a radial maze with eight arms containing food pellets. The first session had four randomly closed arms, and the second session had all arms open. Good performance was associated with more visits as well as a shorter latency to visit the previously unvisited bait-containing arms. D2 overexpression significantly impaired the performance on this task. In the T-maze test, mice learned the rule to visit the previously unvisited arm between the two consecutive choices. Mutants took significantly more trials to accomplish the task. In a task of olfactory discrimination, mice learned an association between an odor and food. After the association was established, mice had to make a new association between a new odor stimulus and food by inhibiting the previous association. Performance in this attention set–shifting task was also impaired in mutant mice, suggesting altered working memory and/or behavioral flexibility (Kellendonk et al., 2006).

In addition, overexpression of D2 receptors affected motivational processes and timing, as measured by award-related operant motivation tests, including interval timing and a progressive ratio task (Drew et al., 2007; Huang et al., 2004). The observed behavioral alterations were associated with decreased DA turnover in the medial prefrontal cortex (mPFC) and an increase in D1 receptor activation. This model provides valuable information regarding the interaction between the mPFC and striatum with relevance to schizophrenia-like dysfunctions (Cazorla et al., 2014; Cazorla et al., 2012; Li et al., 2011).

The catechol-O-methyltransferase (COMT) gene is located on chromosome 22q11, a region that is associated with schizophrenia (Bassett and Chow, 2008). The gene encodes the enzyme that inactivates DA in PFC (Tunbridge et al., 2006). A common functional polymorphism (Val(108/158) Met) in this gene has been associated with variations in working memory in control subjects (Egan et al., 2001). COMT knockout (KO) mice demonstrated increased anxiety in females and increased aggression in males (Gogos et al., 1998). Subsequent studies assessed exploratory and habituation behaviors and found decreased rearing with increased sifting and chewing in heterozygous (HET) COMT mice (Babovic et al., 2007, 2008). The mice also displayed altered responses to DA agonists in locomotor activity assessment (Huotari et al., 2004). In addition, COMT deficiency was associated with increased pain sensitivity (Walsh et al., 2010). Female COMT KO mice also showed an enhanced corticosterone response to acute stress and were more anxious than control littermates (Desbonnet et al., 2012). A transgenic model with overexpression of the human COMT-VAL polymorphism was recently described (Papaleo et al., 2008). The data demonstrated that diminished attention set–shifting and impaired recognition memory were probably related to increased COMT enzyme activity in Val-tg mice. Intriguingly, amphetamine was able to improve recognition memory, consistent with the hypothesis that the optimal levels of DA are necessary for cognitive function (Seamans and Yang, 2004). Val-tg mice also demonstrated lower stress reactivity and pain sensitivity (Papaleo et al., 2008).

Models targeting glutamatergic transmission

The role of the glutamate (GLU) system in schizophrenia has previously been extensively reviewed (Howes et al., 2015; Moghaddam and Javitt, 2012; Poels et al., 2014). The GLU hypothesis of schizophrenia is based on observations that N-methyl-D-aspartate (NMDA) receptor (NMDAR) antagonists, phencyclidine (PCP) or ketamine, can induce behavioral symptoms in healthy individuals or exacerbate existing symptoms in schizophrenic patients (Coyle, 2006). The role of GLU neurotransmission is further supported by neuropathological, neuroimaging, and genetic studies (Howes et al., 2015; Javitt, 2015).

A hypomorphic mouse model characterized by 2–8% expression of the NR1 subunit of the NMDA receptor was generated by inserting a neomycin resistance gene into intron 20 (Mohn et al., 1999). NR1 hypomorphic mice display increased activity and stereotypy and reduced habituation similar to phenotypes induced by MK-801 or PCP administration. Social interaction was decreased in mutants in a resident–intruder behavioral assay. Subsequent studies showed deficits in PPI and increased sensitivity to amphetamine-induced disruption of PPI, as well as reduced metabolic activity in the mPFC, ACC, and hippocampus in these mice (Duncan et al., 2006a,b). Both typical and atypical antipsychotics were able to ameliorate the PPI deficit in mutant mice, suggesting that different transmitter systems might be involved in the complex phenotypes (Miyamoto et al., 2004). In addition to NR1, other subtypes of NMDA, AMPA, or metabotropic GLU receptors have also been targeted in several mouse models (Labrie et al., 2012; Lipina et al., 2007; Wiedholz et al., 2008).

As with the models that target the DA system, GLU models also involve genes beyond the receptors, such as the genes encoding for different factors involved in GLU synaptic transmission. One such factor, D-amino acid oxidase (DAO), is the enzyme that is responsible for degradation of D-serine, which acts as a co-agonist of the NMDA receptor along with D-glycine. DAO expression and activity were increased in postmortem brain samples from patients with schizophrenia and bipolar disorder (Burnet et al., 2008; Madeira et al., 2008). This increase may be related to decreased levels of D-serine, consistent with impaired NMDAR function. A strain of mice that lacks DAO activity due to a single point mutation was identified (Almond et al., 2006). Mice with this mutation had decreased novelty-induced activity and increased performance in spatial learning tests, and the effects of PCP were attenuated in DAO mutants. D-serine in the serum and brain and NMDAR glycine site occupancy were increased (Almond et al., 2006). This model provided information about the role of DAO in behavioral alterations consistent with schizophrenia symptoms.

Another example is PICK1 mutants. Protein interacting with C kinase-1 (PICK-1) is a PDZ domain-containing protein located at presynaptic and postsynaptic sites. PICK1 interacts with AMPA receptor subunits GluR2/3/4 and mGluR7, dopamine, norepinephrine transporters, and serine racemase (SR) enzyme. SR converts L-serine to D-serine (Wolosker et al., 1999). The D-serine levels were decreased in plasma and cerebrospinal fluid samples in schizophrenia patients. In addition, PICK-1 plays a significant role in synaptic stability and plasticity via its effects on AMPAR scaffolding to the synapse and by interacting with GluR2. These interactions, particularly the interaction with SR, suggest a possible role of PICK1 in schizophrenia pathogenesis (Fujii et al., 2006).

In the PICK1 KO mice, SR levels were not changed, but D-serine levels decreased in early postnatal days. The interaction of PICK-1 and SR has been confirmed at the cellular level; D-serine production was increased when only PICK1 and SR were coexpressed in the cells (Tsai et al., 1998, 1999).

Although not based on strong genetic data, the pathophysiological mouse models have advanced our understanding of the involvement of different neurotransmitter systems in the complex phenotypes resembling aspects of schizophrenia. These models have also been useful in elucidating the mechanisms whereby pharmacological compounds ameliorate behavioral abnormalities. We think that these "neurotransmission" models will continue to serve as valuable tools for refining existing and developing novel treatment options.

MODELS BASED ON ETIOLOGICAL FINDINGS

Linkage and association studies have implicated several loci in the genome that harbors genes conferring risk for schizophrenia (Tiwari et al., 2010). Candidate genes identified by a genetic approach have an advantage over genes chosen based on pathophysiological, neuropathological, or neuropharmacological studies, because candidate genes may be directly involved in the disease process (Ross et al., 2006). Recent genome-wide association studies (GWAS) have furthered the progress in psychiatric genetics and highlighted the fact that the risk conferred by each genetic variant is small and combined effects of these variants are required (Gratten et al., 2014; Harrison, 2015). In addition, GWAS have confirmed that structural variations in the genome (copy number variations (CNVs)) may have significant contributions to schizophrenia (Kirov, 2015). One example includes a 22q11 deletion that is associated with an increased risk for psychosis; the largest cohort found more than 40% of individuals had a psychotic disorder (Schneider et al., 2014; Shprintzen et al., 1992). Different groups have developed successful CNV models of neuropsychiatric disorders, and these CNV models are elegantly covered in this book by Nishi and Hiroi.

Models based on genes from genome-wide association studies

One current prominent hypothesis about the genetics of schizophrenia relies on the common gene–common variant hypothesis. Accordingly, GWAS allow researchers to comprehensively search for common variants with low effect sizes throughout the genome of individuals carrying the disorder. To determine the variants with low effect sizes, samples from a substantial number of individuals need to be collected. To collect and analyze the genotypes of a high number of samples with different psychiatric disorders, multicenter collaborations were established. The Schizophrenia Working Group of the Psychiatric Genomics study, which has the largest sample size to date for schizophrenia, included more than 35,000 cases and identified 108 loci contributing to risk of schizophrenia. More than 70% of discovered loci were located in regions encoding proteins involved in dopaminergic and glutamatergic neurotransmission, calcium signaling, synaptic plasticity, potassium channels, and neurodevelopment (Schizophrenia Working Group of the Psychiatric Genomics, 2014). Schizophrenia-associated genes overlapped with neuroimmune diseases and single nucleotide polymorphisms (SNPs) in the extended major histocompatibility complex (MHC) region on chromosome 6. Other immune-related genes were also significantly associated with schizophrenia, suggesting etiological relevance of immune responses and inflammatory pathways (Andreassen et al., 2015; Network & Pathway Analysis Subgroup of Psychiatric Genomics, 2015; Shi et al., 2009). For a thorough discussion on human genetic epidemiology studies and the current best evidence for the genetic hypothesis of schizophrenia, please see the "The Genetic and Epigenetic Basis of Schizophrenia" section.

Generating animal models with these risk variants is anticipated to advance our understanding of the disease mechanisms. However, animal model research in this field has only recently started. Because the variants described with GWAS are common, meaning they can also be found in many control subjects, and because they have a low effect size, there are significant challenges for recreating multiple variants in an animal model.

One of the most replicated schizophrenia GWAS results is the MHC locus. This locus contains multiple genes involved in immunity, neurodevelopment, and synaptic plasticity, presenting a challenge for linking specific genes in this region to a particular domain of the pathophysiology (Corriveau et al., 1998; Huh et al., 2000). Many MHC mouse studies focus on species-specific recognition as a part of mating behavior (Hurst, 2009). A few investigations have evaluated this locus in other behaviors. One study reports deletion of H2-Kb and H2-Db, the two classical MHCI molecules, which are coexpressed in C57BL/6 mice on Purkinje cells (PCs) (McConnell et al., 2009). Mice deficient for both H2-K^b and H2-D^b(K^b $D^{b-/-}$) have a lower threshold for induction of long-term depression (LTD) at fiber parallel to PC synapses and improvement in acquisition and retention of a rotarod behavioral task, suggesting improved motor memory (McConnell et al., 2009). Mice deficient in β-2 microglobulin and transporter associated with antigen processing (β2M$^{-/-}$TAP$^{-/-}$) have greater hypothalamic pituitary adrenal activation with reduced activity after saline injection, suggesting altered stress reactivity (Sankar et al., 2012). Although not specific for schizophrenia, involvement of class I MHC molecules in stress processing supports the potential role of the variants of this genomic region in the genesis of behavioral alterations.

A putative role of one of the most significant schizophrenia susceptibility genes, the basic helix-loop-helix (bHLH) transcription factor TCF4, has been evaluated in transgenic mice that moderately overexpress TCF4 in the postnatal brain. TCF4 is a member of the E-protein subfamily that is involved in neurodevelopment. TCF4 transgenic mice display profound deficits in contextual and cue-dependent fear conditioning and sensorimotor gating. Molecular analyses revealed the dynamic circadian deregulation of neuronal bHLH factors in the adult hippocampus. This study describes the first animal model relating to TCF4, and it suggests that this transcriptional factor contributes to cognitive processing (Brzozka et al., 2010).

Candidate gene models

Although candidate genes may have become less appealing for genetic studies of schizophrenia, some of them can still provide important insights into the underlying biology of the disorder because mutations in some candidate genes have strong effects on neurodevelopment and are highly penetrant in the families that carry those mutations (Farrell et al., 2015). Here, we briefly overview the most recent animal models based on mutations in candidate genes.

Dystrobrevin-binding protein 1, or **Dysbindin**, was discovered as a binding partner of α-dystrobrevin (Benson et al., 2001). Dysbindin is a constituent of the dystrophin-associated protein complex (DPC) and is a part of the biogenesis of lysosome-related organelles complex 1 (BLOCK1) (Nazarian et al., 2006). In humans, dysbindin is encoded by the *DTNBP1* gene (Benson et al., 2001). The Dysbindin locus was first identified in an Irish study (Straub et al., 2002), and research has supported relevance of this gene to schizophrenia (Ross et al., 2006; Tochigi et al., 2006), including haplotypes of the gene linked to the negative symptoms (DeRosse et al., 2006; Fallgatter et al., 2006; Wirgenes et al., 2009). However, negative studies also exist (Joo et al., 2006; Morris et al., 2003; Strohmaier et al., 2010). Studies reported decreased levels of transcription and protein in the hippocampus and PFC (Talbot et al., 2004; Tang et al., 2009; Weickert et al., 2008).

The first mouse model of dysbindin is based on spontaneous deletion in the *Dtnbp1* in the DBA/2 J strain, leading to loss of dysbindin in sandy (sdy) mice (Li et al., 2003). Previous studies reported multiple neuronal, synaptic, and neurotransmitter alterations in the brain of these mice (Kobayashi et al., 2011; Lutkenhoff et al., 2012; Nihonmatsu-Kikuchi et al., 2011). Several behavioral studies have reported deficits in social interaction, increased anxiety-like behavior, and impaired long-term memory, novel object recognition, and working memory (Bhardwaj et al., 2009; Feng et al., 2008; Hattori et al., 2008; Karlsgodt et al., 2011). Due to additional mutations on the DBA/2 J background, it has been suggested that the innate biochemical characteristics in the dopaminergic system and behavioral alterations such as coordination deficits in these mice may make it difficult to interpret the deficits caused by the lack of dysbindin (D'Este, Casini et al., 2007; Papaleo et al., 2012). To eliminate the background effects of DBA/2 J mice, after backcrossing this mutation to the C57BL/6J genetic background, the mutants exhibited hyperactivity and abnormal spatial and working memory, suggesting that the effects observed were due to the dysbindin mutation, not the background (Cox et al., 2009; Papaleo et al., 2012). Dysbindin overexpressing mice were also generated that displayed increased sensitivity to the locomotor effects of acute methamphetamine administration (Shintani et al., 2014).

Neuregulin 1 (NRG1) and its receptors, ErbB2 and ErbB4, have also been associated with schizophrenia in genetic association studies (Benzel et al., 2007; Stefansson et al., 2002, 2003). Neuregulin-1 is encoded by the NRG1 gene located on chromosome 8p13 (Law et al., 2007). NRG1 is a pleiotropic growth factor that has been shown to have diverse functions in the brain (Harrison and Law, 2006). NRG1 acts by activating type I transmembrane ErbB receptor tyrosine kinases (Britsch, 2007; Mei and Xiong, 2008).

Because deletions of Nrg1 and ErbB4 are lethal (Meyer and Birchmeier, 1995), either HET mice or Nrg1 hypomorphic mice were studied. Mice with targeted deletion of the EGF-like domain (NRG1-EGF$^{+/-}$) were reported by Meyer and Birchmeier. Mutant mice exhibited elevated activity, improved rotorod performance, and spontaneous alternation (Gerlai et al., 2000). Stefansson et al. showed that the behavioral phenotypes of the NRG1 hypomorphs could be ameliorated with clozapine (Stefansson et al., 2002), whereas amphetamine impaired PPI (Duffy et al., 2008). Ehrlichman et al. found that Nrg1$^{+/-}$ mice exhibited deficits in contextual fear conditioning and social interaction (Ehrlichman et al., 2009).

NRG-1 KO mice that lack the trans-membrane domain show no gross abnormalities, cognitive abnormalities, or anxiety (O'Tuathaigh et al., 2007), but they exhibit high spontaneous activity and decreased PPI (Karl et al., 2007). No effects of amphetamine, phencyclidine (PCP), or MK-801 were found, and no changes in D2 receptors were observed in Nrg1$^{+/-}$ mice (van den Buuse et al., 2009). John Waddington and his laboratory reported that neuregulin-1 "knockouts" exhibit increased spontaneous incisor chattering and that SKF 83959 induced incisor chattering, vertical jaw movements, and tongue protrusions, consistent with orofacial dyskinesia (Tomiyama et al., 2012). Nrg1 male mutants also showed attenuated responses to PCP (O'Tuathaigh et al., 2010) and mild cognitive abnormalities (Chesworth et al., 2012).

Several studies have examined the behavioral effects of increased Nrg1 expression. Ubiquitous expression of Nrg1 increased locomotor activity, decreased context-dependent fear conditioning, and impaired social interaction (Kato et al., 2010). When overexpression was limited to postmitotic neurons with the help of the murine Thy 1.2 promoter, mutants had a tremor, impaired performance on the accelerating rotarod, and PPI (Deakin et al., 2009). Older (but not younger) mice had impairment in hippocampus-dependent spatial working memory associated with reduction in frequency of carbachol-induced gamma oscillations (Deakin et al., 2012).

NRG1 acts by stimulating a family of single-transmembrane receptor tyrosine kinases called ErbB proteins (Mei and Xiong, 2008). Although ErbB2 and ErbB3 mutant mice displayed no behavioral alterations (Gerlai et al., 2000), ErbB4 HET mice were hyperactive (Stefansson et al., 2002). Conditional deletion of the receptor in PV-positive GABAergic neurons produced phenotypes reminiscent of schizophrenia, such as hyperactivity, impaired contextual and working memory, and disrupted PPI. Some of these phenotypes could be ameliorated by acute treatment with a low dose of diazepam, a drug that enhances GABA transmission, supporting the finding that impaired GABAergic transmission in PV neurons in this model was associated with the behavioral deficits related to schizophrenia (Shamir et al., 2012; Wen et al., 2010).

Disrupted in schizophrenia 1 (*DISC1*) gene was identified in a balanced chromosomal translocation [t(1;11) (q42;q14)] in a Scottish pedigree with a history of various psychiatric disorders, including recurrent major depression, schizophrenia, bipolar disorder, alcohol use disorder, and anxiety disorders (St. Clair et al., 1990). Until the GWAS era, DISC1 was one of the most promising susceptibility genes. The existence of a clear, identifiable mutation put DISC1 in a unique position in schizophrenia research (Chubb et al., 2008; Millar et al., 2000). The DISC1 protein is considered a scaffold protein, with various motifs that bind to multiple proteins to form protein complexes, including many diverse enzymes (Yerabham et al., 2013). Here, we review the characteristics of current DISC1 mouse models (Fig. 20.2) and refer readers to more comprehensive reviews when necessary (Ayhan et al., 2011b; Cash-Padgett and Jaaro-Peled, 2013).

Disc1: risk allele schizophrenia model

Drs. Gogos and Karayiorgou were the first to introduce the Disc1 mouse model based on the 129 strain background that carries a spontaneous 25-bp deletion in exon 6 to induce a frame shift in the

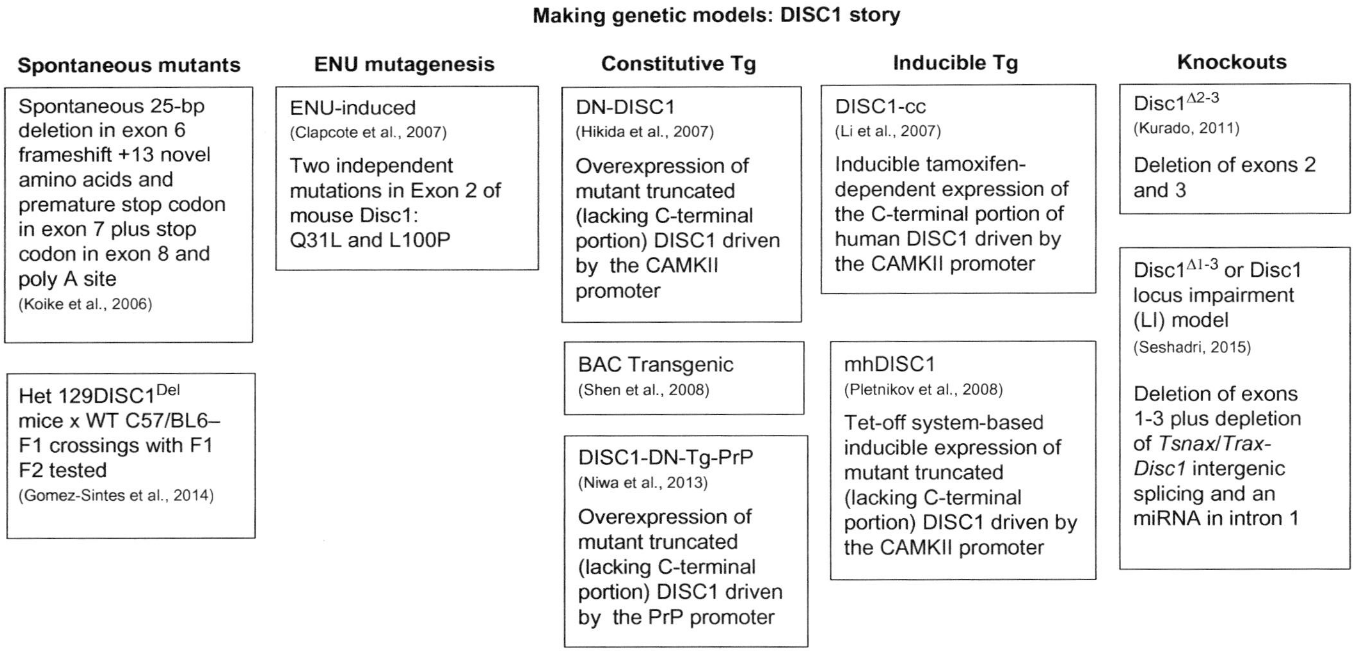

Making genetic models: DISC1 story

Spontaneous mutants

Spontaneous 25-bp deletion in exon 6 frameshift +13 novel amino acids and premature stop codon in exon 7 plus stop codon in exon 8 and poly A site
(Koike et al., 2006)

Het 129DISC1Del mice x WT C57/BL6–F1 crossings with F1 F2 tested
(Gomez-Sintes et al., 2014)

ENU mutagenesis

ENU-induced
(Clapcote et al., 2007)

Two independent mutations in Exon 2 of mouse Disc1: Q31L and L100P

Constitutive Tg

DN-DISC1
(Hikida et al., 2007)

Overexpression of mutant truncated (lacking C-terminal portion) DISC1 driven by the CAMKII promoter

BAC Transgenic
(Shen et al., 2008)

DISC1-DN-Tg-PrP
(Niwa et al., 2013)

Overexpression of mutant truncated (lacking C-terminal portion) DISC1 driven by the PrP promoter

Inducible Tg

DISC1-cc
(Li et al., 2007)

Inducible tamoxifen-dependent expression of the C-terminal portion of human DISC1 driven by the CAMKII promoter

mhDISC1
(Pletnikov et al., 2008)

Tet-off system-based inducible expression of mutant truncated (lacking C-terminal portion) DISC1 driven by the CAMKII promoter

Knockouts

Disc1Δ2-3
(Kurado, 2011)

Deletion of exons 2 and 3

Disc1Δ1-3 or Disc1 locus impairment (LI) model
(Seshadri, 2015)

Deletion of exons 1-3 plus depletion of Tsnax/Trax-Disc1 intergenic splicing and an miRNA in intron 1

FIGURE 20.2 Methodologies used to generate various DISC1 models.

Existing DISC1 mouse models that use various genetic manipulations to perturb DISC1 functions are presented. Importantly, lack of floxed Disc1 mice still presents a major roadblock for studying roles of DISC1 in a cell-type-dependent manner, although the inducible model generated by Pletnikov and colleagues (Pletnikov et al., 2008; Ma et al., 2013) provides an opportunity to selectively affect Disc1 interactome in different cell types.

open reading frame, resulting in 13 novel amino acids, followed by a premature stop codon in exon 7. The mouse strain was further modified by homologous recombination to have a termination codon in exon 8 and a premature polyadenylation site in intron 8 to produce a truncated transcript. Mice were subsequently backcrossed to the C57BL/6J background (Koike et al., 2006). Western blot revealed the elimination of the two major Disc1 isoforms and the preservation of the short N-terminal isoforms (Koike et al., 2006). These mice exhibit impairments in working memory without significant changes in spontaneous activity or PPI. The mutation leads to abnormalities in the organization of newly born and mature neurons of the dentate gyrus and alterations in short-term plasticity. No changes in the number of interneurons were found (Kvajo et al., 2008). Recently, defects in axonal targeting and changes in short-term plasticity at the mossy fiber/CA3 circuit were also reported, along with elevated levels of cAMP (Kvajo et al., 2011). The original 129DISC1 (Del) mutation in a mixed B6/129 strain resulted in a decreased level of full-length DISC1 and sex-specific behavioral alterations. Specifically, hyperactivity in male mice and decreased novelty-induced hyperactivity in females increased immobility in both PFST and tail suspension tests (TST), and decreased cued fear conditioning in males and PPI deficits were noted (Gomez-Sintes et al., 2014). This model demonstrates that a natural polymorphism in the Disc1 gene may be responsible for the neural and behavioral abnormalities consistent with cognitive impairment and neuropathology observed in patients.

Disc1 ENU-induced models

Dr. Roder and his colleagues characterized Disc1 mouse models based on several mutations in the mouse Disc1 gene as a result of N-nitroso-N-ethylurea (ENU) mutagenesis (Clapcote et al., 2007). Two novel missense mutations have led to Q31L and L100P amino acid exchanges in exon 2, respectively. These two mouse models are characterized by distinct phenotypes. Increased immobility in the forced swim test (FST), decreased sociability and social novelty, and decreased sucrose consumption were seen in 31L mice. These 31L mice also had reduced levels of serotonin, norepinephrine, and dopamine levels in the nucleus accumbens (Su et al., 2014). In contrast, increased horizontal activity, decreased PPI, decreased LI, and decreased performance in T-maze were found in 100P mutant mice. Another study assessed DA neurotransmission in Disc1-L100P mice and found that the mutation facilitated the effects of amphetamine and was associated with a 2.1-fold increase in striatal D receptors without altering DA release in the striatum in response to amphetamine (Lipina et al., 2010). A follow-up study analyzed the brain histology of two Disc1 mutant mice and found a reduced neuronal number, decreased neurogenesis, altered neuronal distribution, shorter dendrites, and decreased surface area and spine density in cortical neurons of Disc1 L100P mutant mice (Lee et al., 2011). These ultrastructural alterations could be, at least in part, related to abnormal interaction between Disc1 and Gsk3α in frontal cortical neurons. Pharmacological and genetic inactivation of GSK-3 reversed PPI and normalized hyperactivity of Disc1-L100P mutants and also reduced interaction between DISC1 and GSK-3α and β (Lipina et al., 2011). Further support of an interaction between DISC1, PDE4, and GSK-3 came from a study showing that combined treatment of Disc1-L100P mice with rolipram (0.1 mg/kg, PDE4 inhibitor) and TDZD-8 (2.5 mg/kg, GSK-3 blocker) corrected the PPI deficit and hyperactivity in mutants (Lipina et al., 2012). In the Disc1-Q31L mutant line, TDZD-8 (2.5 mg/kg) was able to correct the PPI deficit, reduce immobility in FST, and increase social motivation and novelty performance. Taken together, these findings suggest that genetic variations in DISC1 can influence the formation of the biochemical complex with PDE4 and GSK-3 and strengthen the possibility of synergistic interactions between these proteins (Lipina et al., 2012). In accordance with this interaction, DISC1 was shown to form a complex with the D2 receptor; in Disc1-L100P mice, this complex is increased similar to the postmortem samples of schizophrenia patients (Su et al., 2014).

These models display how two different mutations in DISC1 result in distinct phenotype profiles associated with two major mental disorders: schizophrenia and depression. In addition, a specific biochemical interaction was targeted as a novel approach for the treatment of mental disorders. These findings, however, were recently challenged (Arime et al., 2014; Shoji et al., 2012).

DISC1 exon 2-3 deletion model

Kuroda and colleagues (2011) developed a mouse model in which exon 2 and exon 3 of DISC1 gene were deleted. The deletion resulted in the loss of DISC1 isoforms containing exons 2 and 3, including a 100-kb full-length isoform; however, other isoforms may still be transcribed. The mutants (called Disc1$^{\Delta 2\text{-}3}$ mice) displayed gender-specific alterations in some behavioral tests. Specifically, in female mutants, PPI was impaired and methamphetamine administration increased locomotor activity. In both sexes, anxiety was reduced, as evidenced by decreased time spent in the closed arms and increased time spent in open arms of the elevated plus maze. Histopathologic analyses revealed shorter or fewer TH-positive neuronal fibers, as well as reduced hippocampal parvalbumin (PV) levels (Iritani et al., 2015; Nakai et al., 2014).

DISC1 transgenic mice

The splicing of DISC1 results in numerous isoforms (Nakata et al., 2009). The chromosomal translocation affects the C-terminal of the three major isoforms (Millar et al., 2001). Possible outcomes of the translocation in the DISC1 gene include haploinsufficiency or the production of a mutant truncated DISC1 protein. The truncated human DISC1 may lose its normal localization and association with interacting proteins and affect the organization of protein interacting complexes via dominant-negative mechanisms. As a result, expression of mutant DISC1 leads to altered levels and distribution of both exogenous full-length DISC1 and endogenous Disc1 in neuronal PC12 rat cells or mouse brain (Kamiya et al., 2005; Pletnikov et al., 2008; Pletnikov et al., 2007). Notably, both dominant-negative and haploinsufficiency mechanisms could perturb similar DISC1-interacting protein complexes, leading to loss of function of DISC1 (Chubb et al., 2008).

Constitutive overexpression of mutant DISC1

In the model of constitutive expression of mutant DISC1, overexpression of mutant truncated DISC1, a putative product of the chromosomal translocation, was driven by the Ca^{2+}/calmodulin-dependent protein kinase II (CAMKII) promoter (Hikida et al., 2007). Mutant DISC1 was expressed in the olfactory bulb, frontal cortex, hippocampus, and basal ganglia starting from the neonatal period. The mutant DISC1 mice (also labeled by the authors as dominant-negative DISC1 [DN-DISC1]) displayed increased activity, deficits in PPI, increased duration of immobility in the Porsolt forced swim test (PFST), and a delay in finding the hidden food. The mice were less sociable and interacted less with a novel animate subject. In addition, decreased reversal learning and mental flexibility in pure cognitive and reward-related paradigms were observed. Increased lateral ventricle size at 6 weeks and decreased PV reactivity were found; however, no difference in the levels of calbindin and calretinin was observed (Hikida et al., 2007). The D2 receptor density was increased, baseline dopamine levels were decreased, and dopamine transporter was increased in the striatum of these mice (Jaaro-Peled et al., 2013). Additionally, oxidative stress parameters were increased in the prefrontal cortices of these animals (Johnson et al., 2013).

Another DISC1 model expresses DN-DISC1 under the PRP promoter. Although no gross behavioral abnormalities were observed, DN-DISC1 mice under the PRP promoter displayed increased spontaneous and methamphetamine-induced locomotor activity, deficits in PPI, and increased immobility in FST after adolescent isolation stress at a level that is not expected to produce prominent changes in normal mice. Tissue dopamine content, extracellular dopamine levels, and tyrosine hydroxylase (TH) expression were low in frontal cortex, which the authors linked to the elevated stress-induced glucocorticoid levels, which in turn altered the methylation pattern of TH projecting to the frontal cortex. This effect, along with the dopamine-associated biochemical and behavioral changes, was long-lasting and could be reversed with a glucocorticoid antagonist (Niwa et al., 2013). The findings demonstrate how stress-related hormonal changes and genetic background interact to generate behavioral abnormalities similar to schizophrenia.

After the generation of the DN-DISC1 model, a bacterial artificial chromosome (BAC) transgenic model was presented in which expression of full-length mouse DISC1 was regulated by the endogenous promoter (Shen et al., 2008). Expression of mutant DISC1 started during embryonic life in the hippocampus and cortex and in cerebellum during adulthood. In adult male mice, morphological alterations included lateral ventricle enlargement, cortical thinning and volumetric reduction, and partial agenesis of the corpus callosum. PV-positive cells were reduced in the hippocampus and medial PFC of these mice. In behavioral tests, mutant mice exhibited impaired latent inhibition (LI), increased immobility in the FST and TST, and significantly fewer stress calls during a maternal separation test. As with the other mutant DISC1 models, dendrite length and complexity were reduced in primary cultures of this model (Shen et al., 2008).

Inducible C-fragment DISC1 model

To assess timing of the effects of mutant human DISC1, Dr. Silva and Dr. Cannon's groups have generated a DISC1 model in which the putative "translocated" C-terminus fragment of DISC1 is expressed under regulation of the CAMKII promoter in forebrain neurons in an inducible manner (Li et al., 2007). Expression of the fragment is induced by injection of tamoxifen that binds to the fragment-associated estrogen receptor ligand-binding domain and releases the bound fragment, leading to its expression. The group reported that early postnatal, but not adult, expression of the fragment impaired performance in the delayed nonmatching to position task (a widely used automated test of spatial memory), attenuated social interaction, and was associated with a shorter latency to immobilization in FST. The behavioral changes were accompanied by reduced dendritic complexity and basal synaptic transmission in the hippocampus. In addition, decreased dendritic spine density and a decreased number of mature spines were observed in cortical neurons during adulthood. These structural changes were accompanied with impaired long-term potentiation and long-term depression, displaying altered plasticity during adulthood dependent on early disruption in DISC1 function (Greenhill et al., 2015). This model was one of the first to demonstrate that early postnatal perturbation in endogenous Disc1 may contribute to neurobehavioral abnormalities, supporting the neurodevelopmental hypothesis of schizophrenia.

Inducible mutant human DISC1 model

An additional mouse model of inducible expression of mutant human DISC1 in forebrain neurons has been generated using the Tet-off system (Pletnikov et al., 2008). In this model, expression of mutant DISC1 is also regulated by the CAMKII promoter and can be turned off by adding tetracycline or a related compound, doxycycline, to food or water. Similar to other DISC1 mouse models, expression of mutant DISC1 produced no gross developmental defects but significantly increased spontaneous locomotor activity in male, but not female, mice, decreased social interaction in male mice, enhanced their aggressive behavior, and was associated with poorer spatial memory in the Morris water maze task in female mice (Pletnikov et al., 2008). The behavioral alterations were accompanied by lateral ventricle enlargement in adult mice, reduced dendritic arborization in primary cortical neurons, and decreased expression of a synaptic protein, SNAP-25. To evaluate the role of DISC1 across neurodevelopment, Ayhan and colleagues (2011a) selectively expressed mutant DISC1 during only the prenatal period, only the postnatal period, or during both periods. All periods of expression similarly led to decreased levels of cortical dopamine (DA) and fewer PV-positive neurons in the cortex. Combined prenatal and postnatal expression produced increased aggression and an enhanced response to psychostimulants in male mice, along with increased linear density of dendritic spines on neurons of the dentate gyrus of the hippocampus and lower levels of endogenous DISC1 and LIS1. Prenatal expression only resulted in smaller brain volume, whereas selective postnatal expression gave rise to decreased social behavior in male mice and depression-like responses in female mice, as well as enlarged lateral ventricles and decreased DA content in the hippocampus of female mice and decreased levels of endogenous DISC1 (Ayhan et al., 2011a). The diverse changes in mutant DISC1 mice are reminiscent of findings in major mental diseases. The transcription profile of this model revealed increased expression of markers of oligodendrocyte (OLG) precursors and mature myelinating OLGs, indicating accelerated differentiation in both neonatal and adult animals (Katsel et al., 2011). Fig. 20.2 summarizes the main approaches to DISC1 mouse models relevant to schizophrenia.

CHALLENGES IN MODELING SCHIZOPHRENIA

The definition of schizophrenia has been variable since the introduction of the term (Tandon et al., 2013). After more than 100 years, the diagnosis is still based on the clinical signs and symptoms, which are not specific to the disease. Even with the new edition of *Diagnostic and Statistical Manual of Mental Disorders* (DSM5), there have only been minor changes in the criteria, except for removing and reassigning the subtypes of the disorder (ie, paranoid, catatonic, hebephrenic) (Tandon et al., 2013). These issues pose a significant challenge for animal models of the disease. Although existing models for schizophrenia have advanced our understanding of the neurobiology of this disease, there are still many caveats that need to be addressed in future research. Different methodological and technological caveats and challenges of the current animal models of schizophrenia are reviewed (Kannan et al., 2013; Ayhan et al., 2015). In this chapter, we discuss a new conceptual framework that is coming to the field to transform animal models of complex psychiatric disorders.

As argued elsewhere (Ayhan et al., 2015), the main goal of a model is to address specific hypotheses. We should try to avoid both overly enthusiastic promotion of an animal model as the best one and needless cynicism regarding the utility of animal models. We strongly believe that the field needs more models to tackle heterogeneity and complexity of the mechanisms of psychotic disorders, including schizophrenia. In fact, given that schizophrenia, schizoaffective disorders, and bipolar disorders with psychoses may have similar underlying genetic bases and pathobiology (Insel, 2010; Hall et al., 2015), the field must move beyond the constrains of the diagnostic categories. It seems that we can achieve more progress by focusing on a dimensional construct and domains of psychopathology, which do not necessarily follow the nosological requirements (O'Tuathaigh and Waddington, 2015).

Currently, the clinical symptoms and signs are the basis for diagnosis of psychiatric disorders, and different constellations of signs and symptoms constitute the criteria set that are the diagnostic parameters for psychiatric disorder categories. Although current diagnostic systems provide good reliability for some diagnoses, validity of the categories and their existence as separate disease entities remain questionable (Hyman, 2010; Kendell and Jablensky, 2003). Thus, "in order to explore mechanisms independent of categories" (Morris and Cuthbert, 2012), the US National Institute of Mental Health has recently implemented Research Domain Criteria Project (RDoC). RDoC's goal is to "develop, for research purposes, new ways of classifying mental disorders based on dimensions of observable behavior and neurobiological measures" (Morris and Cuthbert, 2012). RDoC proposes different domains that may be affected in various disorders through different units of analysis, including genetic, molecular, cellular, circuit level, physiological, and behavioral units. These domains include negative and positive valence systems, cognitive systems, systems for social processes and arousal, and regulatory systems. Importantly, RDoC is not meant to solve the problems associated with schizophrenia diagnosis per se, and not all constructs proposed are affected in schizophrenia. Instead, this approach considers mental disorders as different constellations of domains and could advance our knowledge of the neurobiology of basic processes that are based on phenotypes translatable across different species and perhaps more sensitive to biology-driven treatment (Petrinovic and Künnecke, 2014; Hoffman, 2013; Argyropoulos et al., 2013; Mahoney and Olmstead, 2013; Cuthbert, 2014). Analyzing neurobehavioral abnormalities within the dimensional or RDoC framework is expected to provide a stronger mechanistic link between the model and the specific domains of the psychiatric disease (Cazorla et al., 2014).

Notably, before the RDoC concept, the terms "endophenotype" and "intermediate phenotype" were introduced (Gottesman and Gould, 2003; Kellendonk et al., 2009; Desbonnet et al., 2009; Kaffman and Krystal, 2012; Amann et al., 2010) to stimulate examination of the neurobiological entities readily translatable to humans (Powell, 2006; Young et al., 2015). Novel animal models should expand the use of physiological and neural circuitry intermediate phenotypes, genome-wide gene expression, and epigenetic modification profiling in specific cell types (Kannan et al., 2013). Brain imaging and electrophysiology endophenotypes are also likely to provide additional information, which may be more translatable to human diseases. In addition, we should strive to expand the use of model organisms (Insel, 2009). The value of using multiple model animals is particularly evident when comparing the advantages and disadvantages of mouse and rat models. Genetic manipulations in rats and the availability and reproducibility of social and cognitive behaviors are bringing back rats as a model organism for psychiatric disorders (Ellenbroek et al., 2002; Ratajczak et al., 2013).

CONCLUSIONS

Genetic mouse models will continue to be important tools for psychiatric research. They have been illuminating the functions of candidate genes in neurodevelopment and are stimulating the search for future pathogenic treatments. Table 20.1 provides a brief summary of the major genetic mouse models for schizophrenia. More sophisticated and advanced genetic manipulations in combination with a better understanding of the human clinical phenotypes will be needed to refine the present models to make them more relevant to human psychiatric diseases.

Table 20.1 A Summary of Genetic Models of Schizophrenia

Gene	Approach in Mouse Model	Behavioral Phenotypes	Reference
Physiological Models			
Striatal D2 receptors	Over-expression	Impaired working memory and behavioral flexibility, impaired motivational processes and timing	Kellendonk, 2006 Drew, 2007 Huang, 2004
COMT	COMT-deficient (KO)	Females: Elevated anxiety, enhanced stress response	Gogos, 1998 Walsh, 2010
		Males: Increased aggression	Desbonnet, 2012
		Both: Increased pain sensitivity	Babovic, 2008
	Heterozygous COMT	Decreased rearing, increased sifting and chewing	Babovic, 2007
	Over-expression of human COMT-VAL	Diminished attention set-shifting, impaired recognition memory, lower stress reactivity and reduced pain sensitivity	Papaleo, 2008
NR1	NR1 hypomorphic	Increased activity and stereotypy, decreased habituation, PPI deficits, and reduced social interaction	Mohn, 1999 Duncan, 2006a, 2006b
DAO	ddy/DAO- mice (absence of DAO activity)	Decreased novelty induced activity, increased spatial learning, attenuation of PCP induced effects	Almond, 2006
PICK1	PICK1 KO	Decreased D-Serine during early postnatal period, interaction with serine racemase	G. Tsai, 1998 G.E. Tsai, 1997
Genes based on GWAS findings			
MHCI loci H2-K^b and H2-D^b	H2-K^b and H2-D^b($K^bD^{b/-}$) deficient	Lower LTD induction threshold, improved motor memory	McConnell, 2009
β-2microglobulin and TAP	β-2microglobulin and TAP deficient (β2$M^{-/-}$TAP$^{-/-}$)	Increased stress reactivity	Sankar, 2012
TCF4	Over-expression	Contextual and cue-dependent fear conditioning deficits	Brzozka, 2010
Candidate Genes			
Dysbindin	Spontaneous deletion in *Dtnbp1* leading to loss of dysbindin	Increased anxiety, impaired social interaction, impaired long term memory, novel object recognition, and working memory	Bhardwaj, 2009 Feng, 2008 Hattori, 2008 Karlsgodt, 2011
	Over-expression	Increased sensitivity to psychostimulant induced locomotor activity	Shintani, 2014
Neuregulin 1	heterozygous NRG1	Elevated activity, improved rotorod and spontaneous alternation performance impaired contextual fear conditioning and social interaction, orofacial dyskinesia-like behavior, and cognitive abnormalities	Gerlai, 2000 Ehrlichman, 2009 Tomiyama, 2012 Chesworth, 2012
	Ubiquitous over-expression	Increased locomotor activity, decreased context-dependent fear conditioning, and impaired social interaction	Kato, 2010

(Continued)

Table 20.1 A Summary of Genetic Models of Schizophrenia (Continued)

Gene	Approach in Mouse Model	Behavioral Phenotypes	Reference
ErbB4	Over-expression in post-mitotic neurons	Tremor, impaired rotarod performance, and impaired PPI	Deakin, 2009
	Heterozygous ErB4	Hyperactivity	Stefansson, 2002
	Conditional deletion in PV-positive GABAergic neurons	Hyperactivity, impaired contextual and working memory, and disrupted PPI	Shamir, 2012 Wen, 2010
DISC1	Spontaneous mutant	**Spontaneous 25-bp deletion in exon 6** Impaired working memory **129DISC1 (Del) mutation** Males: hyperactivity Females: hypoactivity Both: deficits in PPI and increased immobility in PFST and TST	Koike, 2006 Kvajo, 2008 Gomez-Sintes, 2014
	ENU mutagenesis	**31L** Increased immobility in FST, decreased social interaction, and decreased sucrose consumption **100P** Increased activity, decreased PPI, decreased LI, and decreased T-maze performance	Su, 2014 Lipina, 2010
	Constitutive Tg	**DN-DISC1** Increased activity, deficits in PPI, increased immobility in FST, decreased social interaction **BAC Transgenic** Impaired latent inhibition, increased immobility in the FST and TST, fewer stress calls during maternal separation **DISC1-DN-Tg-PrP** Increased spontaneous and methamphetamine-induced locomotion, PPI deficits, increased immobility in FST following adolescent isolation stress	Hikida, 2007 Shen, 2008 Niwa, 2013
	Inducible Tg	**DISC1-cc** Impaired spatial memory, attenuated social interaction, reduced latency to immobility in FST **mhDISC1** Males: Increased locomotor activity, decreased social interaction, increased aggression Females: Impaired spatial memory	Li, 2007 Pletnikov, 2008
	Knockout	**Disc1$^{\Delta 2-3}$** Females: Impaired PPI, increased methamphetamine induced locomotor activity Both: Reduced anxiety	Kuroda, 2011

ACKNOWLEDGMENTS

We thank the following funding agencies for the support: MH-083728, MH-094268 Silvo O. Conte center, and the Brain and Behavior Research Foundation (M.V.P.).

REFERENCES

Abakay, A., Gokalp, O., Abakay, O., Evliyaoglu, O., Sezgi, C., Palanci, Y., et al., 2012. Relationships between respiratory function disorders and serum copper levels in copper mineworkers. Biol. Trace. Elem. Res. 145 (2), 151–157. http://dx.doi.org/10.1007/s12011-011-9184-9.

Abi-Dargham, A., 2014. Schizophrenia: overview and dopamine dysfunction. J. Clin. Psychiatry 75 (11), e31. http://dx.doi.org/10.4088/JCP.13078tx2c.

Almond, S.L., Fradley, R.L., Armstrong, E.J., Heavens, R.B., Rutter, A.R., Newman, R.J., et al., 2006. Behavioral and biochemical characterization of a mutant mouse strain lacking D-amino acid oxidase activity and its implications for schizophrenia. Mol. Cell Neurosci. 32 (4), 324–334. http://dx.doi.org/10.1016/j.mcn.2006.05.003.

Amann, L.C., Gandal, M.J., Halene, T.B., Ehrlichman, R.S., White, S.L., McCarren, H.S., et al., 2010. Mouse behavioral endophenotypes for schizophrenia. Brain Res. Bull. 83 (3-4), 147–161. http://dx.doi.org/10.1016/j.brainresbull.2010.04.008.

Andreassen, O.A., Harbo, H.F., Wang, Y., Thompson, W.K., Schork, A.J., Mattingsdal, M., et al., 2015. Genetic pleiotropy between multiple sclerosis and schizophrenia but not bipolar disorder: differential involvement of immune-related gene loci. Mol. Psychiatry 20 (2), 207–214. http://dx.doi.org/10.1038/mp.2013.195.

Argyropoulos, A., Gilby, K.L., Hill-Yardin, E.L., 2013. Studying autism in rodent models: reconciling endophenotypes with comorbidities. Front. Hum. Neurosci. 7, 417. http://dx.doi.org/10.3389/fnhum.2013.00417.

Arime, Y., Fukumura, R., Miura, I., Mekada, K., Yoshiki, A., Wakana, S., et al., 2014. Effects of background mutations and single nucleotide polymorphisms (SNPs) on the Disc1 L100P behavioral phenotype associated with schizophrenia in mice. Behav. Brain Funct. 10, 45. http://dx.doi.org/10.1186/1744-9081-10-45.

Ayhan, Y., Abazyan, B., Nomura, J., Kim, R., Ladenheim, B., Krasnova, I.N., et al., 2011a. Differential effects of prenatal and postnatal expressions of mutant human DISC1 on neurobehavioral phenotypes in transgenic mice: evidence for neurodevelopmental origin of major psychiatric disorders. Mol. Psychiatry 16 (3), 293–306. http://dx.doi.org/10.1038/mp.2009.144.

Ayhan, Y., Jaaro-Peled, H., Sawa, A., Pletnikov, M., 2011b. DISC1 mouse models. In: O'Donell., P. (Ed.), Animal Models of Schizophrenia and Related Disorders. Humana Press, pp. 211–229.

Ayhan, Y., McFarland, R., Pletnikov, M.V., 2015. Animal models of gene-environment interaction in schizophrenia: a dimensional perspective. Prog. Neurobiol. http://dx.doi.org/10.1016/j.pneurobio.2015.10.002.

Babovic, D., O'Tuathaigh, C.M., O'Sullivan, G.J., Clifford, J.J., Tighe, O., Croke, D.T., et al., 2007. Exploratory and habituation phenotype of heterozygous and homozygous COMT knockout mice. Behav. Brain Res. 183 (2), 236–239. http://dx.doi.org/10.1016/j.bbr.2007.07.006.

Babovic, D., O'Tuathaigh, C.M., O'Connor, A.M., O'Sullivan, G.J., Tighe, O., Croke, D.T., et al., 2008. Phenotypic characterization of cognition and social behavior in mice with heterozygous versus homozygous deletion of catechol-O-methyltransferase. Neuroscience 155 (4), 1021–1029. http://dx.doi.org/10.1016/j.neuroscience.2008.07.006.

Barch, D.M., Carter, C.S., Arnsten, A., Buchanan, R.W., Cohen, J.D., Geyer, M., et al., 2009. Selecting paradigms from cognitive neuroscience for translation into use in clinical trials: proceedings of the third CNTRICS meeting. Schizophr. Bull. 35 (1), 109–114. http://dx.doi.org/10.1093/schbul/sbn163.

Bassett, A.S., Chow, E.W., 2008. Schizophrenia and 22q11.2 deletion syndrome. Curr. Psychiatry Rep. 10 (2), 148–157. Retrieved from: http://www.ncbi.nlm.nih.gov/pubmed/18474208.

Benson, M.A., Newey, S.E., Martin-Rendon, E., Hawkes, R., Blake, D.J., 2001. Dysbindin, a novel coiled-coil-containing protein that interacts with the dystrobrevins in muscle and brain. J Biol Chem 276 (26), 24232–24241. http://dx.doi.org/10.1074/jbc.M010418200.

Benzel, I., Bansal, A., Browning, B.L., Galwey, N.W., Maycox, P.R., McGinnis, R., et al., 2007. Interactions among genes in the ErbB-Neuregulin signalling network are associated with increased susceptibility to schizophrenia. Behav. Brain Funct. 3, 31. http://dx.doi.org/10.1186/1744-9081-3-31.

Bhardwaj, S.K., Baharnoori, M., Sharif-Askari, B., Kamath, A., Williams, S., Srivastava, L.K., 2009. Behavioral characterization of dysbindin-1 deficient sandy mice. Behav. Brain Res. 197 (2), 435–441. http://dx.doi.org/10.1016/j.bbr.2008.10.011.

Britsch, S., 2007. The neuregulin-I/ErbB signaling system in development and disease. Adv. Anat. Embryol. Cell Biol. 190, 1–65. Retrieved from: http://www.ncbi.nlm.nih.gov/pubmed/17432114.

Brzozka, M.M., Radyushkin, K., Wichert, S.P., Ehrenreich, H., Rossner, M.J., 2010. Cognitive and sensorimotor gating impairments in transgenic mice overexpressing the schizophrenia susceptibility gene Tcf4 in the brain. Biol. Psychiatry 68 (1), 33–40. http://dx.doi.org/10.1016/j.biopsych.2010.03.015.

Burnet, P.W., Eastwood, S.L., Bristow, G.C., Godlewska, B.R., Sikka, P., Walker, M., et al., 2008. D-amino acid oxidase activity and expression are increased in schizophrenia. Mol. Psychiatry 13 (7), 658–660. http://dx.doi.org/10.1038/mp.2008.47.

Cash-Padgett, T., Jaaro-Peled, H., 2013. DISC1 mouse models as a tool to decipher gene-environment interactions in psychiatric disorders. Front. Behav. Neurosci. 7, 113. http://dx.doi.org/10.3389/fnbeh.2013.00113.

Cazorla, M., Shegda, M., Ramesh, B., Harrison, N.L., Kellendonk, C., 2012. Striatal D2 receptors regulate dendritic morphology of medium spiny neurons via Kir2 channels. J. Neurosci. 32 (7), 2398–2409. http://dx.doi.org/10.1523/JNEUROSCI.6056-11.2012.

Cazorla, M., de Carvalho, F.D., Chohan, M.O., Shegda, M., Chuhma, N., Rayport, S., et al., 2014. Dopamine D2 receptors regulate the anatomical and functional balance of basal ganglia circuitry. Neuron 81 (1), 153–164. http://dx.doi.org/10.1016/j.neuron.2013.10.041.

Chesworth, R., Downey, L., Logge, W., Killcross, S., Karl, T., 2012. Cognition in female transmembrane domain neuregulin 1 mutant mice. Behav. Brain Res. 226 (1), 218–223. http://dx.doi.org/10.1016/j.bbr.2011.09.019.

Chubb, J.E., Bradshaw, N.J., Soares, D.C., Porteous, D.J., Millar, J.K., 2008. The DISC locus in psychiatric illness. Mol. Psychiatry 13 (1), 36–64. http://dx.doi.org/10.1038/sj.mp.4002106.

Clapcote, S.J., Lipina, T.V., Millar, J.K., Mackie, S., Christie, S., Ogawa, F., et al., 2007. Behavioral phenotypes of Disc1 missense mutations in mice. Neuron 54 (3), 387–402. http://dx.doi.org/10.1016/j.neuron.2007.04.015.

Corriveau, R.A., Huh, G.S., Shatz, C.J., 1998. Regulation of class I MHC gene expression in the developing and mature CNS by neural activity. Neuron 21 (3), 505–520. Retrieved from: http://www.ncbi.nlm.nih.gov/pubmed/9768838.

Cox, M.M., Tucker, A.M., Tang, J., Talbot, K., Richer, D.C., Yeh, L., et al., 2009. Neurobehavioral abnormalities in the dysbindin-1 mutant, sandy, on a C57BL/6J genetic background. Genes. Brain Behav. 8 (4), 390–397. http://dx.doi.org/10.1111/j.1601-183X.2009.00477.x.

Coyle, J.T., 2006. Glutamate and schizophrenia: beyond the dopamine hypothesis. Cell Mol. Neurobiol. 26 (4–6), 365–384. http://dx.doi.org/10.1007/s10571-006-9062-8.

Cuthbert, B.N., 2014. The RDoC framework: facilitating transition from ICD/DSM to dimensional approaches that integrate neuroscience and psychopathology. World Psychiatry 13 (1), 28–35. http://dx.doi.org/10.1002/wps.20087.

D'Este, L., Casini, A., Puglisi-Allegra, S., Cabib, S., Renda, T.G., 2007. Comparative immunohistochemical study of the dopaminergic systems in two inbred mouse strains (C57BL/6J and DBA/2J). J. Chem. Neuroanat. 33 (2), 67–74. http://dx.doi.org/10.1016/j.jchemneu.2006.12.005.

Dawe, G.S., Hwang, E.H., Tan, C.H., 2009. Pathophysiology and animal models of schizophrenia. Ann. Acad. Med. Singapore. 38 (5), 425–426. Retrieved from: http://www.ncbi.nlm.nih.gov/pubmed/19521643.

Deakin, I.H., Law, A.J., Oliver, P.L., Schwab, M.H., Nave, K.A., Harrison, P.J., et al., 2009. Behavioural characterization of neuregulin 1 type I overexpressing transgenic mice. Neuroreport 20 (17), 1523–1528. http://dx.doi.org/10.1097/WNR.0b013e328330f6e7.

Deakin, I.H., Nissen, W., Law, A.J., Lane, T., Kanso, R., Schwab, M.H., et al., 2012. Transgenic overexpression of the type I isoform of neuregulin 1 affects working memory and hippocampal oscillations but not long-term potentiation. Cereb. Cortex. 22 (7), 1520–1529. http://dx.doi.org/10.1093/cercor/bhr223.

DeRosse, P., Funke, B., Burdick, K.E., Lencz, T., Ekholm, J.M., Kane, J.M., et al., 2006. Dysbindin genotype and negative symptoms in schizophrenia. Am. J. Psychiatry 163 (3), 532–534. http://dx.doi.org/10.1176/appi.ajp.163.3.532.

Desbonnet, L., Waddington, J.L., Tuathaigh, C.M., 2009. Mice mutant for genes associated with schizophrenia: common phenotype or distinct endophenotypes? Behav. Brain Res. 204 (2), 258–273. Retrieved from: http://www.ncbi.nlm.nih.gov/pubmed/19728400.

Desbonnet, L., Tighe, O., Karayiorgou, M., Gogos, J.A., Waddington, J.L., O'Tuathaigh, C.M., 2012. Physiological and behavioural responsivity to stress and anxiogenic stimuli in COMT-deficient mice. Behav. Brain Res. 228 (2), 351–358. http://dx.doi.org/10.1016/j.bbr.2011.12.014.

Drew, M.R., Simpson, E.H., Kellendonk, C., Herzberg, W.G., Lipatova, O., Fairhurst, S., et al., 2007. Transient overexpression of striatal D2 receptors impairs operant motivation and interval timing. J. Neurosci. 27 (29), 7731–7739. http://dx.doi.org/10.1523/JNEUROSCI.1736-07.2007.

Duffy, L., Cappas, E., Scimone, A., Schofield, P.R., Karl, T., 2008. Behavioral profile of a heterozygous mutant mouse model for EGF-like domain neuregulin 1. Behav. Neurosci. 122 (4), 748–759. http://dx.doi.org/10.1037/0735-7044.122.4.748.

Duncan, G.E., Moy, S.S., Lieberman, J.A., Koller, B.H., 2006a. Effects of haloperidol, clozapine, and quetiapine on sensorimotor gating in a genetic model of reduced NMDA receptor function. Psychopharmacology (Berl) 184 (2), 190–200. http://dx.doi.org/10.1007/s00213-005-0214-1.

Duncan, G.E., Moy, S.S., Lieberman, J.A., Koller, B.H., 2006b. Typical and atypical antipsychotic drug effects on locomotor hyperactivity and deficits in sensorimotor gating in a genetic model of NMDA receptor hypofunction. Pharmacol. Biochem. Behav. 85 (3), 481–491. http://dx.doi.org/10.1016/j.pbb.2006.09.017.

Egan, M.F., Goldberg, T.E., Kolachana, B.S., Callicott, J.H., Mazzanti, C.M., Straub, R.E., et al., 2001. Effect of COMT Val108/158 Met genotype on frontal lobe function and risk for schizophrenia. Proc. Natl. Acad. Sci. U.S.A. 98 (12), 6917–6922. http://dx.doi.org/10.1073/pnas.111134598.

Ehrlichman, R.S., Luminais, S.N., White, S.L., Rudnick, N.D., Ma, N., Dow, H.C., et al., 2009. Neuregulin 1 transgenic mice display reduced mismatch negativity, contextual fear conditioning and social interactions. Brain Res. 1294, 116–127. http://dx.doi.org/10.1016/j.brainres.2009.07.065.

Ellenbroek, B., Cools, A.R., 1995. Animal Models of Psychotic Disturbances. In: Den Boer, J.A., Westenberg, H.G.M., van Praag, H.M. (Eds.), Advances in the Neurobiology of Schizophrenia. John Wiley & Sons Ltd, New York, pp. 89–109.

Ellenbroek, B.A., Cools, A.R., 2002. Apomorphine susceptibility and animal models for psychopathology: genes and environment. Behav. Genet. 32 (5), 349–361. Retrieved from: http://www.ncbi.nlm.nih.gov/pubmed/12405516.

Ellenbroek, B.A., Geyer, M.A., Cools, A.R., 1995. The behavior of APO-SUS rats in animal models with construct validity for schizophrenia. J Neurosci 15 (11), 7604–7611. Retrieved from: http://www.ncbi.nlm.nih.gov/pubmed/7472511.

Fallgatter, A.J., Herrmann, M.J., Hohoff, C., Ehlis, A.C., Jarczok, T.A., Freitag, C.M., et al., 2006. DTNBP1 (dysbindin) gene variants modulate prefrontal brain function in healthy individuals. Neuropsychopharmacology 31 (9), 2002–2010. http://dx.doi.org/10.1038/sj.npp.1301003.

Farrell, M.R., Gruene, T.M., Shansky, R.M., 2015. The influence of stress and gonadal hormones on neuronal structure and function. Horm. Behav. 76, 118–124. http://dx.doi.org/10.1016/j.yhbeh.2015.03.003.

Feng, Y.Q., Zhou, Z.Y., He, X., Wang, H., Guo, X.L., Hao, C.J., et al., 2008. Dysbindin deficiency in sandy mice causes reduction of snapin and displays behaviors related to schizophrenia. Schizophr. Res. 106 (2–3), 218–228. http://dx.doi.org/10.1016/j.schres.2008.07.018.

Francois, J., Ferrandon, A., Koning, E., Angst, M.J., Sandner, G., Nehlig, A., 2009. Selective reorganization of GABAergic transmission in neonatal ventral hippocampal-lesioned rats. Int J Neuropsychopharmacol 12 (8), 1097–1110. http://dx.doi.org/10.1017/S1461145709009985.

Fujii, K., Maeda, K., Hikida, T., Mustafa, A.K., Balkissoon, R., Xia, J., et al., 2006. Serine racemase binds to PICK1: potential relevance to schizophrenia. Mol. Psychiatry 11 (2), 150–157. http://dx.doi.org/10.1038/sj.mp.4001776.

Gerlai, R., Pisacane, P., Erickson, S., 2000. Heregulin, but not ErbB2 or ErbB3, heterozygous mutant mice exhibit hyperactivity in multiple behavioral tasks. Behav. Brain Res. 109 (2), 219–227. Retrieved from: http://www.ncbi.nlm.nih.gov/pubmed/10762692.

Geyer, M.A., McIlwain, K.L., Paylor, R., 2002. Mouse genetic models for prepulse inhibition: an early review. Mol. Psychiatry 7 (10), 1039–1053. http://dx.doi.org/10.1038/sj.mp.4001159.

Gogos, J.A., Morgan, M., Luine, V., Santha, M., Ogawa, S., Pfaff, D., et al., 1998. Catechol-O-methyltransferase-deficient mice exhibit sexually dimorphic changes in catecholamine levels and behavior. Proc. Natl. Acad. Sci. U.S.A. 95 (17), 9991–9996. Retrieved from: http://www.ncbi.nlm.nih.gov/pubmed/9707588.

Goldman-Rakic, P.S., Castner, S.A., Svensson, T.H., Siever, L.J., Williams, G.V., 2004. Targeting the dopamine D1 receptor in schizophrenia: insights for cognitive dysfunction. Psychopharmacology (Berl) 174 (1), 3–16. http://dx.doi.org/10.1007/s00213-004-1793-y.

Gomez-Sintes, R., Kvajo, M., Gogos, J.A., Lucas, J.J., 2014. Mice with a naturally occurring DISC1 mutation display a broad spectrum of behaviors associated to psychiatric disorders. Front. Behav. Neurosci. 8, 253. http://dx.doi.org/10.3389/fnbeh.2014.00253.

Gottesman, I.I., Gould, T.D., 2003. The endophenotype concept in psychiatry: etymology and strategic intentions. Am. J. Psychiatry 160 (4), 636–645. http://dx.doi.org/10.1176/appi.ajp.160.4.636.

Gratten, J., Wray, N.R., Keller, M.C., Visscher, P.M., 2014. Large-scale genomics unveils the genetic architecture of psychiatric disorders. Nat. Neurosci. 17 (6), 782–790. http://dx.doi.org/10.1038/nn.3708.

Greenhill, S.D., Juczewski, K., de Haan, A.M., Seaton, G., Fox, K., Hardingham, N.R., 2015. NEURODEVELOPMENT. Adult cortical plasticity depends on an early postnatal critical period. Science 349 (6246), 424–427. http://dx.doi.org/10.1126/science.aaa8481.

Hall, J., Trent, S., Thomas, K.L., O'Donovan, M.C., Owen, M.J., 2015. Genetic risk for schizophrenia: convergence on synaptic pathways involved in plasticity. Biol. Psychiatry 77 (1), 52–58. http://dx.doi.org/10.1016/j.biopsych.2014.07.011.

Harrison, P.J., 2015. Recent genetic findings in schizophrenia and their therapeutic relevance. J. Psychopharmacol. 29 (2), 85–96. http://dx.doi.org/10.1177/0269881114553647.

Harrison, P.J., Law, A.J., 2006. Neuregulin 1 and schizophrenia: genetics, gene expression, and neurobiology. Biol. Psychiatry 60 (2), 132–140. http://dx.doi.org/10.1016/j.biopsych.2005.11.002.

Hattori, S., Murotani, T., Matsuzaki, S., Ishizuka, T., Kumamoto, N., Takeda, M., et al., 2008. Behavioral abnormalities and dopamine reductions in sdy mutant mice with a deletion in Dtnbp1, a susceptibility gene for schizophrenia. Biochem. Biophys. Res. Commun. 373 (2), 298–302. http://dx.doi.org/10.1016/j.bbrc.2008.06.016.

Hikida, T., Jaaro-Peled, H., Seshadri, S., Oishi, K., Hookway, C., Kong, S., et al., 2007. Dominant-negative DISC1 transgenic mice display schizophrenia-associated phenotypes detected by measures translatable to humans. Proc. Natl. Acad. Sci. U.S.A. 104 (36), 14501–14506. http://dx.doi.org/10.1073/pnas.0704774104.

Hoffman, K.L., 2013. Role of murine models in psychiatric illness drug discovery: a dimensional view. Expert Opin. Drug Discov. 8 (7), 865–877. http://dx.doi.org/10.1517/17460441.2013.797959.

Howes, O., McCutcheon, R., Stone, J., 2015. Glutamate and dopamine in schizophrenia: an update for the 21st century. J. Psychopharmacol. 29 (2), 97–115. http://dx.doi.org/10.1177/0269881114563634.

Huang, Y.Y., Simpson, E., Kellendonk, C., Kandel, E.R., 2004. Genetic evidence for the bidirectional modulation of synaptic plasticity in the prefrontal cortex by D1 receptors. Proc. Natl. Acad. Sci. U.S.A. 101 (9), 3236–3241. http://dx.doi.org/10.1073/pnas.0308280101.

Huh, G.S., Boulanger, L.M., Du, H., Riquelme, P.A., Brotz, T.M., Shatz, C.J., 2000. Functional requirement for class I MHC in CNS development and plasticity. Science 290 (5499), 2155–2159. Retrieved from: http://www.ncbi.nlm.nih.gov/pubmed/11118151.

Huotari, M., Garcia-Horsman, J.A., Karayiorgou, M., Gogos, J.A., Mannisto, P.T., 2004. D-amphetamine responses in catechol-O-methyltransferase (COMT) disrupted mice. Psychopharmacology (Berl) 172 (1), 1–10. http://dx.doi.org/10.1007/s00213-003-1627-3.

Hurst, J.L., 2009. Female recognition and assessment of males through scent. Behav. Brain Res. 200 (2), 295–303. http://dx.doi.org/10.1016/j.bbr.2008.12.020.

Hyman, S.E., 2010. The diagnosis of mental disorders: the problem of reification. Annu. Rev. Clin. Psychol. 6, 155–179. http://dx.doi.org/10.1146/annurev.clinpsy.3.022806.091532.

Insel, T.R., 2009. Disruptive insights in psychiatry: transforming a clinical discipline. J. Clin. Invest. 119 (4), 700–705. http://dx.doi.org/10.1172/JCI38832.

Insel, T.R., 2010. Rethinking schizophrenia. Nature 468 (7321), 187–193. http://dx.doi.org/10.1038/nature09552.

Iritani, S., Sekiguchi, H., Habuchi, C., Torii, Y., Kuroda, K., Kaibuchi, K., et al., 2015. Catecholaminergic neuronal network dysfunction in the frontal lobe of a genetic mouse model of schizophrenia. Acta Neuropsychiatr., 1–7. http://dx.doi.org/10.1017/neu.2015.51.

Jaaro-Peled, H., Ayhan, Y., Pletnikov, M.V., Sawa, A., 2010. Review of pathological hallmarks of schizophrenia: comparison of genetic models with patients and nongenetic models. Schizophr. Bull. 36 (2), 301–313. http://dx.doi.org/10.1093/schbul/sbp133.

Jaaro-Peled, H., Niwa, M., Foss, C.A., Murai, R., de Los Reyes, S., Kamiya, A., et al., 2013. Subcortical dopaminergic deficits in a DISC1 mutant model: a study in direct reference to human molecular brain imaging. Hum. Mol. Genet. 22 (8), 1574–1580. http://dx.doi.org/10.1093/hmg/ddt007.

Javitt, D.C., 2015. Neurophysiological models for new treatment development in schizophrenia: early sensory approaches. Ann. N.Y. Acad. Sci. 1344, 92–104. http://dx.doi.org/10.1111/nyas.12689.

Johnson, A.W., Jaaro-Peled, H., Shahani, N., Sedlak, T.W., Zoubovsky, S., Burruss, D., et al., 2013. Cognitive and motivational deficits together with prefrontal oxidative stress in a mouse model for neuropsychiatric illness. Proc. Natl. Acad. Sci. U.S.A. 110 (30), 12462–12467. http://dx.doi.org/10.1073/pnas.1307925110.

Joo, E.J., Lee, K.Y., Jeong, S.H., Ahn, Y.M., Koo, Y.J., Kim, Y.S., 2006. The dysbindin gene (DTNBP1) and schizophrenia: no support for an association in the Korean population. Neurosci. Lett. 407 (2), 101–106. http://dx.doi.org/10.1016/j.neulet.2006.08.011.

Kaffman, A., Krystal, J.H., 2012. New frontiers in animal research of psychiatric illness. Methods Mol. Biol. 829, 3–30. http://dx.doi.org/10.1007/978-1-61779-458-2_1.

Kamiya, A., Kubo, K., Tomoda, T., Takaki, M., Youn, R., Ozeki, Y., et al., 2005. A schizophrenia-associated mutation of DISC1 perturbs cerebral cortex development. Nat. Cell Biol. 7 (12), 1167–1178. http://dx.doi.org/10.1038/ncb1328.

Kannan, G., Sawa, A., Pletnikov, M.V., 2013. Mouse models of gene-environment interactions in schizophrenia. Neurobiol. Dis. 57, 5–11. http://dx.doi.org/10.1016/j.nbd.2013.05.012.

Karayiorgou, M., Gogos, J.A., 2004. The molecular genetics of the 22q11-associated schizophrenia. Brain Res. Mol. Brain Res. 132 (2), 95–104. http://dx.doi.org/10.1016/j.molbrainres.2004.09.029.

Karl, T., Duffy, L., Scimone, A., Harvey, R.P., Schofield, P.R., 2007. Altered motor activity, exploration and anxiety in heterozygous neuregulin 1 mutant mice: implications for understanding schizophrenia. Genes. Brain Behav. 6 (7), 677–687. http://dx.doi.org/10.1111/j.1601-183X.2006.00298.x.

Karlsgodt, K.H., Robleto, K., Trantham-Davidson, H., Jairl, C., Cannon, T.D., Lavin, A., et al., 2011. Reduced dysbindin expression mediates N-methyl-D-aspartate receptor hypofunction and impaired working memory performance. Biol. Psychiatry 69 (1), 28–34. http://dx.doi.org/10.1016/j.biopsych.2010.09.012.

Kato, T., Kasai, A., Mizuno, M., Fengyi, L., Shintani, N., Maeda, S., et al., 2010. Phenotypic characterization of transgenic mice overexpressing neuregulin-1. PLoS One 5 (12), e14185. http://dx.doi.org/10.1371/journal.pone.0014185.

Katsel, P., Tan, W., Abazyan, B., Davis, K.L., Ross, C., Pletnikov, M.V., et al., 2011. Expression of mutant human DISC1 in mice supports abnormalities in differentiation of oligodendrocytes. Schizophr. Res. 130 (1–3), 238–249. http://dx.doi.org/10.1016/j.schres.2011.04.021.

Kellendonk, C., Simpson, E.H., Polan, H.J., Malleret, G., Vronskaya, S., Winiger, V., et al., 2006. Transient and selective overexpression of dopamine D2 receptors in the striatum causes persistent abnormalities in prefrontal cortex functioning. Neuron 49 (4), 603–615. http://dx.doi.org/10.1016/j.neuron.2006.01.023.

Kellendonk, C., Simpson, E.H., Kandel, E.R., 2009. Modeling cognitive endophenotypes of schizophrenia in mice. Trends Neurosci. 32 (6), 347–358. http://dx.doi.org/10.1016/j.tins.2009.02.003.

Kendell, R., Jablensky, A., 2003. Distinguishing between the validity and utility of psychiatric diagnoses. Am. J. Psychiatry 160 (1), 4–12. http://dx.doi.org/10.1176/appi.ajp.160.1.4.

Kirov, G., 2015. CNVs in neuropsychiatric disorders. Hum. Mol. Genet. 24 (R1), R45–R49. http://dx.doi.org/10.1093/hmg/ddv253.

Kobayashi, K., Umeda-Yano, S., Yamamori, H., Takeda, M., Suzuki, H., Hashimoto, R., 2011. Correlated alterations in serotonergic and dopaminergic modulations at the hippocampal mossy fiber synapse in mice lacking dysbindin. PLoS One 6 (3), e18113. http://dx.doi.org/10.1371/journal.pone.0018113.

Koike, H., Arguello, P.A., Kvajo, M., Karayiorgou, M., Gogos, J.A., 2006. Disc1 is mutated in the 129S6/SvEv strain and modulates working memory in mice. Proc. Natl. Acad. Sci. U.S.A. 103 (10), 3693–3697. http://dx.doi.org/10.1073/pnas.0511189103.

Kuepper, R., Skinbjerg, M., Abi-Dargham, A., 2012. The dopamine dysfunction in schizophrenia revisited: new insights into topography and course. Handb. Exp. Pharmacol. 212, 1–26. http://dx.doi.org/10.1007/978-3-642-25761-2_1.

Kuroda, K., Yamada, S., Tanaka, M., Iizuka, M., Yano, H., Mori, D., et al., 2011. Behavioral alterations associated with targeted disruption of exons 2 and 3 of the Disc1 gene in the mouse. Hum. Mol. Genet. 20 (23), 4666–4683. http://dx.doi.org/10.1093/hmg/ddr400.

Kvajo, M., McKellar, H., Arguello, P.A., Drew, L.J., Moore, H., MacDermott, A.B., et al., 2008. A mutation in mouse Disc1 that models a schizophrenia risk allele leads to specific alterations in neuronal architecture and cognition. Proc. Natl. Acad. Sci. U.S.A. 105 (19), 7076–7081. http://dx.doi.org/10.1073/pnas.0802615105.

Kvajo, M., McKellar, H., Drew, L.J., Lepagnol-Bestel, A.M., Xiao, L., Levy, R.J., et al., 2011. Altered axonal targeting and short-term plasticity in the hippocampus of Disc1 mutant mice. Proc. Natl. Acad. Sci. U.S.A. 108 (49), E1349–E1358. http://dx.doi.org/10.1073/pnas.1114113108.

Labrie, V., Wong, A.H., Roder, J.C., 2012. Contributions of the D-serine pathway to schizophrenia. Neuropharmacology 62 (3), 1484–1503. http://dx.doi.org/10.1016/j.neuropharm.2011.01.030.

Law, A.J., Kleinman, J.E., Weinberger, D.R., Weickert, C.S., 2007. Disease-associated intronic variants in the ErbB4 gene are related to altered ErbB4 splice-variant expression in the brain in schizophrenia. Hum. Mol. Genet. 16 (2), 129–141. http://dx.doi.org/10.1093/hmg/ddl449.

Lee, F.H., Fadel, M.P., Preston-Maher, K., Cordes, S.P., Clapcote, S.J., Price, D.J., et al., 2011. Disc1 point mutations in mice affect development of the cerebral cortex. J. Neurosci. 31 (9), 3197–3206. http://dx.doi.org/10.1523/JNEUROSCI.4219-10.2011.

Li, W., Zhang, Q., Oiso, N., Novak, E.K., Gautam, R., O'Brien, E.P., et al., 2003. Hermansky-Pudlak syndrome type 7 (HPS-7) results from mutant dysbindin, a member of the biogenesis of lysosome-related organelles complex 1 (BLOC-1). Nat. Genet. 35 (1), 84–89. http://dx.doi.org/10.1038/ng1229.

Li, W., Zhou, Y., Jentsch, J.D., Brown, R.A., Tian, X., Ehninger, D., et al., 2007. Specific developmental disruption of disrupted-in-schizophrenia-1 function results in schizophrenia-related phenotypes in mice. Proc. Natl. Acad. Sci. U.S.A. 104 (46), 18280–18285. http://dx.doi.org/10.1073/pnas.0706900104.

Li, Y.C., Kellendonk, C., Simpson, E.H., Kandel, E.R., Gao, W.J., 2011. D2 receptor overexpression in the striatum leads to a deficit in inhibitory transmission and dopamine sensitivity in mouse prefrontal cortex. Proc. Natl. Acad. Sci. U.S.A. 108 (29), 12107–12112. http://dx.doi.org/10.1073/pnas.1109718108.

Lillrank, S.M., Lipska, B.K., Weinberger, D.R., 1995. Neurodevelopmental animal models of schizophrenia. Clin. Neurosci. 3 (2), 98–104. Retrieved from: http://www.ncbi.nlm.nih.gov/pubmed/7583625.

Lipina, T., Weiss, K., Roder, J., 2007. The ampakine CX546 restores the prepulse inhibition and latent inhibition deficits in mGluR5-deficient mice. Neuropsychopharmacology 32 (4), 745–756. http://dx.doi.org/10.1038/sj.npp.1301191.

Lipina, T.V., Niwa, M., Jaaro-Peled, H., Fletcher, P.J., Seeman, P., Sawa, A., et al., 2010. Enhanced dopamine function in DISC1-L100P mutant mice: implications for schizophrenia. Genes. Brain Behav. 9 (7), 777–789. http://dx.doi.org/10.1111/j.1601-183X.2010.00615.x.

Lipina, T.V., Kaidanovich-Beilin, O., Patel, S., Wang, M., Clapcote, S.J., Liu, F., et al., 2011. Genetic and pharmacological evidence for schizophrenia-related Disc1 interaction with GSK-3. Synapse 65 (3), 234–248. http://dx.doi.org/10.1002/syn.20839.

Lipina, T.V., Wang, M., Liu, F., Roder, J.C., 2012. Synergistic interactions between PDE4B and GSK-3: DISC1 mutant mice. Neuropharmacology 62 (3), 1252–1262. http://dx.doi.org/10.1016/j.neuropharm.2011.02.020.

Lipska, B.K., Weinberger, D.R., 1993. Delayed effects of neonatal hippocampal damage on haloperidol-induced catalepsy and apomorphine-induced stereotypic behaviors in the rat. Brain Res. Dev. Brain Res. 75 (2), 213–222. Retrieved from: http://www.ncbi.nlm.nih.gov/pubmed/7903225.

Lipska, B.K., Weinberger, D.R., 2000. To model a psychiatric disorder in animals: schizophrenia as a reality test. Neuropsychopharmacology 23 (3), 223–239. doi:10.1016/S0893-133X(00)00137-8.

Lipska, B.K., Jaskiw, G.E., Weinberger, D.R., 1993. Postpubertal emergence of hyperresponsiveness to stress and to amphetamine after neonatal excitotoxic hippocampal damage: a potential animal model of schizophrenia. Neuropsychopharmacology 9 (1), 67–75. http://dx.doi.org/10.1038/npp.1993.44.

Lu, L., Mamiya, T., Koseki, T., Mouri, A., Nabeshima, T., 2011. Genetic animal models of schizophrenia related with the hypothesis of abnormal neurodevelopment. Biol. Pharm. Bull. 34 (9), 1358–1363. Retrieved from: http://www.ncbi.nlm.nih.gov/pubmed/21881217.

Lutkenhoff, E., Karlsgodt, K.H., Gutman, B., Stein, J.L., Thompson, P.M., Cannon, T.D., et al., 2012. Structural and functional neuroimaging phenotypes in dysbindin mutant mice. Neuroimage 62 (1), 120–129. http://dx.doi.org/10.1016/j.neuroimage.2012.05.008.

Ma, T.M., Abazyan, S., Abazyan, B., Nomura, J., Yang, C., Seshadri, S., et al., 2013. Pathogenic disruption of DISC1-serine racemase binding elicits schizophrenia-like behavior via D-serine depletion. Mol. Psychiatry 18 (5), 557–567. http://dx.doi.org/10.1038/mp.2012.97.

Madeira, C., Freitas, M.E., Vargas-Lopes, C., Wolosker, H., Panizzutti, R., 2008. Increased brain D-amino acid oxidase (DAAO) activity in schizophrenia. Schizophr. Res. 101 (1–3), 76–83. http://dx.doi.org/10.1016/j.schres.2008.02.002.

Mahoney, M.K., Olmstead, M.C., 2013. Neurobiology of an endophenotype: modeling the progression of alcohol addiction in rodents. Curr. Opin. Neurobiol. 23 (4), 607–614. http://dx.doi.org/10.1016/j.conb.2013.03.006.

Matthysse, S., 1986. Animal models in psychiatric research. Prog. Brain Res. 65, 259–270. http://dx.doi.org/10.1016/S0079-6123(08)60655-X.

Meechan, D.W., Maynard, T.M., Tucker, E.S., Fernandez, A., Karpinski, B.A., Rothblat, L.A., et al., 2015. Modeling a model: Mouse genetics, 22q11.2 Deletion Syndrome, and disorders of cortical circuit development. Prog. Neurobiol. 130, 1–28. http://dx.doi.org/10.1016/j.pneurobio.2015.03.004.

McConnell, M.J., Huang, Y.H., Datwani, A., Shatz, C.J., 2009. H2-K(b) and H2-D(b) regulate cerebellar long-term depression and limit motor learning. Proc. Natl. Acad. Sci. U.S.A. 106 (16), 6784–6789. http://dx.doi.org/10.1073/pnas.0902018106.

Mei, L., Xiong, W.C., 2008. Neuregulin 1 in neural development, synaptic plasticity and schizophrenia. Nat. Rev. Neurosci. 9 (6), 437–452. http://dx.doi.org/10.1038/nrn2392.

Meisenzahl, E.M., Schmitt, G.J., Scheuerecker, J., Moller, H.J., 2007. The role of dopamine for the pathophysiology of schizophrenia. Int. Rev. Psychiatry 19 (4), 337–345. http://dx.doi.org/10.1080/09540260701502468.

Meyer, D., Birchmeier, C., 1995. Multiple essential functions of neuregulin in development. Nature 378 (6555), 386–390. http://dx.doi.org/10.1038/378386a0.

Millar, J.K., Wilson-Annan, J.C., Anderson, S., Christie, S., Taylor, M.S., Semple, C.A., et al., 2000. Disruption of two novel genes by a translocation co-segregating with schizophrenia. Hum. Mol. Genet. 9 (9), 1415–1423. Retrieved from: http://www.ncbi.nlm.nih.gov/pubmed/10814723.

Millar, J.K., Christie, S., Anderson, S., Lawson, D., Hsiao-Wei Loh, D., Devon, R.S., et al., 2001. Genomic structure and localisation within a linkage hotspot of disrupted in schizophrenia 1, a gene disrupted by a translocation segregating with schizophrenia. Mol. Psychiatry 6 (2), 173–178. http://dx.doi.org/10.1038/sj.mp.4000784.

Miyamoto, S., Snouwaert, J.N., Koller, B.H., Moy, S.S., Lieberman, J.A., Duncan, G.E., 2004. Amphetamine-induced Fos is reduced in limbic cortical regions but not in the caudate or accumbens in a genetic model of NMDA receptor hypofunction. Neuropsychopharmacology 29 (12), 2180–2188. http://dx.doi.org/10.1038/sj.npp.1300548.

Moghaddam, B., Javitt, D., 2012. From revolution to evolution: the glutamate hypothesis of schizophrenia and its implication for treatment. Neuropsychopharmacology 37 (1), 4–15. http://dx.doi.org/10.1038/npp.2011.181.

Mohn, A.R., Gainetdinov, R.R., Caron, M.G., Koller, B.H., 1999. Mice with reduced NMDA receptor expression display behaviors related to schizophrenia. Cell 98 (4), 427–436. Retrieved from: http://www.ncbi.nlm.nih.gov/pubmed/10481908.

Morris, D.W., McGhee, K.A., Schwaiger, S., Scully, P., Quinn, J., Meagher, D., et al., 2003. No evidence for association of the dysbindin gene [DTNBP1] with schizophrenia in an Irish population-based study. Schizophr. Res. 60 (2–3), 167–172. Retrieved from: http://www.ncbi.nlm.nih.gov/pubmed/12591580.

Morris, S.E., Cuthbert, B.N., 2012. Research domain criteria: cognitive systems, neural circuits, and dimensions of behavior. Dialogues Clin. Neurosci. 14 (1), 29–37. Retrieved from: http://www.ncbi.nlm.nih.gov/pubmed/22577302.

Nakai, T., Nagai, T., Wang, R., Yamada, S., Kuroda, K., Kaibuchi, K., et al., 2014. Alterations of GABAergic and dopaminergic systems in mutant mice with disruption of exons 2 and 3 of the Disc1 gene. Neurochem. Int. 74, 74–83. http://dx.doi.org/10.1016/j.neuint.2014.06.009.

Nakata, K., Lipska, B.K., Hyde, T.M., Ye, T., Newburn, E.N., Morita, Y., et al., 2009. DISC1 splice variants are upregulated in schizophrenia and associated with risk polymorphisms. Proc. Natl. Acad. Sci. U.S.A. 106 (37), 15873–15878. http://dx.doi.org/10.1073/pnas.0903413106.

Nazarian, R., Starcevic, M., Spencer, M.J., Dell'Angelica, E.C., 2006. Reinvestigation of the dysbindin subunit of BLOC-1 (biogenesis of lysosome-related organelles complex-1) as a dystrobrevin-binding protein. Biochem. J. 395 (3), 587–598. http://dx.doi.org/10.1042/BJ20051965.

Network & Pathway Analysis Subgroup of Psychiatric Genomics Consortium, 2015. Psychiatric genome-wide association study analyses implicate neuronal, immune and histone pathways. Nat. Neurosci. 18 (2), 199–209. http://dx.doi.org/10.1038/nn.3922.

Nihonmatsu-Kikuchi, N., Hashimoto, R., Hattori, S., Matsuzaki, S., Shinozaki, T., Miura, H., et al., 2011. Reduced rate of neural differentiation in the dentate gyrus of adult dysbindin null (sandy) mouse. PLoS One 6 (1), e15886. http://dx.doi.org/10.1371/journal.pone.0015886.

Niwa, M., Jaaro-Peled, H., Tankou, S., Seshadri, S., Hikida, T., Matsumoto, Y., et al., 2013. Adolescent stress-induced epigenetic control of dopaminergic neurons via glucocorticoids. Science 339 (6117), 335–339. http://dx.doi.org/10.1126/science.1226931.

O'Tuathaigh, C.M., Waddington, J.L., 2010. Mutant mouse models: phenotypic relationships to domains of psychopathology and pathobiology in schizophrenia. Schizophr. Bull. 36 (2), 243–245. http://dx.doi.org/10.1093/schbul/sbq004.

O'Tuathaigh, C.M., Babovic, D., O'Sullivan, G.J., Clifford, J.J., Tighe, O., Croke, D.T., et al., 2007. Phenotypic characterization of spatial cognition and social behavior in mice with 'knockout' of the schizophrenia risk gene neuregulin 1. Neuroscience 147 (1), 18–27. http://dx.doi.org/10.1016/j.neuroscience.2007.03.051.

O'Tuathaigh, C.M., Harte, M., O'Leary, C., O'Sullivan, G.J., Blau, C., Lai, D., et al., 2010. Schizophrenia-related endophenotypes in heterozygous neuregulin-1 'knockout' mice. Eur. J. Neurosci. 31 (2), 349–358. http://dx.doi.org/10.1111/j.1460-9568.2009.07069.x.

O'Tuathaigh, C.M., Waddington, J.L., 2015. Closing the translational gap between mutant mouse models and the clinical reality of psychotic illness. Neurosci. Biobehav. Rev. 58, 19–35. http://dx.doi.org/10.1016/j.neubiorev.2015.01.016.

Papaleo, F., Crawley, J.N., Song, J., Lipska, B.K., Pickel, J., Weinberger, D.R., et al., 2008. Genetic dissection of the role of catechol-O-methyltransferase in cognition and stress reactivity in mice. J. Neurosci. 28 (35), 8709–8723. http://dx.doi.org/10.1523/JNEUROSCI.2077-08.2008.

Papaleo, F., Yang, F., Garcia, S., Chen, J., Lu, B., Crawley, J.N., et al., 2012. Dysbindin-1 modulates prefrontal cortical activity and schizophrenia-like behaviors via dopamine/D2 pathways. Mol. Psychiatry 17 (1), 85–98. http://dx.doi.org/10.1038/mp.2010.106.

Peleg-Raibstein, D., Feldon, J., Meyer, U., 2012. Behavioral animal models of antipsychotic drug actions. Handb. Exp. Pharmacol. 212, 361–406. http://dx.doi.org/10.1007/978-3-642-25761-2_14.

Petrinovic, M.M., Kunnecke, B., 2014. Neuroimaging endophenotypes in animal models of autism spectrum disorders: lost or found in translation? Psychopharmacology (Berl) 231 (6), 1167–1189. http://dx.doi.org/10.1007/s00213-013-3200-z.

Pletnikov, M.V., Moran, T.H., Carbone, K.M., 2002. Borna disease virus infection of the neonatal rat: developmental brain injury model of autism spectrum disorders. Front. Biosci. 7, d593–d607. Retrieved from: http://www.ncbi.nlm.nih.gov/pubmed/11861216.

Pletnikov, M.V., Xu, Y., Ovanesov, M.V., Kamiya, A., Sawa, A., Ross, C.A., 2007. PC12 cell model of inducible expression of mutant DISC1: new evidence for a dominant-negative mechanism of abnormal neuronal differentiation. Neurosci. Res. 58 (3), 234–244. http://dx.doi.org/10.1016/j.neures.2007.03.003.

Pletnikov, M.V., Ayhan, Y., Nikolskaia, O., Xu, Y., Ovanesov, M.V., Huang, H., et al., 2008. Inducible expression of mutant human DISC1 in mice is associated with brain and behavioral abnormalities reminiscent of schizophrenia. Mol. Psychiatry 13 (2) 173–186, 115. http://dx.doi.org/10.1038/sj.mp.4002079.

Poels, E.M., Kegeles, L.S., Kantrowitz, J.T., Slifstein, M., Javitt, D.C., Lieberman, J.A., et al., 2014. Imaging glutamate in schizophrenia: review of findings and implications for drug discovery. Mol. Psychiatry 19 (1), 20–29. http://dx.doi.org/10.1038/mp.2013.136.

Powell, C.M., Miyakawa, T., 2006. Schizophrenia-relevant behavioral testing in rodent models: a uniquely human disorder? Biol. Psychiatry 59 (12), 1198–1207. http://dx.doi.org/10.1016/j.biopsych.2006.05.008.

Ratajczak, P., Kus, K., Jarmuszkiewicz, Z., Wozniak, A., Cichocki, M., Nowakowska, E., 2013. Influence of aripiprazole and olanzapine on behavioral dysfunctions of adolescent rats exposed to stress in perinatal period. Pharmacol. Rep. 65 (1), 30–43. Retrieved from: http://www.ncbi.nlm.nih.gov/pubmed/23563021.

Ross, C.A., Margolis, R.L., Reading, S.A., Pletnikov, M., Coyle, J.T., 2006. Neurobiology of schizophrenia. Neuron 52 (1), 139–153. http://dx.doi.org/10.1016/j.neuron.2006.09.015.

Sankar, A., MacKenzie, R.N., Foster, J.A., 2012. Loss of class I MHC function alters behavior and stress reactivity. J. Neuroimmunol. 244 (1–2), 8–15. http://dx.doi.org/10.1016/j.jneuroim.2011.12.025.

Seshadri, S., Faust, T., Ishizuka, K., Delevich, K., Chung, Y., Kim, S.H., et al., 2015. Interneuronal DISC1 regulates NRG1-ErbB4 signalling and excitatory-inhibitory synapse formation in the mature cortex. Nat. Commun. 6, 10118. http://dx.doi.org/10.1038/ncomms10118.

Schizophrenia Working Group of the Psychiatric Genomics, C., 2014. Biological insights from 108 schizophrenia-associated genetic loci. Nature 511 (7510), 421–427. http://dx.doi.org/10.1038/nature13595.

Schneider, M., Debbane, M., Bassett, A.S., Chow, E.W., Fung, W.L., van den Bree, M., et al., 2014. Psychiatric disorders from childhood to adulthood in 22q11.2 deletion syndrome: results from the International Consortium on Brain and Behavior in 22q11.2 Deletion Syndrome. Am. J. Psychiatry 171 (6), 627–639. http://dx.doi.org/10.1176/appi.ajp.2013.13070864.

Seamans, J.K., Yang, C.R., 2004. The principal features and mechanisms of dopamine modulation in the prefrontal cortex. Prog. Neurobiol. 74 (1), 1–58. http://dx.doi.org/10.1016/j.pneurobio.2004.05.006.

Shamir, A., Kwon, O.B., Karavanova, I., Vullhorst, D., Leiva-Salcedo, E., Janssen, M.J., et al., 2012. The importance of the NRG-1/ErbB4 pathway for synaptic plasticity and behaviors associated with psychiatric disorders. J. Neurosci. 32 (9), 2988–2997. http://dx.doi.org/10.1523/JNEUROSCI.1899-11.2012.

Shen, S., Lang, B., Nakamoto, C., Zhang, F., Pu, J., Kuan, S.L., et al., 2008. Schizophrenia-related neural and behavioral phenotypes in transgenic mice expressing truncated Disc1. J. Neurosci. 28 (43), 10893–10904. http://dx.doi.org/10.1523/JNEUROSCI.3299-08.2008.

Shi, J., Levinson, D.F., Duan, J., Sanders, A.R., Zheng, Y., Pe'er, I., et al., 2009. Common variants on chromosome 6p22.1 are associated with schizophrenia. Nature 460 (7256), 753–757. http://dx.doi.org/10.1038/nature08192.

Shintani, N., Onaka, Y., Hashimoto, R., Takamura, H., Nagata, T., Umeda-Yano, S., et al., 2014. Behavioral characterization of mice overexpressing human dysbindin-1. Mol. Brain 7, 74. http://dx.doi.org/10.1186/s13041-014-0074-x.

Shoji, H., Toyama, K., Takamiya, Y., Wakana, S., Gondo, Y., Miyakawa, T., 2012. Comprehensive behavioral analysis of ENU-induced Disc1-Q31L and -L100P mutant mice. BMC Res. Notes 5, 108. http://dx.doi.org/10.1186/1756-0500-5-108.

Shprintzen, R.J., Goldberg, R., Golding-Kushner, K.J., Marion, R.W., 1992. Late-onset psychosis in the velo-cardio-facial syndrome. Am. J. Med. Genet. 42 (1), 141–142. http://dx.doi.org/10.1002/ajmg.1320420131.

St Clair, D., Blackwood, D., Muir, W., Carothers, A., Walker, M., Spowart, G., et al., 1990. Association within a family of a balanced autosomal translocation with major mental illness. Lancet 336 (8706), 13–16. Retrieved from: http://www.ncbi.nlm.nih.gov/pubmed/1973210.

Stefansson, H., Sigurdsson, E., Steinthorsdottir, V., Bjornsdottir, S., Sigmundsson, T., Ghosh, S., et al., 2002. Neuregulin 1 and susceptibility to schizophrenia. Am. J. Hum. Genet. 71 (4), 877–892. http://dx.doi.org/10.1086/342734.

Stefansson, H., Sarginson, J., Kong, A., Yates, P., Steinthorsdottir, V., Gudfinnsson, E., et al., 2003. Association of neuregulin 1 with schizophrenia confirmed in a Scottish population. Am. J. Hum. Genet. 72 (1), 83–87. Retrieved from: http://www.ncbi.nlm.nih.gov/pubmed/12478479.

Straub, R.E., Jiang, Y., MacLean, C.J., Ma, Y., Webb, B.T., Myakishev, M.V., et al., 2002. Genetic variation in the 6p22.3 gene DTNBP1, the human ortholog of the mouse dysbindin gene, is associated with schizophrenia. Am. J. Hum. Genet. 71 (2), 337–348. http://dx.doi.org/10.1086/341750.

Strohmaier, J., Frank, J., Wendland, J.R., Schumacher, J., Jamra, R.A., Treutlein, J., et al., 2010. A reappraisal of the association between Dysbindin (DTNBP1) and schizophrenia in a large combined case-control and family-based sample of German ancestry. Schizophr. Res. 118 (1–3), 98–105. http://dx.doi.org/10.1016/j.schres.2009.12.025.

Su, P., Li, S., Chen, S., Lipina, T.V., Wang, M., Lai, T.K., et al., 2014. A dopamine D2 receptor-DISC1 protein complex may contribute to antipsychotic-like effects. Neuron 84 (6), 1302–1316. http://dx.doi.org/10.1016/j.neuron.2014.11.007.

Sullivan, P.F., 2010. The psychiatric GWAS consortium: big science comes to psychiatry. Neuron 68 (2), 182–186. http://dx.doi.org/10.1016/j.neuron.2010.10.003.

Swerdlow, N.R., Geyer, M.A., 1998. Using an animal model of deficient sensorimotor gating to study the pathophysiology and new treatments of schizophrenia. Schizophr. Bull. 24 (2), 285–301. Retrieved from: http://www.ncbi.nlm.nih.gov/pubmed/9613626.

Talbot, K., Eidem, W.L., Tinsley, C.L., Benson, M.A., Thompson, E.W., Smith, R.J., et al., 2004. Dysbindin-1 is reduced in intrinsic, glutamatergic terminals of the hippocampal formation in schizophrenia. J Clin Invest 113 (9), 1353–1363. http://dx.doi.org/10.1172/JCI20425.

Tandon, R., Nasrallah, H.A., Keshavan, M.S., 2009. Schizophrenia, "just the facts" 4. Clinical features and conceptualization. Schizophr. Res. 110 (1–3), 1–23. http://dx.doi.org/10.1016/j.schres.2009.03.005.

Tandon, R., Gaebel, W., Barch, D.M., Bustillo, J., Gur, R.E., Heckers, S., et al., 2013. Definition and description of schizophrenia in the DSM-5. Schizophr. Res. 150 (1), 3–10. http://dx.doi.org/10.1016/j.schres.2013.05.028.

Tang, J., LeGros, R.P., Louneva, N., Yeh, L., Cohen, J.W., Hahn, C.G., et al., 2009. Dysbindin-1 in dorsolateral prefrontal cortex of schizophrenia cases is reduced in an isoform-specific manner unrelated to dysbindin-1 mRNA expression. Hum. Mol. Genet. 18 (20), 3851–3863. http://dx.doi.org/10.1093/hmg/ddp329.

Tiwari, A.K., Zai, C.C., Muller, D.J., Kennedy, J.L., 2010. Genetics in schizophrenia: where are we and what next? Dialogues. Clin. Neurosci. 12 (3), 289–303. Retrieved from: http://www.ncbi.nlm.nih.gov/pubmed/20954426.

Tochigi, M., Zhang, X., Ohashi, J., Hibino, H., Otowa, T., Rogers, M., et al., 2006. Association study of the dysbindin (DTNBP1) gene in schizophrenia from the Japanese population. Neurosci. Res. 56 (2), 154–158. http://dx.doi.org/10.1016/j.neures.2006.06.009.

Tomiyama, K., Drago, J., Waddington, J.L., Koshikawa, N., 2012. Constitutive and conditional mutant mouse models for understanding dopaminergic regulation of orofacial movements: emerging insights and challenges. J. Pharmacol. Sci. 119 (4), 297–301. Retrieved from: http://www.ncbi.nlm.nih.gov/pubmed/22863668.

Tsai, G., Yang, P., Chung, L.C., Lange, N., Coyle, J.T., 1998. D-serine added to antipsychotics for the treatment of schizophrenia. Biol. Psychiatry 44 (11), 1081–1089. Retrieved from: http://www.ncbi.nlm.nih.gov/pubmed/9836012.

Tsai, G.E., Yang, P., Chung, L.C., Tsai, I.C., Tsai, C.W., Coyle, J.T., 1999. D-serine added to clozapine for the treatment of schizophrenia. Am. J. Psychiatry 156 (11), 1822–1825. Retrieved from: http://www.ncbi.nlm.nih.gov/pubmed/10553752.

Tseng, K.Y., Chambers, R.A., Lipska, B.K., 2009. The neonatal ventral hippocampal lesion as a heuristic neurodevelopmental model of schizophrenia. Behav. Brain Res. 204 (2), 295–305. http://dx.doi.org/10.1016/j.bbr.2008.11.039.

Tsuang, M.T., Stone, W.S., Faraone, S.V., 2000. Schizophrenia: family studies and treatment of spectrum disorders. Dialogues Clin. Neurosci. 2 (4), 381–391. Retrieved from: http://www.ncbi.nlm.nih.gov/pubmed/22033752.

Tunbridge, E.M., Harrison, P.J., Weinberger, D.R., 2006. Catechol-o-methyltransferase, cognition, and psychosis: Val158Met and beyond. Biol. Psychiatry 60 (2), 141–151. http://dx.doi.org/10.1016/j.biopsych.2005.10.024.

van den Buuse, M., Wischhof, L., Lee, R.X., Martin, S., Karl, T., 2009. Neuregulin 1 hypomorphic mutant mice: enhanced baseline locomotor activity but normal psychotropic drug-induced hyperlocomotion and prepulse inhibition regulation. Int. J. Neuropsychopharmacol. 12 (10), 1383–1393. http://dx.doi.org/10.1017/S1461145709000388.

Walsh, J., Tighe, O., Lai, D., Harvey, R., Karayiorgou, M., Gogos, J.A., et al., 2010. Disruption of thermal nociceptive behaviour in mice mutant for the schizophrenia-associated genes NRG1, COMT and DISC1. Brain Res. 1348, 114–119. http://dx.doi.org/10.1016/j.brainres.2010.06.027.

Weickert, C.S., Rothmond, D.A., Hyde, T.M., Kleinman, J.E., Straub, R.E., 2008. Reduced DTNBP1 (dysbindin-1) mRNA in the hippocampal formation of schizophrenia patients. Schizophr. Res. 98 (1–3), 105–110. http://dx.doi.org/10.1016/j.schres.2007.05.041.

Weiner, I., Arad, M., 2009. Using the pharmacology of latent inhibition to model domains of pathology in schizophrenia and their treatment. Behav. Brain Res. 204 (2), 369–386. http://dx.doi.org/10.1016/j.bbr.2009.05.004.

Wen, L., Lu, Y.S., Zhu, X.H., Li, X.M., Woo, R.S., Chen, Y.J., et al., 2010. Neuregulin 1 regulates pyramidal neuron activity via ErbB4 in parvalbumin-positive interneurons. Proc. Natl. Acad. Sci. U.S.A. 107 (3), 1211–1216. http://dx.doi.org/10.1073/pnas.0910302107.

Wiedholz, L.M., Owens, W.A., Horton, R.E., Feyder, M., Karlsson, R.M., Hefner, K., et al., 2008. Mice lacking the AMPA GluR1 receptor exhibit striatal hyperdopaminergia and 'schizophrenia-related' behaviors. Mol. Psychiatry 13 (6), 631–640. http://dx.doi.org/10.1038/sj.mp.4002056.

Wirgenes, K.V., Djurovic, S., Agartz, I., Jonsson, E.G., Werge, T., Melle, I., et al., 2009. Dysbindin and d-amino-acid-oxidase gene polymorphisms associated with positive and negative symptoms in schizophrenia. Neuropsychobiology 60 (1), 31–36. http://dx.doi.org/10.1159/000235799.

Wolosker, H., Blackshaw, S., Snyder, S.H., 1999. Serine racemase: a glial enzyme synthesizing D-serine to regulate glutamate-N-methyl-D-aspartate neurotransmission. Proc. Natl. Acad. Sci. U.S.A. 96 (23), 13409–13414. Retrieved from: http://www.ncbi.nlm.nih.gov/pubmed/10557334.

Wood, G.K., Lipska, B.K., Weinberger, D.R., 1997. Behavioral changes in rats with early ventral hippocampal damage vary with age at damage. Brain Res. Dev. Brain Res. 101 (1–2), 17–25. Retrieved from: http://www.ncbi.nlm.nih.gov/pubmed/9263576.

Yerabham, A.S., Weiergraber, O.H., Bradshaw, N.J., Korth, C., 2013. Revisiting disrupted-in-schizophrenia 1 as a scaffold protein. Biol. Chem. 394 (11), 1425–1437. http://dx.doi.org/10.1515/hsz-2013-0178.

Young, J.W., Geyer, M.A., 2015. Developing treatments for cognitive deficits in schizophrenia: the challenge of translation. J. Psychopharmacol. 29 (2), 178–196. http://dx.doi.org/10.1177/0269881114555252.

CHAPTER 21

BEHAVIORAL PHENOTYPES OF GENETIC MOUSE MODELS: CONTRIBUTIONS TO UNDERSTANDING THE CAUSES AND TREATMENT OF SCHIZOPHRENIA

P.M. Moran

School of Psychology, University of Nottingham, Nottingham, United Kingdom

CHAPTER OUTLINE

Introduction	383
How Do You Measure Psychotic Behavior in a Mouse?	384
Models Targeting Dopaminergic Transmission	385
Models Targeting Glutamatergic Transmission	388
Future Directions	389
Conclusion	392
References	392

INTRODUCTION

At the time of this writing, treatment for schizophrenia had not changed substantially since the 1950s and remained based on dopamine D_2 receptor (D_2 receptor) and serotonin 5-HT2 receptor antagonists. Advances in molecular biology in the 1990s included application of genetic modifications in mouse models and revolutionized approaches to identifying new therapeutic targets for a wide range of diseases. In the case of schizophrenia, recent successes in genome-wide association studies (GWAS) (Schizophrenia Working Group of the Psychiatric Genomics Consortium, 2014) following a long period of disillusionment with genetic approaches have given a renewed cause for optimism about the possibility of identifying novel biological treatment targets for schizophrenia using this approach (Harrison, 2015). In this chapter, I consider how behavioral analysis of genetic mouse models has contributed to advancements in identifying the basic biological underpinnings of behaviors relevant to schizophrenia and the mechanism of action of current treatments, concentrating on dopamine and glutamate models, for which there is a significant body of evidence (Howes et al., 2015). For a comprehensive consideration of the important genetic models relevant to schizophrenia, see the chapter by Ayshan and Pletnikov in this volume. Finally, I also consider the implications of new directions in the classification of psychiatric symptoms for interpreting behavioral phenotypes from genetic mouse models in the future and how these can help to identify novel biological targets for therapeutic intervention. The neurophysiological and pharmacological aspects of these models have been reviewed in detail elsewhere and in this volume (Moran et al., 2014; Rafter et al., 2015; Ayshan and Pletnikov, this volume).

HOW DO YOU MEASURE PSYCHOTIC BEHAVIOR IN A MOUSE?

A significant challenge for modeling schizophrenia symptoms in rodents is the psychological nature of the symptoms. Although behavioral hyperactivity or memory might be readily measurable in a mouse, an analogue of hallucinations or unusual use of language is not so obvious. The behavioral endophenotype approach taken in the past approximately 30 years was developed to circumvent this problem. This approach measures a proxy phenotype, in this case a behavior, measurable in both humans and animals that is abnormal in patients compared to controls and relates to a possible causative psychological construct (endophenotype). For example, symptoms such as hallucinations and delusions may result from faulty gating of sensory information (Adler et al., 1982). A measure of this gating is prepulse inhibition (PPI), whereby the startle response to an acoustic stimulus is attenuated when the subject experiences a short warning prepulse. PPI is measurable in both animals and humans and is disrupted in patients and their relatives (Braff and Geyer, 1990). Treatments and drugs that induce psychosis in healthy humans disrupt PPI, and this disruption can be restored by clinically effective antipsychotic drugs (Braff and Geyer, 1990; Geyer et al., 2001). This endophenotype approach has been applied to other relevant behavioral constructs such as learning to ignore irrelevant stimuli (measured in latent inhibition) (Weiner and Arad, 2009) and spatial working memory (Glahn et al., 2003). Because many of these behavioral tests are relatively quick and easy tests to implement, the endophenotype approach was rapidly adopted (Gottesman and Gould, 2003). A large volume of research was generated such that it seemed that any pharmacological or genetic treatment that modulated PPI was almost invariably interpreted as having potential relevance to schizophrenia. The approach has been criticized for syllogism and inability to identify new drug mechanisms as it is validated using existing drugs (Ayshan and Pletnikov, this volume). The lack of success in finding new treatments has been attributed, in part, to limitations with these animal models (Nestler and Hyman, 2010) but is more likely to be attributable to their promiscuous interpretation, as has been suggested for other disorders such as depression (Hendrie and Pickles, 2013). With the benefit of hindsight, it is now clear that indiscriminate interpretation of effects on endophenotypes as synonymous with potential efficacy on all of the symptoms of schizophrenia was going to be limited. Notwithstanding, these data have provided valuable insight into the underlying biology of specific behaviors that seem ready to become increasingly important as new directions emerge in classifying psychiatric illnesses in terms of dimensions and constructs rather than specific diseases. Recent identification of 108 loci in GWAS for schizophrenia (Schizophrenia Working Group of the Psychiatric Genomics Consortium, 2014) has not only yielded potential novel biological targets but also corroborated existing targets that had been identified through behavioral neuroscientific approaches.

MODELS TARGETING DOPAMINERGIC TRANSMISSION

The neurotransmitter dopamine is important for a wide range of behaviors, including motivation (Wise, 2004), reward (Chowdhury et al., 2013), learning (Flagel et al., 2011), action and movement (Glimcher, 2011), decision making (Rogers, 2011), and arousal (Moran et al., 2014; Ueno et al., 2012). Abnormalities in many of these behaviors are part of the wide range of symptoms that characterize schizophrenia (Moran et al., 2014). The D_2 receptor subtype has been considered central to pathophysiology of schizophrenia, particularly the positive symptoms of hallucination and delusions (Howes and Kapur, 2009), and this importance has recently been corroborated by GWAS (Schizophrenia Working Group of the Psychiatric Genomics Consortium, 2014).

An increase in D_2 receptor density, hypersensitivity of the postsynaptic D_2 receptor, and an increase in receptors in a high-affinity state have been considered causal to the symptoms observed in the disorder (Moran et al., 2014). There are several demonstrations of increased receptor density measured in unmedicated patients, although this is, for the most part, not seen in chronically ill recently medicated patients (Kuepper et al., 2012). Clinical studies have limitations for investigating the causal role of such biological factors in symptoms; for example, it may be impossible to manipulate individual genes in humans purely for experimental purposes, but in mouse models these causal hypotheses can be tested directly and specifically. To causally test the hypothesis that overactivity/expression of the D_2 receptor plays a role in schizophrenia, mice have been produced that selectively overexpress the D_2 receptor gene (*Drd2*) in the striatum (Kellendonk et al., 2006). These mice show decreased motivation to work for food reward on a test of motivational behavior, possibly relevant to motivational problems in patients (Drew et al., 2007). Behavioral confounds such as fatigue or reduced palatability of the food were controlled for in these studies. *Drd2* overexpressing mice liked rewards as much as controls but work less hard to obtain them, as shown by their performance in an effort-related choice paradigm. *Drd2* overexpressing mice were shown to be less sensitive to variations in the distribution of rewards than controls, potentially indicative of a problem calculating differences in future reward value or inability to adapt behavior based on those calculations (Simpson et al., 2012). Clearly, more evidence is required, but this is consistent with theoretical ideas that schizophrenia symptoms may arise from striatal dopamine dysregulation producing abnormalities in the calculation of reward prediction error and the computed difference between actual and expected outcomes in learning. One integrative theory suggests that abnormal prediction error underlies delusional and disordered thinking in schizophrenia (Kapur et al., 2005; Moran et al., 2008), and neurophysiological studies are starting to support this and locate neural analogues of prediction error in prefrontal and striatal regions (Boehme et al., 2015; Heinz and Schlagenhauf, 2010; Juckel et al., 2006; Moran et al., 2012).

Catechol-*O*-methyltransferase (COMT) is the enzyme that catalyzes the methylation of catecholamines, including DA, norepinephrine, and epinephrine (Axelrod and Tomchick, 1958). Converging evidence now suggests a role for COMT in the catabolism of cortical dopamine as a consequence of the lack of dopamine transporters at cortical synapses (Sannino et al., 2014). This relatively selective effect is thought to underlie how genetic variations in COMT modulate behavior and, potentially, interactions with potential risk factors for schizophrenia such as cannabis use. COMT knockout mice have partial reduction (heterozygous) or complete absence (homozygous) of COMT enzyme activity (Gogos et al., 1998; Tunbridge et al., 2013). Decrease in DA in the prefrontal cortex has been found in males only, as has blunted psychostimulant sensitivity, but there was no change in sensorimotor gating (Huotari et al., 2002). Male *COMT* KO mice showed abnormal responsiveness to Δ^9-tetrahydrocannabinol (THC; the psychoactive constituent of cannabis) administration during adolescence. This was seen on behavioral measures such as exploration, spatial working memory, and anxiety. This has suggested a possible interaction between *COMT* genotype and cannabis exposure during adolescence (O'Tuathaigh et al., 2010). Mutations in *COMT* have been shown to modulate cortical thickening, neuronal density, and working memory performance with different profiles strongly dependent on sex. The sex dependence of *COMT* effects seen may be of particular importance for understanding sex differences in schizophrenia, which is more prevalent in males (Leung and Chue, 2000).

Mice overexpressing a human COMT-Val158Met polymorphism have been shown to have increased COMT enzyme activity and impairments in extradimensional shift, an attentional task that measures allocation of attentional resources analogous to clinical tests in patients such as Wisconsin cardsorting (Papaleo et al., 2008). Mouse models provide an experimental system to investigate possible interactions between genes. Epistatic interaction between dysbindin (dystrobrevin binding protein 1 (DTNBP1)) and COMT has been suggested in a mouse model where a single disruption of each of these genes in isolation improved working memory, whereas genetic reduction of both in the same mouse produced a disruption (Moran et al., 2014). These findings were mirrored using a human sample from individuals homozygous for the *COMT* rs4680 Met allele that reduces COMT enzyme activity, which showed more efficient prefrontal cortical engagement measured using fMRI; however, those that also had a dysbindin haplotype associated with decreased DTNB1 expression showed less efficient prefrontal processing. This suggests a limitation in studying functional genes in isolation and demonstrates how these models can play a role in the generation and the testing of novel hypotheses. Epistatic interactions between genes are much more likely to underlie schizophrenia than single gene effects, and these models allow investigation of these interactions at biological and behavioral levels.

Genetic mouse models have been important for testing specific hypotheses about the role of the dopamine D_2 receptor (DrD_2) in specific behaviors. In the mid-1990s, mice with deletion of dopamine receptor genes were seized upon as a long-awaited experimental tool to determine the function of the five dopamine receptor subtypes in isolation. The intention was to delete the relevant genes, characterize the behavioral and neural phenotypes, and ascertain the specific function of each receptor (Waddington et al., 2005). Dopamine D_2 receptor gene (Drd_2) knockout mice showed motor symptoms, such as reduced locomotor activity and longer duration wall rears, as anticipated from previous pharmacological studies and observations in Parkinson disease (Fowler et al., 2002; Kelly et al., 1998). *Drd2* knockout mice showed impairment in adjusting responses to previously reinforced stimuli when confronted with unexpected outcomes, suggesting impairment in the ability to regulate associative and reversal learning (Kruzich and Grandy, 2004). These mice also showed reduced ethanol-conditioned place preference compared to controls, confirming that D_2 receptors might be involved in reward (Cunningham et al., 2000). It has also been shown that in wild-type mice, D-amphetamine significantly disrupted PPI; however, in *Drd2* knockouts, D-amphetamine had no effect on PPI. This suggested that *Drd2*, but not *Drd4* or *Drd5*, receptors were necessary for D-amphetamine-induced disruption of PPI in mice (Ralph et al., 1999).

These data corroborated findings of Parkinsonian side effects and sedation in patients treated with antipsychotics and findings of rodent studies that showed reduced locomotor activity and motivation in rodents treated with antipsychotics (Moran et al., 2014). We now appreciate that knockout studies in general have potential influencing variables such as the background strain of the mouse used and the potential for developmental compensatory effects to mask or mimic true behavioral or neural consequences of gene deletion. However, in the case of D_2 receptors, evidence suggests that reduced D_2 receptor function has consistent behavioral effects, whether induced pharmacologically or genetically (Moran et al., 2014).

Combining pharmacological and knockout mouse techniques can elucidate behavioral and neural mechanisms that are not detectable using either approach alone; an example of this is latent inhibition. Latent inhibition is a phenomenon demonstrated in associative learning that measures the ability to ignore irrelevant stimuli (Lubow and Moore, 1959). Latent inhibition is shown as reduced learning about a stimulus in a group receiving preexposure to that stimulus without reinforcement compared to a group without preexposure. Most studies involving schizophrenia patients, relatives of patients, and individuals with high schizotypy have abnormal latent inhibition (either inappropriately high or low depending on symptom profiles) (Weiner and Arad, 2009). Antipsychotic drugs potentiate LI when it is rendered low in controls by reducing the amount of preexposure and attenuate the abolition of the latent inhibition effect by psychotomimetic drugs such as D-amphetamine (Weiner and Arad, 2009). *Drd2* knockout mice show similar potentiated latent inhibition with antipsychotic drugs, but the action of antipsychotic drugs to attenuate latent inhibition disrupted by D-amphetamine is surprisingly unchanged in *Drd2* knockout mice (Bay-Richter et al., 2013). This suggests more than one behavioral mechanism to improve abnormal learning to ignore irrelevant stimuli: one D_2 receptor–dependent and one non-D_2 receptor–dependent. In knockout mouse studies in which gene deletion is present from birth, compensatory developmental changes are likely to occur (Waddington et al., 2005). In these experiments, it is possible that an as-yet-unknown compensatory mechanism may have superseded the habitual D_2 receptor mechanism of the antipsychotic drugs tested to produce an antipsychotic-like behavioral effect (in this case modeled as restored latent inhibition). If this mechanism can be identified, then it could potentially be exploited as an antipsychotic drug target that produces positive behavioral effects but does not interact with D_2 receptors[1].

Two isoforms of the D_2 receptor have been identified, long (D_2L) and short (D_2S) forms, which may play different roles (Giros et al., 1989; Monsma et al., 1989). Evidence in knockout mice specific for each isoform suggests that D_2S acts at the presynaptic site to regulate synthesis and release of dopamine. D_2L acts postsynaptically to mediate G-protein-dependent and G-protein-independent signaling (for review see De Mei et al., 2009). Catalepsy induced by the antipsychotic haloperidol has been shown to be prevented in D_2L knockout mice, suggesting that D_2L specifically may be required for the motor cataleptic effects of the drug (Usiello et al., 2000). In the absence of specific ligands for the isoforms of the D_2 receptor, genetic approaches are an excellent tool to understand their role in behavior.

Adenoviral delivery of microRNA or other short regulatory RNAs provides an alternative approach to achieve knockdown of gene expression in mouse models. Combined with microinjection techniques, this can be used to target specific brain regions with some spatial precision. This approach was used to suppress *Drd2* expression in the nucleus accumbens core, which prevented methamphetamine-induced hyperactivity (Miyamoto et al., 2015). Overexpression of the dopamine D_3 receptor has been shown to reduce locomotor activity (Cote et al., 2014). Recently, this approach has been combined with optogenetic technology. Optogenetics is a combination of genetic and optical techniques that introduce microbial opsin genes sensitive to different kinds of light into specific neurons. D_2 receptors may play a role in anatomical plasticity (Cazorla et al., 2014; Kupchik et al., 2015). It was found in mice that by modulating neuronal excitability, striatal D_2 receptors bidirectionally control the density of direct pathway collaterals in the globus pallidus that bridge between direct and indirect pathways. An increase in bridging collaterals and abnormal locomotor activity was induced by optogenetic stimulation of the main pathway, and this effect was reversed by chronic treatment with the antipsychotic drug haloperidol. This makes the novel suggestion that regulation of the output from the striatum may involve these bridging collaterals, suggesting it corroborated a mechanism for etiology of symptoms as well as a further possible mechanism of action of antipsychotic drugs.

---

[1] Antipsychotics induce severe motor side effects via D_2 receptor antagonism.

MODELS TARGETING GLUTAMATERGIC TRANSMISSION

Glutamate is the major excitatory neurotransmitter in the mammalian brain (Fonnum, 1984). Since its relatively recent identification in the 1950s, a role for glutamate has been established in a variety of functions, including neuro-plasticity, neurotoxicity, development, and learning and memory (Mcdonald and Johnston, 1990; Nakanishi, 1994; Riedel et al., 2003). Presynaptically released glutamate binds different subclasses of ionotropic glutamate receptors (iGluRs), including α-amino-3-hydroxy-5-methyl-4-isoxazolepropionic acid (AMPA) receptors (AMPARs), kainate receptors, and N-methyl-D-aspartate (NMDA) receptors (for recent review, see Karakas et al., 2015). As for dopamine, genetic mouse models have built on data generated using pharmacological and physiological models to identify specific roles for subtypes of receptors in neural transmission and behavior (for review of pharmacological models, see Rafter et al., 2015). Since the early formulations of the hypothesis (eg, Olney and Farber, 1995), considerable evidence has converged to suggest that glutamate plays a role in the pathophysiology of schizophrenia (Marsman et al., 2013; Olney and Farber, 1995). Evidence for this hypothesis was based on pharmacological evidence that glutamate antagonist drugs induced psychotic symptoms (Javitt, 2007). A number of risk genes were subsequently identified that affect NMDA receptor function or glutamatergic neurotransmission (Harrison et al., 1991; Harrison and Weinberger, 2005; Coyle, 2006), whereas postmortem studies indicated decreased expression of glutamate receptor genes in hippocampal and temporal lobe regions (Harrison et al., 1991). More recently, in vivo magnetic resonance spectroscopy methods for measuring glutamatergic function in vivo have confirmed reduced glutamate in medial frontal regions of patients (Marsman et al., 2013).

Mouse genetic models of reduced NMDA receptor function have been developed in which the expression of the NMDA R1 subunit (*GRIN1*[2]) of the GluN1 receptor has been reduced to 5–10% (Mohn et al., 1999; van den Buuse, 2010). These NMDA receptor hypofunctioning mice have a phenotype that includes increased locomotor activity and deficits in both PPI and social interactions. Restriction of *GRIN1* hypofunction to the striatum using CreLox conditional gene targeting induced developmental abnormalities and locomotor hyperactivity, but mice survived to only postnatal day 21 (Mohn et al., 1999). This was replicated by Ohtsuka et al. (2008), who also showed that this hyperactivity was reversed by dopamine D_1 receptor and D_2 receptor agonists. This suggested that the NMDA receptor played an essential role in the integration of dopamine signaling controlling behavior mediated by the striatum (Ohtsuka et al., 2008). Mice lacking the *GRIN1* subunit selectively in medium spiny neurons display impaired motor learning measured on an accelerating rotarod and failed to learn to press a lever for food reward or acquire an active avoidance task (Beutler et al., 2011).

GRM3 is a metabotropic glutamate receptor–modulating synaptic glutamate. A common haplotype has been suggested to increase risk for schizophrenia, and evidence suggests that GRM3 genotype alters prefrontal levels of the glial glutamate transporter N-acetylaspartate, suggesting alteration in glutamatergic transmission (Egan et al., 2004). Mice with combined GRM2/3 deletion have been found to display spatial and other memory impairments when motivated by food reward, but not on aversively motivated tasks, suggesting interaction between anxiety/arousal and cognition. However, mice with independent deletions of GRM2 or GRM3, although showing a spatial learning phenotype, did not replicate the pattern of impairment of combined GRM2/GRM3 mice (De Filippis et al., 2015). Other studies using targeted disruption of the GRM3 gene have suggested a role for mGLuR3

---

[2] Nomenclature recommendations have changed since these studies and were formerly called NR1.

in contextual fear learning, without effects on reactivity to stress. Reduced hippocampal long-term potentiation, a synaptic correlate of learning, and increased sensitivity to methamphetamine-induced dopamine release from the nucleus accumbens (a striatal region implicated in hallucinatory symptoms of schizophrenia) accompanied these behavioral changes. This study suggests that working memory may also be disrupted, but the delayed alternation test used is a very imprecise measure of this behavior (Fujioka et al., 2014). Activation of mGLuR3 has been found to induce long-term depression of excitatory transmission in medial prefrontal cortex, suggesting a synaptic mechanism for these effects (Walker et al., 2015). It has been suggested that potentiators of mGLuR3 may alleviate behavioral impairments associated with reduced prefrontal activity, such as memory and planning, which are common in schizophrenia (Walker et al., 2015).

NMDA receptors are regulated by the glycine transporter1 (GlyT1), and it has been suggested that inhibition of GlyT1 might prove to be an effective means to increase NMDA receptor function therapeutically. Selective disruption of forebrain neuronal GlyT1 increases NMDA receptor function and enhances aversive conditioning and latent inhibition (Yee et al., 2006). These effects may be behaviorally specific to cues for threat (Dubroqua et al., 2014).

FUTURE DIRECTIONS

Developments in genetics, neuroscience, and cognitive neuropsychology in the past 30 years have driven a shift in thinking about how psychiatric disorders should be classified, suggesting that a dimensional approach is more likely to be adopted in the future. A well-developed example of this is the RDoC research domains criteria (Cuthbert, 2015). This is proposed by the National Institute for Mental Health in the United States and represents a classification matrix of scientific knowledge. It groups different aspects of human function (called constructs) into five overarching categories (called domains) (see Table 21.1). Each of these constructs are defined in terms of eight different levels of analysis (genes, molecules, cells, circuits, physiology, behavior, self-report, and paradigms) used to test them (see Table 21.2). Such an approach has implications for drug discovery; therefore, rather than searching for a drug

Table 21.1 Higher-Order Domains (1–5) and Associated Constructs in the RDoC Framework

1. Negative Valence	2. Positive Valence	3. Cognitive Systems	4. Social Processes	5. Arousal/Regulatory
Acute threat (fear)	Reward learning	Declarative memory	Affiliation and attachment	Arousal
Potential threat (anxiety)	Approach motivation	Perception	Social communication	Biological rhythms
Sustained threat	Habit	Attention	Self-perception and understanding	Sleep/wake cycle
Loss	Initial responsiveness to reward	Working memory	Perception and understanding of others	
Frustrating nonreward	Sustained responsiveness to reward	Language Cognitive control		

Source: Adapted from Badcock and Hugdahl, 2014.

Table 21.2 An Example of One Construct From Each of the Higher-Order Domains of the Research Criteria (RDoC)

Domain Name (Subconstruct)/ Units of Analysis	1. Negative Valence (Acute Threat/Fear)	2. Positive Valence (Reward Learning)	3. Cognitive (Declarative Memory)	4. Social Processes (Affiliation and Attachment [Attachment Formation and Maintenance])	5. Arousal Regulatory (Arousal)
1. Genes	BDNF, serotonin receptors, CRF, glutamate, NMDA, COMT, dopamine	Genes involved in dopamine synthesis, clearance, and signaling, DARP32, COMT, plasticity-related genes (eg, CREB, FosB, NMDA receptors on DRD1 neurons)	BDNF, KIBRA	OXTR, AVPR1A, MOR, 3OXT, tyrosine hydroxylase, *DRD2*, MOR, 3CRF, KOR, CRFR2, DRD1	Acetylcholine, serotonin, tryptophan hydroxylase, dopamine, GABA, glutamate adenosine NET, DAT, leptin, ghrelin cytokine
2. Molecules	NMDAr, glutamate, dopamine, cortisol, BDNF, GABA	Dopamine-related molecules, acetylcholine, co-released neuromodular glutamate	Acetylcholine, glutamate noradrenaline, opioid	Oxytocin, vasopressin, oxytocin receptor, vasopressin 1a receptor, dopamine, Mu opioid receptor, 3CRF, KOR, CRFR2, D1	Glutamate, norepinephrine, acetylcholine, histamine, dopamine, hypocretin/orexin, CRF, serotonin, leptin, ghrelin, opioids, oxytocin, GABA, cytokines
3. Cells	Neurons, glia, pyramidal GABAergic	Medium spiny neurons dopaminergic neurons	Pyramidal cells, granule cells, interneurons	Magnocellular OT	Locus coeruleus, tuberomammillary nucleus, LDT, PPT, basal forebrain nuclei, lateral, periforical, and dorsomedial hypothalamus, dorsal raphe, VTA, amygdala
4. Circuits	Amygdala (central, basal, lateral nuclei), hippocampus, PFC-insula, ACC, pons	Dorsal striatum, medial prefrontal cortex, F, VT, SN, amygdala	Intrinsic hippocampal circuitry (eg, DG, CA1, CA3, subiculum); extrinsic hippocampal circuitry (connections with higher-order cortical areas and the parahippocampal region; PFC and PPC interactions with multiple association cortices)	VTA- NAcc- VP- amygdala, PVN, OFC, FF gyrus, VMPFC, 3Amygdala, BNST, PVN, NAcc	Cholinergic and monoaminergic nuclei projections to thalamic and cortical circuits, hypothalamic to thalamic and cortical circuits, basal forebrain nuclei to cortical circuits, brainstem monoaminergic and cholinergic projections to basal forebrain, amygdala to monoaminergic and basal forebrain cholinergic nuclei

5. Physiology	Fear-potentiated startle, skin conductance, respiration	Error-related negativity, feedback-related negativity, midline theta	LTP/LTD, NMDA-related synaptic plasticity, AMPA-related synaptic plasticity, place cell activity, conjunction codes, up/down states, frontal/temporal coordinated oscillations, subsequent memory effect (fMRI, ERP)	Sex steroid changes; HPA downregulation; vagal tone; immune markers, HPA axis activation; immune responses ("sickness"); activation of sympathetic activity; vagal withdrawal	EEGEMG, autonomic ERPs: heart rate; blood pressure; pupil size; galvanic skin response; breathing, HPA axis: glucocorticoids; ACTH; CRF, sex-specific differences in arousal, brain activation as measured by fMRI, neural activity
6. Behavior	Freezing, response time, open field, social approach, facial expressions	Approach behaviors, consummatory behaviors toward any goal object	Learning, recall, discrimination, familiarity, recognition	Attachment formation — maintaining proximity; preference for individual, attachment maintenance: distress on separation	Waking, startle, eye blink, motor activity, learning, memory; attention; executive function, affective states: agitation, sensory reactivity, motivated behavior
7. Self-reports	Fear questionnaire, STAI, structured diagnostic assessment scales	Ecological momentary assessment, ambulatory assessment and monitoring	Cognitive assessment interview	Inventory of parent and peer attachment scale; attachment interview, attachment style interview; QSORT parent attachment interview, bereavement scales	Arousal, self-report scales (eg, ADACL, POMS arousal subscale, etc.); self-assessment mannequin
8. Paradigms	Fear conditioning, viewing aversive pictures or films, emotional imagery	Probabilistic reinforcement learning; deterministic reinforcement learning, pavlovian conditioning, instrumental conditioning, prediction error tasks	Paired associate learning; delayed recall; transitive inference; acquired equivalence; list and story learning	Social buffering of stress, strange situation, separation	EEG and EMG recording local field potentials and single neuron recordings; fMRI/PET – Psycho-motor vigilance, continuous performance tasks – Eye-blink – Startle – Odd-ball tasks

Source: Adapted from: www.nimh.nig.gov/research-priorities/rdoc.

Abbreviations: BDNF, brain-derived neurotrophic factor; COMT, catechol-O-methyl transferase; ACH, acetylcholine; NMDA, N-methyl-D-aspartate; OXTR, oxytocin receptor; AVPR1A, arginine vasopressin receptor 1A; TH, tyrosine hydroxylase; DRD2, dopamine D2 receptor; CRFR, corticotropin releasing factor receptor; KOR, kappa opioid receptor; DRD1, dopamine D1 receptor; CREB, cAMP response element binding protein; FOSB, FBJ murine osteosarcoma viral oncogene homolog B; DARPP 32, dopamine and cAMP-regulated phosphoprotein; EEG, electroencephalography; PF, prefrontal; PP, pedunculopontine; ACC, anterior cingulate; fMRI, functional magnetic resonance imaging; SN, substantia nigra; DG, dentate gyrus; LTD, long-term depression; ERP, event-related potential.

Table 21.3 Criteria Relevant to Prioritizing Drug Targets Based on Genetic Findings

1. The gene contains a causal variant unequivocally associated with the disorder.
2. The biological function of the gene, and the causal variant within it, is known.
3. The gene harbors multiple causal variants of known biological function.
4. The gene has a gain-of-function allele that protects against the disorder or a loss-of-function allele that increases risk.
5. The gene must be related to the clinical indications targeted for treatment.
6. The genetic variant is associated with an intermediate phenotype that can serve as a biomarker.
7. The gene is "druggable."
8. The genetic variant is not associated with other adverse phenotypes.
9. Corroborating biological data support the genetic findings.

Source: The criteria are reproduced from Harrison (2015) and adapted from Plenge (2013), and listed in order of priority.

to treat *all* of the symptoms of schizophrenia, in the future a drug might target a specific construct such as motivation or attention via basic biological knowledge of its underpinning genetic, neurochemical, and physiological circuitry (Insel et al., 2010). This makes understanding biology of specific behavioral constructs from a number of different scientific levels of enquiry of primary importance. It should be noted that the RDoC approach has also been criticized, for example, for over-reliance on clinical psychology (eg, Lieblich et al., 2015). At present, these domains and constructs are provisional starting points to update current classifications in line with new discoveries in psychology and biology and are not strict criteria. Behavioral neuroscience has the means to investigate the underlying biology of many behavioral constructs quite specifically; consequently, it should play an important role in the development of the new approach. We could never measure schizophrenia in mice, but we can measure specific behaviors such as reward, memory, or attention and integrate findings with other scientific techniques. Harrison (2015) has suggested a set of criteria relevant to prioritizing drug targets based on genetic findings (Table 21.2), and behavioral evidence will be crucial for attaining sufficient evidence to fulfill these criteria (Table 21.3).

CONCLUSION

Genetic mouse models have played and will continue to play an important role in understanding basic biological underpinnings of behaviors relevant to schizophrenia and its treatment. New dimensional approaches are being adopted for the classification of psychiatric symptoms, allowing the co-option of a rich repository of information about the neural bases of specific behaviors in the drive toward developing improved treatment for patients.

REFERENCES

Adler, L.E., Pachtman, E., Franks, R.D., Pecevich, M., Waldo, M.C., Freedman, R., 1982. Neurophysiological evidence for a defect in neuronal mechanisms involved in sensory gating in schizophrenia. Biol. Psychiatry 17 (6), 639–654.

REFERENCES

Axelrod, J., Tomchick, R., 1958. Enzymatic O-methylation of epinephrine and other catechols. J. Biol. Chem. 233 (3), 702–705.

Badcock, J.C., Hugdahl, K., 2014. A synthesis of evidence on inhibitory control and auditory hallucinations based on the Research Domain Criteria (RDoC) framework. Front. Hum. Neurosci. 8. http://dx.doi.org/10.3389/fnhum.2014.00180.

Bay-Richter, C., O'Callaghan, M.J., Mathur, N., O'Tuathaigh, C.M., Heery, D.M., Fone, K.C., et al., 2013. D-amphetamine and antipsychotic drug effects on latent inhibition in mice lacking dopamine D2 receptors. Neuropsychopharmacology 38 (8), 1512–1520. http://dx.doi.org/10.1038/npp.2013.50.

Beutler, L.R., Eldred, K.C., Quintana, A., Keene, C.D., Rose, S.E., Postupna, N., et al., 2011. Severely impaired learning and altered neuronal morphology in mice lacking NMDA receptors in medium spiny neurons. PLoS ONE 6 (11). http://dx.doi.org/10.1371/journal.pone.0028168.

Boehme, R., Deserno, L., Gleich, T., Katthagen, T., Pankow, A., Behr, J., et al., 2015. Aberrant salience is related to reduced reinforcement learning signals and elevated dopamine synthesis capacity in healthy adults. J. Neurosci. 35 (28), 10103–10111. http://dx.doi.org/10.1523/JNEUROSCI.0805-15.2015.

Braff, D.L., Geyer, M.A., 1990. Sensorimotor gating and schizophrenia - human and animal-model studies. Arch. Gen. Psychiatry 47 (2), 181–188.

Cazorla, M., de Carvalho, F.D., Chohan, M.O., Shegda, M., Chuhma, N., Rayport, S., et al., 2014. Dopamine D2 receptors regulate the anatomical and functional balance of basal ganglia circuitry. Neuron 81 (1), 153–164. http://dx.doi.org/10.1016/j.neuron.2013.10.041.

Chowdhury, R., Guitart-Masip, M., Lambert, C., Dayan, P., Huys, Q., Duzel, E., et al., 2013. Dopamine restores reward prediction errors in old age. Nat. Neurosci. 16 (5), 648–653. http://dx.doi.org/10.1038/nn.3364. http://www.nature.com/neuro/journal/v16/n5/abs/nn.3364.html#supplementary-information.

Cote, S.R., Chitravanshi, V.C., Bleickardt, C., Sapru, H.N., Kuzhikandathil, E.V., 2014. Overexpression of the dopamine D3 receptor in the rat dorsal striatum induces dyskinetic behaviors. Behav. Brain Res. 263, 46–50. <http://dx.doi.org/10.1016/j.bbr.2014.01.011>.

Coyle, J.T., 2006. Glutamate and schizophrenia: beyond the dopamine hypothesis. Cell Mol. Neurobiol. 26 (4–6), 365–384.

Cunningham, C.L., Howard, M.A., Gill, S.J., Rubinstein, M., Low, M.J., Grandy, D.K., 2000. Ethanol-conditioned place preference is reduced in dopamine D2 receptor-deficient mice. Pharmacol. Biochem. Behav. 67 (4), 693–699.

Cuthbert, B.N., 2015. Research domain criteria: toward future psychiatric nosologies. Dialogues. Clin. Neurosci. 17 (1), 89–97.

De Filippis, B., Lyon, L., Taylor, A., Lane, T., Burnet, P.W.J., Harrison, P.J., et al., 2015. The role of group II metabotropic glutamate receptors in cognition and anxiety: comparative studies in GRM2$^{-/-}$, GRM3$^{-/-}$ and GRM2/3$^{-/-}$ knockout mice. Neuropharmacology 89, 19–32. <http://dx.doi.org/10.1016/j.neuropharm.2014.08.010>.

De Mei, C., Ramos, M., Iitaka, C., Borrelli, E., 2009. Getting specialized: presynaptic and postsynaptic dopamine D2 receptors. Curr. Opin. Pharmacol. 9 (1), 53–58. http://dx.doi.org/10.1016/j.coph.2008.12.002.

Drew, M.R., Simpson, E.H., Kellendonk, C., Herzberg, W.G., Lipatova, O., Fairhurst, S., et al., 2007. Transient overexpression of striatal D-2 receptors impairs operant motivation and interval timing. J. Neurosci. 27 (29), 7731–7739. http://dx.doi.org/10.1523/Jneurosci.1736-07.2007.

Dubroqua, S., Singer, P., Yee, B.K., 2014. Deletion of forebrain glycine transporter 1 enhances conditioned freezing to a reliable, but not an ambiguous, cue for threat in a conditioned freezing paradigm. Behav. Brain Res. 273, 1–7.

Egan, M.F., Straub, R.E., Goldberg, T.E., Yakub, I., Callicott, J.H., Hariri, A.R., et al., 2004. Variation in GRM3 affects cognition, prefrontal glutamate, and risk for schizophrenia. Proc. Natl. Acad. Sci. U. S. A. 101 (34), 12604–12609. http://dx.doi.org/10.1073/pnas.0405077101.

Flagel, S.B., Clark, J.J., Robinson, T.E., Mayo, L., Czuj, A., Willuhn, I., et al., 2011. A selective role for dopamine in stimulus-reward learning. Nature 469 (7328), 53–57. <http://www.nature.com/nature/journal/v469/n7328/abs/nature09588.html#supplementary-information>.

Fonnum, F., 1984. Glutamate: a neurotransmitter in mammalian brain. J. Neurochem. 42 (1), 1–11.

Fowler, S.C., Zarcone, T.J., Vorontsova, E., Chen, R., 2002. Motor and associative deficits in D2 dopamine receptor knockout mice. Int. J. Dev. Neurosci. 20 (3–5), 309–321.

Fujioka, R., Nii, T., Iwaki, A., Shibata, A., Ito, I., Kitaichi, K., et al., 2014. Comprehensive behavioral study of mGluR3 knockout mice: implication in schizophrenia related endophenotypes. Mol. Brain 7 (1), 31.

Geyer, M.A., Krebs-Thomson, K., Braff, D.L., Swerdlow, N.R., 2001. Pharmacological studies of prepulse inhibition models of sensorimotor gating deficits in schizophrenia: a decade in review. Psychopharmacology 156 (2–3), 117–154.

Giros, B., Sokoloff, P., Martres, M.P., Riou, J.F., Emorine, L.J., Schwartz, J.C., 1989. Alternative splicing directs the expression of two D2 dopamine receptor isoforms. Nature 342 (6252), 923–926. http://dx.doi.org/10.1038/342923a0.

Glahn, D.C., Therman, S., Manninen, M., Huttunen, M., Kaprio, J., Lonnqvist, J., et al., 2003. Spatial working memory as an endophenotype for schizophrenia. Biol. Psychiatry 53 (7), 624–626.

Glimcher, P.W., 2011. Understanding dopamine and reinforcement learning: the dopamine reward prediction error hypothesis. Proc. Natl. Acad. Sci. 108 (Suppl. 3), 15647–15654. http://dx.doi.org/10.1073/pnas.1014269108.

Gogos, J.A., Morgan, M., Luine, V., Santha, M., Ogawa, S., Pfaff, D., et al., 1998. Catechol-O-methyltransferase-deficient mice exhibit sexually dimorphic changes in catecholamine levels and behavior. Proc. Natl. Acad. Sci. U. S. A. 95 (17), 9991–9996.

Gottesman, I.I., Gould, T.D., 2003. The endophenotype concept in psychiatry: etymology and strategic intentions. Am. J. Psychiatry 160 (4), 636–645.

Harrison, P.J., 2015. Recent genetic findings in schizophrenia and their therapeutic relevance. J. Psychopharmacol. 29 (2), 85–96. http://dx.doi.org/10.1177/0269881114553647.

Harrison, P.J., Weinberger, D.R., 2005. Schizophrenia genes, gene expression, and neuropathology: on the matter of their convergence (vol 10, pg 420, 2005). Mol. Psychiatry 10 (8), 804.

Harrison, P.J., Mclaughlin, D., Kerwin, R.W., 1991. Decreased hippocampal expression of a glutamate receptor gene in schizophrenia. Lancet 337 (8739), 450–452.

Heinz, A., Schlagenhauf, F., 2010. Dopaminergic dysfunction in schizophrenia: salience attribution revisited. Schizophr. Bull. 36 (3), 472–485. http://dx.doi.org/10.1093/schbul/sbq031.

Hendrie, C., Pickles, A., 2013. The failure of the antidepressant drug discovery process is systemic. J. Psychopharmacol. 27 (5), 407–416.

Howes, O., McCutcheon, R., Stone, J., 2015. Glutamate and dopamine in schizophrenia: an update for the 21st century. J. Psychopharmacol. 29 (2), 97–115. http://dx.doi.org/10.1177/0269881114563634.

Howes, O.D., Kapur, S., 2009. The dopamine hypothesis of schizophrenia: version III - the final common pathway. Schizophr. Bull. 35 (3), 549–562.

Huotari, M., Gogos, J.A., Karayiorgou, M., Koponen, I., Forsberg, M., Raasmaja, A., et al., 2002. Brain catecholamine metabolism in catechol-O-methyltransferase (COMT)-deficient mice. Eur. J. Neurosci. 15 (2), 246–256. http://dx.doi.org/10.1046/j.0953-816x.2001.01856.x.

Insel, T., Cuthbert, B., Garvey, M., Heinssen, R., Pine, D.S., Quinn, K., et al., 2010. Research domain criteria (RDoC): toward a new classification framework for research on mental disorders. Am. J. Psychiatry 167 (7), 748–751. http://dx.doi.org/10.1176/appi.ajp.2010.09091379.

Javitt, D.C., 2007. Glutamate and schizophrenia: phencyclidine, N-methyl-D-aspartate receptors, and dopamine-glutamate interactions Integrating the Neurobiology of Schizophrenia, vol. 78. Elsevier Academic Press Inc., San Diego, CA.69–108

Juckel, G., Schlagenhauf, F., Koslowski, M., Wustenberg, T., Villringer, A., Knutson, B., et al., 2006. Dysfunction of ventral striatal reward prediction in schizophrenia. Neuroimage 29 (2), 409–416. http://dx.doi.org/10.1016/j.neuroimage.2005.07.051.

Kapur, S., Mizrahi, R., Li, M., 2005. From dopamine to salience to psychosis - linking biology, pharmacology and phenomenology of psychosis. Schizophr. Res. 79 (1), 59–68. http://dx.doi.org/10.1016/j.schres.2005.01.003.

Karakas, E., Regan, M.C., Furukawa, H., 2015. Emerging structural insights into the function of ionotropic glutamate receptors. Trends Biochem. Sci. 40 (6), 328–337. <http://dx.doi.org/10.1016/j.tibs.2015.04.002>.

Kellendonk, C., Simpson, E.H., Polan, H.J., Malleret, G., Vronskaya, S., Winiger, V., et al., 2006. Transient and selective overexpression of dopamine D2 receptors in the striatum causes persistent abnormalities in prefrontal cortex functioning. Neuron 49 (4), 603–615. http://dx.doi.org/10.1016/j.neuron.2006.01.023.

Kelly, M.A., Rubinstein, M., Phillips, T.J., Lessov, C.N., Burkhart-Kasch, S., Zhang, G., et al., 1998. Locomotor activity in D2 dopamine receptor-deficient mice is determined by gene dosage, genetic background, and developmental adaptations. J. Neurosci. 18 (9), 3470–3479.

Kruzich, P.J., Grandy, D.K., 2004. Dopamine D2 receptors mediate two-odor discrimination and reversal learning in C57BL/6 mice. BMC Neurosci. 5, 12. http://dx.doi.org/10.1186/1471-2202-5-12.

Kuepper, R., Skinbjerg, M., Abi-Dargham, A., 2012. The dopamine dysfunction in schizophrenia revisited: new insights into topography and course. Handb. Exp. Pharmacol. 212, 1–26. http://dx.doi.org/10.1007/978-3-642-25761-2_1.

Kupchik, Y.M., Brown, R.M., Heinsbroek, J.A., Lobo, M.K., Schwartz, D.J., Kalivas, P.W., 2015. Coding the direct/indirect pathways by D1 and D2 receptors is not valid for accumbens projections. Nat. Neurosci. 18 (9), 1230–1232. http://dx.doi.org/10.1038/nn.4068.

Leung, A., Chue, P., 2000. Sex differences in schizophrenia, a review of the literature. Acta Psychiatry Scand. 101, 3–38.

Lieblich, S.M., Castle, D.J., Everall, I.P., 2015. RDoC: we should look before we leap. Aust. N. Z. J. Psychiatry 49 (9), 770–771. http://dx.doi.org/10.1177/0004867415592956.

Lubow, R.E., Moore, A.U., 1959. Latent inhibition: the effect of nonreinforced pre-exposure to the conditional stimulus. J. Comp. Physiol. Psychol. 52, 415–419.

Marsman, A., van den Heuvel, M.P., Klomp, D.W.J., Kahn, R.S., Luijten, P.R., Pol, H.E.H., 2013. Glutamate in schizophrenia: a focused review and meta-analysis of H-1-MRS studies. Schizophr. Bull. 39 (1), 120–129.

Mcdonald, J.W., Johnston, M.V., 1990. Physiological and pathophysiological roles of excitatory amino-acids during central-nervous-system development. Brain Res. Rev. 15 (1), 41–70. http://dx.doi.org/10.1016/0165-0173(90)90011-C.

Miyamoto, Y., Iida, A., Sato, K., Muramatsu, S.-I., Nitta, A., 2015. Knockdown of dopamine D2 receptors in the nucleus accumbens core suppresses methamphetamine-induced behaviors and signal transduction in mice. Int. J. Neuropsychopharmacol. 18 (4), pyu038.

Mohn, A.R., Gainetdinov, R.R., Caron, M.G., Koller, B.H., 1999. Mice with reduced NMDA receptor expression display behaviors related to schizophrenia. Cell 98 (4), 427–436.

Monsma Jr., F.J., McVittie, L.D., Gerfen, C.R., Mahan, L.C., Sibley, D.R., 1989. Multiple D2 dopamine receptors produced by alternative RNA splicing. Nature 342 (6252), 926–929. http://dx.doi.org/10.1038/342926a0.

Moran, P.M., Owen, L., Crookes, A.E., Al-Uzri, M.M., Reveley, M.A., 2008. Abnormal prediction error is associated with negative and depressive symptoms in schizophrenia. Prog. Neuropsychopharmacol. Biol. Psychiatry 32 (1), 116–123. http://dx.doi.org/10.1016/j.pnpbp.2007.07.021.

Moran, P.M., Rouse, J.L., Cross, B., Corcoran, R., Schurmann, M., 2012. Kamin blocking is associated with reduced medial-frontal gyrus activation: implications for prediction error abnormality in schizophrenia. PLoS One 7 (8), e43905. http://dx.doi.org/10.1371/journal.pone.0043905.

Moran, P.M., O'Tuathaigh, C.M., Papaleo, F., Waddington, J.L., 2014. Dopaminergic function in relation to genes associated with risk for schizophrenia: translational mutant mouse models. Prog. Brain Res. 211, 79–112. http://dx.doi.org/10.1016/B978-0-444-63425-2.00004-0.

Nakanishi, S., 1994. Metabotropic glutamate receptors: synaptic transmission, modulation, and plasticity. Neuron 13 (5), 1031–1037.

Nestler, E.J., Hyman, S.E., 2010. Animal models of neuropsychiatric disorders. Nat. Neurosci. 13 (10), 1161–1169.

Ohtsuka, N., Tansky, M.F., Kuang, H., Kourrich, S., Thomas, M.J., Rubenstein, J.L.R., et al., 2008. Functional disturbances in the striatum by region-specific ablation of NMDA receptors. Proc. Natl. Acad. Sci. 105 (35), 12961–12966. http://dx.doi.org/10.1073/pnas.0806180105.

Olney, J.W., Farber, N.B., 1995. Glutamate receptor dysfunction and schizophrenia. Arch. Gen. Psychiatry 52 (12), 998–1007.

O'Tuathaigh, C.M.P., Hryniewiecka, M., Behan, A., Tighe, O., Coughlan, C., Desbonnet, L., et al., 2010. Chronic adolescent exposure to Δ-9-tetrahydrocannabinol in COMT mutant mice: impact on psychosis-related and other phenotypes. Neuropsychopharmacology 35 (11), 2262–2273. http://dx.doi.org/10.1038/npp.2010.100.

Papaleo, F., Crawley, J.N., Song, J., Lipska, B.K., Pickel, J., Weinberger, D.R., et al., 2008. Genetic dissection of the role of catechol-O-methyltransferase in cognition and stress reactivity in mice. J. Neurosci. 28 (35), 8709–8723. http://dx.doi.org/10.1523/JNEUROSCI.2077-08.2008.

Rafter, M., Fone, K.C., Moran, P.M., 2015. Glutamate Pharmacological models relevant to schizophrenia and psychosis: can a receptor occupancy normalisation approach reduce the gap between animal and human experiments ? Pletnikov, M. Waddington, J. (Eds.),. In: Modeling Psychopathogical dimensions of Schizophrenia: From molecules to Behaviour, 23. Elsevier, pp. 140–184.

Ralph, R.J., Varty, G.B., Kelly, M.A., Wang, Y.M., Caron, M.G., Rubinstein, M., et al., 1999. The dopamine D2, but not D3 or D4, receptor subtype is essential for the disruption of prepulse inhibition produced by amphetamine in mice. J. Neurosci. 19 (11), 4627–4633.

Riedel, G., Platt, B., Micheau, J., 2003. Glutamate receptor function in learning and memory. Behav. Brain Res. 140 (1–2), 1–47.

Rogers, R.D., 2011. The roles of dopamine and serotonin in decision making: evidence from pharmacological experiments in humans. Neuropsychopharmacology 36 (1), 114–132.

Sannino, S., Gozzi, A., Cerasa, A., Piras, F., Scheggia, D., Manago, F., et al., 2014. COMT genetic reduction produces sexually divergent effects on cortical anatomy and working memory in mice and humans. Cereb. Cortex. http://dx.doi.org/10.1093/cercor/bhu053.

Schizophrenia Working Group of the Psychiatric Genomics, Consortium, 2014. Biological insights from 108 schizophrenia-associated genetic loci. Nature 511 (7510), 421–427. http://dx.doi.org/10.1038/nature13595, http://www.nature.com/nature/journal/v511/n7510/abs/nature13595.html#supplementary-information.

Simpson, E.H., Waltz, J.A., Kellendonk, C., Balsam, P.D., 2012. Schizophrenia in translation: dissecting motivation in schizophrenia and rodents. Schizophr. Bull. 38 (6), 1111–1117. http://dx.doi.org/10.1093/schbul/sbs114.

Tunbridge, E.M., Farrell, S.M., Harrison, P.J., Mackay, C.E., 2013. Catechol-O-methyltransferase (COMT) influences the connectivity of the prefrontal cortex at rest. Neuroimage 68, 49–54.

Ueno, T., Tomita, J., Tanimoto, H., Endo, K., Ito, K., Kume, S., et al., 2012. Identification of a dopamine pathway that regulates sleep and arousal in Drosophila. Nat. Neurosci. 15 (11), 1516–1523. <http://www.nature.com/neuro/journal/v15/n11/abs/nn.3238.html#supplementary-information>.

Usiello, A., Baik, J.H., Rouge-Pont, F., Picetti, R., Dierich, A., LeMeur, M., et al., 2000. Distinct functions of the two isoforms of dopamine D-2 receptors. Nature 408 (6809), 199–203.

van den Buuse, M., 2010. Modeling the positive symptoms of schizophrenia in genetically modified mice: pharmacology and methodology aspects. Schizophr. Bull. 36 (2), 246–270. http://dx.doi.org/10.1093/schbul/sbp132.

Waddington, J.L., O'Tuathaigh, C., O'Sullivan, G., Tomiyama, K., Koshikawa, N., Croke, D.T., 2005. Phenotypic studies on dopamine receptor subtype and associated signal transduction mutants: insights and challenges from 10 years at the psychopharmacology-molecular biology interface. Psychopharmacology (Berl) 181 (4), 611–638. http://dx.doi.org/10.1007/s00213-005-0058-8.

Walker, A.G., Wenthur, C.J., Xiang, Z., Rook, J.M., Emmitte, K.A., Niswender, C.M., et al., 2015. Metabotropic glutamate receptor 3 activation is required for long-term depression in medial prefrontal cortex and fear extinction. Proc. Natl. Acad. Sci. U. S. A. 112 (4), 1196–1201. http://dx.doi.org/10.1073/pnas.1416196112.

Weiner, I., Arad, M., 2009. Using the pharmacology of latent inhibition to model domains of pathology in schizophrenia and their treatment. Behav. Brain Res. 204 (2), 369–386. http://dx.doi.org/10.1016/j.bbr.2009.05.004.

Wise, R.A., 2004. Dopamine, learning and motivation. Nat. Rev. Neurosci. 5 (6), 483–494.

Yee, B.K., Balic, E., Singer, P., Schwerdel, C., Grampp, T., Gabernet, L., et al., 2006. Disruption of glycine transporter 1 restricted to forebrain neurons is associated with a procognitive and antipsychotic phenotypic profile. J. Neurosci. 26 (12), 3169–3181.

CHAPTER 22

GENETIC MECHANISMS EMERGING FROM MOUSE MODELS OF CNV-ASSOCIATED NEUROPSYCHIATRIC DISORDERS

A. Nishi[1,2] and N. Hiroi[1,3,4]

[1]*Department of Psychiatry and Behavioral Sciences, Albert Einstein College of Medicine, Bronx, NY, United States*
[2]*Department of Psychiatry, Course of Integrated Brain Sciences and Medical Informatics, Institute of Biomedical Sciences, Tokushima University Graduate School, Tokushima, Japan* [3]*Dominick P. Purpura Department of Neuroscience, Albert Einstein College of Medicine, Bronx, NY, United States* [4]*Department of Genetics, Albert Einstein College of Medicine, Bronx, NY, United States*

CHAPTER OUTLINE

Copy Number Variants and Diverse Neuropsychiatric Disorders 398
Dimensional Behavioral Models of Neuropsychiatric Disorders 400
Mechanisms by Which CNV-Encoded Genes Cause Behavioral Dimensions of Neuropsychiatric
Disorders 402
    Noncontiguous Effects 406
    Mass Action and Net Effects 407
    Pleiotropy 407
    Developmental Trajectories 407
    Phenotypic Variation 408
Acknowledgment 409
References 409

Various genetic variants have been explored as potential risk factors for schizophrenia. Single nucleotide polymorphisms (SNPs) have been extensively studied as variants found in more than 1% of schizophrenia samples (ie, common variants). Although SNPs collectively account for one-third to one-half of genetic liability for schizophrenia, each allele has very weak effects (odds ratio (OR) <1.2) (Lee et al., 2012; Ripke et al., 2013; Purcell et al., 2009). Because of their very weak effects, genes with common SNPs do not easily translate to genetic mouse models. Recently discovered variants, termed copy number variants (CNVs), include multiple genes in most cases. Each of these CNVs is found in less than 1% of the schizophrenia population; therefore, they are called "rare variants." Despite their rarity, each of these CNVs confers considerable risk for schizophrenia; two of these CNVs have ORs higher than ten, six have ORs higher than five, and four have ORs higher than two (Malhotra and Sebat, 2012). Given their robust and reproducible association with schizophrenia, these CNVs have been recapitulated in several genetic mouse models. In this chapter, we describe what we have learned about the genetic mechanisms through which CNVs cause dimensional aspects of schizophrenia and other neuropsychiatric disorders. There are also pharmacological and single gene models of schizophrenia, and the reader is referred to the chapter by Ayhan and Pletnikov in this work.

COPY NUMBER VARIANTS AND DIVERSE NEUROPSYCHIATRIC DISORDERS

Recent, concerted, large-scale studies have identified many robust, reliable genetic correlates of neuropsychiatric disorders. Among them are deletions or duplications of kilobase to megabase chromosomal segments, each representing less than 1% of disease cases. One remarkable feature of these rare CNVs is the robustness of their association with neuropsychiatric disorders. According to an estimate by Kirov and colleagues (see Fig. 22.1; Kirov et al., 2013), 3q29 deletion and 22q11.2 deletion confer extraordinarily high penetrance rates for schizophrenia. Other CNVs confer considerable risk, such as 16p11.2 duplication, 7q11.23 duplication, 1q21.1 deletion, and 15q13.3 deletion. Similarly, penetrance of CNVs for developmental delay, congenital malformation, and autism spectrum disorders (ASDs) is high with 7q11.23 deletion, 22q11.2 deletion, 15q11-13 duplication, 15q13.3 deletion,

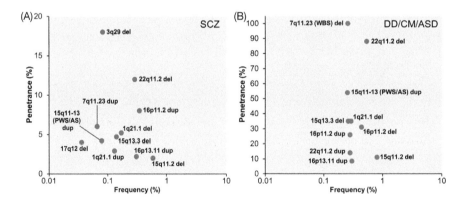

FIGURE 22.1

Risk for CNV carriers of developing disorders. Some representative high-risk cases of rare (<1%) variants are shown. For developmental delay (DD), congenital malformation (CM), and ASDs, only those cases with frequencies at 0.25% or higher are shown. Penetrance estimates for schizophrenia (SCZ) (A) and DD, CM, and ASDs (B) are based on published data (Kirov et al., 2013). Frequency indicates an estimated frequency of each CNV in a disease population.

This figure was taken from Hiroi, N., Nishi, A., 2016. Dimensional deconstruction and reconstruction of CNV-associated neuropsychiatric disorders. In: Modeling the psychopathological dimensions of schizophrenia and related psychoses. Handbook of Behaviorial Neuroscience, Vol. 23. Elsevier, San Diego, CA, pp. 285–302, Chapter 17, with permission.

1q21.1 deletion, 16p11.2 deletion, 16p11.2 duplication, 22q11.2 duplication, 15q11.2 deletion, and 16p13.11 duplication.

Several CNVs contain both deletion and duplication at the same chromosomal loci, including 1q21.1, 7q11.23, 16p11.2, and 22q11.2. Both deletion and duplication of all of these CNVs are associated with ID; deletion and duplication at 16p11.2 and 22q11.2 are associated with high rates of ASD; and deletion and duplication at 1q21.1 are associated with schizophrenia. However, there are cases in which high and low gene doses do not confer the same disorder. Duplication, but not deletion, at 1q21.1 and 7q11.23 is associated with ASD, and duplication at 16p11.2 is associated with schizophrenia. Although 22q11.2 deletion is associated with schizophrenia, duplication is not (Malhotra and Sebat, 2012).

How genes encoded in a CNV are expressed in the brain is still not well understood. Normally, blood samples are used to determine the presence of CNVs. It should be noted, however, that gene doses in blood cells do not necessarily indicate those in the brain. In fact, copy number variation could differ between individual neurons in humans, a phenomenon termed mosaicism (Gole et al., 2013; McConnell et al., 2013), whereby actual DNA content varies among individual neurons (Westra et al., 2008; Fischer et al., 2012).

Another feature of CNVs is pleiotropy. Some CNVs are associated with not only schizophrenia and ASDs but also intellectual disability (ID), developmental delay, congenital malformations, bipolar disorder, and recurrent depression (Kirov et al., 2013; Malhotra and Sebat, 2012; Rees et al., 2013; Szatkiewicz et al., 2014). Some CNVs are associated with ID, ASDs, and schizophrenia (eg, 15q11-13 duplication, 15q13.3 deletion, 16p11.2 duplication, and 22q11.2 deletion), but others are associated with only some of these disorders (eg, 16p11.2 deletion and 22q11.2 duplication) (Malhotra and Sebat, 2012), indicating that some CNV-encoded genes preferentially affect certain neuropsychiatric disorders.

The mode of inheritance varies among CNVs; 22q11.2 deletion, 7q11.23 deletion, and 16p11.2 deletion are predominantly de novo, whereas other CNVs are predominantly inherited, including 16p11.2 duplication, 15q11.2 deletion, 16p13.11 duplication, 15q13.3 deletion, 22q11.2 duplication, and 1q21.1 deletion duplication (Girirajan et al., 2012).

Since its association with schizophrenia was established (Driscoll et al., 1992a,b; Scambler et al., 1991, 1992; Shprintzen et al., 1992), hemizygous deletion at human chromosome 22q11.2 has been extensively studied and reliable statistics are available. With 22q11.2 hemizygous deletion, the percentages of carriers who exhibit disorders reach as high as 30% for schizophrenia, 27% for ASDs, 37% for attention deficit hyperactivity disorder (ADHD), and 36% for anxiety disorders (Schneider et al., 2014). Carriers of 22q11.2 deletion have an average IQ of approximately 70 and are typically diagnosed with mild to borderline ID (Swillen et al., 1999; Butcher et al., 2012; De Smedt et al., 2007; Niklasson et al., 2009; Schneider et al., 2014). Although some claimed that 22q11.2 hemizygosity was not associated with ASDs, it is now established that carriers exhibit ASDs at high rates (Hiroi et al., 2012; Hiroi and Nishi, 2016; Hiroi et al., 2013; Schneider et al., 2014).

Duplication is also found at this chromosomal locus, and most carriers exhibit developmental delays in cognitive, socioemotional, and motor function. Some reach the diagnostic criteria of ASDs, mild to borderline ID, and ADHD (Mukaddes and Herguner, 2007; Alberti et al., 2007; Brunet et al., 2006; Courtens et al., 2008; Descartes et al., 2008; Edelmann et al., 1999; Ensenauer et al., 2003; Hassed et al., 2004; Lo-Castro et al., 2009; Ou et al., 2008; Portnoi et al., 2005, 2009; Ramelli et al., 2008; van Campenhout et al., 2012; Wentzel et al., 2008; Hiroi et al., 2013; de La Rochebrochard et al., 2006).

When a disease population is examined, 22q11.2 deletion and duplication cases are found in less than 1% of cases (ie, rare variants). Although previous small-scale studies suggested a much higher rate, a recent study of 12,202 schizophrenic patients established the rate of 22q11.2 deletion in this population as being between 0.2 and 0.3% (Malhotra and Sebat, 2012; Rees et al., 2014). Duplications of this locus are also found among individuals with ASDs, intellectual disability, and developmental delay (Kirov et al., 2013; Malhotra and Sebat, 2012).

DIMENSIONAL BEHAVIORAL MODELS OF NEUROPSYCHIATRIC DISORDERS

Neuropsychiatric disorders are defined using clinical diagnostic scales that include the presumably essential core feature(s) of a given psychiatric disorder. Psychopathological features of schizophrenia include positive symptoms (ie, delusions and hallucinations), negative symptoms (ie, lack of motivation, reduction in spontaneous speech, social withdrawal, and decreased social functioning and self-care). Further, schizophrenia is characterized by decreased attention, speed of processing, working and long-term memory, executive function, and social cognition (Fioravanti et al., 2005).

These pathological features are not specific to schizophrenia. For instance, delusions and hallucinations may also be present in patients with bipolar disorder. Defective social cognition is seen in patients with ASDs, ADHD, and developmental language disorders (Korkmaz, 2011), as well as in those with schizophrenia (Sprong et al., 2007). Working memory deficits are found in patients with schizophrenia (Piskulic et al., 2007), ASDs (Bennett and Heaton, 2012; Pennington and Ozonoff, 1996; Russo et al., 2007; Luna et al., 2007; O'Hearn et al., 2008), and ADHD (Martinussen et al., 2005; Willcutt et al., 2005). It should be noted, however, that a dimensional measure may have distinct mechanisms among distinct disorders. For example, social impairments seen in ASDs and ADHD seem to have different underlying causes (Walcott and Landau, 2004).

Given that each CNV is associated with multiple disorders, it might be futile to attempt to model a single neuropsychiatric disorder in CNV mouse models. Rodent behavioral tasks are dimensional and quantitative in nature and are designed to evaluate specific perceptional, cognitive, social, affective, emotional, and motor processes separately. Because these dimensional measures are not specific to a particular mental disorder in humans, rodent behavioral tasks are not intended to model a mental disorder. However, these dimensional measures might more faithfully represent the way in which a CNV affects a phenotype (Hiroi et al., 2013; Hiroi and Nishi, 2016). Here, we describe some of the commonly used tasks for analysis of mouse models of CNVs.

Prepulse inhibition (PPI). When a weak acoustic auditory stimulus is presented immediately before a loud sound, a startle response induced by the loud sound is inhibited in rodents and humans. This phenomenon, termed PPI, is attenuated not only in schizophrenic patients but also in individuals with schizotypal personality disorder, obsessive compulsive disorder, Tourette syndrome, Huntington disorder, bipolar disorder, seizures, Lewy body dementia, and ADHD (Geyer, 2006). PPI deficits are not normally seen in individuals with ASDs when tested using standard procedures, but a subtle deficit appears at a high prepulse intensity under specific parameters (Belmonte et al., 2004; McAlonan et al., 2002; Perry et al., 2007; Yuhas et al., 2010; Ornitz et al., 1993)

Social behaviors. Patients with schizophrenia or ASDs exhibit defective social motivation and cognition (Penn et al., 1997; Sigman, 1998). Social behaviors are modeled in mice in a naturalistic cage set-up (Silverman et al., 2010) and a three-chamber "sociability" test. The former is designed to evaluate the ethological details of reciprocal social interaction. Two mice are placed into a home cage setting in which neither mouse has previously resided; because there is no "resident" mouse, affiliative social behaviors, but not aggressive behaviors, are optimally measured. The three-chamber sociability task is automated; a stimulus mouse is confined in a small cage or compartment in one of the three chambers. A test mouse's approach to the cage or compartment is measured. Several technical issues of the three-chamber apparatus recently have been pointed out, indicating that data should be interpreted with caution. First, time spent near the caged stimulus mouse widely varies across cohorts and is not appropriate for parametric quantitative analyses (Crawley, 2014). Second, scores from the three-chamber test do not correlate well with those of genuine reciprocal social interaction in the naturalistic test setup, suggesting that "sociability" measured in the three-chamber apparatus is different from genuine reciprocal social interaction (Fairless et al., 2013; Spencer et al., 2005, 2011).

Working memory and flexibility. Both humans and rodents have the capacity to temporarily hold incoming information to aid decisions about the next course of action based on that memory. This memory or working memory is impaired in patients with schizophrenia (Piskulic et al., 2007) and in adolescents and adults with ASDs (Bennett and Heaton, 2012; Pennington and Ozonoff, 1996; Russo et al., 2007). Spontaneous alternation in the T-maze or Y-maze is designed to measure working memory and memory-based repetitive behavioral tendencies (Lalonde, 2002). Spontaneous alternation takes advantage of a mouse's natural behavioral tendency to alternate behavior based on working memory. Animals remember which of the two goal arms they visited at a previous trial and tend to visit the other goal arm at the next trial. Rewarded alternation also is used to evaluate working memory, but this task requires extensive pretraining with reward; if motivation for reward is altered, then it is difficult to isolate and evaluate working memory deficits. These two alternation tasks do not distinguish a working memory deficit from a repetitive behavioral tendency, and a reduction in alternation could reflect either or both. Reversal of a learned behavior is used as an index of cognitive flexibility in attentional set shifting, Morris water maze, rewarded alternation, and discriminative operant learning (Brigman et al., 2010).

Anxiety-like behaviors. Anxiety is heightened in patients with schizophrenia or ASDs (van Steensel et al., 2011, 2013). The elevated plus maze and an inescapable open field are used to measure anxiety-related behaviors under varying degrees of stress. Stress levels are considered much lower when mice have a choice to escape from an open arm to a closed arm of the elevated plus maze compared to when they do not have such a choice in the inescapable open field (Misslin et al., 1982; Zhu et al., 2007). Anxiety-like behaviors seen in various anxiety tasks have nonidentical behavioral components and distinct genetic underpinnings (Trullas and Skolnick, 1993; Rogers et al., 1999; Ramos and Mormede, 1998; Carola et al., 2002; Henderson et al., 2004; Takahashi et al., 2008; Turri et al., 2001; Ramos et al., 1999; O'Leary et al., 2013). In fact, measures taken from an open field apparatus and an elevated plus maze yield contradictory phenotypes in many cases. In other words, it is unreasonable to assume that anxiety-related phenotypes should be "corroborated" in multiple tasks. A phenotype in one anxiety-related task, but not another, should be instead interpreted as an indication that the gene differentially contributes to specific aspects of anxiety. However, if a phenotype appears in more than one anxiety-related tasks, then such data should be interpreted as an indication that the gene nonspecifically contributes to various aspects of anxiety-related behaviors or to a commonly shared aspect of various anxiety-related behaviors.

Neonatal vocalization. Atypical preverbal vocalizations are prognostic of ASDs (Ozonoff et al., 2010, 2014; Esposito and Venuti, 2010). Atypical cries do not facilitate bonding or reciprocity between babies and mothers, probably because the emotional state of these infants is not easily understood or is negatively perceived by mothers (Esposito et al., 2013; Esposito and Venuti, 2009). When separated from dams, mouse pups emit ultrasonic vocal calls, and these calls elicit a maternal approach (Sewell, 1970; Okabe et al., 2013; Uematsu et al., 2007). Atypical pup calls have been described in many mouse models of ASDs (Lai et al., 2014; Takahashi et al., 2015). Neonatal preverbal vocalization in humans and pup vocalization in mice have an innate component (Arriaga and Jarvis, 2013; Kikusui et al., 2011). Early neonatal vocalization emitted by human infants and rodent pups occurs without auditory feedback, because deaf infants and deaf mouse pups spontaneously emit various vocal signals (Brors et al., 2008; Hammerschmidt et al., 2012; Scheiner and Fischer, 2011; Volkenstein et al., 2009). Because of temporal proximity between atypical vocalizations in infants and clinical diagnosis of ASDs, this association has been well studied; however, the association between atypical vocalization and late-onset disorders (eg, schizophrenia) is technically difficult to establish due to more than a 10-year gap between atypical vocalization and symptom onset in humans.

MECHANISMS BY WHICH CNV-ENCODED GENES CAUSE BEHAVIORAL DIMENSIONS OF NEUROPSYCHIATRIC DISORDERS

Mouse models of human CNVs have been developed. These include 7q11.23 deletion, paternal and maternal 15q11-13 duplication, 15q13.3 deletion, 16p11.2 deletion and duplication, and 22q11.2 duplication and deletion. These model mice have been examined for their behavioral phenotypes; some, but not all, of these CNV models exhibit dimensional features of human neuropsychiatric disorders (see Table 22.1). Data from these studies indicate promises as well as limitations of these genetic mouse models (Hiroi and Nishi, 2016).

Although CNV mouse models recapitulate the extent of genomic deletion and duplication, mouse models with smaller segmental overexpression and deletion of murine chromosomes within a CNV are useful for identifying chromosomal segments critical for behavioral phenotypes. For instance, overexpression of a 200-kb segment of human 22q11.2, which includes the four protein-coding human genes *SEPT5*, *GP1BB*, *TBX1*, and *GNB1L* (Table 22.1; Fig. 22.2), induces social behavioral deficits and spontaneous exacerbation of repetitive hyperactivity. Spontaneous exacerbation of hyperactivity was attenuated by chronic treatment with the antipsychotic drug clozapine, suggesting relevance of this segment to neuropsychiatric disorders (Hiroi et al., 2005). In contrast, overexpression of an adjacent 190-kb human chromosomal region (*TXNRD2*, *COMT*, and *ARVCF*) selectively delayed developmental maturation of working memory capacity from 1 month to 2 months of age (Table 22.1; Fig. 22.2) but had no effect on PPI, social interactions, or anxiety-like behaviors (Suzuki et al., 2009a). Overexpression of other segmental regions outside the 200-kb and 190-kb region had no effect on PPI (Tg-2; Table 22.1; Fig. 22.2) or even increased PPI (Tg-1; Table 22.1; Fig. 22.2; Stark et al., 2009). Taken together, these data provided evidence that overexpression of genes encoded within this CNV may not contribute equally to a given phenotype.

Segmental deletions similarly supported the contribution of the 200-kb segment to a phenotype relevant to neuropsychiatric disorders. PPI was reduced in mice with large, partially overlapping deletions as long as the deletion included the 200-kb critical region (Fig. 22.2; Table 22.1; Df(16)1/+, Df(16)3/+, Df(16)4/+). Note, however, that mice with a large 22q11.2 hemizygous deletion have hearing impairments, and PPI deficit might reflect this peripheral deficit (Fuchs et al., 2013). Hemizygous deletions outside the 200-kb region do not recapitulate the PPI reduction seen in 22q11.2 hemizygous carriers in humans (Sobin et al., 2005; Fig. 22.2; Table 22.1; Df(16)2/+ and Df(16)5/+) (Paylor et al., 2006) or

Table 22.1 Mouse Models of CNVs and Dimensional Behavioral Phenotypes

CNV	Designation	ES or Zygote Cell Background	Additional Background	PPI	WM	RL	SI	Ax	LA	Pup Voc	Reference
7q11.23 del	PD *Limk1-Gtf2i*	129S7/SvEvBrd-Hprt	C57BL/6J and 129SvEv	↓	-		↑	(↑)	-		Li et al. (2009)
	DD *Trim50-Limk1*			-			-		↓		
	D/P *Trim50-Gtf2i*			-			(↑)	(↑)	↓		
15q11-13 dup	patDp/+ *Herc2-Mkrn3*	129S7/SvEvBrd-Hprt<b-m2>	129S6/SvEvTac or C57BL/6J	-	-	↓	↓	↑	↓	↑	Nakatani et al. (2009)
	matDp/+ *Herc2-Mkrn3*			-	-	-	-	-	-	-	
15q11-13 dup	patDp/+ *Herc2-Mkrn3*	129S7/SvEvBrd-Hprt<b-m2>	C57BL/6J, congenic >10 backcrosses	-	-			↑	↓		Tamada et al. (2010)
15q13.3 del	Df(h15q13)/+ *Fan1-Chrna7*	C57BL/6NTac	C57BL/6NTac, co-isogenic	-	-		-	-	-		Fejgin et al. (2013)
16p11.2 del	df/+ *Spn-Coro1a*	129S7/SvEvBrd-Hprt<b-m2>	F2 C57BL/6N	-	-		-	-	↑		Horev et al. (2011)
16p11.2 dup	dp/+ *Spn-Coro1a*			-	-		-	-	↓		
16p11.2 del	df/+ *Spn-Coro1a*	129S7/SvEvBrd-Hprt<b-m2>	C57BL/6N, N8	-	-		-	↑	-		Pucilowska et al. (2015)
16p11.2 del	16p11+/- *Spn-Coro1a*	129/Ola, 129S1/SvImJ	CD1; C57BL/6N, at least 5–7 backcrosses	-	-		-	-	↑		Portmann et al. (2014)
16p11.2 del	16p11+/- *Spn-Coro1a*	129/Ola, 129S1/SvImJ	94% C57BL/6N 6% 129P2/Ola and CD-1	↓	-		↓	(↓)	-	Sim ↑ Com ↓	Yang et al. (2015b)
22q11.2 dup	200 kb Tg *Gnb1l -Sept5* High copy	FVB	FVB, co-isogenic C57BL/6J; N4 FVB	-	-		↓	-	↑		Hiroi et al. (2005)
	Low copy			-	-		↓	-	↑		
22q11.2 dup	190 kb Tg *Arvcf-Txnrd2*	FVB	C57BL/6J, congenic 10 backcrosses	-	↓		-	-	↓		Suzuki et al. (2009a)
22q11.2 dup	Tg-1 *Vpreb2, Prod* Tg-2 *Zdhhc8, Ranbp1, Htf9c, T10, Arvcf, Comt*	FVB/N	FVB/N × SW × C57BL/6J, N2 backcross	-	-		-	-	↓		Stark et al. (2009)

(Continued)

Table 22.1 Mouse Models of CNVs and Dimensional Behavioral Phenotypes (Continued)

CNV	Designation	ES or Zygote Cell Background	Additional Background	PPI	WM	RL	SI	Ax	LA	Pup Voc	Reference
22q11.2 del	Znf520-ps-Slc25a1	129S6/SvEvTac	129S6/SvEvTac co-isogenic or NIH Black Swiss	↑				-	-		Kimber et al. (1999)
22q11.2 del	Df(16)1/+ Dgcr14-Ufd1l	129S7/SvEvBrd-Hprt<b-m2>	129S5/SvEvBrd; C57BL/6^{c-/c-}, 4 or 5 backcrosses	↓				-	-		Paylor et al. (2001)
22q11.2 del	Df(16)1/+ Dgcr14-Ufd1l	129S7/SvEvBrd-Hprt<b-m2>	129S5/SvEvBrd; C57BL/6^{c-/c-}, 5–6 backcrosses	↓							Paylor et al. (2006)
	Df(16)2/+ Dgcr14-Txnrd2			-							
	Df(16)3/+ Dgcr14-Sept5			↓							
	Df(16)4/+ Tango2-Hira			↓							
	Df(16)5/+ Cldn5-Hira			-							
22q11.2 del	Lgdel/+ Dgcr2-Hira	129/Sv, C57BL/6J, SJL; 129S6/SvEvTac; FVB/N	C57BL/6; 129Sv; CD1; C57BL/6J, >5 backcrosses	↓				-	-		Long et al. (2006)
22q11.2 del	Df(16)A^{+/−} Dgcr2-Hira	129S7/SvEvBrd-Hprt<b-m2>	C57BL/6J, 3 backcrosses	↓	↓			↑	↑		Stark et al. (2008)
22q11.2 del	Df(16)A^{+/−} Dgcr2-Hira	129S7/SvEvBrd-Hprt<b-m2>	C57BL/6J, congenic		-						Sigurdsson et al. (2010)
22q11.2 del	Lgdel/+ Dgcr2-Hira	129/Sv, C57BL/6J, SJL; 129S6/SvEvTac; FVB/N	C57BL/6; 129Sv; CD1; C57BL/6J, >25 backcrosses			↓					Meechan et al., 2015

This Table is taken from our previous publication with permission Hiroi, N., Nishi, A., 2016. Dimensional deconstruction and reconstruction of CNV-associated neuropsychiatric disorders. In: Modeling the psychopathological dimensions of schizophrenia and related psychoses. Handbook of Behavioral Neuroscience, Elsevier, San Diego, CA, Vol. 23, pp. 285–302, Chapter 17. PPI, prepulse inhibition; WM, working memory; RL, reversal learning; SI, social interaction; Ax, Anxiety; LA, locomotor activity; Pup Voc, neonatal vocalization; Del, deletion; dup, duplication; F#, generation of intercrossed line; N#, generation of backcrossed line; red arrow, effects consistent with human phenotypes; black arrow and bar, effects inconsistent with phenotypes or not reported in humans; black bar, nonsignificant genetic effects on phenotypes. Sim, simple call types; Com, complicated call types; ( ) indicates mixed phenotypes.

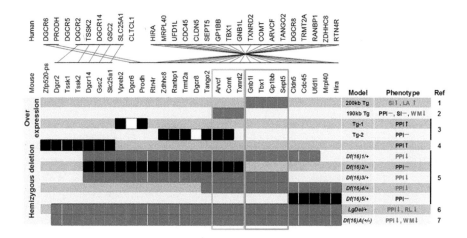

FIGURE 22.2

Mouse models of 22q11.2 CNVs. Segmental overexpression and deletions induce behavioral phenotypes consistent (red) and inconsistent (black) with human variant carriers. Only protein-coding genes are shown. Blue frame shows a 200-kb critical region in which overexpression induces defective social behaviors and clozapine-responsive compulsive hyperactivity, and outside which hemizygous deletions do not induce PPI deficits. Green frame indicates an adjacent 190-kb region critical for working memory. (1, Hiroi et al., 2005; 2, Suzuki et al., 2009a; 3, Stark et al., 2009; 4, Kimber et al., 1999; 5, Paylor et al., 2001; Paylor et al., 2006; 6, Long et al., 2006; 7, Stark et al., 2008).

Taken from our previous publications with permission Hiroi, N., Takahashi, T., Hishimoto, A., Izumi, T., Boku, S., Hiramoto, T., 2013. Copy number variation at 22q11.2: from rare variants to common mechanisms of developmental neuropsychiatric disorders. Mol. Psychiatry 18, 1153–1165; Hiroi, N., Nishi, A., 2016. Dimensional deconstruction and reconstruction of CNV-associated neuropsychiatric disorders. In: Modeling the psychopathological dimensions of schizophrenia and related psychoses. Handbook of Behavioral Neuroscience, Vol. 23. Elsevier, San Diego, CA, pp. 285–302, Chapter 17.

resulted in increased PPI (see Fig. 22.2; Table 22.1; *Zfp520-ps* to *Slc25a1*) (Kimber et al., 1999). These negative data nonetheless are critical for eliminating segmental regions irrelevant for a phenotype.

Other CNVs have not been as extensively analyzed as 22q11.2 CNVs. Nonetheless, when available, data are consistent with the nonequal segmental contributions to phenotypes. 7q11.23 CNVs are associated with ID and developmental delays (see Fig. 22.1). A deletion mouse model with the ortholog of the entire 7q11.23 from *Gtf2i* to *Fkbp6* exhibits impairments in motor coordination and spatial memory, an increase in startle reaction to an acoustic stimulus, and an increase in social behavior, reminiscent of hypersociality in 7q11.23 deletion patients (Segura-Puimedon et al., 2014). These phenotypes are not equally induced by deletions of different segments within the 7q11.23 ortholog. Deletions of the proximal segment from *Gtf2i* to *Link1* and the distal segment from *Link1* to *Trim50* of murine 7q11.23 ortholog result in nonidentical phenotypes. The proximal deletion results in increased reciprocal social interaction and acoustic startle response to pain and impairs prepulse inhibition, whereas the distal deletion results in deficits in contextual and cued fear conditioning (Li et al., 2009). Anxiety-related behaviors were affected by both proximal and distal deletions (Li et al., 2009).

Studies have further examined individual genes encoded in critical segments. We previously proposed a set of hypothetical rules that govern gene–phenotype relation based on individual 22q11.2 genes (Hiroi and Nishi, 2016; Hiroi et al., 2013). Since then, more data have become available to reevaluate the generality of some of the hypothetical rules in other CNV mouse models. Here, we provide an updated re-evaluation of those hypotheses.

NONCONTIGUOUS EFFECTS

For any given phenotype, some (but not all) encoded genes, not necessarily located contiguously, are responsible. The noncontiguous effect of CNV-encoded genes was most elegantly revealed in mouse models of 22q11.2 genes thanks to extensive behavioral screenings (Fig. 22.3). Within the 200-kb segment of murine 22q11.2 ortholog we identified (Hiroi et al., 2005), heterozygous deletion of *Tbx1* induced a PPI reduction (Paylor et al., 2006), but homozygous—not heterozygous—deletion of another gene, namely *Sept5*, potentiated PPI (Suzuki et al., 2009b). Single genes located in a segment with no PPI deficit (Fig. 22.3; Table 22.1; Df(16)2/+, Tg-1, and Tg-2) do not have much effect on this phenotype. Deletions of *Zdhhc8* and *Dgcr8* had very weak effects on PPI at only one or two prepulse levels. Deletions of *Gsc2*, *Rtn4r*, or *Comt* had no effect on this phenotype. Noncongenic *Prodh* mutant mice exhibited a PPI reduction (Gogos et al., 1999), but congenic *Prodh* mutant mice did not (Paterlini et al., 2005).

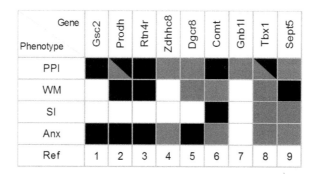

FIGURE 22.3

Impact of homozygous and heterozygous deletions of individual 22q11.2 gene on various behavioral phenotypes. Genes cause mouse phenotypes consistent with (red) and opposite to (gray) large deletions in humans. Deletions of some genes apparently cause no effect on a phenotype (black). Phenotypes that have not been studied in relation to given genes are indicated by blank squares. *PPI*, prepulse inhibition; *WM*, working memory; *SI*, social behaviors; *Anx*, anxiety-like behaviors. (1, Long et al., 2006; 2, Gogos et al., 1999; Paterlini et al., 2005; 3, Gogos et al., 1999; Paterlini et al., 2005; 4, Mukai et al., 2004; 5, Stark et al., 2008; Ouchi et al., 2013; Chun et al., 2014; 6, Gogos et al., 1998; Papaleo et al., 2008; Babovic et al., 2008; O'Tuathaigh et al., 2010; O'Tuathaigh et al., 2012; 7, Paylor et al., 2006; 8, Long et al., 2006; Paylor et al., 2006; Hiramoto et al., 2011; 9, Paylor et al., 2006; Suzuki et al., 2009b; Harper et al., 2012).

Taken from Hiroi, N., Takahashi, T., Hishimoto, A., Izumi, T., Boku, S., Hiramoto, T., 2013. Copy number variation at 22q11.2: from rare variants to common mechanisms of developmental neuropsychiatric disorders. Mol. Psychiatry 18, 1153–1165 and Hiroi, N., Nishi, A., 2016. Dimensional deconstruction and reconstruction of CNV-associated neuropsychiatric disorders. In: Modeling the psychopathological dimensions of schizophrenia and related psychoses. Handbook of Behavioral Neuroscience, Vol. 23. Elsevier, San Diego, CA, pp. 285–302, Chapter 17, with permission.

For working memory impairments, deficiencies of *Dgcr8* and *Tbx1* are contributory, but those of other genes are not (eg, *Prodh*, *Rtn4r*, and *Sept5*). Dose deficiencies of *Zdhhc8*, *Comt*, and *Tbx1* contribute to heightened anxiety traits, but those in *Gsc2*, *Prodh*, *Rtn4r*, *Dgcr8*, or *Sept5* do not.

Genes encoded in 7q11.23 CNV similarly have noncontiguous effects. For instance, *Gtf2ird1* deletion results in increased anxiety-like behaviors (Schneider et al., 2012), but *eif4h* deletion does not (Capossela et al., 2012).

Such noncontiguous effects suggest that caution is warranted in the search to identify neuronal and molecular mechanisms underlying neuropsychiatric disorders simply based on the presence of a gene in a CNV. Some CNV genes whose gene dose alteration causes anatomical, neuronal, and molecular phenotypes might have little to do with dimensional behavioral phenotypes relevant to mental disorders.

Although CNVs on the human genome also include nonprotein-coding genes, such as microRNAs, their murine orthologs are still poorly characterized.

MASS ACTION AND NET EFFECTS

Some phenotypes are collectively affected by more than one gene encoded within a CNV. Among 22q11.2-encoded genes, deficiencies of *Prodh*, *Zdhhc8*, *Dgcr8*, *Gnb1l*, and *Tbx1* contribute to PPI deficits (see Fig. 22.3). Deficiencies of *Dgcr8* and *Tbx1* contribute to working memory impairments. Social interaction is impaired by *Tbx1* and *Sept5* deficiencies (Hiramoto et al., 2011; Harper et al., 2012; Suzuki et al., 2009b). Anxiety-related behaviors are potentiated by gene dose reduction in *Zdhhc8*, *Comt*, and *Tbx1*.

When deleted or overexpressed, some individual CNV-encoded genes produce behavioral phenotypes opposite to what is seen in carriers of that CNV. For example, deletion of *Sept5* increases PPI and reduces anxiety-related behavior (see Fig. 22.3). Deletion of *Comt* improves working memory (see Fig. 22.3). Given that these effects are opposite to those seen in 22q11.2 deletion carriers, a plausible explanation is that these effects of some genes are masked by much larger net effects of other genes.

PLEIOTROPY

Some individual genes contribute to more than one phenotype. Such pleiotropic action is supported by the diverse phenotypic effects of some of CNV-encoded genes (see Fig. 22.3). *Tbx1* heterozygosity of 22q11.2 CNV induces defective PPI, social interaction, working memory, and heightened anxiety traits (Hiramoto et al., 2011). *Dgcr8* deficiency causes PPI and working memory deficits (Stark et al., 2008; Ouchi et al., 2013; Chun et al., 2014). *Gtf2i* deficiency of 7q11.23 CNV contributes to impaired social interaction and response to novelty (Sakurai et al., 2011) and anxiety-related behaviors (Lucena et al., 2010). Deficiency of cholinergic receptor neuronal α7 subunit (*Chrna7*), a gene encoded in 15q13.3 CNV, impairs attentional processing, working memory, impulsivity control, and accuracy in spatial discrimination learning and decreases anxiety-like thigmotaxis (Young et al., 2004, 2007; Hoyle et al., 2006; Keller et al., 2005; Levin et al., 2009).

DEVELOPMENTAL TRAJECTORIES

Gene dose alteration causes phenotypes to appear at specific time points along the postnatal developmental course. We demonstrated that mice at 2 months of age have better working memory performance than at 1 month of age, and transgenic mice constitutively overexpressing the 190-kb 22q11.2 segment (*TXNRD2*, *COMT*, and *ARVCF*; see Fig. 22.2; Table 22.1) exhibited impaired working memory performance in the rewarded alteration at 2 months but not at 1 month (Suzuki et al., 2009a). This finding parallels the developmental trajectory of working memory capacity and IQ in humans. Working memory performance shows a gradual but steady improvement from childhood (eg, 6 years old) to adulthood (ie, 20 years old) (Brocki and Bohlin, 2004; De Luca et al., 2003; Demetriou et al., 2002; Gathercole et al., 2004; Luna et al., 2004; Luna, 2009; Swanson, 1999). A high-activity *COMT* allele is associated with poor visuospatial working memory after, but not before, 10 years of age, compared to a low-activity *COMT* allele (Dumontheil et al., 2011). For 22q11.2 deletion carriers, those who develop psychosis start to show a decline in verbal IQ at approximately 11 years of age compared to those who do not develop psychosis (Vorstman et al., 2015).

Other phenotypes appear much earlier. Neonatal vocalization in pups under maternal separation is atypical in several mouse models of CNVs and a CNV-encoded gene. They include paternal 15q11-13 duplication (Nakatani et al., 2009), *Tbx1* heterozygosity of 22q11.2 CNV (Hiramoto et al., 2011; Takahashi et al., 2015), and 16p11.2 hemizygosity (Yang et al., 2015b; see Table 22.1). This atypical behavioral appears within the first week of life in pups of mouse models of ASDs (Lai et al., 2014) and is one of the earliest phenotypes associated with ASDs in humans (Ozonoff et al., 2010, 2014; Esposito and Venuti, 2010). Infants with ASDs and ID have distinctly atypical vocalizations (Esposito and Venuti, 2008, 2009). However, whether atypical preverbal vocalizations are present in other neuropsychiatric disorders, including schizophrenia and ADHD, is not well understood.

PHENOTYPIC VARIATION

Phenotypic features of carriers of a given CNV vary (Hiroi et al., 2013; Malhotra and Sebat, 2012). For example, up to 30% and 27% of 22q11.2 hemizygosity carriers are diagnosed with schizophrenia and ASDs, respectively (Hiroi et al., 2013; Schneider et al., 2014). This might be one of the reasons why some mouse models of CNVs do not differ from control mice in some behavioral dimensions (see Table 22.1). Mouse models of 16p11.2 deletion and duplication show no detectable phenotypic features of working memory, "sociability," or anxiety (Horev et al., 2011; Portmann et al., 2014; see Table 22.1). One congenic mouse model of large 22q11.2 deletion (Df(16)A/+) is not impaired in spatial working memory (Sigurdsson et al., 2010), whereas noncongenic Df(16)A/+ mice are thus impaired (Stark et al., 2008). A coisogenic mouse model of 15q13.3 deletion has normal PPI, no deficits in working memory, or no anxiety-like behaviors (Fejgin et al., 2013). A congenic mouse model of paternal 15q11-13 duplication is indistinguishable from wild-type mice regarding PPI and working memory (Tamada et al., 2010). Like human CNV carriers with incomplete penetrance for a given neuropsychiatric disorder, a subset of mice might exhibit expected phenotypes.

Another possible explanation for the apparent absence of phenotype is that genetic background modifies phenotypic expression (Hiroi, 2014; Hiroi et al., 2013). We showed that when deletion of *Sept5*, a 22q11.2 gene, is placed on different genetic backgrounds of inbred mouse lines, the phenotypic expression in social interaction is either amplified or attenuated (Hiroi et al., 2012; Hiroi et al., 2013; Suzuki et al., 2009b; Harper et al., 2012). Similarly, others demonstrated that mouse models of fragile X syndrome (Spencer et al., 2011), *Nlgn3* mutants (Jaramillo et al., 2014), and *Shank3* mutation (Drapeau et al., 2014) result in varying degrees of ASD-related behavioral phenotypes on different genetic backgrounds. These data suggest that alleles in the genetic background have modulatory effects on the phenotype caused by CNVs. Such action of background alleles could explain why a CNV does not cause 100% penetrance and why phenotypic expression varies among carriers of the same CNVs (Hiroi et al., 2012; Hiroi, 2014; Hiroi and Nishi, 2016; Hiroi et al., 2013; Suzuki et al., 2009b).

The impacts of genetic background might also include those of a second CNV among carriers of the primary CNV in humans (Girirajan et al., 2012). For example, some CNVs are accompanied by a second CNV at statistically significant rates, including 16p11.2 duplication and 15q11.2 deletion (Girirajan et al., 2012). In those cases, genetic mouse models of a single CNV might not show phenotypes.

Environmental factors are another possible cause for varying phenotypic expression. For instance, we reported that individual housing, a condition known to reduce stress in male mice (Abramov et al., 2004; Arndt et al., 2009; Gresack et al., 2010; Panksepp et al., 2007; Voikar et al., 2005), selectively elevated expression levels of *Sept5*, a 22q11.2 gene, in the amygdala and increased active affiliative social interactions (Harper et al., 2012). Litter composition is another such environmental factor. Recently, Yang and colleagues elegantly demonstrated that some ASD-related phenotype characteristics disappeared in mice of a 16p11.2 deletion model separated from wild-type littermates and raised with mice of the same deletion genotype only (Yang et al., 2015a).

In summary, mouse models of CNVs present an unprecedented opportunity to delve into the precise manners by which CNV-encoded genes cause dimensional phenotypes relevant to neuropsychiatric disorders. Many questions remain. If CNVs alter dimensional phenotypes that are common among more than one clinically defined neuropsychiatric disorder, then why is the same CNV associated with various clinically distinct diagnoses? We suspect that modulatory forces, such as individually diverse genetic background and environmental factors, determine ultimate phenotypic expression, but more work is needed. Moreover, hypothetical gene–phenotype relations have been identified in mouse models of 22q11.2 CNVs (Hiroi and Nishi, 2016; Hiroi et al., 2013), but it remains unclear how these hypothetical rules can be generalized to other CNV models. Further, in the human genome, many noncoding genes, microRNA, and pseudogenes (in addition to protein-coding genes) are encoded in CNVs, but their conservation in the murine genome and their functional contributions to dimensional aspects of neuropsychiatric disorders are still poorly understood. Despite caveats, genetic mouse models are likely to greatly improve our understanding of the gene–phenotype relation in psychiatric disorders. However, what we learn from mouse models of CNVs is not specific to schizophrenia, but rather imparts information about the relationship between a genotype and dimensional aspects of many mental disorders in general. If CNVs indeed cause mental disorders by acting on dimensional measures, then human genetic association studies are also likely to benefit from incorporation of dimensional measures rather than clinical categorical classification.

ACKNOWLEDGMENT

This work was supported by the NIH (R21HD05311 and R01MH099660), an NARSAD Independent Investigator Award, and the Maltz Foundation (to N.H.).

REFERENCES

Abramov, U., Raud, S., Koks, S., Innos, J., Kurrikoff, K., Matsui, T., et al., 2004. Targeted mutation of CCK(2) receptor gene antagonises behavioural changes induced by social isolation in female, but not in male mice. Behav. Brain Res. 155, 1–11.

Alberti, A., Romano, C., Falco, M., Cali, F., Schinocca, P., Galesi, O., et al., 2007. 1.5 Mb de novo 22q11.21 microduplication in a patient with cognitive deficits and dysmorphic facial features. Clin. Genet. 71, 177–182.

Arndt, S.S., Laarakker, M.C., van Lith, H.A., van der Staay, F.J., Gieling, E., Salomons, A.R., et al., 2009. Individual housing of mice – impact on behaviour and stress responses. Physiol. Behav. 97, 385–393.

Arriaga, G., Jarvis, E.D., 2013. Mouse vocal communication system: are ultrasounds learned or innate? Brain Lang. 124, 96–116.

Babovic, D., O'Tuathaigh, C.M., O'Connor, A.M., O'Sullivan, G.J., Tighe, O., Croke, D.T., et al., 2008. Phenotypic characterization of cognition and social behavior in mice with heterozygous versus homozygous deletion of catechol-O-methyltransferase. Neuroscience 155, 1021–1029.

Belmonte, M.K., Cook Jr., E.H., Anderson, G.M., Rubenstein, J.L., Greenough, W.T., Beckel-Mitchener, A., et al., 2004. Autism as a disorder of neural information processing: directions for research and targets for therapy. Mol. Psychiatry 9, 646–663.

Bennett, E., Heaton, P., 2012. Is talent in autism spectrum disorders associated with a specific cognitive and behavioural phenotype? J. Autism Dev. Disord. 42, 2739–2753.

Brigman, J.L., Graybeal, C., Holmes, A., 2010. Predictably irrational: assaying cognitive inflexibility in mouse models of schizophrenia. Front. Neurosci. 4.

Brocki, K.C., Bohlin, G., 2004. Executive functions in children aged 6 to 13: a dimensional and developmental study. Dev. Neuropsychol. 26, 571–593.

Brors, D., Hansen, S., Mlynski, R., Volkenstein, S., Aletsee, C., Sendtner, M., et al., 2008. Spiral ganglion outgrowth and hearing development in p75-deficient mice. Audiol. Neurootol. 13, 388–395.

Brunet, A., Gabau, E., Perich, R.M., Valdesoiro, L., Brun, C., Caballin, M.R., et al., 2006. Microdeletion and microduplication 22q11.2 screening in 295 patients with clinical features of DiGeorge/Velocardiofacial syndrome. Am. J. Med. Genet. A 140, 2426–2432.

Butcher, N.J., Chow, E.W., Costain, G., Karas, D., Ho, A., Bassett, A.S., 2012. Functional outcomes of adults with 22q11.2 deletion syndrome. Genet. Med. 14, 836–843.

Capossela, S., Muzio, L., Bertolo, A., Bianchi, V., Dati, G., Chaabane, L., et al., 2012. Growth defects and impaired cognitive-behavioral abilities in mice with knockout for Eif4h, a gene located in the mouse homolog of the Williams-Beuren syndrome critical region. Am. J. Pathol. 180, 1121–1135.

Carola, V., D'Olimpio, F., Brunamonti, E., Mangia, F., Renzi, P., 2002. Evaluation of the elevated plus-maze and open-field tests for the assessment of anxiety-related behaviour in inbred mice. Behav. Brain Res. 134, 49–57.

Chun, S., Westmoreland, J.J., Bayazitov, I.T., Eddins, D., Pani, A.K., Smeyne, R.J., et al., 2014. Specific disruption of thalamic inputs to the auditory cortex in schizophrenia models. Science 344, 1178–1182.

Courtens, W., Schramme, I., Laridon, A., 2008. Microduplication 22q11.2: a benign polymorphism or a syndrome with a very large clinical variability and reduced penetrance? – report of two families. Am. J. Med. Genet. A 146A, 758–763.

Crawley, J., 2014. Optimizing behavioral assays for mouse models of autism. SFARI News and Opinion.

de La Rochebrochard, C., Joly-Helas, G., Goldenberg, A., Durand, I., Laquerriere, A., Ickowicz, V., et al., 2006. The intrafamilial variability of the 22q11.2 microduplication encompasses a spectrum from minor cognitive deficits to severe congenital anomalies. Am. J. Med. Genet. A 140, 1608–1613.

De Luca, C.R., Wood, S.J., Anderson, V., Buchanan, J.A., Proffitt, T.M., Mahony, K., et al., 2003. Normative data from the CANTAB. I: development of executive function over the lifespan. J. Clin. Exp. Neuropsychol. 25, 242–254.

De Smedt, B., Devriendt, K., Fryns, J.P., Vogels, A., Gewillig, M., Swillen, A., 2007. Intellectual abilities in a large sample of children with Velo-Cardio-Facial Syndrome: an update. J. Intellect. Disabil. Res. 51, 666–670.

Demetriou, A., Christou, C., Spanoudis, G., Platsidou, M., 2002. The development of mental processing: efficiency, working memory, and thinking. Monogr Soc. Res. Child Dev. 67 i–155.

Descartes, M., Franklin, J., de Stahl, T.D., Piotrowski, A., Bruder, C.E., Dumanski, J.P., et al., 2008. Distal 22q11.2 microduplication encompassing the BCR gene. Am. J. Med. Genet. A 146A, 3075–3081.

Drapeau, E., Dorr, N.P., Elder, G.A., Buxbaum, J.D., 2014. Absence of strong strain effects in behavioral analyses of Shank3-deficient mice. Dis. Model. Mech. 7, 667–681.

Driscoll, D.A., Budarf, M.L., Emanuel, B.S., 1992a. A genetic etiology for DiGeorge syndrome: consistent deletions and microdeletions of 22q11. Am. J. Hum. Genet. 50, 924–933.

Driscoll, D.A., Spinner, N.B., Budarf, M.L., Donald-McGinn, D.M., Zackai, E.H., Goldberg, R.B., et al., 1992b. Deletions and microdeletions of 22q11.2 in velo-cardio-facial syndrome. Am. J. Med. Genet. 44, 261–268.

Dumontheil, I., Roggeman, C., Ziermans, T., Peyrard-Janvid, M., Matsson, H., Kere, J., et al., 2011. Influence of the COMT genotype on working memory and brain activity changes during development. Biol. Psychiatry 70, 222–229.

Edelmann, L., Pandita, R.K., Spiteri, E., Funke, B., Goldberg, R., Palanisamy, N., et al., 1999. A common molecular basis for rearrangement disorders on chromosome 22q11. Hum. Mol. Genet. 8, 1157–1167.

Ensenauer, R.E., Adeyinka, A., Flynn, H.C., Michels, V.V., Lindor, N.M., Dawson, D.B., et al., 2003. Microduplication 22q11.2, an emerging syndrome: clinical, cytogenetic, and molecular analysis of thirteen patients. Am. J. Hum. Genet. 73, 1027–1040.

Esposito, G., Venuti, P., 2008. How is crying perceived in children with Autistic Spectrum Disorder. Res. Autism Spectr. Disord. 2, 371–384.

Esposito, G., Venuti, P., 2009. Comparative analysis of crying in children with autism, developmental delays, and typical development. Focus Autism Other Dev. Disabil. 24, 240–247.

Esposito, G., Venuti, P., 2010. Understanding early communication signals in autism: a study of the perception of infants' cry. J. Intellect. Disabil. Res. 54, 216–223.

Esposito, G., Nakazawa, J., Venuti, P., Bornstein, M.H., 2013. Componential deconstruction of infant distress vocalizations via tree-based models: a study of cry in autism spectrum disorder and typical development. Res. Dev. Disabil. 34, 2717–2724.

Fairless, A.H., Katz, J.M., Vijayvargiya, N., Dow, H.C., Kreibich, A.S., Berrettini, W.H., et al., 2013. Development of home cage social behaviors in BALB/cJ vs. C57BL/6J mice. Behav. Brain Res. 237, 338–347.

Fejgin, K., Nielsen, J., Birknow, M.R., Bastlund, J.F., Nielsen, V., Lauridsen, J.B., et al., 2013. A mouse model that recapitulates cardinal features of the 15q13.3 microdeletion syndrome including schizophrenia- and epilepsy-related alterations. Biol. Psychiatry 76, 128–137.

Fioravanti, M., Carlone, O., Vitale, B., Cinti, M.E., Clare, L., 2005. A meta-analysis of cognitive deficits in adults with a diagnosis of schizophrenia. Neuropsychol. Rev. 15, 73–95.

Fischer, H.G., Morawski, M., Bruckner, M.K., Mittag, A., Tarnok, A., Arendt, T., 2012. Changes in neuronal DNA content variation in the human brain during aging. Aging Cell 11, 628–633.

Fuchs, J.C., Zinnamon, F.A., Taylor, R.R., Ivins, S., Scambler, P.J., Forge, A., et al., 2013. Hearing loss in a mouse model of 22q11.2 deletion syndrome. PLoS One 8, e80104.

Gathercole, S.E., Pickering, S.J., Ambridge, B., Wearing, H., 2004. The structure of working memory from 4 to 15 years of age. Dev. Psychol. 40, 177–190.

Geyer, M.A., 2006. The family of sensorimotor gating disorders: comorbidities or diagnostic overlaps? Neurotox. Res. 10, 211–220.

Girirajan, S., Rosenfeld, J.A., Coe, B.P., Parikh, S., Friedman, N., Goldstein, A., et al., 2012. Phenotypic heterogeneity of genomic disorders and rare copy-number variants. N. Engl. J. Med. 367, 1321–1331.

Gogos, J.A., Morgan, M., Luine, V., Santha, M., Ogawa, S., Pfaff, D., et al., 1998. Catechol-O-methyltransferase-deficient mice exhibit sexually dimorphic changes in catecholamine levels and behavior. Proc. Natl. Acad. Sci. U.S.A. 95, 9991–9996.

Gogos, J.A., Santha, M., Takacs, Z., Beck, K.D., Luine, V., Lucas, L.R., et al., 1999. The gene encoding proline dehydrogenase modulates sensorimotor gating in mice. Nat. Genet. 21, 434–439.

Gole, J., Gore, A., Richards, A., Chiu, Y.J., Fung, H.L., Bushman, D., et al., 2013. Massively parallel polymerase cloning and genome sequencing of single cells using nanoliter microwells. Nat. Biotechnol. 31, 1126–1132.

Gresack, J.E., Risbrough, V.B., Scott, C.N., Coste, S., Stenzel-Poore, M., Geyer, M.A., et al., 2010. Isolation rearing-induced deficits in contextual fear learning do not require CRF(2) receptors. Behav. Brain Res. 209, 80–84.

Hammerschmidt, K., Reisinger, E., Westekemper, K., Ehrenreich, L., Strenzke, N., Fischer, J., 2012. Mice do not require auditory input for the normal development of their ultrasonic vocalizations. BMC Neurosci. 13, 40.

Harper, K.M., Hiramoto, T., Tanigaki, K., Kang, G., Suzuki, G., Trimble, W., et al., 2012. Alterations of social interaction through genetic and environmental manipulation of the 22q11.2 gene *Sept5* in the mouse brain. Hum. Mol. Genet. 21, 3489–3499.

Hassed, S.J., Hopcus-Niccum, D., Zhang, L., Li, S., Mulvihill, J.J., 2004. A new genomic duplication syndrome complementary to the velocardiofacial (22q11 deletion) syndrome. Clinical Genetics 65, 400–404.

Henderson, N.D., Turri, M.G., DeFries, J.C., Flint, J., 2004. QTL analysis of multiple behavioral measures of anxiety in mice. Behav. Genet. 34, 267–293.

Hiramoto, T., Kang, G., Suzuki, G., Satoh, Y., Kucherlapati, R., Watanabe, Y., et al., 2011. Tbx1: identification of a 22q11.2 gene as a risk factor for autism spectrum disorder in a mouse model. Hum. Mol. Genet. 20, 4775–4785.

Hiroi, N., 2014. Small cracks in the dam: rare genetic variants provide opportunities to delve into mechanisms of neuropsychiatric disorders. Biol. Psychiatry 76, 91–92.

Hiroi, N., Nishi, A., 2016. Dimensional deconstruction and reconstruction of CNV-associated neuropsychiatric disorders. In: Modeling the psychopathological dimensions of schizophrenia and related psychoses. Handbook of Behavioral Neuroscience, Vol. 23. Elsevier, San Diego, CA, pp. 285–302, Chapter 17.

Hiroi, N., Zhu, H., Lee, M., Funke, B., Arai, M., Itokawa, M., et al., 2005. A 200-kb region of human chromosome 22q11.2 confers antipsychotic-responsive behavioral abnormalities in mice. Proc. Natl. Acad. Sci. U.S.A. 102, 19132–19137.

Hiroi, N., Hiramoto, T., Harper, K.M., Suzuki, G., Boku, S., 2012. Mouse models of 22q11.2-associated autism spectrum disorder. Autism S1, 1–9.

Hiroi, N., Takahashi, T., Hishimoto, A., Izumi, T., Boku, S., Hiramoto, T., 2013. Copy number variation at 22q11.2: from rare variants to common mechanisms of developmental neuropsychiatric disorders. Mol. Psychiatry 18, 1153–1165.

Horev, G., Ellegood, J., Lerch, J.P., Son, Y.E., Muthuswamy, L., Vogel, H., et al., 2011. Dosage-dependent phenotypes in models of 16p11.2 lesions found in autism. Proc. Natl. Acad. Sci. U.S.A. 108, 17076–17081.

Hoyle, E., Genn, R.F., Fernandes, C., Stolerman, I.P., 2006. Impaired performance of α7 nicotinic receptor knock-out mice in the five-choice serial reaction time task. Psychopharmacology (Berl) 189, 211–223.

Jaramillo, T.C., Liu, S., Pettersen, A., Birnbaum, S.G., Powell, C.M., 2014. Autism-related neuroligin-3 mutation alters social behavior and spatial learning. Autism Res. 7, 264–272.

Keller, J.J., Keller, A.B., Bowers, B.J., Wehner, J.M., 2005. Performance of α7 nicotinic receptor null mutants is impaired in appetitive learning measured in a signaled nose poke task. Behav. Brain Res. 162, 143–152.

Kikusui, T., Nakanishi, K., Nakagawa, R., Nagasawa, M., Mogi, K., Okanoya, K., 2011. Cross fostering experiments suggest that mice songs are innate. PLoS One 6, e17721.

Kimber, W.L., Hsieh, P., Hirotsune, S., Yuva-Paylor, L., Sutherland, H.F., Chen, A., et al., 1999. Deletion of 150 kb in the minimal DiGeorge/velocardiofacial syndrome critical region in mouse. Hum. Mol. Genet. 8, 2229–2237.

Kirov, G., Rees, E., Walters, J.T., Escott-Price, V., Georgieva, L., Richards, A.L., et al., 2013. The penetrance of copy number variations for schizophrenia and developmental delay. Biol. Psychiatry 75, 378–385.

Korkmaz, B., 2011. Theory of mind and neurodevelopmental disorders of childhood. Pediatr. Res. 69, 101R–108R.

Lai, J.K., Sobala-Drozdowski, M., Zhou, L., Doering, L.C., Faure, P.A., Foster, J.A., 2014. Temporal and spectral differences in the ultrasonic vocalizations of fragile X knock out mice during postnatal development. Behav. Brain Res. 259, 119–130.

Lalonde, R., 2002. The neurobiological basis of spontaneous alternation. Neurosci. Biobehav. Rev. 26, 91–104.

Lee, S.H., DeCandia, T.R., Ripke, S., Yang, J., Sullivan, P.F., Goddard, M.E., et al., 2012. Estimating the proportion of variation in susceptibility to schizophrenia captured by common SNPs. Nat. Genet. 44, 247–250.

Levin, E.D., Petro, A., Rezvani, A.H., Pollard, N., Christopher, N.C., Strauss, M., et al., 2009. Nicotinic α7- or β2-containing receptor knockout: effects on radial-arm maze learning and long-term nicotine consumption in mice. Behav. Brain Res. 196, 207–213.

Li, H.H., Roy, M., Kuscuoglu, U., Spencer, C.M., Halm, B., Harrison, K.C., et al., 2009. Induced chromosome deletions cause hypersociability and other features of Williams-Beuren syndrome in mice. EMBO Mol. Med. 1, 50–65.

Lo-Castro, A., Galasso, C., Cerminara, C., El-Malhany, N., Benedetti, S., Nardone, A.M., et al., 2009. Association of syndromic mental retardation and autism with 22q11.2 duplication. Neuropediatrics 40, 137–140.

Long, J.M., Laporte, P., Merscher, S., Funke, B., Saint-Jore, B., Puech, A., et al., 2006. Behavior of mice with mutations in the conserved region deleted in velocardiofacial/DiGeorge syndrome. Neurogenetics 7, 247–257.

Lucena, J., Pezzi, S., Aso, E., Valero, M.C., Carreiro, C., Dubus, P., et al., 2010. Essential role of the N-terminal region of TFII-I in viability and behavior. BMC Med. Genet. 11, 61.

Luna, B., 2009. Developmental changes in cognitive control through adolescence. Adv. Child Dev. Behav. 37, 233–278.

Luna, B., Garver, K.E., Urban, T.A., Lazar, N.A., Sweeney, J.A., 2004. Maturation of cognitive processes from late childhood to adulthood. Child Dev. 75, 1357–1372.

Luna, B., Doll, S.K., Hegedus, S.J., Minshew, N.J., Sweeney, J.A., 2007. Maturation of executive function in autism. Biol. Psychiatry 61, 474–481.

Malhotra, D., Sebat, J., 2012. CNVs: harbingers of a rare variant revolution in psychiatric genetics. Cell 148, 1223–1241.

Martinussen, R., Hayden, J., Hogg-Johnson, S., Tannock, R., 2005. A meta-analysis of working memory impairments in children with attention-deficit/hyperactivity disorder. J. Am. Acad. Child Adolesc. Psychiatry 44, 377–384.

McAlonan, G.M., Daly, E., Kumari, V., Critchley, H.D., van, A.T., Suckling, J., et al., 2002. Brain anatomy and sensorimotor gating in Asperger's syndrome. Brain 125, 1594–1606.

McConnell, M.J., Lindberg, M.R., Brennand, K.J., Piper, J.C., Voet, T., Cowing-Zitron, C., et al., 2013. Mosaic copy number variation in human neurons. Science 342, 632–637.

Meechan, D.W., Rutz, H.L., Fralish, M.S., Maynard, T.M., Rothblat, L.A., LaMantia, A.S., 2015. Cognitive ability is associated with altered medial frontal cortical circuits in the LgDel mouse model of 22q11.2DS. Cereb. Cortex 25, 1143–1151.

Misslin, R., Herzog, F., Koch, B., Ropartz, P., 1982. Effects of isolation, handling and novelty on the pituitary – adrenal response in the mouse. Psychoneuroendocrinology 7, 217–221.

Mukaddes, N.M., Herguner, S., 2007. Autistic disorder and 22q11.2 duplication. World J. Biol. Psychiatry 8, 127–130.

Mukai, J., Liu, H., Burt, R.A., Swor, D.E., Lai, W.S., Karayiorgou, M., et al., 2004. Evidence that the gene encoding ZDHHC8 contributes to the risk of schizophrenia. Nat. Genet. 36, 725–731.

Nakatani, J., Tamada, K., Hatanaka, F., Ise, S., Ohta, H., Inoue, K., et al., 2009. Abnormal behavior in a chromosome-engineered mouse model for human 15q11-13 duplication seen in autism. Cell 137, 1235–1246.

Niklasson, L., Rasmussen, P., Oskarsdottir, S., Gillberg, C., 2009. Autism, ADHD, mental retardation and behavior problems in 100 individuals with 22q11 deletion syndrome. Res. Dev. Disabil. 30, 763–773.

O'Hearn, K., Asato, M., Ordaz, S., Luna, B., 2008. Neurodevelopment and executive function in autism. Dev. Psychopathol. 20, 1103–1132.

O'Leary, T.P., Gunn, R.K., Brown, R.E., 2013. What are we measuring when we test strain differences in anxiety in mice? Behav. Genet. 43, 34–50.

O'Tuathaigh, C.M., Hryniewiecka, M., Behan, A., Tighe, O., Coughlan, C., Desbonnet, L., et al., 2010. Chronic adolescent exposure to delta-9-tetrahydrocannabinol in COMT mutant mice: impact on psychosis-related and other phenotypes. Neuropsychopharmacology 35, 2262–2273.

O'Tuathaigh, C.M., Clarke, G., Walsh, J., Desbonnet, L., Petit, E., O'Leary, C., et al., 2012. Genetic vs. pharmacological inactivation of COMT influences cannabinoid-induced expression of schizophrenia-related phenotypes. Int. J. Neuropsychopharmacol. 15, 1331–1342.

Okabe, S., Nagasawa, M., Kihara, T., Kato, M., Harada, T., Koshida, N., et al., 2013. Pup odor and ultrasonic vocalizations synergistically stimulate maternal attention in mice. Behav. Neurosci. 127, 432–438.

Ornitz, E.M., Lane, S.J., Sugiyama, T., deTraversay, J., 1993. Startle modulation studies in autism. J. Autism Dev. Disord. 23, 619–637.

Ou, Z., Berg, J.S., Yonath, H., Enciso, V.B., Miller, D.T., Picker, J., et al., 2008. Microduplications of 22q11.2 are frequently inherited and are associated with variable phenotypes. Genet. Med. 10, 267–277.

Ouchi, Y., Banno, Y., Shimizu, Y., Ando, S., Hasegawa, H., Adachi, K., et al., 2013. Reduced adult hippocampal neurogenesis and working memory deficits in the Dgcr8-deficient mouse model of 22q11.2 deletion-associated schizophrenia can be rescued by IGF2. J. Neurosci. 33, 9408–9419.

Ozonoff, S., Iosif, A.M., Baguio, F., Cook, I.C., Hill, M.M., Hutman, T., et al., 2010. A prospective study of the emergence of early behavioral signs of autism. J. Am. Acad. Child Adolesc. Psychiatry 49, 256–266.

Ozonoff, S., Young, G.S., Belding, A., Hill, M., Hill, A., Hutman, T., et al., 2014. The broader autism phenotype in infancy: when does it emerge? J. Am. Acad. Child Adolesc. Psychiatry 53, 398–407.

Panksepp, J.B., Jochman, K.A., Kim, J.U., Koy, J.J., Wilson, E.D., Chen, Q., et al., 2007. Affiliative behavior, ultrasonic communication and social reward are influenced by genetic variation in adolescent mice. PLoS One 2, e351.

Papaleo, F., Crawley, J.N., Song, J., Lipska, B.K., Pickel, J., Weinberger, D.R., et al., 2008. Genetic dissection of the role of catechol-O-methyltransferase in cognition and stress reactivity in mice. J. Neurosci. 28, 8709–8723.

Paterlini, M., Zakharenko, S.S., Lai, W.S., Qin, J., Zhang, H., Mukai, J., et al., 2005. Transcriptional and behavioral interaction between 22q11.2 orthologs modulates schizophrenia-related phenotypes in mice. Nat. Neurosci. 8, 1586–1594.

Paylor, R., McIlwain, K.L., McAninch, R., Nellis, A., Yuva-Paylor, L.A., Baldini, A., et al., 2001. Mice deleted for the DiGeorge/velocardiofacial syndrome region show abnormal sensorimotor gating and learning and memory impairments. Hum. Mol. Genet. 10, 2645–2650.

Paylor, R., Glaser, B., Mupo, A., Ataliotis, P., Spencer, C., Sobotka, A., et al., 2006. Tbx1 haploinsufficiency is linked to behavioral disorders in mice and humans: implications for 22q11 deletion syndrome. Proc. Natl. Acad. Sci. U.S.A. 103, 7729–7734.

Penn, D.L., Corrigan, P.W., Bentall, R.P., Racenstein, J.M., Newman, L., 1997. Social cognition in schizophrenia. Psychol. Bull. 121, 114–132.

Pennington, B.F., Ozonoff, S., 1996. Executive functions and developmental psychopathology. J. Child Psychol. Psychiatry 37, 51–87.

Perry, W., Minassian, A., Lopez, B., Maron, L., Lincoln, A., 2007. Sensorimotor gating deficits in adults with autism. Biol. Psychiatry 61, 482–486.

Piskulic, D., Olver, J.S., Norman, T.R., Maruff, P., 2007. Behavioural studies of spatial working memory dysfunction in schizophrenia: a quantitative literature review. Psychiatry Res. 150, 111–121.

Portmann, T., Yang, M., Mao, R., Panagiotakos, G., Ellegood, J., Dolen, G., et al., 2014. Behavioral abnormalities and circuit defects in the basal ganglia of a mouse model of 16p11.2 deletion syndrome. Cell Rep. 7, 1077–1092.

Portnoi, M.F., 2009. Microduplication 22q11.2: a new chromosomal syndrome. Eur. J. Med. Genet. 52, 88–93.

Portnoi, M.F., Lebas, F., Gruchy, N., Ardalan, A., Biran-Mucignat, V., Malan, V., et al., 2005. 22q11.2 duplication syndrome: two new familial cases with some overlapping features with DiGeorge/velocardiofacial syndromes. Am. J. Med. Genet. A 137, 47–51.

Pucilowska, J., Vithayathil, J., Tavares, E.J., Kelly, C., Karlo, J.C., Landreth, G.E., 2015. The 16p11.2 deletion mouse model of autism exhibits altered cortical progenitor proliferation and brain cytoarchitecture linked to the ERK MAPK pathway. J. Neurosci. 35, 3190–3200.

Purcell, S.M., Wray, N.R., Stone, J.L., Visscher, P.M., O'Donovan, M.C., Sullivan, P.F., et al., 2009. Common polygenic variation contributes to risk of schizophrenia and bipolar disorder. Nature 460, 748–752.

Ramelli, G.P., Silacci, C., Ferrarini, A., Cattaneo, C., Visconti, P., Pescia, G., 2008. Microduplication 22q11.2 in a child with autism spectrum disorder: clinical and genetic study. Dev. Med. Child Neurol. 50, 953–955.

Ramos, A., Mormede, P., 1998. Stress and emotionality: a multidimensional and genetic approach. Neurosci. Biobehav. Rev. 22, 33–57.

Ramos, A., Moisan, M.P., Chaouloff, F., Mormede, C., Mormede, P., 1999. Identification of female-specific QTLs affecting an emotionality-related behavior in rats. Mol. Psychiatry 4, 453–462.

Rees, E., Walters, J.T., Chambert, K.D., O'Dushlaine, C., Szatkiewicz, J., Richards, A.L., et al., 2013. CNV analysis in a large schizophrenia sample implicates deletions at 16p12.1 and SLC1A1 and duplications at 1p36.33 and CGNL1. Hum. Mol. Genet.

Rees, E., Walters, J.T., Georgieva, L., Isles, A.R., Chambert, K.D., Richards, A.L., et al., 2014. Analysis of copy number variations at 15 schizophrenia-associated loci. Br. J. Psychiatry 204, 108–114.

Ripke, S., O'Dushlaine, C., Chambert, K., Moran, J.L., Kahler, A.K., Akterin, S., et al., 2013. Genome-wide association analysis identifies 13 new risk loci for schizophrenia. Nat. Genet. 45, 1150–1159.

Rogers, D.C., Jones, D.N., Nelson, P.R., Jones, C.M., Quilter, C.A., Robinson, T.L., et al., 1999. Use of SHIRPA and discriminant analysis to characterise marked differences in the behavioural phenotype of six inbred mouse strains. Behav. Brain Res. 105, 207–217.

Russo, N., Flanagan, T., Iarocci, G., Berringer, D., Zelazo, P.D., Burack, J.A., 2007. Deconstructing executive deficits among persons with autism: implications for cognitive neuroscience. Brain Cogn. 65, 77–86.

Sakurai, T., Dorr, N.P., Takahashi, N., McInnes, L.A., Elder, G.A., Buxbaum, J.D., 2011. Haploinsufficiency of Gtf2i, a gene deleted in Williams Syndrome, leads to increases in social interactions. Autism Res. 4, 28–39.

Scambler, P.J., Carey, A.H., Wyse, R.K., Roach, S., Dumanski, J.P., Nordenskjold, M., et al., 1991. Microdeletions within 22q11 associated with sporadic and familial DiGeorge syndrome. Genomics 10, 201–206.

Scambler, P.J., Kelly, D., Lindsay, E., Williamson, R., Goldberg, R., Shprintzen, R., et al., 1992. Velo-cardio-facial syndrome associated with chromosome 22 deletions encompassing the DiGeorge locus. Lancet 339, 1138–1139.

Scheiner, E., Fischer, J., 2011. Emotion expression: the evolutionary heritage in the human voice. In: Welsch, W. (Ed.), Interdisciplinary Anthropology Springer-Verlag, Berlin, Heidelberg, pp. 105–129.

Schneider, M., Debbane, M., Bassett, A.S., Chow, E.W., Fung, W.L., van den Bree, M.B., et al., 2014. Psychiatric disorders from childhood to adulthood in 22q11.2 deletion syndrome: results from the International Consortium on Brain and Behavior in 22q11.2 deletion syndrome. Am. J. Psychiatry 171, 627–639.

Schneider, T., Skitt, Z., Liu, Y., Deacon, R.M., Flint, J., Karmiloff-Smith, A., et al., 2012. Anxious, hypoactive phenotype combined with motor deficits in Gtf2ird1 null mouse model relevant to Williams syndrome. Behav. Brain Res. 233, 458–473.

Segura-Puimedon, M., Sahun, I., Velot, E., Dubus, P., Borralleras, C., Rodrigues, A.J., et al., 2014. Heterozygous deletion of the Williams-Beuren syndrome critical interval in mice recapitulates most features of the human disorder. Hum. Mol. Genet. 23, 6481–6494.

Sewell, G.D., 1970. Ultrasonic communication in rodents. Nature 227, 410.

Shprintzen, R.J., Goldberg, R., Golding-Kushner, K.J., Marion, R.W., 1992. Late-onset psychosis in the velo-cardio-facial syndrome. Am. J. Med. Genet. 42, 141–142.

Sigman, M., 1998. The Emanuel Miller Memorial Lecture 1997. Change and continuity in the development of children with autism. J. Child Psychol. Psychiatry 39, 817–827.

Sigurdsson, T., Stark, K.L., Karayiorgou, M., Gogos, J.A., Gordon, J.A., 2010. Impaired hippocampal-prefrontal synchrony in a genetic mouse model of schizophrenia. Nature 464, 763–767.

Silverman, J.L., Yang, M., Lord, C., Crawley, J.N., 2010. Behavioural phenotyping assays for mouse models of autism. Nat. Rev. Neurosci. 11, 490–502.

Sobin, C., Kiley-Brabeck, K., Karayiorgou, M., 2005. Associations between prepulse inhibition and executive visual attention in children with the 22q11 deletion syndrome. Mol. Psychiatry 10, 553–562.

Spencer, C.M., Alekseyenko, O., Serysheva, E., Yuva-Paylor, L.A., Paylor, R., 2005. Altered anxiety-related and social behaviors in the Fmr1 knockout mouse model of fragile X syndrome. Genes Brain Behav. 4, 420–430.

Spencer, C.M., Alekseyenko, O., Hamilton, S.M., Thomas, A.M., Serysheva, E., Yuva-Paylor, L.A., et al., 2011. Modifying behavioral phenotypes in Fmr1KO mice: genetic background differences reveal autistic-like responses. Autism Res. 4, 40–56.

Sprong, M., Schothorst, P., Vos, E., Hox, J., van, E.H., 2007. Theory of mind in schizophrenia: meta-analysis. Br. J. Psychiatry 191, 5–13.

Stark, K.L., Xu, B., Bagchi, A., Lai, W.S., Liu, H., Hsu, R., et al., 2008. Altered brain microRNA biogenesis contributes to phenotypic deficits in a 22q11-deletion mouse model. Nat. Genet. 40, 751–760.

Stark, K.L., Burt, R.A., Gogos, J.A., Karayiorgou, M., 2009. Analysis of prepulse inhibition in mouse lines overexpressing 22q11.2 orthologues. Int. J. Neuropsychopharmacol. 12, 983–989.

Suzuki, G., Harper, K.M., Hiramoto, T., Funke, B., Lee, M., Kang, G., et al., 2009a. Over-expression of a human chromosome 22q11.2 segment including TXNRD2, COMT, and ARVCF developmentally affects incentive learning and working memory in mice. Hum. Mol. Genet. 18, 3914–3925.

Suzuki, G., Harper, K.M., Hiramoto, T., Sawamura, T., Lee, M., Kang, G., et al., 2009b. Sept5 deficiency exerts pleiotropic influence on affective behaviors and cognitive functions in mice. Hum. Mol. Genet. 18, 1652–1660.

Swanson, H.L., 1999. What develops in working memory? A life span perspective. Dev. Psychol. 35, 986–1000.

Swillen, A., Vandeputte, L., Cracco, J., Maes, B., Ghesquiere, P., Devriendt, K., et al., 1999. Neuropsychological, learning and psychosocial profile of primary school aged children with the velo-cardio-facial syndrome (22q11 deletion): evidence for a nonverbal learning disability? Child Neuropsychol. 5, 230–241.

Szatkiewicz, J.P., O'Dushlaine, C., Chen, G., Chambert, K., Moran, J.L., Neale, B.M., et al., 2014. Copy number variation in schizophrenia in Sweden. Mol. Psychiatry 19, 762–773.

Takahashi, A., Nishi, A., Ishii, A., Shiroishi, T., Koide, T., 2008. Systematic analysis of emotionality in consomic mouse strains established from C57BL/6J and wild-derived MSM/Ms. Genes Brain Behav. 7, 849–858.

Takahashi, T., Okabe, S., Broin, P.Ó., Nishi, A., Ye, K., Beckert, M.V., et al., 2015. Structure and function of neonatal social communication in a genetic mouse model of autism. Mol. Psychiatry 2015 Dec 15. [Epub ahead of print].

Tamada, K., Tomonaga, S., Hatanaka, F., Nakai, N., Takao, K., Miyakawa, T., et al., 2010. Decreased exploratory activity in a mouse model of 15q duplication syndrome; implications for disturbance of serotonin signaling. PLoS One 5, e15126.

Trullas, R., Skolnick, P., 1993. Differences in fear motivated behaviors among inbred mouse strains. Psychopharmacology (Berl) 111, 323–331.

Turri, M.G., Datta, S.R., DeFries, J., Henderson, N.D., Flint, J., 2001. QTL analysis identifies multiple behavioral dimensions in ethological tests of anxiety in laboratory mice. Curr. Biol. 11, 725–734.

Uematsu, A., Kikusui, T., Kihara, T., Harada, T., Kato, M., Nakano, K., et al., 2007. Maternal approaches to pup ultrasonic vocalizations produced by a nanocrystalline silicon thermo-acoustic emitter. Brain Res. 1163, 91–99.

van Campenhout, S., Devriendt, K., Breckpot, J., Frijns, J.-P., Peters, H., van Buggenhout, G., et al., 2012. Microduplication 22q11.2: a description of the clinical, developmental and behavioral characteristics during childhood. Genet. Couns. 23, 135–147.

van Steensel, F.J., Bogels, S.M., Perrin, S., 2011. Anxiety disorders in children and adolescents with autistic spectrum disorders: a meta-analysis. Clin. Child Fam. Psychol. Rev. 14, 302–317.

van Steensel, F.J., Bogels, S.M., de Bruin, E.I., 2013. Psychiatric comorbidity in children with autism spectrum disorders: a comparison with children with ADHD. J. Child Fam. Stud. 22, 368–376.

Voikar, V., Polus, A., Vasar, E., Rauvala, H., 2005. Long-term individual housing in C57BL/6J and DBA/2 mice: assessment of behavioral consequences. Genes Brain Behav. 4, 240–252.

Volkenstein, S., Brors, D., Hansen, S., Berend, A., Mlynski, R., Aletsee, C., et al., 2009. Auditory development in progressive motor neuronopathy mouse mutants. Neurosci. Lett. 465, 45–49.

Vorstman, J.A., Breetvelt, E.J., Duijff, S.N., Eliez, S., Schneider, M., Jalbrzikowski, M., et al., 2015. Cognitive decline preceding the onset of psychosis in patients with 22q11.2 deletion syndrome. JAMA Psychiatry 72, 377–385.

Walcott, C.M., Landau, S., 2004. The relation between disinhibition and emotion regulation in boys with attention deficit hyperactivity disorder. J. Clin. Child Adolesc. Psychol. 33, 772–782.

Wentzel, C., Fernstrom, M., Ohrner, Y., Anneren, G., Thuresson, A.C., 2008. Clinical variability of the 22q11.2 duplication syndrome. Eur. J. Med. Genet. 51, 501–510.

Westra, J.W., Peterson, S.E., Yung, Y.C., Mutoh, T., Barral, S., Chun, J., 2008. Aneuploid mosaicism in the developing and adult cerebellar cortex. J. Comp. Neurol. 507, 1944–1951.

Willcutt, E.G., Doyle, A.E., Nigg, J.T., Faraone, S.V., Pennington, B.F., 2005. Validity of the executive function theory of attention-deficit/hyperactivity disorder: a meta-analytic review. Biol. Psychiatry 57, 1336–1346.

Yang, M., Lewis, F., Foley, G., Crawley, J.N., 2015a. In tribute to Bob Blanchard: divergent behavioral phenotypes of 16p11.2 deletion mice reared in same-genotype versus mixed-genotype cages. Physiol. Behav. 146, 16–27.

Yang, M., Mahrt, E.J., Lewis, F., Foley, G., Portmann, T., Dolmetsch, R.E., et al., 2015b. 16p11.2 deletion syndrome mice display sensory and ultrasonic vocalization deficits during social interactions. Autism Res. 8, 507–521.

Young, J.W., Finlayson, K., Spratt, C., Marston, H.M., Crawford, N., Kelly, J.S., et al., 2004. Nicotine improves sustained attention in mice: evidence for involvement of the α7 nicotinic acetylcholine receptor. Neuropsychopharmacology 29, 891–900.

Young, J.W., Crawford, N., Kelly, J.S., Kerr, L.E., Marston, H.M., Spratt, C., et al., 2007. Impaired attention is central to the cognitive deficits observed in α7 deficient mice. Eur. Neuropsychopharmacol. 17, 145–155.

Yuhas, J., Cordeiro, L., Tassone, F., Ballinger, E., Schneider, A., Long, J.M., et al., 2010. Brief report: sensorimotor gating in idiopathic autism and autism associated with fragile X syndrome. J. Autism Dev. Disord. 41, 248–253.

Zhu, H., Lee, M., Agatsuma, S., Hiroi, N., 2007. Pleiotropic impact of constitutive fosB inactivation on nicotine-induced behavioral alterations and stress-related traits in mice. Hum. Mol. Genet. 16, 820–836.

Author Index

Note: Page numbers followed by "*f*" and "*t*" refer to figures and tables, respectively.

A

Aalto, S., 118
Abdolmaleky, H.M., 66–67
Abel, T., 200
Aberg, K.A., 27–28, 66–67
Abi-Dargham, A., 113–114, 116–117
Abu-Akel, A., 302–303
Abukmeil, S., 271
Adams, R.A., 283, 288, 291
Adelstein, J.S., 314
Adler, L.E., 20*t*, 215
Adrian, M., 126
Agid, O., 111
Ahearn, T., 304*f*
Ahuja, N., 185
Aimone, J.B., 202–204
Akbarian, S., 62–63, 67–68, 71–73, 134–135
Aleman, A., 303
Alexander-Bloch, A.F., 322–324
Allen, P., 306–307
American Psychiatric Association, 3–4, 9
Anderson, A., 322–323
Anderson, A.K., 300*f*
Anderson, G., 180
Andreasen, N.C., 263–264, 266–267
Andreassen, O.A., 62
Andreou, C., 241–242, 290–291
Angrist, B., 109
Anisman, H., 181
Anticevic, A., 288, 303
Antonova, E., 263–264
Apic, G., 198–199
Appasani, K., 62–63
Argyelan, M., 241
Asai, Y., 158
Ascher-Svanum, H., 167–168
Ashburner, J., 264–265, 267
Assaf, Y., 313–314
Aston, C., 67–68
Atack, J.R., 154–156
Auta, J., 158
Averbeck, B.B., 290–291

B

Babulas, V., 183
Backasch, B., 302–303
Baggio, H.-C., 314, 329
Bahn, S., 196
Bailey, A.L., 302–303
Baker, L.A., 19
Bakken, T.E., 101–102
Bale, T.L., 64–65
Balla, A., 117
Ballard, D.H., 285
Ballard, T.M., 159–160
Banerjee, A., 126, 129, 131, 135, 137
Baran, B., 9
Barch, D.M., 116–117
Barkus, C., 299–300
Baron, M.K., 126
Barr, M.S., 228
Bartos, M., 242–243
Basar, E., 225–227, 238
Basser, P.J., 315
Bassett, D.S., 314, 324–329
Bayes, A., 127, 197–198
Beaulieu, C., 315
Beaulieu, J.-M., 288
Becerril, K.E., 324
Bechter, K., 180, 188
Bedford, N.J., 304–305
Belforte, J.E., 155–156, 248–249
Belger, A., 220–221
Belvederi Murri, M., 110
Bender, H.U., 173
Benes, F.M., 67–68, 71–72, 243
Beneyto, M., 67–68, 134–135
Benros, M.E., 184
Berdah-Tordjman, D., 159
Berenbaum, H., 18
Berger, G.E., 169
Berger, H., 238
Bernet, B., 313–314
Bernstein, B.E., 73
Berridge, K.C., 290
Bertolino, A., 117–118
Bharadwaj, R., 69–71
Bhattacharyya, S., 299–300
Bien, C.G., 183
Bigos, K.L., 243
Billingslea, E.N., 243, 245–249, 247*f*
Bini, L., 9
Bird, A.P., 65–66

419

Biswal, B., 316
Biswal, B.B., 316
Blackstock, W.P., 131
Blackwood, N.J., 299, 301, 304–305
Bleuler, E., 7, 263–264, 313–314
Blomberg, F., 126–127
Bodatsch, M., 222–223
Bohacek, J., 27
Bohland, J.W., 85
Bohlken, M.M., 20t
Boileau, I., 115
Boksa, P., 183
Bollati, V., 68–69
Bonneau, R.H., 181
Borgdorff, A.J., 129–130
Borglum, A.D., 51
Borgwardt, S.J., 275–276
Bourne, J.N., 126
Boutros, N.N., 217
Bowie, C.R., 111–112
Bradbury, T.N., 26–27
Bradshaw, N.J., 202–204
Braff, D.L., 19, 125–126, 218–220, 222
Bramon, E., 20t, 217, 224
Brandon, N.J., 137, 202–204
Braudeau, J., 159–160
Bredy, T.W., 68–69
Breier, A., 113, 118
Brennand, K., 202–204
Brent, B.J., 304–305
Brockhaus-Dumke, A., 239–241
Brown, A.S., 44, 64, 183–184
Brunet-Gouet, E., 304–305, 308
Buchanan, R.W., 156
Buchsbaum, M.S., 172
Buckner, R.L., 241
Buka, S.L., 183
Bullmore, E., 314, 317, 321–322, 324–328
Bullmore, E.T., 30
Bullock, T.H., 226–227, 238
Bumgarner, R., 42
Burns, J., 313–314
Buxbaum, J.D., 202
Buzsáki, G., 29, 152–154, 238, 242–243

C

Cadenhead, K.S., 218
Calhoun, V.D., 20t, 28–30
Calkins, M.E., 20t
Camargo, L.M., 137, 202
Camchong, J., 30, 319
Cammoun, L., 315

Cannon, T.D., 264–267
Cao, P., 71–72
Carlén, M., 245–246
Carlin, R.K., 127, 131
Carlson, G.C., 29, 244–248
Carlsson, A., 9–10, 109, 182
Carr, D.B., 289
Carrard, A., 66–67
Carrasco, M.A., 314–315
Carroll, B.T., 185
Carter, C.S., 116–117
Casaccia, P., 65–66
Castner, S.A., 155–156
Catts, V.S., 62
Ceaser, A., 116–117
Ceresoli-Borroni, G., 183
Chafee, M.V., 133
Chaim, T.M., 265–266
Chakos, M.H., 184
Chakravarthy, B., 129
Champagne, F.A., 64
Chan, R.C., 20t
Charych, E.I., 67–68
Chaudhry, I.B., 185
Chen, X., 127–128
Cheng, D., 128, 131–132
Cheng, M.C., 68–69
Cherlyn, S.Y., 173
Cheung, I., 73–74
Chiveri, L., 180
Cho, R.Y., 228
Chong, V.Z., 135
Choquet, D., 129–130
Chung, H.J., 200–201
Cingolani, P., 53–54
Claes, S., 182
Clatworthy, P.L., 289
Clelland, C.D., 202–204
Clinton, S.M., 134–135
Coba, M.P., 197–198, 200–202
Cocchi, L., 319
Cohen, A.S., 303
Cohen, M., 271
Cohen, M.S., 322–323
Cohen, P., 200
Cohen, S., 68–69
Cole, M.W., 314, 329
Colledge, M., 128
Collin, G., 314, 318–319, 322–329
Collins, A.G.E., 290
Collins, M.O., 128, 197–198, 201–202
Collinson, N., 157, 159–160

Condray, R., 183
Conn, P.J., 130–131
Cookson, W., 70f
Corcoran, R., 301–302
Cordes, D., 316
Corlett, P.R., 290
Costa, E., 158–159
Costain, G., 195
Covington III, H.E., 71–72
Cowden, R.C., 109
Coyle, J.T., 125–126, 135, 155–156, 244
Cremer, C., 69–71
Cremer, T., 69–71
Crestani, F., 154–155, 157, 159–160
Criss, T.B., 263–264
Crocker, L.D., 21
Cross-Disorder Group of the Psychiatric Genomics Consortium, 23, 46–47, 52, 243
Crossley, N.A., 328–329
Crow, T.J., 110
Cumming, P., 113, 115
Cunningham, C., 181
Cunningham, M.O., 226–227, 243
Cuthbert, B.N., 21–22, 25, 26f

D

Dale, A.M., 264–265
Dalman, C., 183
Damoiseaux, J., 241, 316
Dandash, O., 318–319
Dani, A., 128
Dantzer, R., 180–181
Davatzikos, C., 263–266, 268, 270f, 274–275
Davies, M.N., 66–67
Davis, K.L., 110
Davis, P.A., 239–241
Day, J.J., 65–66, 68–69
Dayan, P., 285
de Bartolomeis, A., 125–126
De Haan, W., 328–329
de Ligt, J., 55
De Reus, M.A., 322
De Rojas, J.O., 250
Deakin, J.F., 202–204
Deardorff, M.A., 69–71
Decety, J., 304–305, 308
Deco, G., 29–30
Degner, J.F., 70f
Dehorter, N., 151–152
Del Pino, I., 245–249
Delay, J., 9–10, 109
Delini-Stula, A., 159

Demiralp, T., 241–242
Demjaha, A., 287
Dempster, E.L., 27–28
Den Ouden, H.E.M., 289
Deniker, P., 9–10
Dennis, E.L., 328–329
Derkits, E.J., 44
Derks, E.M., 314
Desaulniers, D., 68–69
Deserno, L., 290
Deutch, A.Y., 134
Dhanak, D., 71–72
Dharuri, H., 168
Dickinson, D., 288
Diederen, K.M., 306–307
Dierks, T., 306
Diwadkar, V.A., 24–25, 27
Dokmanovic, M., 71–72
Dong, E., 65, 68–69, 71
Dong, H.W., 86–87
Doorduin, J., 184
Dorph-Petersen, K.A., 62
Dosemeci, A., 128, 131
Dracheva, S., 67–68, 134–135
Draguhn, A., 238
Dreyer, J.K., 289
Dringen, R., 173–174
Duan, Z., 69–71
Duncan, C.E., 67–68
Durbin, R., 53–54
Durbin, R.M., 54
Durstewitz, D., 287
Duval, F., 110

E

Eberle, A.L., 314–315
Edgar, J.C., 228–229
Ehlers, M.D., 127–128
Ehrlichman, R.S., 245–246
Ellenbroek, B.A., 218
Ellison-Wright, I., 317, 328
Ellman, L.M., 184
Elston, R.C., 54–55
The ENCODE Project Consortium, 53–54
Endres, M., 68–69
Engin, E., 158–159
Enomoto, T., 243
Erhardt, S., 183
Erk, S., 23–24
Ermakova, A.O., 290–291
Errico, F., 134–135
Erwin, R.J., 217

Esslinger, C., 23–24, 299–300
Esteller, M., 62f–63f
Ethridge, L.E., 20t, 29
Ettinger, U., 264–265
Evans, D.R., 169–170
Ewing, S.G., 250

F

Fagiolini, M., 152
Fan, G., 65–66
Fan, Y., 265–266, 269–270, 274–276
Farde, L., 110–111
Farley, M.M., 127
Farrer, C., 305–306
Fazzari, P., 150
Fejgin, K., 227–228
Feldman, H., 287
Feldon, J., 183
Felgenhauer, K., 180
Felsenfeld, G., 69–71
Fenton, G.W., 239–241
Fernandes, C.P., 24
Fernandez, E., 132, 197–198
Fernandez, F., 159–160
Fernandez-Egea, E., 172
Ferran, J.L., 91–92
Ferrarelli, F., 152–154
Fifkova, E., 127
Filippakopoulos, P., 71–72
Fillman, S.G., 151–152
Fine, C., 290–291
Finley, K.H., 239–241
Finnema, S.J., 113
First, M., 271
Fischer, B.D., 157–158
Fischl, B., 264–265
Fisher, M., 28–29
Fitzsimmons, J., 317
Flagel, S.B., 290
Flagstad, P., 248–249
Flanagan, R.J., 168
Flint, J., 19, 22–23
Focking, M., 196–197
Fonteh, A.N., 169–170
Foong, J., 317
Ford, J.M., 11, 19, 225–226, 241
Fornito, A., 29–30, 314, 317–319, 324–328
Fox, M.D., 316
Fraga, M.F., 27
Francois, J., 247–248
Frank, J., 23
Frank, M.G., 181
Frank, M.J., 287, 289

Frankel, W.N., 244–245
Franklin, K.B.J., 86
Franklin, T.B., 27
Freedman, R., 215, 219–220, 222, 230
Fries, P., 287
Friston, K., 285, 289–290
Friston, K.J., 264–265, 267, 287, 291, 314, 316, 328
Frith, C., 291
Frith, C.D., 314, 328
Fritschy, J.M., 151–152, 157
Fromer, M., 55, 126, 133, 150, 195, 197–198
Fryland, T., 71–72
Fung, S.J., 243
Funk, A.J., 135
Furukawa, H., 181
Fusar-Poli, P., 7, 115, 116f, 117, 264–265, 275–276, 289

G

Gainetdinov, R.R., 288
Gallagher, S., 291
Gallinat, J., 307
Gandal, M.J., 226, 228–230, 238, 240f, 244–249, 248f, 251f
Gao, Q., 314
Gapp, K., 27
Garety, P.A., 303
Garey, L.J., 134
Gargus, J.J., 243
Garrity, A.G., 241
Gaser, C., 263–264
Gassab, L., 224
Gaszner, M., 69–71
Gattaz, W.F., 183
Gauthier, A.S., 65–66
Gejman, P.V., 51
Genome of the Netherlands Consortium, 54
Georgieva, L., 202
Gerdjikov, T.V., 157
Gervasini, C., 69–71
Ghadirivasfi, M., 66–67
Giannitrapani, D., 239–241
Giegling, I., 168
Gil-da-Costa, R., 220–221
Gill, I., 128–129
Gill, K.M., 157–158, 247–248
Girard, S.L., 55
Girgis, R.R., 111
Glahn, D.C., 17, 20t, 22, 29
Glantz, L.A., 134
Gleeson, J.G., 202–204
Gleich, T., 117–118
Glykys, J., 157
Godbout, J.P., 181
Goebel-Goody, S.M., 131

Gogtay, N., 184
Gold, J.M., 290
Goldman, D., 19
Goldman-Rakic, P.S., 134
Goldszal, A.F., 268
Golebiewska, A., 65–66
Gomperts, S.N., 128–129
Gonzalez-Burgos, G., 150, 154, 227–228, 243–244
Good, C.D., 267, 271–272
Gordon, J.A., 241
Goto, Y., 245–246, 288
Gottesman, I., 20t, 30
Gottesman, I.I., 17, 19, 19t, 20t, 180
Gottschalk, M.G., 196
Gould, T.D., 17, 19
Grace, A.A., 115–116, 157–158, 245–248, 250, 288–289
Gradin, V.B., 271–272, 303, 304f
Graff, A., 300f
Graff-Guerrero, A., 110
Grant, S.F., 73
Grant, S.G., 131, 133, 138
Gratten, J., 102
Grayson, D.R., 66–67, 71–72
Green, E.K., 243
Green, M.F., 308
Greenwood, T.A., 202, 219–220, 286–287
Greicius, M.D., 316, 319
Griesinger, W., 5–6
Griffa, A., 314, 322–327
Grimm, O., 299–300
Groman, S.M., 289
Gross, J., 238–239
Gross, O., 313–314
Gross, R.W., 170
Grüber, L., 182, 185
Gründer, G., 9–10, 110–111, 115–116
Grunwald, T., 215
Gryglewski, G., 115
Gudbjartsson, D.F., 54–55
Guest, P.C., 172, 196–197
Guidotti, A., 67–68, 159, 243
Guillozet-Bongaarts, A.L., 100t, 103
Gulsuner, S., 55, 202
Guo, J.U., 65–66, 68–69
Guo, W., 202–204
Gur, R., 271
Gur, R.E., 20t, 263–264

H

Habel, U., 308
Hagmann, P., 313–315, 328–329
Hahn, C.G., 126, 131, 133, 135–137
Hakak, Y., 67–68
Halgren, E., 224
Hall, J., 125–126, 133
Hall, M.H., 222, 229
Hamm, J.P., 228
Han, X., 170
Hansen, T., 172
Hanson, D.R., 180
Harris, K.M., 126–127
Harrison, P.J., 133–134, 136–137, 150, 154, 286–287
Hartley, B.J., 202–204
Harvey, P.D., 19
Hasan, A., 10, 71–72
Hashimoto, R., 125–126, 135
Hashimoto, T., 67–68, 154
Hasnain, M., 196
Hattori, S., 248–249
Hauser, J., 157
Hawrylycz, M., 97
Hawrylycz, M.J., 97
Hayashi, M.K., 127
Hayashi, Y., 126
Hayley, S., 181
He, Y., 168, 173, 322
Healy, D.J., 67–68
Heckers, S., 291–292, 313–314
Heijmans, B.T., 64, 73–74
Heinrichs, R.W., 20t
Heintz, N., 62f–63f
Heinz, A., 303
Helin, K., 71–72
Hemby, S.E., 67–68
Henley, J.M., 129–130
Hennekens, C.H., 62
Hensch, T.K., 152
Hering, H., 126
Herrmann, C.S., 238, 241–242
Higginbotham, H.R., 202–204
Hilgetag, C.C., 322
Hines, R.M., 155–156
Hiroshi, H., 180
Hirrlinger, J., 173–174
Hoen, W.P., 172
Hoff, P., 5–6
Hoffman, R.E., 306–307, 318–319
Hohlfeld, R., 180
Holt, D.J., 304–305
Holtmaat, A., 126
Homayoun, H., 244
Honea, R., 263–264
Honey, C.J., 319, 321
Hong, L.E., 228–229, 239–241, 288
Hoogenraad, C.C., 126, 128, 131, 202–204
Hoopmann, M.R., 197–198

Horga, G., 307–308
Horrobin, D.F., 169–170
Horton, A.C., 127
Horvath, S., 125–126
Hosak, L., 22–24
Hosobuchi, Y., 224
Houpt, T.A., 68–69
Howes, O.D., 287
Hu, B.R., 128
Huang, H.S., 67–68
Huang, Y.S., 169–170
Huettel, S.A., 301–302
Huffaker, S.J., 243
Huganir, R.L., 200
Hughes, V., 314–315
Humphries, C., 134–135
Hurley, R.A., 307
Husi, H., 131
Hyman, S.E., 150

I

Iacono, W.G., 17, 19, 20t, 22–23
Iacovelli, L., 130
Ikegame, T., 66–67
Innocenti, G.M., 329
Inoue, A., 128
Insel, T.R., 21, 25, 103
The International HapMap Consortium, 41
International Schizophrenia Consortium, 22–23, 52
Iossifov, I., 55
Ishizuka, K., 202–204
Itil, T.M., 239–241
Itsara, A., 52
Ivleva, E.I., 20t
Iwamoto, K., 66, 73–74
Iyengar, S.K., 54–55

J

Jaaro-Peled, H., 150
Jablensky, A., 51
Jacobson, L., 250
Jakovcevski, M., 71–73
Jardri, R., 306–307
Jarskog, L.F., 182–183
Javitt, D.C., 117, 135, 152–154, 215, 219–223, 230, 244
Jbabdi, S., 315
Jensen, J., 299–300, 301f
Jentsch, J.D., 244
Jessen, F., 222
Jin, J., 198–199
Jin, S.G., 62f–63f
Job, D.E., 184

Johansen-Berg, H., 318–319
Johnson, G.A., 86
Johnson, R.W., 181–182
Jones, C., 249
Joober, R., 219
Juckel, G., 290, 299–300
Jurado, S., 128

K

Kaas, G.A., 65–66
Kaas, J., 271
Kabiersch, A., 182–183
Kaddurah-Daouk, R., 169–171, 174
Kaech, S., 127
Kagey, M.H., 69–71
Kahn, R.S., 303, 329
Kaiser, M., 322
Kalia, L.V., 129
Kambeitz, J., 115–117, 265–266
Kandel, E.R., 200
Kappenman, E.S., 214
Kapur, S., 110–111, 287, 290, 299–300
Karayiorgou, M., 52
Karbasforoushan, H., 241
Karlsgodt, K.H., 248–249
Kasai, H., 126–127
Katagiri, H., 152
Kathmann, N., 20t
Katsel, P., 67–68
Kawakubo, Y., 222–223
Kay, S.R., 271
Kaynig, V., 314–315
Kayton, L., 239–241
Keane, B.P., 288
Keefe, R.S., 19, 111–112
Keefe, R.S.E., 329
Keeser, D., 250
Kegeles, L.S., 114, 118
Keller, M.C., 55
Keller, J., 18
Kempf, L., 173
Kendler, K.S., 24
Kennard, M.A., 239–241
Kennedy, M.B., 125–131
Kerbeshian, J., 180
Kerschensteiner, M., 180
Keshavan, M.S., 167, 169–170, 263–264
Kew, J.N., 128–129
Khaitovich, P., 172
Kharazia, V.N., 128
Khashan, A.S., 64
Kim, D.H., 62

Kim, H.Y., 170
Kim, J.S., 239, 242
Kim, M.J., 127–129
Kim, S.Y., 314–315
Kinon, B.J., 114
Kircher, T.T., 222
Kirihara, K., 244
Kirkbride, J.B., 64
Kirkpatrick, B., 172
Kirov, G., 52, 55, 133, 195, 197–198
Kissler, J., 239–241
Klausberger, T., 150–154
Klein, J., 250
Kloppel, S., 265–266
Klose, R.J., 65–66
Knabl, J., 154–155
Knable, M.B., 110
Knoflach, F., 150–151
Kocsis, B., 245–246
Kok, A., 223
Kolluri, N., 134
Kolomeets, N.S., 134
Kompus, K., 307
Konradi, C., 243
Koreen, A.R., 169
Korostil, M., 300f
Korotkova, T.M., 247–249
Körschenhausen, D.A., 182
Kotronen, A., 172
Koutsouleris, N., 265–266, 269–276
Kouzarides, T., 62f–63f
Kozak, M.J., 22, 25, 26f
Kraepelin, E., 6, 313–314
Krause, D., 183
Kriaucionis, S., 62f–63f
Kringelbach, M.L., 29–30
Krishnan, G.P., 239–241
Krishnan, K.R., 169
Kristal, B.S., 168
Kristiansen, L.V., 134–135
Kritis, A.A., 130–131
Kritzer, M.F., 134
Krljes, S., 222
Krug, A., 23–24, 301–302
Krumm, N., 202
Krystal, J.H., 150, 244, 246f, 286–287
Kubicki, M., 317
Kuhn, S., 307
Kuijpers, M., 202–204
Kumakura, Y., 115–116
Kumar, P., 304f
Kumari, V., 219, 230

Kundakovic, M., 64–69
Kurita, M., 71–72
Kutas, M., 239–241
Kutzelnigg, A., 180
Kwon, B., 68–69
Kwon, J.S., 228

L

Laan, W., 184–185
Lahti, A.C., 244
Lammel, S., 289
Lander, E.S., 54–55
Landin-Romero, R., 241
Lao, Z., 265–266
LaPlant, Q., 68–69
Laruelle, M., 110, 112–113
Lau, C., 90
Laufs, H., 239
Laursen, T.M., 62
Lavin, A., 245–246
Lawrie, S., 271
Lazarewicz, M.T., 245–246
Le Pen, G., 248–249
Leal, S., 54–55
Leaver, A.M., 324
Lee, D.A., 303
Lee, H., 248–249
Lee, S.H., 46
Leeson, V.C., 289
Lein, E., 84
Lencz, T., 44, 83–84
Lenzenweger, M.F., 17, 19t
Letunic, I., 201f
Levenson, J.M., 68–69
Levitt, P., 150
Levkovitz, Y., 185
Leweke, F.M., 183
Lewensohn, R., 168, 174
Lewis, C.M., 40, 44
Lewis, D.A., 62, 134, 150, 154, 156, 228–229, 243–244
Li, B., 54–55
Li, G., 62f–63f
Li, H., 53–54, 68–69
Li, J., 68–69, 71
Li, K.W., 131
Li, Y., 314, 329
Lichtman, J.W., 314–315
Liddle, P.F., 264–265
Lieberman, J.A., 62, 114, 169
Light, G.A., 20t, 220–222
Lilienfeld, S.O., 18
Lim, E.T., 55

Lin, H., 71–72
Lin, S.P., 64
Linden, D.E., 224
Linderholm, K.R., 183
Lindqvist, M., 109
Lindström, L.H., 115–116
Ling, Z.D., 182–183
Lipska, B.K., 247–249
Lisman, J.E., 244
Lister, R., 65–66, 73–74
Liu, H., 173
Liu, J., 65–66, 152
Liu, Y., 172, 322–324
Liu, Z., 150
Lodge, D.J., 157–158, 245–248
Loh, N., 9
Lopes da Silva, F., 238–239
López-Muñoz, F., 9–10
Low, K., 154–155
Lowe, M.J., 316
Lu, J.C., 151–152
Lubin, F.D., 28
Luck, S.J., 214
Lykken, D.T., 17, 22–23
Lynall, M.-E., 30, 318–319, 324
Lyu, H., 20*t*

M
Ma, D.K., 68–69
Maas, J.W., 110
MacDonald, M.L., 131
MacDonald 3rd, A.W., 133
MacGillavry, H.D., 126
Madisen, L., 85
Mah, S., 44
Mahadik, S.P., 169–170
Maia, T.V., 289
Mainen, Z.F., 127
Majewska, A.K., 126
Malenka, R.C., 128
Malhotra, A.K., 83–84
Malhotra, D., 52
Malinow, R., 128
Malone, S.M., 20*t*
Mamah, D., 241
Mamo, D., 110–111, 300*f*
Mandl, R.C.W., 315, 317
Manoliu, A., 241–242
Mansuy, I.M., 27
Mantini, D., 239
Marín, O., 244
Markham, J.A., 127

Martinez-Cue, C., 159–160
Martinez-Lozada, Z., 130
Martinowich, K., 68–69
Martins-de-Souza, D., 67–68, 196–197
Marutha Ravindran, C.R., 68–69
Mathalon, D.H., 223–224, 228
Mathew, I., 20*t*
Mathiak, K., 217
Mathys, C., 285
Matsuzaki, M., 126
Mattheisen, M., 202
Maunakea, A.K., 62*f*–63*f*
Maze, I., 73
Mazhari, S., 20*t*
McCarley, R.W., 214, 223–225
McCarroll, S.A., 150
McCarthy, S.E., 52
McCullumsmith, R.E., 196–197
McDannald, M.A., 248–249
McEvoy, J., 171
McGowan, P.O., 67–68
McKernan, R.M., 154–155, 158–159
Meador-Woodruff, J.H., 67–68, 134–135
Meda, S.A., 30
Meduna, L., 9
Meinl, E., 180
Meisenzahl, E.M., 184
Melchitzky, D.S., 134
Mellios, N., 67
Meltzer, H.Y., 109
Menon, M., 299–300, 300*f*, 301*f*
Merali, Z., 181
Mercadante, M.T., 180
Meunier, D., 322
Meyer, U., 183–184
Meyer-Lindenberg, A., 24–25, 30, 115, 116*f*, 117
Meyer-Lindenberg, A.S., 291–292
Middleton, F.A., 67–68
Milders, M., 304*f*
Mill, J., 62–63, 66, 68–69, 73–74
Miller, A.H., 182
Miller, C.A., 68–69
Miller, F.D., 65–66
Miller, G.A., 18–19, 20*t*, 21–23, 25, 30
Miller, J.A., 100
Miller, P., 249
Miller, R.L., 19, 22–23, 30
Minor, K.S., 303
Minzenberg, M.J., 116–117, 308
Mirnics, K., 67–68, 125–126
Misra, C., 265–266
Miyauchi, T., 239–241

Mizoguchi, H., 185
Mizrahi, R., 111, 115
Modinos, G., 125–126
Moghaddam, B., 150, 244
Mohler, H., 150–151, 157
Moore, H., 248–249
Morabito, M.A., 200
Moreno-De-Luca, D., 52
Mori, S., 313–315
Morishita, H., 152
Moritz, S., 303
Morris, J.A., 202–204
Morris, M.J., 71–72
Moss, J.F., 69–71
Moutoussis, M., 290–291
Muddyman, D., 54
Mueller, B.R., 65
Mueller, H.T., 134–135
Müller, N., 180, 182–185, 188, 196
Munafò, M.R., 19, 22–23
Munton, R.P., 201–202
Murgatroyd, C., 65
Murray, C.J., 167
Murray, G.K., 290, 299–300
Murray, J.D., 222
Myint, A.M., 183

N

Näätänen, R., 220–221
Nair, A., 181
Nakagawa, T.T., 30, 319
Nakao, K., 245–246
Nakazawa, K., 245–246
Nanda, P., 264–265
Narayan, S., 71–72
Narayanan, B., 29
Narendran, R., 113
Narr, K.L., 324
Nasrallah, H., 167
Neale, B.M., 54–55
Neelam, K., 20t
Nelson, B., 305–306
Nelson, E.D., 68–69
Nelson, E.E., 104
Nelson, M.D., 271
Nestler, E.J., 62–63, 65
Network and Pathway Analysis Subgroup of Psychiatric Genomics Consortium, 62f–63f
Neuner, I., 239
Ng, L.L., 84, 86
Nickl-Jockschat, T., 24–25
Nicolae, D.L., 70f

Nielsen, M.O., 299–300
Nieman, D.H., 224, 230
Nieuwenhuis, S., 224
Nikolich-Zugich, J., 180, 182–183
Nishioka, M., 27–28
Niswender, C.M., 130
Nitta, M., 184–185
Niv, Y., 289
Niwa, M., 65
Nohesara, S., 66–67
Nordström, A., 168, 174
Nordström, A.L., 110–111
Northoff, G., 242, 250
Numachi, Y., 68–69
Nyberg, S., 111

O

O'Brien, R.J., 128–129
O'Carroll, D., 67
O'Connor, S., 224
O'Dell, T.J., 200
O'Donnell, P., 249
O'Donovan, M.C., 44
Oh, G., 26–27, 64
Oh, S.W., 85, 93
Ohi, K., 243
Okabe, S., 128
Okaty, B.W., 152–154
Okubo, Y., 116–117
Olincy, A., 215, 217
Olsen, R.W., 151
Olsson, S.K., 183
O'Neill, L.A., 179
Ongür, D., 241
Oommen, K.J., 180
Orešic, M., 170–173
Orliac, F., 241
O'Roak, B.J., 55, 202
Ortega, A., 130
Ott, T., 289
Ouchi, Y., 202–204
Owen, M.J., 133, 136–137
Ozerdem, A., 29
Oztas, E., 314–315

P

Pak, D.T., 128
Palaniyappan, L., 264–265, 271–272, 318–319
Pantelis, C., 264–265, 275–276
Panzanelli, P., 151–152
Papaleo, F., 248–249

Paparelli, A., 62f–63f
Park, P.J., 73
Pasternak, O., 313–314
Patrick, C.J., 19, 22
Patti, G.J., 168
Paulus, M.P., 301–302
Pauly, K., 303–305, 308
Pawson, T., 198–200
Paxinos, G., 86
Pearce, B.D., 183
Pearlson, G.D., 17, 20t, 23
Peedicayil, J., 27–28
Peet, M., 169–170
Peng, J., 131–132
Pennington, K., 196
Penny, W.D., 291–292
Perez, S.M., 250
Perez, V.B., 222–223, 230
Perry, T.L., 243
Perry, V.H., 181
Petralia, R.S., 127
Petronis, A., 26–27, 62–64
Petryshen, T.L., 152
Pettersson-Yeo, W., 318–319
Phillips, G.R., 131
Phillips, K.G., 226–228, 245–246
Picardi, A., 314
Pierpaoli, C., 315
Pin, J.P., 130
Pinault, D., 227–228
Pinto, D., 202
Pishva, E., 73–74
Poels, E.M., 117
Polan, M.B., 152
Polich, J., 223–224
Politi, P., 7
Popov, T., 28–30, 217
Popov, T.G., 28–29
Popova, P., 28–29
Port, R.G., 238
Porteous, D.J., 51
Potter, D., 215, 217
Potter, E.D., 182–183
Potvin, S., 182
Potwarka, J.J., 169–170
Powell, S.B., 248–249
Prasad, K.M., 29
Price, A.L., 54–55
Price, D.J., 329
Puckett, R.E., 28
Puelles, L., 91–92
Purcell, S.M., 44–45, 57, 69, 70f, 126, 133, 150, 195

Q
Quednow, B.B., 112
Quinones, M.P., 174

R
Racca, C., 128
Radtke, K.M., 28
Radulescu, E., 20t
Ragland, J.D., 308
Raichle, M.E., 316
Raison, C.L., 182
Raj, A., 328–329
Rakic, P., 101–102
Ramyead, A., 239
Ranlund, S., 29
Rao, R.P., 285
Rauch, A., 55
Rausch, F., 118
Rees, E., 52
Regenold, W.T., 67–68
Reich, D.E., 54–55
Reif, A., 202–204
Reijmer, Y.D., 314, 329
Reinberg, D., 62f–63f
Reite, M., 215
Reith, J., 115
Repovs, G., 318–319
Reynolds, G.P., 243
Richiardi, J., 103–104
Richter, A., 118
Riecker, A., 318–319
Ripke, S., 69, 83–84
Ritter, P., 239
Roach, B.J., 228
Roalf, D.R., 20t
Roberts, R.C., 134
Robinson, C.V., 197–198
Rockstroh, B., 18–19, 20t, 22, 25
Rodin, E., 239–241
Rodriguez-Murillo, L., 62
Rodriguez-Paredes, M., 62f–63f
Roiser, J.P., 290, 318–319
Rolland, B., 242
Rondard, P., 130
Ross, C.A., 125–126
Ross, R.M., 301–302
Rotarska-Jagiela, A., 242
Roth, C., 64
Roth, T.L., 65
Rouse, B.T., 182–183
Roussos, P., 69, 70f, 71, 83–84
Roux, L., 152–154

Rowlett, J.K., 158–159
Rubenstein, J.L., 91–92
Rubinov, M., 321–328
Rudenko, A., 65–66
Ruderfer, D.M., 55, 195
Rudie, J.D., 322
Rudolph, U., 150–151, 154–155, 158–159
Rujescu, D., 52, 173
Rutherford, B.R., 185–186
Rutter, L., 239–242
Ryan, M.C., 196
Ryman, S.G., 314

S

Sabri, O., 299
Saha, S., 62
Sala, C., 127
Salnikow, K., 68–69
Salter, M.W., 129
Sams-Dodd, F., 249
Sanders, S.J., 55
Sanfilipo, M., 263–264
Saperstein, A.M., 20t
Sathyasaikumar, K.V., 183
Satta, R., 68–69
Saunders, J.A., 240f, 245–246
Saunders, R.C., 117–118
Savina, I., 301f
Sawa, A., 202–204
Scannevin, R.H., 200
Schaefer, A., 67
The Schizophrenia Psychiatric Genome-Wide Association Study (GWAS) Consortium, 23, 45
Schizophrenia Working Group of the Psychiatric Genomics Consortium, 23, 45, 46f, 48, 52, 126, 133–134, 150
Schlagenhauf, F., 299–300, 303
Schluesener, H.J., 71–72
Schmidt, C., 197–198
Schmidt, R., 328–329
Schmitt, A., 26–27
Schmitz, T.W., 300f
Schneider, K., 7–8
Schnitzler, A., 239
Schork, N.J., 40–41
Schreiber, H., 222
Schroeder, F.A., 71–73
Schultz, W., 289
Schulze, T.G., 23
Schuman, E.M., 128
Schumann, G., 17, 19, 23
Schwarcz, R., 183
Schwarz, E., 196
Schwarz, M., 196
Schwarz, M.J., 182–183
Schwarzenbacher, R., 71–72
Scott, J.D., 198–200
Sebat, J., 22–23, 52
Seeley, W.W., 299–300
Seeman, P., 110
Segal, M., 127
Sehrawat, S., 182–183
Seidman, L.J., 20t, 22
Sejnowski, T.J., 242–243
Senden, M., 322
Seneca, N., 113
Sesack, S.R., 289
Severinsen, J.E., 71–72
Shannon, P., 200f, 203f
Sharma, R.P., 62f–63f, 71–72
Shelley, A.M., 222
Shen, D.G., 268
Sheng, M., 126–131
Shenton, M.E., 263–264, 266–267, 271, 317
Shergill, S.S., 288
Shi, F., 324–327
Shi, J., 44, 69
Shields, J., 17, 19, 19t
Shih, C.-T., 314–315
Shimabukuro, M., 68–69
Siegel, C., 217–218
Siegel, S.J., 239–241
Sieghart, W., 151
Siegmund, K.D., 66, 73–74
Siekevitz, P., 127
Sigmundsson, T., 263–264
Sigurdsson, T., 29
Silva, A.J., 200
Silverstein, S.M., 288
Singer, W., 20t, 29, 214–215, 226, 228–229, 238
Sivarao, D.V., 227–228
Skinner, M.K., 26–27
Sklar, P., 69
Skudlarski, P., 241, 317–319
Slifstein, M., 116–117
Smith, A.J., 301f
Smith, A.K., 217
Smith, G.S., 118
Smith, Y., 289
Smucny, J., 217
Sohal, V.S., 226–227
Sokolov, B.P., 134–135
Sommer, I.E., 184–185
Somogyi, P., 150–154
Song, C.X., 62f–63f

Sorensen, H.J., 183
Sorra, K.E., 126–127
Sotiras, A., 267
Sowell, E.R., 263–264
Spacek, J., 127
Sparkman, N.L., 181–182
Spellman, T.J., 241
Spencer, K.M., 152–154, 153f, 228
Sporns, O., 28–29, 313–316, 319, 321–328
Spring, B., 181–182
St Onge, J.R., 289
Stahl, S.M., 62, 109
Stam, C.J., 328–329
Stanzione, P., 224
Steen, R.G., 184
Stefansson, H., 44, 69, 299–300
Steffek, A.E., 135
Stein, Z., 64
Steinberg, E.E., 289
Steiner, J., 183
Stephan, K.E., 103–104, 287, 314, 319, 328
Stephens, M., 54–55
Stevens, K.E., 215
Stickle, C., 304f
Stone, J.M., 244
Stone, T.W., 183
Stoner, R., 100t
Stransky, E., 313–314
Strauss, G.P., 290
Strelnikov, K., 308
Strogatz, S.H., 322
Strzelecki, D., 117–118
Stubbs, B., 172
Sugranyes, G., 303
Sullivan, P.F., 44, 51
Sun, J., 242, 263–264
Sun, L., 173
Sunkin, S.M., 83–84
Sur, M., 67
Suren, P., 64
Suridjan, I., 115
Susser, E.S., 64
Svoboda, K., 127
Svrakic, D.M., 22–24, 27–28
Swartz, M.S., 62
Sweatt, J.D., 62–63, 65–66, 68–69
Sweet, R.A., 134
Swerdlow, N.R., 20t, 218–220, 249
Szatkiewicz, J.P., 52

T

Taber, K.H., 307
Takano, A., 184
Takao, K., 248–249
Takata, A., 62f–63f
Talbot, K., 136–137
Talbot, P.S., 112
Taly, A., 62
Tam, G.W., 195
Tan, K.R., 156–159
Tan, M., 62f–63f
Tandon, R., 4, 298, 314
Tang, J., 241
Tansey, K.E., 47
Tao-Cheng, J.H., 130
Tatard-Leitman, V.M., 244–249, 247f
Taverna, S.D., 62f–63f
Taylor, S.F., 115, 303
Teixeira Jr., A.L., 180
Theodoridou, A., 5–6
Thermenos, H.W., 264–265
Thoma, R.J., 217
Thompson, C.L., 84, 91–92
Thompson, P.M., 329
Thönnessen, H., 222
Thornicroft, G., 71
The 1000 Genomes Project Consortium, 41–42
Thurman, R.E., 73
Ticku, M.K., 68–69
Tiesinga, P., 242–243
Tkachev, D., 67–68
Tobi, E.W., 64
Toro, C.T., 202–204
Torrealba, F., 314–315
Torrey, E.F., 183
Towers, S.K., 157
Toyooka, K., 135
Traub, R.D., 244
Trinidad, J.C., 128, 133, 201–202
Tsankova, N., 64
Tsuboi, D., 85f
Tu, J.C., 130
Tu, P.C., 241, 318–319
Tunquist, B.J., 128
Turetsky, B., 228–229, 263–264
Turetsky, B.I., 20t, 214–215, 223–224, 228–229, 239–241, 288
Tuulio-Henriksson, A., 20t

U

Uddin, L.Q., 328–329
Uhlhaas, P.J., 20t, 29, 214–215, 226–229, 238
Umbricht, D., 220–222
Uranova, N.A., 317
Urbach, T.P., 239–241

V

Vacic, V., 52
Vaidyanathan, U., 20*t*
Valtschanoff, J.G., 127
van Berckel, B.N., 184
van Dam, A.P., 180–181
Van Deal, F., 301–302
Van den Heuvel, M.P., 29, 314–315, 318–319, 322–329
van der Kemp, W.J., 172
Van der Meer, L., 304–305
Van der Velde, J., 29
Van Diessen, E., 250
van Os, J., 26–27, 64, 102
Van Rossum, J.M., 109
van Scheltinga, A.F.T., 23
van Tricht, M.J., 224
van Zijl, P.C.M., 315
Vapnik, V.N., 265–266
Varshney, L.R., 314–315
Velakoulis, D., 263–264
Vemuri, P., 265–266
Venables, N.C., 239–241
Vercelli, A., 314–315
Verhoeff, N.P., 112–113
Verkhratsky, A., 127
Vernaleken, I., 110–111, 115–116, 118
Versijpt, J.J., 184
Verstraete, E., 315
Villringer, A., 239
Vincent, A., 183
Vincent, S.L., 134
Vinson, P.N., 130–131
Vohs, J.L., 245–246
Voineskos, A.N., 317
Volk, D.W., 154, 243
Volkow, N.D., 112–113
Vollenweider, F.X., 118
Vos, T., 167
Vrieze, S.I., 22–23

W

Waiter, G., 304*f*
Walikonis, R.S., 131
Walker, D.M., 65
Walter, H., 302–303
Waltz, J.A., 289
Wang, H.D., 134
Wang, J., 250
Wang, K., 53–54
Wang, L., 324
Wang, Q., 202–204, 324, 327–328
Wang, X.-J., 242–243
Wang, Z., 83–84
Watanabe, Y., 87
Watkins, S.M., 170–171
Watson, B.O., 29
Watson, T.D., 224
Watts, D.J., 322
Weaver, I.C., 65
Wei, H., 67
Weickert, C.S., 134–135, 137, 196
Weinberg, R.J., 127–128
Weinberger, D.R., 110, 134, 136–137, 150, 249
Weisbrod, C.R., 197–198
Wenk, M.R., 170
Wernicke, C., 313–314
Wesseling, H., 173
Westergaard, T., 183
Wheeler, A.L., 317
White, J.G., 314–315
White, T., 20*t*, 30
White, T.P., 242
Whitfield-Gabrieli, S., 241, 318–319, 327–328
Whitford, T.J., 317
Whittington, M.A., 227–228, 242–244
Wildenauer, D.B., 182
Wilhelmsen, K.C., 19
Wilkins, M.R., 195–196
Wilkinson, K.A., 129–130
Willeit, M., 113, 301*f*
Williams, L.M., 303
Winblad, B., 110
Winkler, A.M., 264–265
Winslow, J.T., 104
Winter, C., 182–183
Winterer, G., 150, 214, 224–225, 239–241
Winton-Brown, T.T., 290, 299–302
Wockner, L.F., 27–28, 66–67
Wolkin, A., 111, 263–264
Wolters, D.A., 195–196
Wong, D.F., 110–112
Wong, E.H., 11
Woo, T.U., 67–68, 134–135
Wood, A.J., 69–71
Wood, P.L., 171
Wood, S.J., 264–265
Woodward, N., 241
Workman, A.D., 101
World Health Organization, 3–4, 9
Wotruba, D., 30
Wright, I.C., 271
Wu, H., 65–66
Wuchty, S., 199–200

X

Xiao, B., 130
Xu, B., 55, 195
Xu, G., 151–152
Xu, X., 24
Xuan, J., 172–173
Xue, J.-G., 244

Y

Yan, H., 319
Yang, J., 46, 173–174
Yang, J.M., 150
Yao, J.K., 169–170
Yates, J.R., 195–196
Yee, B.K., 150, 156–157
Yee, C.M., 19, 25
Yeragani, V.K., 241–242
Yingling, C.D., 224
Yki-Jarvinen, H., 172
Yolken, R.H., 183
Yu, A.J., 285
Yu, Q., 324–328
Yun-Hong, Y., 127
Yushkevich, P., 86

Z

Zaehle, T., 250
Zai, G., 152
Zalesky, A., 314–315, 319, 322–324, 327–329
Zamora-López, G., 322
Zanetti, M.V., 265–266
Zemankovics, R., 244
Zeng, H., 100*t*
Zhang, J., 313–315
Zhang, T.H., 269, 272–274
Zhang, T.Y., 67–68
Zhang, W., 128
Zhang, Y., 324–327
Zhang, Z.J., 243
Zhang, Z.Y., 71–72
Zhao, J., 73
Zhao, X., 152, 202–204
Zhitkovich, A., 68–69
Zhong, H., 70*f*
Zhou, D., 181
Zhou, J., 328–329
Zhou, V.W., 62*f*–63*f*
Zhou, X., 54–55
Zhou, Y., 318–319
Zhu, H., 71–72
Zhu, J., 73
Zhu, J.J., 129–130
Zipursky, R.B., 301*f*
Zubin, J., 181–182
Zucchi, F.C., 27–28
Zukin, S.R., 244
Zweig, A., 313–314

Subject Index

Note: Page numbers followed by "*b*," "*f*," and "*t*" refer to boxes, figures, and tables, respectively.

A

A priori hypothesis, 266–267
AA. *See* Arachidonic acid (AA)
Aberrant salience, 242, 290
   attribution, 299–301
   hypothesis, 299–300
   processing, 303
Aberrant salience detection measurement
   MMN, 220
      measurement, 220–221
      neurobiology, 221–222
      and schizophrenia, 221*f*, 222–223
   P300, 223
      methodology, 223
      neurobiology, 224
      and schizophrenia, 224–225
   Abnormal gamma band oscillations measurement, 225–226
      gamma oscillations neurobiology, 226–228
      neural oscillations measurement, 226
      and schizophrenia, 228–229
ACC. *See* Acetylcysteine (ACC)Anterior cingulate cortex (ACC)
Acetylcysteine (ACC), 185
Activation likelihood estimation (ALE), 338–339
Activity-regulated cytoskeleton-associated protein (ARC-associated protein), 52, 133
Acute auditory hallucinations, neural correlation, 306–307
Adenylate cyclase 5 (*Adcy*5), 89
ADHD. *See* Attention deficit hyperactivity disorder (ADHD)
Adverse drug reactions (ADRs), 167–168
Affymetrix, 42
AIS. *See* Axon initial segment (AIS)
ALE. *See* Activation likelihood estimation (ALE)
Allele discrimination by hybridization, 42
Allen Brain Atlas portal, 83–84
Allen Developing Mouse Brain Atlas, 84, 91–92
Allen Human Brain Atlas, 97–98, 98*f*
Allen Institute for brain science resources, 83–84
   applications in schizophrenia research, 102–104
   atlases, 102
   human atlas resources, 97–102
   mouse atlas resources, 84–97
   portal, 84
Allen Mouse Brain Atlas, 84
Allen Mouse Brain Connectivity Atlas, 84–85, 93, 96*f*
Allen Spinal Cord Atlas, 84–85

α-amino-3-hydroxy-5-methyl-4-isoxazolepropionic acid (AMPA), 388
α3-containing $GABA_A$ receptors, dopaminergic neurons inhibition by, 156–157
$α_5$(H105R) mice, 157
α5-containing $GABA_A$ receptors, indirect inhibition of dopaminergic neurons by, 157–158
α5-negative allosteric modulators and cognition enhancement, 159–160
α-methyl-para-tyrosine (AMPT), 112–113
   dopamine depletion paradigm, 113–114
Ambitious clinical studies, 73–74
AMPA. *See* α-amino-3-hydroxy-5-methyl-4-isoxazolepropionic acid (AMPA)
AMPA receptors (AMPARs), 127, 129–130, 388
AMPT. *See* α-methyl-para-tyrosine (AMPT)
Animal models of noisy brain
   baseline gamma in animal models, 244–245
   EEG/LFP validation of models, 248–249
   NR1 hypomorphic animals, 248*f*
   SCZ with relation to resting gamma activity, 245–246
   in vitro examination of animal models of increased resting gamma, 247–248
Annotation, 54
Anterior cingulate cortex (ACC), 134–135
Anti-inflammatory therapeutic approach, COX-2 inhibition as, 184–185
Antipsychotic(s), 10
   drugs, 71, 387
   exposure, 134
   medications, 62, 168
Anxiety-like behaviors, 401
API. *See* Application programming interface (API)
Application programming interface (API), 102
Arachidonic acid (AA), 172
Arc protein complex, 197–198
ARC-associated protein. *See* Activity-regulated cytoskeleton-associated protein (ARC-associated protein)
Area under the receiver-operating characteristic curve (AUC), 173–174
ASDs. *See* Autism spectrum disorders (ASDs)
Atlas space, 267
Attention deficit hyperactivity disorder (ADHD), 46–47, 52, 399
Atypical antipsychotics, 168
AUC. *See* Area under the receiver-operating characteristic curve (AUC)

433

Audio-visual hallucinations (AVH), 341
Auditory oddball-induced time-frequency alterations, 29
Autism spectrum disorders (ASDs), 46–47, 398–399
Automated voxel-based methods, 264–265
"Average reward rate" model, 289
AVH. *See* Audio-visual hallucinations (AVH)
Axon initial segment (AIS), 155–156

B

Bacterial artificial chromosome (BAC) transgenic model, 365
Baseline gamma in animal models, 244–245
Basic helix-loop-helix (bHLH), 359–360
Bayesian brain, precision, and hierarchical models, 283–285
BDNF. *See* Brain-derived neurotrophic factor (*BDNF*)
Behavioral phenotypes of genetic mouse models
    future directions, 389–392
    higher-order domains
        and associated constructs, 389t
        of research criteria, 390t–391t
    measuring psychotic behavior in mouse, 384
    models targeting dopaminergic transmission, 385
        COMT knockout, 385–386
    models targeting glutamatergic transmission, 388–389
    schizophrenia, 383–384
Benzodiazepines, 154–155
    benzodiazepine-sensitive GABA$_A$ receptors, 154–155, 155f
β2 subunit of GABA$_A$ receptor (GABRAB2), 152
Between-group differences, 114
bHLH. *See* Basic helix-loop-helix (bHLH)
Binding potential (BP), 112
Biogenesis of lysosome-related organelles complex 1 (BLOCK1), 360
Bioinformatics, 53–54
Biomarkers, 168, 196
Bipolar disorder (BPD), 46–47
Bizarre delusional beliefs, 8–9
Bleuler, Eugen, 6, 7f
BLOCK1. *See* Biogenesis of lysosome-related organelles complex 1 (BLOCK1)
Blocks, 41
Blood oxygenation level–dependent (BOLD), 297
BOLD. *See* Blood oxygenation level–dependent (BOLD)
BoMa consortium. *See* Bonn–Mannheim (BoMa) consortium
Bonn–Mannheim (BoMa) consortium, 44
BP. *See* Binding potential (BP)
BPD. *See* Bipolar disorder (BPD)
BPRS. *See* Brief psychiatric rating scale (BPRS)
Brain
    comprehensive, region-specific, and cell type–specific epigenome mappings, 73
    hubs, 322, 324–329
    network, 314

Brain computations in schizophrenia. *See also* Brain volumes in schizophrenia; Computational neuroanatomy
    Bayesian brain, precision, and hierarchical models, 283–285
    computational implications of striatal presynaptic dopamine elevation, 288
        aberrant salience, 290
        incentive salience, 290
        overweighting evidence *vs.* lowered decision threshold, 290–291
        phasic dopamine signaling, 289–290
        tonic dopamine signaling, 289
    hierarchical probabilistic model, 286f
    precision, 284f
    psychosis, 285–287
    synaptic gain, 285–287
        computational implications of decreased high-level, 287–288
Brain connectivity dynamics, 28–29
    anatomic alterations in schizophrenia, 29
    features of brain chronnectomics, 30
    large-scale functional neuronal networks, 30
    neural oscillations, 29
    neuromagnetic oscillatory activity aspects, 29
Brain Explorer window, 90
Brain volumes in schizophrenia, 270–271. *See also* Brain computations in schizophrenia
    morphometric differences, 272–274
    MRI data, 271–272
    ODVBA, 271f
    reduced gray matter volumes, 274f
    ROI-based approaches, 271
    SCZ-like neuroantomical patterns, 276, 277f
    single-subject classification, 274–275
    single-subject differentiation, 275–276
    VBM comparison of gray matter density, 273f
Brain-derived neurotrophic factor (*BDNF*), 66–67
Brain's connectional anatomy, 314–315
    brain network, 314
    functional brain connectivity, 316
    structural brain connectivity, 315b–316b
Brain's connections, schizophrenia and
    affected white matter tracts in, 316f
    connectome disorganization and cognitive deficits relationship in, 329
    connectomics in long-term outcome and clinical practice, 329
    functional connectivity in, 317f, 318–319
    hub pathology to, 328–329
    intersection of structural and functional connectivity, 319
    structural connections underlying functional connectivity, 318f
    structural connectivity in, 317
BrainSpan atlas expression viewer, 101f

BrainSpan Atlas of Developing Human Brain, 97, 99–100
Brief psychiatric rating scale (BPRS), 156
Bromodomains, 71–72
Burrows-Wheeler transform aligner (BWA), 53–54

C

C57Bl/6J mouse brain, 86
Ca^{2+}/calmodulin-dependent protein kinase II (CAMKII), 129, 364
CACNA1C gene, 23–24, 243
Caenorhabditis elegans, 314–315
CAMKII. *See* Ca^{2+}/calmodulin-dependent protein kinase II (CAMKII)
cAMP. *See* Cyclic adenosine monophosphate (cAMP)
Candidate genes, 358
Catechol-*O*-methyltransferase (COMT), 28, 66, 357, 385–386
    COMT-Val158Met polymorphism, 386
    knockout mice, 385–386
CCF. *See* Common Coordinate Framework (CCF)
CDCV hypothesis. *See* Common Disease, Common Variant (CDCV) hypothesis
CdLS. *See* Cornelia de Lange syndrome (CdLS)
CDP. *See* Chlordiazepoxide (CDP)
CDRV hypothesis. *See* Common Disease, Rare Variant (CDRV) hypothesis
Central nervous system (CNS), 180
    inflammation in, 180
    volume loss in imaging studies, 184
Cerebellar (CER), 241
Cerebral correlates
    of delusions, 298–299
        aberrant salience attribution, 299–301
        neural correlates of impaired self-monitoring and self-reflection, 304–306
        neuronal correlates of cognitive and emotional biases, 301–303
    of hallucinations, 306
        neural correlates of acute auditory hallucinations, 306–307
        neural correlates of hallucination proneness, 307–308
Cerebrospinal fluid (CSF), 168
Chlordiazepoxide (CDP), 156
Chlorpromazine, 9–10
Cholinergic receptor neuronal α7 subunit (*Chrna*7), 407
CHR. *See* Clinical high-risk (CHR)
*Chrna*7. *See* Cholinergic receptor neuronal α7 subunit (*Chrna*7)
Chromatin, 62–64
    chromatin-modifying drugs, 72–73
Chromosomal loopings, 69–71
Chronnectome, 28–29
Cingulo-opercular (CO), 241
Circuit models and evidence in humans, 242–243

Class III HDACs, 71–72
Clinical high-risk (CHR), 264–265
Clozapine, 110–111, 168
CMC test. *See* Combined Multivariate and Collapsing (CMC) test
Cnksr2. *See* Connector enhancer of kinase suppressor of ras 2 (Cnksr2)
CNS. *See* Central nervous system (CNS)
CNV-associated neuropsychiatric disorders, 398–400. *See also* Copy number variants (CNVs)
    CNV-encoded genes causing behavioral dimensions, 402
        developmental trajectories, 407–408
        mass action and net effects, 407
        noncontiguous effects, 406–407
        phenotypic variation, 408–409
        pleiotropy, 407
        segmental deletions, 402–405
    dimensional behavioral models, 400–402
CNVs. *See* Copy number variants (CNVs)
CO. *See* Cingulo-opercular (CO)
Cognitive and emotional biases, neuronal correlates of, 301–303
Cognitive dysmetria, 263–264
Cognitive subtraction, 338
COGS. *See* Consortium on the Genetics of Schizophrenia Endophenotypes (COGS)
Combined Multivariate and Collapsing (CMC) test, 54–55
Common Coordinate Framework (CCF), 86–87
Common Disease, Common Variant (CDCV) hypothesis, 40–41
Common Disease, Rare Variant (CDRV) hypothesis, 40–41
Common disease rare variant hypothesis, 54–55
Common reference space in mouse brain, 86–87
Common risk variant, 42–43, 49
Common variant common disease hypothesis, 54–55
Complex GWAS, 54–55
Computational neuroanatomy
    machine learning, 269–270
    MVPA, 269–270
    ODVBA, 269
    region of interest *vs.* voxel-based analysis, 266–267
    regional tissue volumetrics, 267–268
    topography of reduced brain volumes in SCZ, 270–271
        morphometric differences, 272–274
        MRI data, 271–272
        ODVBA, 271*f*
        reduced gray matter volumes, 274*f*
        ROI-based approaches, 271
        SCZ-like neuroantomical patterns, 276, 277*f*
        single-subject classification, 274–275
        single-subject differentiation, 275–276
        VBM comparison of gray matter density, 273*f*

COMT. *See* Catechol-*O*-methyltransferase (COMT)
Conditioned stimulus (CS), 157
Connectivity in mouse brain, 93–96
Connectome, 321
  hubs, 341
  schizophrenia and, 321–328
    affected hub connectivity, 326*f*
    from connections to connectome, 320*f*
    graph theoretical studies of human, 322
    integration, 321, 324–327, 325*f*
    organization in relation to clinical symptoms and outcome in, 327–328
    segregation, 321–324, 323*f*
Connector enhancer of kinase suppressor of ras 2 (Cnksr2), 198
Consortium on the Genetics of Schizophrenia Endophenotypes (COGS), 228
Constitutive overexpression of mutant DISC1, 364–365
Coordinate-based meta-analysis, 339
Copy number variants (CNVs), 11, 42, 52, 102, 133, 195, 197–198, 358, 397–398. *See also* CNV-associated neuropsychiatric disorders
  and diverse neuropsychiatric disorders, 398–400
  mouse models
    and dimensional behavioral phenotypes, 403*t*–404*t*
    of 22q11.2 CNVs, 405*f*
  risk for carriers of developing disorders, 398*f*
Cornelia de Lange syndrome (CdLS), 69–71
Correlative Search utility, 89
Cortical connectivity, 3D reconstruction of, 95*f*
Cortical period plasticity, GABA$_A$ receptor subtypes in, 152
Cortico-striato-pallido-pontine (CSPP) circuitry, 218
Cortico-striato-thalamo-cortical loops, 289
COX-2. *See* Cyclooxygenase-2 (COX-2)
CpG dinucleotides. *See* Cytosine–guanine dinucleotides (CpG dinucleotides)
Cross-Disorder Working Group of PGC, 46–47
CS. *See* Conditioned stimulus (CS)
CSF. *See* Cerebrospinal fluid (CSF)
CSPP circuitry. *See* Cortico-striato-pallido-pontine (CSPP) circuitry
Cyclic adenosine monophosphate (cAMP), 287
Cyclooxygenase-2 (COX-2), 184–185
  inhibition as anti-inflammatory therapeutic approach, 184–185
Cystine, 173–174
Cytosine methylation, 65–66
Cytosine–guanine dinucleotides (CpG dinucleotides), 65–66

D

D-amino acid oxidase (DAO), 357–358
D$_2$ receptor gene (*Drd2*), 385
D$_2$ receptor long form (D$_2$L), 387
D$_2$ receptor short form (D$_2$S), 387
D$_{2/3}$ receptor occupancy, 110–111
D$_2$L. *See* D$_2$ receptor long form (D$_2$L)
D$_2$Rs. *See* Dopamine 2 receptors (D$_2$Rs)
D$_2$S. *See* D$_2$ receptor short form (D$_2$S)
DA. *See* Dopamine (DA)
DALY. *See* Disability-adjusted life-year (DALY)
DAO. *See* D-amino acid oxidase (DAO)
Data-driven approaches, 341–342
De novo LOF mutations, 55
De novo mutations, 55
Default mode network (DMN), 241
Delayed nonmatch to sample (DNMTS), 356
Delusions
  cerebral correlates, 298–299
    aberrant salience attribution, 299–301
    neural correlates of impaired self-monitoring and self-reflection, 304–306
    neuronal correlates of cognitive and emotional biases, 301–303
  delusional perceptions, 7–8, 285
  delusional symptoms, 299–300
  of reference, 299
Dementia praecox, 6–7
Dendritic spines, 126–127
  alterations, 134
Deutsche Forschungsanstalt für Psychiatrie, 10
Developmental trajectories, 407–408
Developmental transcriptome, 99
DHA. *See* Docosahexaenoic acid (DHA)
Diagnostic and Statistical Manual of Mental Disorders (DSM-V), 3–4, 9, 366
Differential Search, 87–89
Diffusion MRI (dMRI), 341
Diffusion-weighted imaging (DWI), 315
Dimensional behavioral models of neuropsychiatric disorders, 400
Dimensional measure, 400
Disability-adjusted life-year (DALY), 167
DISC1. *See* Disrupted in schizophrenia 1 (DISC1)
Disconnection syndrome, 263–264
Discriminative patterns, 344
Disk large-associated homologues 1–4 (DLGs), 198
Disk large-associated protein 1 (Dlgap1), 198
Disorder of dysconnectivity, 103–104
Disorganized speech, 7
Disrupted in schizophrenia 1 (DISC1), 65, 202, 361
  constitutive overexpression of mutant DISC1, 364–365
  ENU-induced models, 363–364
  exon 2–3 deletion model, 364
  gene, 51
  inducible C-fragment DISC1 model, 365
  inducible mutant human DISC1 model, 365–366
  interactome, 202
  methodologies to generation, 361–363
  risk allele schizophrenia model, 361–363
  transgenic mice, 364

Dissociation psychosis. *See* Sejunktionspsychose
Diverse neuropsychiatric disorders, 398–400
Dlgap1. *See* Disk large-associated protein 1 (Dlgap1)
DLGs. *See* Disk large-associated homologues 1–4 (DLGs)
DLPFC. *See* Dorso-lateral prefrontal cortex (DLPFC)
dlPFC. *See* Dorsolateral prefrontal cortex (dlPFC)
DMN. *See* Default mode network (DMN)
dMRI. *See* Diffusion MRI (dMRI)
DNA methylation, 64
   changes, 66
   and gene regulation in brain, 65–66
   studies in peripheral tissue and epigenetic biomarkers in SCZ, 66–67
DNA microarrays, 42
DNMTS. *See* Delayed nonmatch to sample (DNMTS)
Docosahexaenoic acid (DHA), 171–172
Docosapentaenoic acid (DPA), 172
DOPA decarboxylase, 115
Dopamine (DA), 112, 115, 289, 356, 365–366
   hypothesis of schizophrenia, 109
      history, 109–110
      ketamine psychosis, 118
      molecular imaging studies of dopaminergic neurotransmission, 112–116
      prefrontal–subcortical dopamine dysregulation, 116–118
      receptor occupancy studies, 110
         $D_{2/3}$ receptor occupancy, 110–111
         PET studies, 111
         positive symptoms, 111–112
Dopamine 2 receptors (D_2Rs), 287, 383–384
Dopaminergic neurons inhibition by α3-containing $GABA_A$ receptors, 156–157
Dopaminergic system, 9–10, 360
Dopaminergic transmission, models targeting, 356–357, 385
   COMT knockout, 385–386
Dorso-lateral prefrontal cortex (DLPFC), 134–135
Dorsolateral prefrontal cortex (dlPFC), 344
DPA. *See* Docosapentaenoic acid (DPA)
DPC. *See* Dystrophin-associated protein complex (DPC)
Drag-and-drop feature, 87
Drd2. *See* D_2 receptor gene (*Drd2*)
Drd2 dopamine receptor, 87, 92f
DSM-V. *See* Diagnostic and Statistical Manual of Mental Disorders (DSM-V)
DTNBP1. *See* Dystrobrevin binding protein 1 (DTNBP1)
Dutch Famine Study, 64
DWI. *See* Diffusion-weighted imaging (DWI)
Dysbindin. *See* Dystrobrevin binding protein 1 (DTNBP1)
Dysfunctional brain activation, 299
Dystrobrevin binding protein 1 (DTNBP1), 245–246, 360, 386
   Dysbindin-1, 137
Dystrophin-associated protein complex (DPC), 360

E

EEG. *See* Electroencephalography (EEG)
EEG/LFP validation of models, 248–249
EGFP. *See* Enhanced green fluorescent protein (EGFP)
Electroencephalography (EEG), 213, 226, 238–239, 297, 316
   electrophysiological research in schizophrenia, 214
   measures of aberrant salience detection
      MMN, 220–223
      P300, 223–225
   measures of abnormal gamma band oscillations, 225–226
      gamma oscillations neurobiology, 226–228
      neural oscillations measurement, 226
      and schizophrenia, 228–229
   measures of inhibitory failure
      P50 auditory sensory gating, 215–218
      PPI of startle, 218–220
   noninvasive methodology, 214
Electron microscopy (EM), 126
Electrophysiological studies, 288
EM. *See* Electron microscopy (EM)
Embryogenesis, 27
Embryonic brain, 202–204
Encyclopedia of DNA Elements (ENCODE), 53–54, 73
Endophenotypes, 17, 18f, 367
   assumptions, 18
   cognitive and social functions, 21
   conceptual issues, 18
   criteria for, 19t
   emerging frontier for, 26–28
   limitations and future directions, 30–31
   mediation model, 18
   mental illness, 19
   new frontier for, 28–30
   numerous inputs to endophenotype and feedback pathways, 21f
   potential, 19
   psychopathology literature, 19
   in schizophrenia, 20t, 22
      ambitious recent studies, 22
      characterization of biomarkers, 24
      cognitive factors, 22
      gene set enrichment, 24
      gene-hunting, 25
      GWAS, 22–24
      limitations, 26
      multiple and variable genetic and environmental factors, 24
      neural risk architecture, 25
      NIMH Research Domain Criteria matrix, 26f
      NRGN gene, 23–24
      psychological and biological endophenotypes, 24–25
      RDoC, 25

Energy metabolism in schizophrenia, 172
  AUC, 173–174
  global changes in metabolic signatures, 172–173
  metabolites, 173
  MK-801-treated mice, 173
  serum glutamate levels, 173
Enhanced green fluorescent protein (EGFP), 93
Entrained oscillations, 226
ENU. *See* N-Nitroso-N-ethylurea (ENU)
ENU-induced models, 363–364
Epidemiological continuum, 264–265
Epigenetic approaches to molecular and genetic risk architectures, 62. *See also* Sequencing approaches to mapping genes
  ambitious clinical studies, 73–74
  building blocks of epigenome, 62f–63f
  chromatin-modifying drugs, 72–73
  comprehensive, region-specific, and cell type–specific epigenome mappings, 73
  DNA methylation in schizophrenia, 65–67
  epigenetic link between environmental risk factors and schizophrenia, 64
    evidence from animal studies, 65
    evidence from human studies, 64
  epigenetic mechanisms, 67–69
  functional neuroepigenomics, 69
  higher-order chromatin and genetic risk architecture of SCZ, 69–71
  neuroepigenetic approaches, 62–63
  potential contributions, 63
Epigenetic(s), 26–27
  alterations, 28
  differential DNA methylation in schizophrenia etiology, 27–28
  effects, 27
  gene expression, 27
  markings, 62–63
  mechanisms, 27
  neurotransmitter-related epigenetic alterations, 28
  schizophrenia, 28
Epigenome, building blocks of, 62f–63f
EPS effects. *See* Extrapyramidal side (EPS) effects
EPSC. *See* Excitatory postsynaptic current (EPSC)
ERK pathway. *See* Extracellular receptor kinase (ERK) pathway
ERPs. *See* Event-related potentials (ERPs)
ESP. *See* Exome Sequencing Project (ESP)
Etiological findings, models based on, 358. *See also* Modeling schizophrenia in animals
  candidate gene models, 360
    constitutive overexpression of mutant DISC1, 364–365
    DISC1, 361–363

Disc1 ENU-induced models, 363–364
DISC1 exon 2–3 deletion model, 364
DISC1 transgenic mice, 364
inducible C-fragment DISC1 model, 365
inducible mutant human DISC1 model, 365–366
NRG-1 KO mice, 361
NRG1, 360
  models based on genes from GWAS, 359–360
Event-related potentials (ERPs), 214
EVH1 domain, 130
Evoked oscillations, 226
ExAC. *See* Exome Aggregation Consortium (ExAC)
Excitatory postsynaptic current (EPSC), 247
Excitatory tuning, 150
Exome Aggregation Consortium (ExAC), 54
Exome sequencing, 53
Exome Sequencing Project (ESP), 54
Experiment Image Viewer, 87
Extended endophenotypes, 29
Extracellular receptor kinase (ERK) pathway, 130
Extrapyramidal side (EPS) effects, 110–111

F

FA. *See* Fractional anisotropy (FA)
FACS. *See* Fluorescence-activated cell sorting (FACS)
Family-based analysis, 54
Fast rhythmic bursting (FRB) neurons, 226–227
Fast-Fourier transform (FFT), 226
Fast-spiking PV-expressing GABA neurons, 152–154
FC. *See* Functional connectivity (FC)
[^{18}F]FDOPA, 115
  PET, 115–116
FFT. *See* Fast-Fourier transform (FFT)
First-generation
  antipsychotics, 110–111
  treatments, 168
Fisher exact test, 54–55
Fluorescence-activated cell sorting (FACS), 195–196
fMRI. *See* Functional magnetic resonance imaging (fMRI)
FMRP. *See* Fragile X mental retardation protein (FMRP)
Folate, 64
Folic acid, 64
Forced swim test (FST), 363–364
Formal thought disorder, 7
FP. *See* Fronto-parietal (FP)
Fractional anisotropy (FA), 315, 344
Fragile X mental retardation protein (FMRP), 45–46
FRB neurons. *See* Fast rhythmic bursting (FRB) neurons
Fronto-parietal (FP), 241
FST. *See* Forced swim test (FST)
Functional brain connectivity, 316
Functional connectivity (FC), 319

Subject Index

intersection of structural and functional connectivity, 319
in schizophrenia, 317f, 318–319
Functional endophenotype, 24–25
Functional imaging of schizophrenia symptom clusters, 301
  cerebral correlates of delusions, 298–299
    aberrant salience attribution, 299–301
    neural correlates of impaired self-monitoring and self-reflection, 304–306
    neuronal correlates of cognitive and emotional biases, 301–303
  cerebral correlates of hallucinations, 306
    neural correlates of acute auditory hallucinations, 306–307
    neural correlates of hallucination proneness, 307–308
Functional magnetic resonance imaging (fMRI), 10–11, 238–239, 297, 316
  paradigms, 290
  in schizophrenia, 298
Functional meta-analyses of schizophrenia, 341
Functional neuroepigenomics, 69, 70f
Functional prediction, 54

G

G × E interactions. *See* Gene × environment (G × E) interactions
G-protein-coupled receptors (GPCRs), 87, 92–93, 130
  kinase, 135
GABA. *See* γ-amino butyric acid (GABA)
GABA$_A$ receptors, 150
  α$_1$ GABA$_A$ receptors, 151–152
  α$_2$/α$_3$ GABA$_A$ receptor modulation, attempts to enhancing cognition via, 155–156
  α$_5$-negative allosteric modulators and cognition enhancement, 159–160
  benzodiazepine-sensitive, 155f
  deficits of GABAergic cortical interneurons, 152–154
  dopaminergic neurons inhibition by α3-containing, 156–157
  excitatory and inhibitory tuning, 150
  genetics and SCZ, 152
  impaired cortical oscillations in SCZ, 152–154
  indirect inhibition of dopaminergic neurons by α5-containing, 157–158
  maturation, 151–152
  multiplicity, 150–151
  multispecific modulation of α2-, α3-, and α5-containing, 158–159
  perception and cognition indexed by neuronal synchrony, 153f
  physiological and pharmacological functions, 154–155
  proof-of-concept clinical study, 159
  subtypes in cortical period plasticity, 152
GABA$_B$ receptor 1 gene (GABRB1), 152
GABAergic neurons, 117

GABRAB2. *See* β2 subunit of GABA$_A$ receptor (GABRAB2)
GABRB1. *See* GABA$_B$ receptor 1 gene (GABRB1)
GAD1 gene, 28
GAD$_{67}$. *See* Glutamic acid decarboxylase 67 (GAD$_{67}$)
Gametogenesis, 27
Gamma activity, 242
  membrane properties of conditional NR1 knockout mice, 247f
  schizophrenia with relation to resting, 245–246
  in vitro examination of animal models of increased resting gamma, 247–248
Gamma oscillations, 152–154, 226, 238
  neurobiology, 226–228
  in schizophrenia, 239–241
γ-amino butyric acid (GABA), 28, 286–287
Gaussian filter, 267
GCTA. *See* Genome-wide Complex Trait Analysis (GCTA)
GD17. *See* Gestational day 17 (GD17)
Gene × environment (G × E) interactions, 19
Gene expression, 83–84, 87–89
Gene Search, 87–89
Gene-hunting, 25
Genetic(s), 359
  evidence, 133–134
  Generic P300, 224
  mouse models, 386
  variants, 134, 397–398
Genome of the Netherlands Consortium (Go-NL), 54
Genome-wide association studies (GWAS), 22–24, 41, 52, 83–84, 134, 195, 243, 353, 358, 383–384
  developments, 41–42
  findings in schizophrenia, 44
    Manhattan plot, 46f
    novel gene loci, 45
    PGC, 45
    polygenic risk score analysis, 44–45
    preliminary biological pathway analyses, 45
    schizophrenia-associated loci, 45–46
  future directions, 49
  genetic relationship between schizophrenia and psychiatric disorders, 46–47
  models based on genes from, 359–360
  number of genome-wide significant findings, 48f
  pre-GWAS era, 40–41
  principle of, 42–43
  of schizophrenia, 48
  SNP-based heritabilities of psychiatric disorders, 47f
  SNPs associated with traits per chromosome, 43f
Genome-wide Complex Trait Analysis (GCTA), 46
Genomic-restricted maximum likelihood estimation (GREML), 46
Gestational day 17 (GD17), 245–246

$G_{i/o}$ proteins, 130
Glutamate (GLU), 28, 117–118, 182, 357, 388
Glutamatergic function, 117–118
Glutamatergic transmission, models targeting, 357–358, 388–389
Glutamic acid decarboxylase 67 (GAD_{67}), 247–248
Glutathione (GSH), 173–174
Glycine transporter1 (GlyT1), 389
GM. *See* Gray matter (GM)
Go-NL. *See* Genome of the Netherlands Consortium (Go-NL)
GPCRs. *See* G-protein-coupled receptors (GPCRs)
Gprin family member 3 (*Gprin*3), 89
Graph theory, 313–314, 321
Gray matter (GM), 338
GREML. *See* Genomic-restricted maximum likelihood estimation (GREML)
Griesinger, Wilhelm, 5–6, 5*f*
GRM3 genotype, 388–389
Group I mGluRs, 130–131
Group II mGluRs, 130–131
Group of schizophrenias, 263–264
GSH. *See* Glutathione (GSH)
GWAS. *See* Genome-wide association studies (GWAS)

H

"Hallmark of neural signals", 28–29
Hallucinations, 242
  cerebral correlates, 306
    neural correlates of acute auditory hallucinations, 306–307
    neural correlates of hallucination proneness, 307–308
Haloperidol, 168
Haplotype tagging SNP, 41
HATs. *See* Histone acetyltansferases (HATs)
HC. *See* Hippocampus (HC)
HDAC inhibitor (HDACi), 71–72
HDACs. *See* Histone deacetylases (HDACs)
Healthy controls (HC), 270–271, 273*f*
Hemodynamic response, 297
Herpes simplex virus type 1 (HSV-1), 180
Heterozygous (HET), 357
Hidden Markov field model (HMRF model), 271–272
Hierarchical models, 285
Hierarchical probabilistic model, 286*f*
High displacers, 114
Higher-order chromatin, 69–71
Hippocampal CA3-CA4 interneurons, 215
Hippocampus (HC), 286–287
hiPSC-derived neurons, 202–204
Histone acetylation, 71–72
Histone acetyltansferases (HATs), 71–72
Histone deacetylases (HDACs), 71–72
  *HDAC*1, 71–72

Histone methylation, 69, 73
HMRF model. *See* Hidden Markov field model (HMRF model)
Homovanillic acid (HVA), 110
HPA axis. *See* Hypothalamic–pituitary–adrenal (HPA) axis
HSV-1. *See* Herpes simplex virus type 1 (HSV-1)
*HTR*2A. *See* Serotonin receptor type-2 (*HTR*2A)
Hub pathology to schizophrenia, 328–329
Human atlas resources, 97. *See also* Mouse atlas resources
  Allen Human Brain Atlas, 97–98, 98*f*
  BrainSpan atlas expression viewer, 101*f*
  BrainSpan Atlas of Developing Human Brain, 99–100
  genetic topography of human brain, 99*f*
  ISH gene expression studies in human brain, 100*t*
  NIH blueprint nonhuman primate atlas, 101–102
Human connectome, 314
Human genetics, 41
  studies, 354–355
Human Genome Project, 52–53
Human studies, evidence from, 64
HVA. *See* Homovanillic acid (HVA)
Hyper-ToM, 302–303
Hyperactivity, 249
Hypomorphic mouse model, 357
Hypothalamic–pituitary–adrenal (HPA) axis, 196

I

[^{123}I]IBZM SPECT studies of patients, 113
ICD-10. *See* International Statistical Classification of Diseases and Related Health Problems (ICD-10)
Ich-Störung, 7–8
ID. *See* Intellectual disability (ID)
IFN-γ. *See* Interferon-gamma (IFN-γ)
iGluRs. *See* Ionotropic glutamate receptors (iGluRs)
IL-1β, 182–183
IL-6, 182–183
Illuminia's HiSeq, 52–53
Image series, 87
Imidazenil, 158
Immune dysbalance in SCZ, 182
Immune response, 181
  kindling and sensitization of, 181
Imputation, 42
In situ hybridization (ISH), 84
In vitro examination of animal models, 247–248
In vivo competition paradigm, 113
In vivo multimodal imaging studies, 117
Incentive salience, 290
INDELs. *See* Insertions or deletions (INDELs)
Indirect inhibition of dopaminergic neurons by α5-containing $GABA_A$ receptors, 157–158
Inducible C-fragment DISC1 model, 365
Inducible mutant human DISC1 model, 365–366

Infection role in SCZ, 183–184
Infinium Assay of Illumina, 42
Inflammation, 179
   in CNS, 180
   CNS volume loss in imaging studies, 184
   immune dysbalance in SCZ, 182
   impact on neurotransmitters in SCZ, 182–183
   kindling and sensitization of immune response, 181
   microglia as cellular basis, 180–181
   vulnerability-stress-inflammation model of SCZ, 181–182
ING model. See Interneuron network gamma (ING) model
Inhibitory postsynaptic potentials (IPSPs), 226–227
Inhibitory tuning, 150
Inositol 1, 4, 5-trisphosphate (IP_3), 130
Insertions or deletions (INDELs), 53–54
Integration, connectome and schizophrenia, 321, 324–327, 325f
Intellectual disability (ID), 55
Intellectual disability, 399, 405
Interferon-gamma (IFN-γ), 182, 185
Intermediate phenotype, 367
International HapMap Project, 41
International Schizophrenia Consortium (ISC), 44
International Statistical Classification of Diseases and Related Health Problems (ICD-10), 3–4, 9
Interneuron network gamma (ING) model, 243
Ionotropic glutamate receptors (iGluRs), 388
IP_3. See Inositol 1, 4, 5-trisphosphate (IP_3)
IP3 receptor (IP3R), 130
IPSPs. See Inhibitory postsynaptic potentials (IPSPs)
ISC. See International Schizophrenia Consortium (ISC)
ISH. See In situ hybridization (ISH)

K

KCC2, 151–152
Ketamine psychosis, 118
Kindling, 181
Knockout mice (KO mice), 357
Kraepelin, Emil, 6, 6f

L

LA. See Linoleic acid (LA)
Large-scale functional neuronal networks, 30
Laser Microdissection (LMD) Microarray, 100
Latent inhibition (LI), 365
LC–MS/MS. See Liquid chromatography coupled with tandem MS (LC–MS/MS)
LD. See Linkage disequilibrium (LD)
LDA. See Linear discriminant analysis (LDA)
LFP. See Local field potentials (LFP)
LI. See Latent inhibition (LI)
Linear discriminant analysis (LDA), 344

Linkage disequilibrium (LD), 41
Linoleic acid (LA), 172
Lipidomics in schizophrenia, 169
   analytical systems, 170
   metaanalyses, 172
   MS, 170
   phospholipids groups, 170
   plasmalogens, 171
   shotgun lipidomic analysis, 171
   VLDLs, 172
Lipids, 169–170
   metabolic signatures, 171
Liquid chromatography coupled with tandem MS (LC–MS/MS), 131–132, 197
LMD Microarray. See Laser Microdissection (LMD) Microarray
Local field potentials (LFP), 244–245
LOF. See Loss-of-function (LOF)
Long-term depression (LTD), 128, 159–160, 359
Long-term potentiation (LTP), 128, 159–160
Loss-of-function (LOF), 55
Low penetrance, 41, 49
Lowered decision threshold, 290–291
Lowered decision threshold vs. overweighting evidence, 290–291
LTD. See Long-term depression (LTD)
LTP. See Long-term potentiation (LTP)

M

Machine-learning (ML), 269–270, 343
   advanced ML studies in schizophrenia, 346
   applications in schizophrenia, 346
   brain–behavior architecture, 347
   discriminative features, 346–347
   hypothesis-guided ML studies in schizophrenia, 345–346
   structural ML studies in schizophrenia, 344
   supervised ML, 343
MAF. See Minor allele frequency (MAF)
Magnetic resonance imaging (MRI), 10–11, 86
Magnetic resonance spectroscopy (MRS), 117–118
Magnetic transfer ratio (MTR), 315
Magnetoencephalography (MEG), 214, 226, 238–239, 316. See also Electroencephalography (EEG)
MAGUK. See Membrane-associated guanylate kinase (MAGUK)
Major depressive disorder (MDD), 46–47
Major disease genes, 40
Major histocompatibility complex (MHC), 44, 69, 359
MALDI-TOF. See Matrix-assisted laser desorption/ionization time-of-flight (MALDI-TOF)
MAM. See Methylazoxymethanol (MAM)
Mammalian target of rapamycin (MTOR), 130

Manisch-depressives Irreseyn, 6
MAPK pathway. *See* Mitogen-activated protein kinase (MAPK) pathway
Mass action and net effects, 407
Mass spectrometry (MS), 131, 170, 195–196
Maternal stress, 64
Matrix-assisted laser desorption/ionization time-of-flight (MALDI-TOF), 131
MD. *See* Mean diffusivity (MD)
MDD. *See* Major depressive disorder (MDD)
Mean diffusivity (MD), 344
Medial frontal cortex, 241
Medial prefrontal cortex (mPFC), 356–357
Mediation model, 18
Meduna, Ladislas, 9
MEG. *See* Magnetoencephalography (MEG)
Membrane receptors, 128
Membrane-associated guanylate kinase (MAGUK), 127–129, 198
Mental disorders, 303
Mental illnesses, 5–6
Meta-analyses, 298
Metabolic markers in diagnosis, 168
Metabolites, 168
Metabolomics of schizophrenia, 168. *See also* Proteomics of schizophrenia
    biomarkers, 168
    energy metabolism in schizophrenia, 172–174
    lipidomics in schizophrenia, 169–172
    metabolic markers in diagnosis, 168
    "omics" approaches in schizophrenia, 169f
Metabotropic glutamate receptors (mGluRs), 128, 130–131, 287
Methylazoxymethanol (MAM), 157–158, 245–246
mGluRs. *See* Metabotropic glutamate receptors (mGluRs)
MGS. *See* Molecular Genetics of Schizophrenia (MGS)
MHC. *See* Major histocompatibility complex (MHC)
Microarrays, 42, 87, 97
Microglia as cellular basis, 180–181
Minor allele frequency (MAF), 41
*MIR*137, 45
Mismatch negativity (MMN), 214–215, 220
    measurement, 220–221
    neurobiology, 221–222
    and schizophrenia, 221f, 222–223
Mitogen-activated protein kinase (MAPK) pathway, 130
MK-0777. *See* TPA023
MK-801-treated mice, 173
ML. *See* Machine-learning (ML)
MMN. *See* Mismatch negativity (MMN)
Modeling schizophrenia in animals, 354. *See also* Etiological findings, models based on
    approaches, 355f

challenges, 366–367
    genetic animal models in SCZ, 355
        models based on etiological findings, 358–366
        pathophysiological models, 356–358
    human genetics studies, 354–355
    pathogenic mechanisms of disease, 354
"Modulated VBM" method, 267
Molecular Genetics of Schizophrenia (MGS), 44
Molecular imaging studies of dopaminergic neurotransmission, 112
    amphetamine-induced dopamine release, 114f
    AMPT dopamine depletion paradigm, 113–114
    between-group differences, 114
    [^{18}F]FDOPA, 115
    [^{18}F]FDOPA PET, 115–116
    high displacers, 114
    meta-analysis
        of serotonin transporter imaging studies, 115
        of striatal dopamine synthesis capacity, 116f
    pharmacological challenge paradigms, 113
    [^{11}C]-(+)-PNHO binding, 115
    psychostimulant-evoked dopamine release, 112–113
    in vivo competition paradigm, 113
Morphometric gap, 265
Mouse atlas resources, 84. *See also* Human atlas resources
    common reference space in mouse brain, 86–87
    connectivity in mouse brain, 93–96
    integrated search and visualization in Allen brain atlas, 87
        Brain Explorer window, 90
        Correlative Search utility, 89
        detailed correlative search return, 89f
        Experiment Image Viewer, 87
        G-protein-coupled receptor, 87
        image series, 87, 88f
        individual genes from correlative search with *Drd*2, 90f
        3D viewing of gene expression in Brain Explorer, 91f
    local connectivity in Allen Mouse Brain Connectivity Atlas, 95f
    mouse developmental atlas, 91–93
    nervous system, 85
    in situ hybridization images from Allen Mouse Brain Atlas, 85f
    3D expression summaries and heatmap for developmental mouse, 94f
    3D reconstruction of cortical connectivity, 95f
Mouse brain
    connectivity in, 93–96
    reference space in, 86–87
Mouse developmental atlas, 91–93
Mouse models, 245–246, 250
    of 22q11.2 CNVs, 405f
    and dimensional behavioral phenotypes, 403t–404t
mPFC. *See* Medial prefrontal cortex (mPFC)
MRI. *See* Magnetic resonance imaging (MRI)
MRS. *See* Magnetic resonance spectroscopy (MRS)

MS. *See* Mass spectrometry (MS)Multiple sclerosis (MS)
MTOR. *See* Mammalian target of rapamycin (MTOR)
MTR. *See* Magnetic transfer ratio (MTR)
Multiple agglomerative gene-based association tests, 54–55
Multiple sclerosis (MS), 180
Multispecific "dirty" drugs, 158–159
Multispecific modulation of α2-, α3-, and α5-containing GABA$_A$ receptors, 158–159
Multivariate pattern analysis (MVPA), 269–270
Multivariate voxel pattern analysis (MVPA) methods, 265
MVPA. *See* Multivariate pattern analysis (MVPA)
MVPA methods. *See* Multivariate voxel pattern analysis (MVPA) methods

N

N-acetylaspartate (NAA), 117–118
N-methyl-D-aspartate (NMDA), 220, 357, 388
N-methyl-D-asparate receptors (NMDARs), 52, 117, 127, 286–287, 357
  protein complex, 197–198, 201–202
  signaling complex, 128–129
N-methyl-D-aspartate receptors (NMDA-R), 243
NAA. *See* N-acetylaspartate (NAA)
National Institute of Mental Health (NIMH), 11
National Institutes of Health (NIH), 97
Negative emotional biases, 303
Negative symptoms, 7
Neonatal ventral hippocampus lesions (NVHL), 245–246
Neonatal vocalization, 401–402
Nervous system, 85
Neural correlates
  of acute auditory hallucinations, 306–307
  of hallucination proneness, 307–308
  of impaired self-monitoring and self-reflection, 304–306
Neural oscillations, 29, 225–226
  measurement, 226
Neural risk architecture, 25
Neural traffic, 322
Neural-rich club connections, 322
Neuregulin 1 (NRG1), 360
Neuroepigenetic approaches, 62–63
Neurofilament-L (NF-L), 135
Neurogranin (*NRGN*), 23–24, 44
Neuroinflammation, 180
Neuronal correlates of cognitive and emotional biases, 301–303
Neuronal progenitor cells (NPC), 202–204
Neuropsychiatric disorders
  CNV-encoded genes causing behavioral dimensions, 402–409
  CNVs and, 398–400
  dimensional behavioral models, 400–402
Neurotransmission models, 358
Neurotransmitter-related epigenetic alterations, 28

Next-generation sequencing, 52–53
NF-L. *See* Neurofilament-L (NF-L)
NHP Atlas. *See* Nonhuman Primate Atlas (NHP Atlas)
Nicotinic antagonists, 215
NIH. *See* National Institutes of Health (NIH)
NIH blueprint nonhuman primate atlas, 101–102
NIMH. *See* National Institute of Mental Health (NIMH)
NIMH Research Domain Criteria matrix, 26f
N-Nitroso-N-ethylurea (ENU), 363–364
NMDA. *See* *N*-methyl-D-aspartate (NMDA)
NMDA-R. *See* *N*-methyl-D-aspartate receptors (NMDA-R)
NMDA-R subunit 1 (NR1), 245–246
NMDARs. *See* *N*-methyl-D-asparate receptors (NMDARs)
[^{11}C]NMSP, 110
Noisy brain, animal models of, 244–250. *See also* Computational neuroanatomy
Noncontiguous effects, 406–407
Nonhuman Primate Atlas (NHP Atlas), 97
Novel gene loci, 45
NPC. *See* Neuronal progenitor cells (NPC)
NR1. *See* NMDA-R subunit 1 (NR1)
NRG1. *See* Neuregulin 1 (NRG1)
NRGL–ERBB4 complex, 219–220
NRGN. *See* Neurogranin (*NRGN*)
NVHL. *See* Neonatal ventral hippocampus lesions (NVHL)

O

Odds ratio (OR), 397–398
ODVBA. *See* Optimally discriminant voxel-based analysis (ODVBA)
Olanzapine, 168
OLG. *See* Oligodendrocyte (OLG)
Oligodendrocyte (OLG), 365–366
Online Mendelian Inheritance of Man (OMIM), 69–71, 202–204
Optimally discriminant voxel-based analysis (ODVBA), 269
Optogentics, 387
OR. *See* Odds ratio (OR)
Outcome uncertainty, 285
Overweighting evidence, 290–291

P

P50 auditory sensory gating, 215
  P50 suppression, 222
    neurobiology, 215–217
    and schizophrenia, 217–218
  responses to paired clicks, 216f
p70 S6 kinase pathway, 130
P300, 223
  methodology, 223
  neurobiology, 224
  and schizophrenia, 224–225

Pacemaker cells, 226–227
PANSS. *See* Positive and negative symptom scale (PANSS)
Paranoia. *See* Persecutory delusions
Parkinson disease, 386
Parvalbumin (PV), 243, 364
　PV-positive basket cells, 151–152
Passivity symptoms, 304–305
Pathophysiological models, 356
　models targeting
　　dopaminergic transmission, 356–357
　　glutamatergic transmission, 357–358
PC. *See* Phosphotidylcholine (PC)
PCA. *See* Principal component analysis (PCA)
PCP. *See* Phencyclidine (PCP)
PCs. *See* Purkinje cells (PCs)
Pde1b. *See* Phosphodiesterase 1B (*Pde1b*)
PDZ-binding domains. *See* Pentylenetetrazole-binding domains (PDZ-binding domains)
PE. *See* Phosphotidylethanolamine (PE)
Pedunculopontine (PPTg), 219
Pentylenetetrazole (PTZ), 159–160
Pentylenetetrazole-binding domains (PDZ-binding domains), 128–129
Periconceptional exposure to famine, 64
Persecutory delusions, 299
PEs. *See* Prediction errors (PEs)
PET. *See* Positron emission tomography (PET)
PFC. *See* Prefrontal cortex (PFC)
PFST. *See* Porsolt forced swim test (PFST)
PGC. *See* Psychiatric Genomics Consortium (PGC)
Pharmacologic models, 354
Pharmacological challenge paradigms, 113
Pharmacological studies, 221–222
Phasic dopamine signaling, 289–290
Phencyclidine (PCP), 173, 354, 357, 361
Phenotypic variation, 408–409
Phosphodiesterase 1B (*Pde1b*), 89
Phospholipids, 169–170
Phosphotidylcholine (PC), 170
Phosphotidylethanolamine (PE), 170
PICK-1. *See* Protein interacting with C kinase-1 (PICK-1)
PING. *See* Pyramidal and interneuron model of generating gamma oscillations (PING)
PKA. *See* Protein kinase A (PKA)
PKC. *See* Protein kinase C (PKC)
Plasmalogens, 171
Pleiotropy, 47, 407
[^{11}C]-(+)-PNHO binding, 115
Polygenic risk score analysis, 44–45
Polygenicity, 45
Polyunsaturated fatty acids (PUFAs), 170
Porsolt forced swim test (PFST), 364
Portal, 84

Positive and negative symptom scale (PANSS), 271
Positive symptoms, 7
Positron emission tomography (PET), 110, 184, 316
Post-translational modifications (PTMs), 198–200
Postmortem findings, 28
Postmortem studies, 196–197
Postsynaptic density (PSD), 126–127, 197–198
　abundant protein domains in, 201*f*
　clustering of PSD protein complexes, 200*f*
　constellation and function of proteins, 128
　dendritic spines, 126–127
　dysregulations in PSD proteins and signaling in SCZ, 134–135
　evidence supporting role of PSD in schizophrenia, 133–136
　extracellular domains of NMDAR and AMPAR, 128
　membrane receptors, 128
　as microdomain for converging molecular alterations in schizophrenia, 136–137
　molecular abnormalities in PSD converge, 136*f*
　molecular architecture, 132*f*
　postmortem evidence for PSD dysregulation in SCZ, 134
　proteomic landscape, 131, 133
　　MS analyses, 131
　　NMDAR complexes, 131
　　purified PSD fractions, 131–132
　　subcellular fractions, 132
　scaffold protein, 198, 199*f*
　signaling pathways in, 128
　　AMPA receptors, 129–130
　　mGluRs, 130–131
　　NMDA receptor signaling complex, 128–129
Postsynaptic proteome (PSP), 197–198
Power and coherence of brain activity, 241–242
PPI. *See* Prepulse inhibition (PPI)
PPTg. *See* Pedunculopontine (PPTg)
Pre-GWAS era, 40–41
Prediction errors (PEs), 285
Predictive coding, 283, 285
Prefrontal cortex (PFC), 130–131, 286–287, 356
Prefrontal–subcortical dopamine dysregulation, 116–117
　glutamatergic function, 117–118
　NMDA receptor, 117
　in vivo multimodal imaging studies, 117
Preimputation, 27
Premature dementia, 7
Prenatal expression, 365–366
Prenatal immune activation, 184
Prepulse inhibition (PPI), 218, 354, 384, 400
　startle response
　　neurobiology of PPI, 218–219
　　PPI and schizophrenia, 219–220
　　PPI methodology, 218
Principal component analysis (PCA), 344

PRODH gene, 173
Prodromal state, 305–306
Proof-of-concept clinical study, 159
Protein
    expression of serine racemase, 135
    interaction pathways, 197–202
    phosphorylation, 200–201
    protein–protein associations, 137
    protein–protein interaction studies, 198–199
Protein interacting with C kinase-1 (PICK-1), 358
Protein kinase A (PKA), 128
Protein kinase C (PKC), 128
Protein kinases (PTKs), 129
Proteome, 195–196
Proteomic landscape of PSD, 131, 133
    MS analyses, 131
    NMDAR complexes, 131
    purified PSD fractions, 131–132
    subcellular fractions, 132
Proteomics of schizophrenia, 196
    biomarkers, 196
    circular clustering of protein complexes, 203*f*
    embryonic brain, 202–204
    postmortem studies, 196–197
    protein interaction pathways, 197–202
    signaling networks, 197–202
PSD. *See* Postsynaptic density (PSD)
PSD-95, 128–129
Psd95 protein complex, 197–198
PSP. *See* Postsynaptic proteome (PSP)
PsychENCODE, 73
Psychiatric disorders
    implications for, 186*f*
    stress-induced inflammatory response in, 181
    vulnerability–stress model, 181–182
Psychiatric Genomics Consortium (PGC), 44–45, 55
Psychiatric Genomics study, 359
Psychopathology, 17–18, 25
Psychosis, 285–287
Psychotic disorders, 6
PTKs. *See* Protein kinases (PTKs)
PTMs. *See* Post-translational modifications (PTMs)
PTZ. *See* Pentylenetetrazole (PTZ)
PUFAs. *See* Polyunsaturated fatty acids (PUFAs)
Purkinje cells (PCs), 359
PV. *See* Parvalbumin (PV)
Pyramidal and interneuron model of generating gamma oscillations (PING), 244

Q

Quantitative connectivity map, 93
Quantitative meta-analysis, 338–339
Quetiapine, 110–111

R

[^{11}C]raclopride PET study, 113
Rare variant analysis, 54
Rare variants, 54–55, 397–398
RAVENS approach, 268
RDoCs. *See* Research Domain Criteria (RDoCs)
Reelin (*RELN*), 27–28, 66
Reference Atlas, 86, 100
Regions of homozygosity (ROH), 55
Regions of interest (ROI), 263–264, 266–267, 276
    limitations, 266*f*
    voxel-based analysis *vs.*, 266–267
RELN. *See* Reelin (*RELN*)
Research Domain Criteria (RDoCs), 11, 18, 25, 367
Resting state brain activity, 238
Resting state gamma activity
    schizophrenia with relation to, 245–246
    in vitro examination of animal models of increasing, 247–248
Resting state network in schizophrenia, 239–241
Rich club organization, 321, 327–328
Risk allele schizophrenia model, 361–363
Ro4938581 cognitive enhancer, 159–160
ROH. *See* Regions of homozygosity (ROH)
ROI. *See* Regions of interest (ROI)

S

SVM. *See* Support vector machines (SVM)
Salient network (SN), 241
Sandy mice (sdy mice), 360
SAP-97 protein level, 135
SC. *See* Structural connectivity (SC)
Scaffold protein, 198, 199*f*
Schizophrenia (SCZ), 3, 51, 62, 83–84, 125–126, 167, 195, 270–271, 313–314, 353, 383–384, 399
    brain network, 314
    burden on human societies, 337–338
    COX-2 inhibition, 184–185
    de novo mutations, 55
    diagnostic criteria for, 4*t*, 8*t*
    epigenetic mechanisms and treatment of, 71–72
    first-and second-rank symptoms, 9*t*
    functional meta-analyses, 341
    heterogeneity, 4
    higher-order chromatin and genetic risk architecture of, 69–71
    history, 5–9
    immune dysbalance, 182
    immune-related substances in therapy, 185
    impact on neurotransmitters, 182–183
    infection role, 183–184
    lessons from therapeutic approaches, 9–10

Schizophrenia (SCZ) (*Continued*)
   methodological aspects of response to immune-based therapy, 185–186
   neurobiology, 10–12
   neuropsychiatric disorder, 213
   proteomics, 196
      biomarkers, 196
      embryonic brain, 202–204
      postmortem studies, 196–197
      protein interaction pathways, 197–202
      signaling networks, 197–202
   psychopathological features, 400
   with relation to resting gamma activity, 245–246
   Schizophrenia Working Group, 359
   SCZ-associated genes, 359
   SCZ-like neuroanatomical patterns, 276, 277f
   structural meta-analyses in, 339–341
   studies in peripheral tissue and epigenetic biomarkers in, 66–67
   studies in
      brain activity in gamma frequency range, 240f
      circuit models and evidence in humans, 242–243
      EEG, 238–239
      fMRI, 239
      gamma oscillations in, 239–241
      ING model, 243
      inherent noise and relationship to symptoms in, 242
      MEG, 239
      PING, 244
      power and coherence of brain activity, 241–242
      resting state brain activity, 238
      resting state network in, 239–241
      similarities among schizophrenia, 251f
   tryptophan/kynurenine metabolism, 186f
   vulnerability-stress-inflammation model, 181–182, 187f
Schneider, Kurt, 6–8, 8f
SCZ. *See* Schizophrenia (SCZ)
sdy mice. *See* Sandy mice (sdy mice)
Second-generation "atypical" antipsychotics, 110–111
Second-generation antipsychotics, 168
Segregation, connectome and schizophrenia, 321–324, 323f
Sejunktionspsychose, 313–314
Self-monitoring and self-reflection, neural correlates of impaired, 304–306
Self-related information processing, 305–306
"Self-tag", 306–307
Sensitization, 181
Sequence Kernel Association Test (SKAT), 54–55
Sequencing approaches to mapping genes, 51. *See also* Epigenetic approaches to molecular and genetic risk architectures
   CNV, 52
   exome *vs.* whole genome, 53–55
   GWAS, 52
   multiple environmental factors, 51
   next-generation sequencing, 52–53
   schizophrenia risk genes converge on functional pathways, 56f
   schizophrenia-associated CNVs, 52
   sequencing studies to date, 55–57
   single gene mutation, 51
Serine racemase (SR) enzyme, 358
D-Serine, 135, 357–358
Serotonin receptor type-2 (*HTR2A*), 66
Sex-determining region Y-box containing gene 10 (*SOX*10), 27–28, 66
SFKs. *See* Src/Fyn family kinases (SFKs)
SH-053–2'F-R-CH$_3$, 157–158
SH3 and multiple ankyrin repeat domain proteins (SHANKs 1–3), 198
Shotgun lipidomic analysis, 171
Signaling networks, 197–202
SILAC. *See* Stable isotope labeling by amino acids in cell culture (SILAC)
Single gene mutation, 51
Single nucleotide polymorphisms (SNPs), 22–23, 41–42, 64, 346, 359, 397–398
   associated with traits per chromosome, 43f
   SNP-based heritabilities of psychiatric disorders, 47f
Single photo emission computer tomography (SPECT), 110
Single-nucleotide variants (SNVs), 55, 195, 197–198
Single-subject
   classification, 274–275
   differentiation, 275–276
Sirtuins. *See* Class III HDACs
SKAT. *See* Sequence Kernel Association Test (SKAT)
Small-world topology, 322
Smooth endoplasmic reticulum, 127
SN. *See* Salient network (SN)
"Snapshots" of cellular states, 195–196
SNPs. *See* Single nucleotide polymorphisms (SNPs)
SNVs. *See* Single-nucleotide variants (SNVs)
Social behaviors, 400–401
Social-cognitive schemes, 301
*SOX*10. *See* Sex-determining region Y-box containing gene 10 (*SOX*10)
Spatial resolution, 297
SPECT. *See* Single photo emission computer tomography (SPECT)
Spine apparatus, 127
SQUID. *See* Supra-conducting quantum interference device (SQUID)
SR enzyme. *See* Serine racemase (SR) enzyme
Src/Fyn family kinases (SFKs), 129

SSEPs. *See* Steady-state evoked potentials (SSEPs)
ST6GALNAC1, 27–28
Stable isotope labeling by amino acids in cell culture (SILAC), 202–204
Startle response of PPI, 218
  neurobiology of PPI, 218–219
  PPI and schizophrenia, 219–220
  PPI methodology, 218
State uncertainty, 285
Statistical learning of neurobiology
  histopathology, 338
  ML, 343
    advanced ML studies in schizophrenia, 346
    applications in schizophrenia, 346
    brain–behavior architecture, 347
    discriminative features, 346–347
    hypothesis-guided ML studies in schizophrenia, 345–346
    structural ML studies in schizophrenia, 344
  quantitative meta-analyses, 342
  quantitative meta-analysis, 338–339
    functional meta-analyses of schizophrenia, 341
    structural meta-analyses in schizophrenia, 339–341
  schizophrenia, 337–338
    gray matter volume alterations in, 340f
    pathophysiology, 342
Steady-state evoked potentials (SSEPs), 226
STG. *See* Superior temporal gyrus (STG)
Stress-induced inflammatory response in psychiatric disorders, 181
Striatal presynaptic dopamine elevation
  aberrant salience, 290
  computational implications, 288
  incentive salience, 290
  overweighting evidence *vs.* lowered decision threshold, 290–291
  phasic dopamine signaling, 289–290
  tonic dopamine signaling, 289
Striatum, 130–131
Structural brain connectivity, 315b–316b
Structural connectivity (SC), 319
  analyses, 29
  intersection of structural and functional connectivity, 319
  in schizophrenia, 317
Structural endophenotypes, 29
Structural meta-analyses in schizophrenia, 339–341
Subcellular fractions, 132
Subcortical hyperdopaminergia in psychosis, 116–117
Subgroup of delusions, 7–8
Subtraction, 298
Superior temporal gyrus (STG), 225
Support vector machine (SVM) algorithms, 344
Support vector machines (SVM), 269–270

Supra-conducting quantum interference device (SQUID), 214
SVM algorithms. *See* Support vector machine (SVM) algorithms
Synaptic gain, 285–287
  computational implications of decreased high-level, 287–288
"Synaptic" scenario, 202
Synaptogenesis, 151–152

T

Tachykinn, precursor 1 (*Tac*1), 89
Tail suspension tests (TST), 361–363
Tandem affinity purification (TAP) tags, 132
TAP tags. *See* Tandem affinity purification (TAP) tags
Targeted cognitive training (TCT), 222–223
Task-based neuroimaging studies, 338
*TCF*4. *See* Transcription factor 4 (*TCF*4)
TCT. *See* Targeted cognitive training (TCT)
Testing phase, 343
Δ^9-Tetrahydrocannabinol (THC), 385–386
TH. *See* Tyrosine hydroxylase (TH)
THC. *See* Δ^9-Tetrahydrocannabinol (THC)
Theory of mind (ToM), 302–303
Three-chamber "sociability" test, 400–401
3D Allen Reference Atlas model, 93
3D models, 86
Tissue density maps, 267
TNiK. *See* Traf and Nck interacting kinase (TNiK)
ToM. *See* Theory of mind (ToM)
Tonic dopamine signaling, 289
TPA023, 155–156
Traf and Nck interacting kinase (TNiK), 200
Training phase, 343
Transcription factor 4 (*TCF*4), 44, 359–360
Triton-X 100, 131
TST. *See* Tail suspension tests (TST)
2D Allen Reference Atlas coronal plates, 93
Typical antipsychotics, 168
Tyrosine hydroxylase (TH), 365

U

Ultra-high-risk (UHR), 219
Umbrella construct, 263–264
Unitary psychosis, 5–6
Unsupervised ML, 343
UPenn study, 276

V

Valproate, 71–72
Variant Effect Predictor (VEP), 53–54
Variants, 53–54
VBM. *See* Voxel-based morphometry (VBM)

VBMA. *See* Voxel-based morphometric analyses (VBMA)
Ventral hippocampus (VHP), 354
VEP. *See* Variant Effect Predictor (VEP)
Very-low-density lipoproteins (VLDLs), 172
VHP. *See* Ventral hippocampus (VHP)
VLDLs. *See* Very-low-density lipoproteins (VLDLs)
Voxel-based analysis, 266–267
Voxel-based morphometric analyses (VBMA), 267, 269
Voxel-based morphometry (VBM), 339
Vulnerability-stress-inflammation model of SCZ, 181–182

W

White matter (WM), 339–341
Whole-exome sequencing (WES), 53
Whole-genome sequencing (WGS), 53

Y

Years of life lived with disability (YLDs), 167

Z

ZNF804A
  allele, 23–24
  transcription factor, 44